Applied Stochastic Processes and Control for Jump-Diffusions

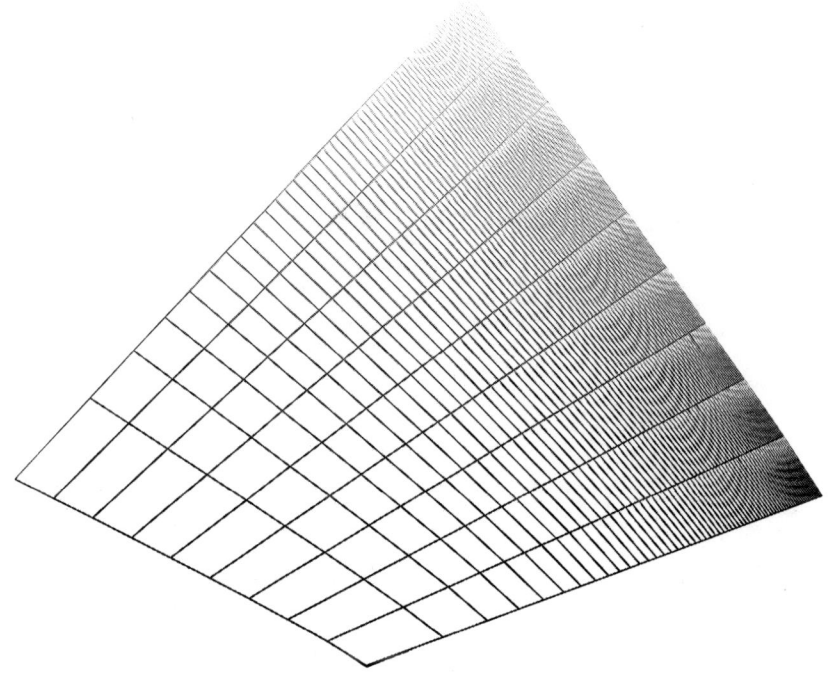

Advances in Design and Control

SIAM's Advances in Design and Control series consists of texts and monographs dealing with all areas of design and control and their applications. Topics of interest include shape optimization, multidisciplinary design, trajectory optimization, feedback, and optimal control. The series focuses on the mathematical and computational aspects of engineering design and control that are usable in a wide variety of scientific and engineering disciplines.

Editor-in-Chief
Ralph C. Smith, North Carolina State University

Editorial Board

Series Volumes

Hanson, Floyd B., *Applied Stochastic Processes and Control for Jump-Diffusions: Modeling, Analysis, and Computation*

Michiels, Wim and Niculescu, Silviu-Iulian, *Stability and Stabilization of Time-Delay Systems: An Eigenvalue-Based Approach*

Ioannou, Petros and Fidan, Baris, *Adaptive Control Tutorial*

Bhaya, Amit and Kaszkurewicz, Eugenius, *Control Perspectives on Numerical Algorithms and Matrix Problems*

Robinett III, Rush D., Wilson, David G., Eisler, G. Richard, and Hurtado, John E., *Applied Dynamic Programming for Optimization of Dynamical Systems*

Huang, J., *Nonlinear Output Regulation: Theory and Applications*

Haslinger, J. and Mäkinen, R. A. E., *Introduction to Shape Optimization: Theory, Approximation, and Computation*

Antoulas, Athanasios C., *Approximation of Large-Scale Dynamical Systems*

Gunzburger, Max D., *Perspectives in Flow Control and Optimization*

Delfour, M. C. and Zolésio, J.-P., *Shapes and Geometries: Analysis, Differential Calculus, and Optimization*

Betts, John T., *Practical Methods for Optimal Control Using Nonlinear Programming*

El Ghaoui, Laurent and Niculescu, Silviu-Iulian, eds., *Advances in Linear Matrix Inequality Methods in Control*

Helton, J. William and James, Matthew R., *Extending H^∞ Control to Nonlinear Systems: Control of Nonlinear Systems to Achieve Performance Objectives*

Applied Stochastic Processes and Control for Jump-Diffusions

Modeling, Analysis, and Computation

Floyd B. Hanson
University of Illinois at Chicago
Chicago, Illinois

Society for Industrial and Applied Mathematics
Philadelphia

Library of Congress Cataloging-in-Publication Data

Hanson, Floyd B.
      Applied stochastic processes and control for jump-diffusions : modeling, analysis, and computation / Floyd B. Hanson.
    p. cm. — (Advances in design and control)
    Includes bibliographical references and index.
    ISBN 978-0-898716-33-7 (alk. paper)
     1. Stochastic processes. 2. Markov processes. 3. Diffusion processes. I. Title.

QA274.H36 2007
519.2'3—dc22   2007061739

Partial royalties from the sale of this book are placed in a fund to help students attend SIAM meetings and other SIAM-related activities. This fund is administered by SIAM, and qualified individuals are encouraged to write directly to SIAM for guidelines.

 is a registered trademark.

To five generations of women in my life:

Margaret Geiger Bliss, Violet Bliss Hanson,
Ethel Hutchins Hanson, Lisa Hanson, and
Chiara Hanson Whitehurst

Contents

List of Figures   xv

List of Tables   xxi

Preface   xxiii

1   Stochastic Jump and Diffusion Processes: Introduction   1
    1.1   Poisson and Wiener Processes Basics 1
    1.2   Wiener Process Basic Properties 3
    1.3   More Wiener Process Moments 6
    1.4   Wiener Process Nondifferentiability 9
    1.5   Wiener Process Expectations Conditioned on the Past 9
    1.6   Poisson Process Basic Properties 11
    1.7   Poisson Process Moments 16
    1.8   Poisson Zero-One Jump Law 18
    1.9   Temporal, Nonstationary Poisson Process 20
    1.10   Poisson Process Expectations Conditioned on the Past 24
    1.11   Exercises 25

2   Stochastic Integration for Diffusions   31
    2.1   Ordinary or Riemann Integration 32
    2.2   Stochastic Integration in $W(t)$: The Foundations 34
    2.3   Stratonovich and Other Stochastic Integration Rules 55
    2.4   Conclusion 57
    2.5   Exercises 57

3   Stochastic Integration for Jumps   63
    3.1   Stochastic Integration in $P(t)$: The Foundations 63
    3.2   Stochastic Jump Integration Rules and Expectations 74
    3.3   Conclusion 77
    3.4   Exercises 77

4   Stochastic Calculus for Jump-Diffusions: Elementary SDEs   81
    4.1   Diffusion Process Calculus Rules 82
        4.1.1   Functions of Diffusions Alone, $G(W(t))$ 82

		4.1.2	Functions of Diffusions and Time $G(W(t), t)$	86
		4.1.3	Itô Stochastic Natural Exponential Construction	88
		4.1.4	Transformations of Linear Diffusion SDEs	94
		4.1.5	Functions of General Diffusion States and Time $F(X(t), t)$	98
	4.2		Poisson Jump Process Calculus Rules	99
		4.2.1	Jump Calculus Rule for $h(dP(t))$	99
		4.2.2	Jump Calculus Rule for $\mathcal{H}(P(t), t)$	100
		4.2.3	Jump Calculus Rule with General State $Y(t) = F(X(t), t)$	103
		4.2.4	Transformations of Linear Jump with Drift SDEs	104
	4.3		Jump-Diffusion Rules and SDEs	106
		4.3.1	Jump-Diffusion Conditional Infinitesimal Moments	106
		4.3.2	Stochastic Jump-Diffusion Chain Rule	107
		4.3.3	Linear Jump-Diffusion SDEs	109
		4.3.4	SDE Models Exactly Transformable to Purely Time-Varying Coefficients	119
	4.4		Poisson Noise Is White Noise Too!	120
	4.5		Exercises	122

5   Stochastic Calculus for General Markov SDEs: Space-Time Poisson, State-Dependent Noise, and Multidimensions — **129**

	5.1		Space-Time Poisson Process	130
	5.2		State-Dependent Generalization of Jump-Diffusion SDEs	139
		5.2.1	State-Dependent Generalization for Space-Time Poisson Processes	139
		5.2.2	State-Dependent Jump-Diffusion SDEs	141
		5.2.3	Linear State-Dependent SDEs	142
	5.3		Multidimensional Markov SDE	158
		5.3.1	Conditional Infinitesimal Moments in Multidimensions	160
		5.3.2	Stochastic Chain Rule in Multidimensions	162
	5.4		Distributed Jump SDE Models Exactly Transformable	163
		5.4.1	Distributed Jump SDE Models Exactly Transformable	164
		5.4.2	Vector Distributed Jump SDE Models Exactly Transformable	164
	5.5		Exercises	165

6   Stochastic Optimal Control: Stochastic Dynamic Programming — **169**

	6.1		Stochastic Optimal Control Problem	169
	6.2		Bellman's Principle of Optimality	172
	6.3		Hamilton–Jacobi–Bellman (HJB) Equation of Stochastic Dynamic Programming (SDP)	176
	6.4		Linear Quadratic Jump-Diffusion (LQJD) Problem	179
		6.4.1	LQJD in Control Only (LQJD/U) Problem	180
		6.4.2	LLJD/U or the Case $C_2 \equiv 0$	183
		6.4.3	Canonical LQJD Problem	184
	6.5		Exercises	188

7   Kolmogorov Forward and Backward Equations and Their Applications   193
   7.1   Dynkin's Formula and the Backward Operator 193
   7.2   Backward Kolmogorov Equations 196
   7.3   Forward Kolmogorov Equations 198
   7.4   Multidimensional Backward and Forward Equations 202
   7.5   Chapman–Kolmogorov Equation for Markov Processes
         in Continuous Time 205
   7.6   Jump-Diffusion Boundary Conditions 205
       7.6.1   Absorbing Boundary Conditions 205
       7.6.2   Reflecting Boundary Conditions 206
   7.7   Stopping Times: Expected Exit and First Passage Times 206
       7.7.1   Expected Stochastic Exit Time 208
   7.8   Diffusion Approximation Basis 213
   7.9   Exercises 215

8   Computational Stochastic Control Methods   219
   8.1   Finite Difference PDE Methods of SDP 220
       8.1.1   Linear Dynamics and Quadratic Control Costs 221
       8.1.2   Crank–Nicolson, Extrapolation-Predictor-Corrector Finite
             Difference Algorithm for SDP 222
       8.1.3   Upwinding Finite Differences If Not Diffusion-Dominated . 228
       8.1.4   Multistate Systems and Bellman's Curse of Dimensionality . 229
   8.2   Markov Chain Approximation for SDP 231
       8.2.1   MCA Formulation for Stochastic Diffusions 232
       8.2.2   MCA Local Diffusion Consistency Conditions 233
       8.2.3   MCA Numerical Finite Differences for State Derivatives and
             Construction of Transition Probabilities 233
       8.2.4   MCA Extensions to Include Jump Processes 236

9   Stochastic Simulations   241
   9.1   SDE Simulation Methods 241
       9.1.1   Convergence and Stability for Stochastic Problems
             and Simulations 242
       9.1.2   Stochastic Diffusion Euler Simulations 244
       9.1.3   Milstein's Higher Order Diffusion Simulations 248
       9.1.4   Convergence and Stability of Jump-Diffusion Euler
             Simulations 251
       9.1.5   Jump-Diffusion Euler Simulation Procedures 255
   9.2   Monte Carlo Methods 258
       9.2.1   Basic Monte Carlo Simulations 260
       9.2.2   Inverse Method for Generating Nonuniform Variates . . . 268
       9.2.3   Acceptance and Rejection Method of von Neumann 270
       9.2.4   Importance Sampling 274
       9.2.5   Stratified Sampling 276
       9.2.6   Antithetic Variates 279
       9.2.7   Control Variates 281

10   Applications in Financial Engineering   **287**
   10.1   Classical Black–Scholes Option Pricing Model 288
   10.2   Merton's Three Asset Option Pricing Model Version of Black–Scholes 291
         10.2.1   PDE of Option Pricing 299
         10.2.2   Final and Boundary Conditions for Option Pricing PDE . . . 301
         10.2.3   Transforming PDE to Standard Diffusion PDE 304
   10.3   Jump-Diffusion Option Pricing 309
         10.3.1   Jump-Diffusions with Normal Jump-Amplitudes 310
         10.3.2   Risk-Neutral Option Pricing for Jump-Diffusions 311
   10.4   Optimal Portfolio and Consumption Models 317
         10.4.1   Log-Uniform Amplitude Jump-Diffusion for Log-Returns . . 318
         10.4.2   Log-Uniform Jump-Amplitude Model 319
         10.4.3   Optimal Portfolio and Consumption Policies Application . . 321
         10.4.4   CRRA Utility and Canonical Solution Reduction 325
   10.5   Important Financial Events Model: The Greenspan Process 327
         10.5.1   Stochastic Scheduled and Unscheduled Events Model with
                 Stochastic Parameter Processes 328
         10.5.2   Further Properties of Quasi-Deterministic or Scheduled Event
                 Processes: $K(\hat{q}; A(t))dQ(t)$ 330
         10.5.3   Optimal Portfolio Utility, Stock Fraction, and Consumption . 330
         10.5.4   Canonical CRRA Model Solution 333
   10.6   Exercises 335

11   Applications in Mathematical Biology and Medicine   **339**
   11.1   Stochastic Bioeconomics: Optimal Harvesting Applications 339
         11.1.1   Optimal Harvesting of Logistically Growing Population
                 Undergoing Random Jumps 340
         11.1.2   Optimal Harvesting with Both Price and Population
                 Random Dynamics 344
   11.2   Stochastic Biomedical Applications 347
         11.2.1   Diffusion Approximation of Tumor Growth and Tumor
                 Doubling Time Application 347
         11.2.2   Optimal Drug Delivery to Brain PDE Model 353

12   Applied Guide to Abstract Theory of Stochastic Processes   **361**
   12.1   Very Basic Probability Measure Background 362
         12.1.1   Mathematical Measure Theory Basics 362
         12.1.2   Change of Measure: Radon–Nikodým Theorem
                 and Derivative 367
         12.1.3   Probability Measure Basics 368
         12.1.4   Stochastic Processes in Continuous Time on Filtered
                 Probability Spaces 371
         12.1.5   Martingales in Continuous Time 372
         12.1.6   Jump-Diffusion Martingale Representation 375
   12.2   Change in Probability Measure: Radon–Nikodým Derivatives
         and Girsanov's Theorem 376

		12.2.1	Radon–Nikodým Theorem and Derivative for Change of Probability Measure 376
		12.2.2	Change in Measure for Stochastic Processes: Girsanov's Theorem 382
	12.3	Itô, Lévy, and Jump-Diffusion Comparisons 389
		12.3.1	Itô Processes and Jump-Diffusion Processes 389
		12.3.2	Lévy Processes and Jump-Diffusion Processes 390
	12.4	Exercise 401

Bibliography   **403**

Index   **423**

A   Online Appendix: Deterministic Optimal Control   **A1**
   A.1   Hamilton's Equations: Hamiltonian and Lagrange Multiplier Formulation of Deterministic Optimal Control A2
        A.1.1   Deterministic Computation and Computational Complexity . A11
   A.2   Optimum Principles: The Basic Principles Approach A12
   A.3   Linear Quadratic (LQ) Canonical Models A22
        A.3.1   Scalar, Linear Dynamics, Quadratic Costs (LQ) A22
        A.3.2   Matrix, Linear Dynamics, Quadratic Costs (LQ) A24
   A.4   Deterministic Dynamic Programming (DDP) A28
        A.4.1   Deterministic Principle of Optimality A29
        A.4.2   Hamilton–Jacobi–Bellman (HJB) Equation of Deterministic Dynamic Programming A30
        A.4.3   Computational Complexity for Deterministic Dynamic Programming A31
        A.4.4   Linear Quadratic (LQ) Problem by Deterministic Dynamic Programming A32
   A.5   Control of PDE Driven Dynamics: Distributed Parameter Systems (DPS) A34
        A.5.1   DPS Optimal Control Problem A34
        A.5.2   DPS Hamiltonian Extended Space Formulation A35
        A.5.3   DPS Optimal State, Costate, and Control PDEs A37
   A.6   Exercises A39

B   Online Appendix: Preliminaries in Probability and Analysis   **B1**
   B.1   Distributions for Continuous Random Variables B2
        B.1.1   Probability Distribution and Density Functions B2
        B.1.2   Expectations and Higher Moments B4
        B.1.3   Uniform Distribution B5
        B.1.4   Normal Distribution and Gaussian Processes B8
        B.1.5   Simple Gaussian Processes B9
        B.1.6   Lognormal Distribution B11
        B.1.7   Exponential Distribution B14
   B.2   Distributions of Discrete Random Variables B17
        B.2.1   Poisson Distribution and Poisson Process B18

B.3   Joint and Conditional Distribution Definitions B20
       B.3.1   Conditional Distributions and Expectations B25
       B.3.2   Law of Total Probability B29
B.4   Probability Distribution of a Sum: Convolutions B30
B.5   Characteristic Functions B33
B.6   Sample Mean and Variance: Sums of Independent, Identically
       Distributed (IID) Random Variables B36
B.7   Law of Large Numbers B38
       B.7.1   Weak Law of Large Numbers (WLLN) B38
       B.7.2   Strong Law of Large Numbers (SLLN) B38
B.8   Central Limit Theorem B39
B.9   Matrix Algebra and Analysis B39
B.10   Some Multivariate Distributions B45
       B.10.1   Multivariate Normal Distribution B45
       B.10.2   Multinomial Distribution B46
B.11   Basic Asymptotic Notation and Results B49
B.12   Generalized Functions: Combined Continuous and
       Discrete Processes B52
B.13   Fundamental Properties of Stochastic and Markov Processes B59
       B.13.1   Basic Classification of Stochastic Processes B59
       B.13.2   Markov Processes and Markov Chains B59
       B.13.3   Stationary Markov Processes and Markov Chains B61
B.14   Continuity, Jump Discontinuity, and Nonsmoothness Approximations . B61
       B.14.1   Beyond Continuity Properties B61
       B.14.2   Taylor Approximations of Composite Functions B63
B.15   Extremal Principles B67
B.16   Exercises B69

C   Online Appendix: MATLAB Programs   C1
C.1   Program: Uniform Distribution Simulation Histograms C1
C.2   Program: Normal Distribution Simulation Histograms C2
C.3   Program: Lognormal Distribution Simulation Histograms C3
C.4   Program: Exponential Distribution Simulation Histograms C4
C.5   Program: Poisson Distribution Versus Jump Counter k C5
C.6   Program: Binomial Distribution Versus Binomial Frequency f_1 . . . C6
C.7   Program: Simulated Diffusion $W(t)$ Sample Paths C7
C.8   Program: Simulated Diffusion $W(t)$ Sample Paths Showing Variation
       with Time Step Size C8
C.9   Program: Simulated Simple Poisson $P(t)$ Sample Paths C9
C.10   Program: Simulated Simple Incremental Poisson $\Delta P(t)$ Sample Paths . C10
C.11   Program: Simulated Diffusion Integrals $\int (dW)^2(t)$ by Itô Partial Sums C12
C.12   Program: Simulated Diffusion Integrals $\int g(W,t)dW$: Direct Case by
       Itô Partial Sums C13
C.13   Program: Simulated Diffusion Integrals $\int g(X,t)dW$: Chain Rule . . . C14
C.14   Program: Simulated Linear Jump-Diffusion Sample Paths C16
C.15   Program: Simulated Linear Mark-Jump-Diffusion Sample Paths . . . C18

C.16   Program: Curse of Dimensionality C21
C.17   Program: Euler–Maruyama Simulations for Linear Diffusion SDE . . C23
C.18   Program: Milstein Simulations for Linear Diffusion SDE C25
C.19   Program: Monte Carlo Simulation Comparing Uniform and
        Normal Errors C27
C.20   Program: Monte Carlo Simulation Testing Uniform Distribution . . . C29
C.21   Program: Monte Carlo Acceptance-Rejection Technique C30
C.22   Program: Monte Carlo Multidimensional Integration C32
C.23   Program: Regular and Bang Control Examples C34
C.24   Program: Simple Optimal Control Example C37
C.25   Program: Bang-Bang Control with Control Switching Example C38
C.26   Program: Singular Control Examples C40

List of Figures

Note: Figures A1–A5 and B1–B6 appear online only and can be found at
www.siam.org/books/dc13

1.1   (a) Paths were simulated using MATLAB [210] with $N = 1000$ sample points, four randn states, and maximum time $T = 1.0$. (b) Paths were simulated using subsets of the same random state of randn used for the finer grids using $N = 1000, 100, 10$.   6

1.2   (a) Simulated sample paths for the simple Poisson process $P(t)$ versus the dimensionless time λt using four different MATLAB [210] random states for four different sample paths and using the exponential distribution of the time between jumps. (b) Similar illustration for the simple Poisson process increment simulations versus t with $\lambda = 1.0$ and $\Delta t = 0.05$, based upon the zero-one jump law implemented with uniform distribution paths and simulated using subsets of the same random state of rand used with a finer grid of 501 time-steps so the zero-one jump law was a good approximation. 15

2.1   Simulated sample path for the Itô forward integration approximating sum of $\int (dW)^2(t) \overset{\text{ims}}{=} t \simeq \sum_i (\Delta W_i)^2$ for $n = 10^4$ MATLAB randn sample size. 43

4.1   Example of a simulated Itô discrete approximation to the stochastic diffusion integral $I_n[g](t_{i+1}) = \sum_{j=0}^{i} g_j \Delta W_j$ for $i = 0 : n$, using MATLAB randn with sample size $n = 10,000$ on $0 \leq t \leq 2.0$. Presented are the simulated Itô partial sums S_{i+1}, the simulated noise W_{i+1}, and the error E_{i+1} relative to the exact integral, $I^{(\text{ims})}[g](t_{i+1}) \overset{\text{ims}}{=} \exp(W_{i+1} - t_{i+1}/2) - 1$, in the Itô mean square sense. 92

4.2   Example of a simulated Itô discrete approximation to the stochastic diffusion integral $I_n[g](t_{i+1}) = \sum_{j=0}^{i} g_j \Delta W_j$ for $i = 0 : n$, using MATLAB randn with sample size $n + 1 = 10,001$ on $0 \leq t \leq 2.0$. Presented are the simulated Itô partial sums S_{i+1}, the simulated noise W_{i+1}, and the error E_{i+1} relative to the stochastic chain rule partially integrated form I_{i+1} given in (4.23). 93

4.3   Four linear jump-diffusion sample paths for constant coefficients are simulated using MATLAB [210] with $N = 1000$ sample points, maximum time $T = 1.0$, and four randn and four rand states. Parameter values are $\mu_0 = 0.5$, $\sigma_0 = 0.10$, $\nu_0 = -0.10$, $\lambda_0 = 3.0$, and $x_0 = 1.0$. In addition to the four simulated states, the expected state $E[X(t)]$ and two deviation measures $E[X(t)] * V(t)$ and $E[X(t)]/V(t)$, are displayed where the factor $V(t)$ is based on the standard deviation of the state exponent $Y(t)$. 114

4.4   Four linear pure-diffusion sample paths for constant coefficients are simulated using MATLAB [210] with $N = 1000$ sample points, maximum time $T = 1.0$, and four randn states. Parameter values are $\mu_0 = 0.5$, $\sigma_0 = 0.10$, $\nu_0 = 0.0$, and $x_0 = 1.0$. In addition to the four simulated states, the expected state $E[X(t)]$ and two deviation measures $E[X(t)] * V(t)$ and $E[X(t)]/V(t)$ are displayed, where the factor $V(t)$ is based on the standard deviation of the state exponent $Y(t)$. 115

4.5   Four linear pure jump with drift sample paths for constant coefficients are simulated using MATLAB [210] with $N = 1000$ sample points, maximum time $T = 1.0$, and four randn states. Parameter values are $\mu_0 = 0.5$, $\sigma_0 = 0.0$, $\nu_0 = -0.10$, and $x_0 = 1.0$. In addition to the four simulated states, the expected state $E[X(t)]$ and two deviation measures $E[X(t)] * V(t)$ and $E[X(t)]/V(t)$ are displayed, where the factor $V(t)$ is based on the standard deviation of the state exponent $Y(t)$. 116

5.1   Four linear mark-jump-diffusion sample paths for time-dependent coefficients are simulated using MATLAB [210] with $N = 1000$ timesteps, maximum time $T = 2.0$, and four randn and four rand states. Initially, $x_0 = 1.0$. Parameter values are given in vectorized functions using vector functions and dot-element operations, $\mu_d(t) = 0.1 * \sin(t)$, $\sigma_d(t) = 1.5 * \exp(-0.01 * t)$, and $\lambda = 3.0 * \exp(-t. * t)$. The marks are uniformly distributed on $[-2.0, +1.0]$. In addition to the four simulated states, the expected state $E[X(t)]$ is presented using the quasi-deterministic equivalence (5.54) of Hanson and Ryan [115], but also presented are the sample mean of the four sample paths. . . . 158

6.1   Multibody stochastic dynamical system under feedback control. . . . 171

8.1   Estimate of the logarithm to base 2 of the order of the growth of memory and computing demands using 8 byte words to illustrate the curse of dimensionality in the diagonal Hessian case for $n_x = 1:10$ dimensions and $N_x = 1:64 = 1:2^6$ nodes per dimension. Note that 1 KB has a base 2 exponent of $10 = \log_2(2^{10})$, while the base 2 exponent is 20 for 1 MB, 30 for 1 GB, 40 for 1 TB, 50 for 1 PB, and 60 for 1 EB. 231

9.1   EM SDE simulations code. 245

9.2   Comparison of coarse Euler–Maruyama and fine exact paths, simulated
       using MATLAB with $N_t = 1024$ fine sample points for the exact
       path (9.15) and $N_t/8 = 128$ coarse points for the Euler path (9.13),
       initial time $t_0 = 0$, final time $t_f = 5$, and initial state $x_0 = 1.0$.
       Time-dependent parameter values are $\mu(t) = 0.5/(1 + 0.5t)^2$ and
       $\sigma(t) = 0.5$. 246

9.3   Error in coarse Euler–Maruyama and fine exact paths using the coarse
       discrete time points. The simulations use MATLAB with the same
       values and time-dependent coefficients as in Figure 9.2. The Euler
       maximal-absolute error for this example is $1.3 \simeq 34\Delta t/8$, while for
       $N_t = 4096$ the maximal error is better at $0.28 \simeq 29\Delta t/8$. 246

9.4   Comparison of coarse Milstein and fine exact paths, simulated using
       MATLAB with $N_t = 1024$ fine sample points for the exact path (9.15)
       and $N_t/8 = 128$ coarse points for the Milstein path (9.23), initial time
       $t_0 = 0$, final time $t_f = 5$, and initial state $x_0 = 1.0$ as in Figure 9.2.
       Time-dependent parameter values are $\mu(t) = 0.5/(1 + 0.5t)^2$ and
       $\sigma(t) = 0.5$. 250

9.5   Error in coarse Milstein and fine exact paths using the coarse discrete
       time points. The simulations use MATLAB with the same values and
       time-dependent coefficients as in Figure 9.2. The Milstein maximal-
       absolute error for this example is 1.2, while for $N_t = 4096$ the maximal
       error is better at 0.95. 250

9.6   Difference in coarse Milstein and Euler paths using the coarse discrete
       time points. The simulations use MATLAB with the same values
       and time-dependent coefficients as in Figure 9.2. The Milstein–Euler
       maximal-absolute difference for this example is 0.19, while for $N_t =$
       4096 the maximal difference is comparable at 0.24. 251

9.7   Jump-adaptive code fragment. 259

9.8   Monte Carlo simulations for testing use of the uniform distribution to
       approximate the integral of the integrand $F(x) = \sqrt{1 - x^2}$ on $(a, b) =$
       $(0, 1)$ using MATLAB code C.20, called mcm1test.m, on p. C29 in
       Online Appendix C, for $n = 10^k$, $k = 1:7$. 265

9.9   Inverse Poisson method code to generate jump-counts using the uni-
       form distribution [97, *Figure* 3.9]. 270

9.10   Monte Carlo simulations shown apply the acceptance and rejection
       technique and the normal distribution to compute the estimates for the
       mean $\widehat{\mu}_n$ and the magnified standard error $10 \cdot \widehat{\sigma}_n/\sqrt{n}$ for the integral
       of the truncated normal distribution with $F(x) = \phi_n(x)$ on $[a, b] =$
       $[-2, 2]$ using MATLAB code C.21, mcm2acceptreject.m, on
       p. C30 in Online Appendix C, for $n = 10^k$, $k = 1:7$. 272

9.11   Monte Carlo simulations for estimating multi-dimensional integrals for the n_x-dimension normal integrand $F(\mathbf{x}) = \phi_n(\mathbf{x})$ on $[\mathbf{a}, \mathbf{b}] = [-2, 2]^{n_x}$ using MATLAB code C.22, mcm3multidim.m, on p. C32 in Online Appendix C, for $n = 10^k$, $k = 1:6$. The acceptance-rejection technique is used to handle the finite domain. 273

10.1   Optimal portfolio stock fraction policy $u^*(t)$ on $t \in [0, 12]$ subject to the control constraint set $[U_0^{(\min)}, U_0^{(\max)}] = [-10, 10]$. 327
10.2   Optimal consumption policy $c^*(t, w)$ for $(t, w) \in [0, 12] \times [0, 100]$. . 328

11.1   Optimal tumor density $Y_1^*(x_1, t)$ in the one-dimensional case with time as a parameter rounded at quartile values $\{0, t_{q_1} = t_f/4, t_{\mathrm{mid}} = t_f/2, t_{q_3} = 3t_f/4, t_f\}$, where $t_f = 5$ days. The total tumor density integral is reduced by 29% in the 5-day simulated drug treatment trial. 358

A.1   Hamitonian and optimal solutions for regular control problem example from (A.30) for $X^*(t)$ and (A.31) for $\lambda^*(t)$. Note that the $\gamma = 0.5$ power utility is only for illustration purposes. A10
A.2   Hamiltonian and optimal solutions for bang control problem example from (A.30) for $X^*(t)$ and (A.31) for $\lambda^*(t)$. Note that the $\gamma = 2.0$ power utility is only for illustration purposes. A11
A.3   Optimal solutions for a simple, static optimal control problem represented by (A.35) and (A.36), respectively. A13
A.4   Optimal control, state, and switch time multiplier sum are shown for bang-bang control example with sample parameter values $t_0 = 0$, $t_f = 2.0$, $a = 0.6$, $M = 2$, $K = 2.4$, and $x_0 = 1.0$. The computed switch time t_s is also indicated. A18
A.5   Optimal state solutions for singular control example leading to a bang-singular-bang trajectory represented by (A.60). Subfigure (a) yields a maximal bang trajectory from x_0 using $U^{(\max)}$, whereas Subfigure (b) yields a minimal bang trajectory from x_0 using $U^{(\min)}$. A22

B.1   Histograms of simulations of uniform distribution on $(0, 1)$ using MATLAB [210] for two different sample sizes N. B7
B.2   Histograms of simulations of the standard normal distribution with mean 0 and variance 1 using MATLAB [210] with 50 bins for two sample sizes N. The histogram for the large sample size of $N = 10^5$ in (b) exhibits a better approximation to the theoretical normal density $\phi_n(x; 0, 1)$. B10
B.3   Histograms of simulations of the lognormal distribution with mean $\mu_n = 0$ and variance $\sigma_n = 0.5$ using MATLAB [210] normal distribution simulations, x = exp(mu*ones(N,1) + sigma*randn(N,1)) with 150 bins for two sample sizes. The histogram for the large sample size of $N = 10^5$ in (b) exhibits a better approximation to the theoretical lognormal density $\phi_n(x; 0, 1)$ than the one in (a). B14

B.4   Histograms of simulations of the standard exponential distribution, with mean taken to be $\mathrm{mu} = 1$, using MATLAB's hist function [210] with 50 bins for two sample sizes N, generated by $\mathrm{x} = -\mathrm{mu} * \log(\mathrm{rand(N, 1)})$ in MATLAB. The histogram for the large sample size of $N = 10^5$ in (b) exhibits a better approximation to the standard theoretical exponential density $\phi_e(x; 1)$. B16

B.5   Poisson distributions with respect to the Poisson counter variable k for parameter values $\Lambda = 0.2, 1.0, 2.0,$ and 5.0. These represent discrete distributions, but discrete values are connected by dashed, dotted, and dash-dotted lines only to help visualize the distribution form for each parameter value. B20

B.6   Binomial distributions with respect to the binomial frequency f_1 with $N = 10$ for values of the probability parameter, $\pi_1 = 0.25, 0.5,$ and 0.75. These represent discrete distributions, but discrete values are connected by dashed, dotted, and dash-dotted lines only to help visualize the distribution form for each parameter value. B49

List of Tables

Note: Tables A1 *and* B1 *appear online only and can be found at*
`www.siam.org/books/dc13`

1.1   Some expected moments (powers) of absolute value of the Wiener increments.   7

1.2   Some expected moments (powers) of Poisson increments and their deviations.   18

2.1   Some Itô stochastic diffusion differentials with an accuracy with error $o(dt)$ as $dt \to 0^+$. 54

3.1   Some stochastic jump-integrals of powers with an accuracy with error $o(dt)$ as $dt \to 0^+$. 68

3.2   Some Itô stochastic jump differentials with an accuracy with error $o(dt)$ as $dt \to 0^+$. 74

4.1   Example transforms listing original coefficients in terms of target and transform coefficients. 120

7.1   Some simple jump-amplitude models and inverses. 202

A.1   Some final conditions for deterministic optimal control. A4

B.1   Some expected moments of bivariate normal distribution. B46

Preface

Everything should be as simple as it is,
but not simpler.
—Albert Einstein (1879–1955)

A mathematical theory is not to be considered complete
until you have made it so clear that you can explain it
to the first man whom you meet on the street.
—David Hilbert (1862–1943)

Always take a pragmatic view in applied mathematics:
the proof of the pudding is in the eating.
—N. H. Bingham and Rüdiger Kiesel [33]

Overview of This Book

The aim of this book is to be a self-contained, practical, entry-level text on stochastic processes and control for jump diffusions in continuous time, technically Markov processes in continuous time.

The book is intended to be accessible to graduate students and to be a research monograph useful to researchers in applied mathematics, computational science, and engineering. In fact, a number of colleagues have said that they would like to learn about stochastic processes but have found it difficult to learn from the existing literature. Also, the book may be useful for practitioners of financial engineering who need fast and efficient answers to stochastic financial problems. Hence, the exposition is based upon integrated basic principles of applied mathematics, applied probability, and computational science. The target audience includes mathematical modelers and students in many areas of science and engineering seeking to construct models for scientific applications subject to uncertain environments. The prime focus of the text is on modeling and problem solving. The utility of the exposition, based upon systematic derivations along with essential proofs in the spirit of classical applied mathematics, is related more to setting up a stochastic model of an application than to abstract theory. However, a lengthy last chapter is intended to bridge the gap between the applied world and the abstract world so that students and readers of applied mathematics may understand the more abstract literature.

The formulation and proof of more rigorous theorems is not of immediate importance compared to modeling and solving an applied problem, although many proofs are given

xxiii

here. Many research problems deal with new applications, and often these new applications require models beyond those in the existing literature. So, it is important to have a reasonably understandable derivation for a nearby model that can be perturbed to obtain a proper new model. The level of rigor here is embodied in correct and systematic derivations, with many proofs and results not available elsewhere, under reasonable conditions, though not necessarily the tightest possible conditions. In fact, much of this book and the theory of Markov processes in continuous time are based upon modifying the formulations for continuous functions in regular and advanced calculus to extend them to the discontinuous and nonsmooth functions of stochastic calculus.

Origin of the Book

The book is based upon the author's courses *Math 574: Applied Optimal Control*; *Math 590: Special Topics: Applied Stochastic Control*; *MCS 507: Mathematical, Statistical and Scientific Software for Industry*; and (in part) on *MCS 571: Numerical Methods for Partial Differential Equations*. In addition, the results from research papers on computational stochastic dynamic programming, computational finance, and computational biomedicine are included. Courses in asymptotic analysis and numerical analysis play a role as well. However, as with lectures, every attempt is made to keep the book self-contained through an integrated approach, without depending heavily on prerequisites, so that the book can be of use to a diverse interdisciplinary audience.

This book integrates many of the research and exposition advances made in computational stochastic dynamic programming and stochastic modeling. These advances exhibit the broader impact of the book's applications and its computationally oriented approach. The stochastic applications are wide ranging, including the optimal economics of biological populations in uncertain and disastrous environments, bio-medical applications in cancer modeling and optimal treatment, and financial engineering with applications in option pricing and optimal portfolios.

How This Book Is Organized and How to Use It

- The prerequisites are too numerous to expect of any reader of a wide-ranging interdisciplinary book such as this one, so a single online appendix, Appendix B, on the essentials of probability theory, mathematical analysis, matrix algebra, and other topics can be found at

  www.siam.org/books/dc13

  Overspecification of prerequisites tends to filter out too many students who could benefit from this material. Appendix B is intended to bring all readers up to the same level by self-study, where necessary, of the basic concepts and notations of probability and analysis needed for jump-diffusion processes and their deviations from continuity. The book is not meant to be taught or read in sequence but to include relevant results when needed and to make the presentation as self-contained as possible.

- Chapters 1, 2, 3, and 4 cover the basics for simple jump-diffusions, i.e., stochastic diffusion (Wiener or Brownian motion) and simple Poisson driven processes, including stochastic integration based upon Itô's computationally motivated mean square convergence for Markov processes and stochastic calculus for transformations of stochastic differential equations (SDEs). The speed and depth of coverage for students or readers will depend on their level of expertise, particularly with respect to prior knowledge of probability and diffusion processes which are more well known. The presentation is more elementary than that of later chapters to reduce the likelihood that readers will get lost at the basic level.

- Advanced and special topics are found in Chapters 5 through 12 and can be selected according to the instructor's or reader's interests. There are more chapters than can be covered in any one course.

  - Chapter 5, Stochastic Calculus for General Markov SDEs, covers more advanced and general topics for SDEs. These include jumps driven by compound Poisson or Poisson random measure processes that allow randomly distributed jump-amplitudes, state-dependent jump-diffusions, and multidimensional jump-diffusions.

  - Chapter 6, Stochastic Optimal Control: Stochastic Dynamic Programming, and Chapter 8, Computational Stochastic Control Methods, can form a stochastic control theory and the computational components of a course. Also, if a chapter on deterministic control theory is desired for introduction and contrast to stochastic control, Online Appendix A, Deterministic Optimal Control, is available at the above listed URL. Appendix A gives a summary of deterministic optimal control results to provide a background for comparison to the stochastic optimal control results, but this could be skipped if a deterministic control course is a prerequisite or if only stochastic optimal control is of interest. In Chapter 6, stochastic optimal control problems are introduced and the equation of stochastic dynamic programming is systematically derived from the basic principles of applied mathematics. Chapter 8 has treatments using either modified finite difference methods for optimal control problems or the Markov chain approximation methods. Computational methods are important for stochastic optimal control problems because there are so few exact analytical solutions.

  - Chapter 7, Kolmogorov Equations, covers PDE methods for solving stochastic problems using the forward and backward Kolmogorov equations, Dynkin's integral formulas (also Feynman–Kac as Dynkin's with an integrating factor) that help provide PDE solutions without directly solving the PDE, boundary conditions, and stopping time problems. Knowledge of partial derivatives from advanced courses in calculus is all that should be needed; a course in PDEs will be of little help, since such a course is not essential and only these integral formulas are used in this chapter. PDE methods are an applied alternative to the abstract method of using martingales to solve stochastic problems, such as those in finance. (See Chapter 12 for martingale and other abstract approaches.)

  - Chapter 9, Stochastic Simulations, contains treatments for direct simulations of SDEs and general simulations by the Monte Carlo method. This chapter, along with Chapter 8 on computational dynamic programming, could form a computational component of a course.

  - Chapter 10 on financial applications and Chapter 11 on biomedical applications provide substantial examples of application of the theory and techniques treated in this book. Chapter 10 explains Merton's mathematical justification and generalization of the classical Black–Scholes option pricing problem in sufficient detail for those familiar with the diffusion processes properties in Chapters 1–4 and is a good motivating application for Chapter 5. Also treated are option pricing models for jump-diffusions, optimal portfolio and consumption models, and an important events model that modifies the jump-diffusion model with a quasi-deterministic jump model for scheduled announcements and random responses. Chapter 11 includes applications to stochastic optimal control or bioeconomic models, diffusion approximation models of tumor growth, and a deterministic optimal control model of a PDE-driven drug delivery model for the brain.

- Chapter 12 is an applied description of abstract probability methods, including probability measure, probability space, martingales, and change in probability measure using either Radyn–Nikodým or Girsanov theorems. The last section of this chapter is a generalization of jump-diffusions called Lévy processes that permit the jump rate to be infinite. This chapter is meant to be a bridge between the applied view of stochastic processes and the abstract view to ease the transition to reading some of the more abstract literature on stochastic processes. However, depending on the instructor or reader, parts of this chapter can be woven into the coverage of the earlier chapters. For instance, a colleague has said that Girsanov's measure change transformation was needed in his financial applications course, and there are a pure diffusion version and a jump-diffusion version of the Girsanov theorem in this chapter.

- As mentioned previously, Online Appendix A covers deterministic optimal control and analysis, and Online Appendix B presents background preliminaries in probability and analysis. A third Online Appendix, C, contains example MATLAB programs that were used to generate many of the figures. All three Online Appendices can be found at

  www.siam.org/books/dc13

  The MATLAB source files can be found at

  http://www.math.uic.edu/~hanson/pub/SIAMbook/MATLABCodes/

Distinct Features of This Book

- Both analytical and computational methods are emphasized based on the utility, with respect to the computational complexity, of the problems. Exercises and examples in the elementary chapters are both computational and analytic. Students need to have good analytic and computational skills to do well, since diverse skills are needed for many jobs.

- The treatment of jump and diffusion processes is balanced as well, rather than there being a stronger or nearly exclusive emphasis on diffusion processes. This is a unique feature of this book. This treatment of jump-diffusions is important for training graduate students to do research on stochastic processes; since the analysis of diffusion processes is so well developed, there are many opportunities for open problems on jump diffusions.

- The book clearly shows the strong role that discontinuous as well as nonsmooth properties of stochastic processes play compared to the random properties by emphasizing a concrete jump calculus, without much reliance on measure-theoretic constructs, except for the very useful random Poisson measure concept.

- Basic principles of probability theory in the spirit of classical applied mathematics are used to set up the practical foundations through clear and systematic derivations, making the book an accessible research tool for many who work with applications.

- The book shows how analytical-canonical control problem models, such as the linear-quadratic jump-diffusion (LQJD) problem and financial risk-adverse power utilities, can be used to reduce computational dimensional complexity of approximate solutions along with other computational techniques.

- Insightful and useful materials are presented so that the book can be readily used to model realistic applications and even modify the derivations when new applications do not quite fit the old stochastic model.

- Clear explanations are given for the entry-level student. In particular, clear and consistent notation is used, such that the notation is clearly identified with the quantity it symbolizes rather than arbitrarily selected. Sometimes this has meant some compromise on some standard notation—for instance, P is used for the Poisson process to be consistent with the W used for the Wiener process. This means that P could not be used for probability, so Prob is used in place of P (or Pr) and is clearer to a diverse audience. Similarly, probability distributions are denoted by Φ and densities by ϕ since P is used for Poisson and F is used for transformation functions throughout the book.

Target Audience

Colleagues and students have requested a more accessible, practical treatment of these topics. They are interested in learning about stochastic calculus and optimal stochastic control in continuous time but are reluctant to invest time to learn it from more advanced treatments relying heavily on abstract concepts. Hence, this book should be of interest to an interdisciplinary audience of applied mathematicians, applied probabilists, engineers (including control engineers dealing with deterministic problems and financial engineers needing fast, useful methods for modeling rapidly changing market developments), statisticians, and other scientists. After this primary audience, a secondary audience would be mathematicians, engineers, and scientists, who may use this book as a research monograph when seeking to more intuitively and more fully understand stochastic processes and how the more advanced analytical approaches fit in with important applications, such as financial market modeling.

Prerequisites

For optimal use of this book, it would be helpful for students to have had prior introduction to applied probability theory, including continuous random variables and mathematical analysis at least at the level of advanced calculus. Ordinary differential equations, partial differential equations, and basic computational methods would be helpful, but the book does not rely on prior knowledge of these topics; rather, it uses basic calculus-style motivations. In other words, the more-or-less usual preparation for students of applied mathematics, science, and engineering should be sufficient. However, the author has striven to make this book as self-contained as practical, not strongly relying on readers' prior knowledge, but instead explaining or reviewing the prerequisite knowledge at the point it is needed to justify a step in the systematic derivation of some mathematical result. Online Appendix B supplies essential preliminaries.

MATLAB® Computation

As part of the theme of balancing computation and analysis, MATLAB the matrix laboratory computation system, is used for almost all computational examples and figure illustrations. Simple MATLAB codes are described in class, and the code for all text figures is given in Online Appendix C in both listing and source forms. MATLAB greatly facilitates the development of code and is ideally suited to stochastic processes and control problems. Also, MATLAB now comes with the Maple kernel built into the MATLAB student package for including elementary symbolic computations with numeric computations. Beyond the initial elementary assignments of my courses, students are required to submit their assignments with professionally done illustrations, for which they can find examples in Online Appendix C. Many students surveyed at the end of the class list MATLAB with the other topics that they were happy to learn. MATLAB is also helpful for producing professional research papers and theses.

Acknowledgments

I am grateful to a number of coworkers and students, as well as other authors, who helped as reviewers or who contributed to this applied stochastic book through research contributions, and agencies who have provided grant support for computational stochastic dynamic programming:

- My research assistants and graduate students, Siddhartha Pratim Chakrabarty, Zongwu Zhu, Guoqing Yan, and Jinchun Ye, have helped review drafts of this book with the keen eyes of applied mathematics and computer science graduate students to make sure that it would be useful and understandable to other graduate students.

- Over the years many have helped develop pieces of the underlying applied theory or model applications: Abdul Majid Wazwaz, Dennis Ryan, Kumarss Naimipour, Siu-Leung Chung, Huihuang (Howard) Xu, Dennis J. Jarvis, Christopher J. Pratico, Michael S. Vetter, John J. Westman, Raghib abu-Saris, and Daniel L. Kern.

- Some of my students have used the SIAM Travel Fund, so half the royalties of this book are to be donated to this fund.

- My wife Ethel did the major job of the final proofread.

- This work has been influenced, consciously and subconsciously, by books and related works by many authors, such as

Applebaum [12], Arnold [13], Bingham and Kiesel [33], Bliss [36], Çinlar [56], Clark [57], Cont and Tankov [60], Feller [84, 85], Fleming and Rishel [86], Gihman and Skorohod [95, 96], Goel and Richter-Dyn [99], Glasserman [97], Hammersley and Handscomb [105], D. Higham [140, 141], Hull [148], Itô [150], Jäckel [151], Jazwinski [155], Karlin and Taylor [162, 163, 265], Kirk [164], Kloeden and Platen [166], Kushner [174, 176], Kushner and Dupuis [179], Ludwig [187], Merton [203], Mikosch [209], Øksendal [222], Øksendal and Sulem, [223], Parzen [224], Protter [232], Runggaldier [239], Schuss [244], Snyder and Miller [252], Tuckwell [270], Steele [256], Wonham [286], and others.

Although their influence may not be directly apparent here, some have shown how to make the presentation much clearer, while others have supplied the motivation to simplify the presentation, making it accessible to a more general audience and to use with other applications.

• This material is based upon work supported by the National Science Foundation (NSF) under Grants 02-07081, 99-73231, 96-26692, 93-01107, 91-02343, and 88-0699 in the Computational Mathematics Program *Advanced Computational Stochastic Dynamic Programming for Continuous Time Problems* at the University of Illinois at Chicago. Any opinions, findings, and conclusions or recommendations expressed in this material are those of the author and do not necessarily reflect the views of the NSF. In addition, the NSF supplied Research Experience for Undergraduates support for Mike Vetter to develop a portable object-oriented version of our multidimensional computational control visualization system.

• The Argonne National Laboratory Advanced Computing Research Facility supplied parallel processing training through summer and sabbatical support from 1985 through 1988 that enabled the development of large-scale computational stochastic applications.

• Many of our national supercomputing centers provided supercomputing time during 1987 through 2003 on the most powerful supercomputers for continuing research for solving large-scale stochastic control problems in advanced computational stochastic dynamic programming and also for computational science education. In addition to Argonne National Laboratory, these were the National Center for Supercomputing Applications, Los Alamos National Laboratory's Advanced Computing Laboratory, the Cornell Theory Center, the Pittsburgh Supercomputing Center, and the San Diego Supercomputing Center.

• At the University of Illinois at Chicago, the Laboratory of Advanced Computing and associate centers have supplied us with cluster computing, and the Electronic Visualization Laboratory supplied a most capable graduate student, Chris Pratico, and facilities for developing a multidimensional computational control visualization system using a real-time socket feed from our Los Alamos National Laboratory account.

Floyd B. Hanson

Chapter 1

Stochastic Jump and Diffusion Processes: Introduction

Life is good for only two things, discovering mathematics and teaching mathematics.
—Siméon Denis Poisson (1781–1840)

I do not regret my attempts, for it is only by trying problems that exceed his powers that the mathematician can ever learn to use these powers to their full extent.
—Norbert Wiener (1894–1964) in *Ex-Prodigy*

The generation of random numbers is too important to be left to chance.
—Robert Coveyou at http://www.xs4all.nl/˜jcdverha/ scijokes/1_5.html#subindex

1.1 Poisson and Wiener Processes Basics

This chapter introduces Wiener processes $W(t)$ and simple Poisson jump processes $P(t)$ in differential and integral forms. The Wiener and Poisson processes form the tools of a toolbox to create jump-diffusion process models. Wiener processes are also called diffusion or, loosely, Brownian motion.

The processes $W(t)$ and $P(t)$ are **continuous-time stochastic processes**, which basically means they are continuous time-dependent random variables.[1] They are also a special form of stochastic processes called a Markov process that is without memory of all but the prior state and that can be simply defined [56], repeating the essential definition given in Online Appendix B, as follows.

---

[1] In this book, the words stochastic and random have the same meaning, involving probability or chance.

Definition 1.1. *The stochastic process $X(t)$ is a **Markov process** provided the conditional probability satisfies*

$$\text{Prob}[X(t + \Delta t) = x \mid X(s), 0 \le s \le t] = \text{Prob}[X(t + \Delta t) = x | X(t)]$$

for any $t \ge 0$ and any $\Delta t \ge 0$ and x is in the state space, $\mathcal{D}_x$.

The stochastic processes serve as useful concepts for modeling random changes in time with stochastic differential equations, similar to the use of ordinary differential equations to model deterministic (nonstochastic) problems. These standard processes have basic infinitesimal moments

$$\text{E}[dW(t)] = 0 \ \text{ and } \ \text{Var}[dW(t)] = dt \tag{1.1}$$

for the **differential Wiener process** with initial condition $W(0^+) = 0$ **with probability one** (w.p.o.), while

$$\text{E}[dP(t)] = \lambda dt = \text{Var}[dP(t)] \tag{1.2}$$

for the **differential of the simple Poisson counting process** with rate $\lambda > 0$ and initial condition $P(0^+) = 0$ with probability one. The Wiener process is a mathematical idealization of **Brownian motion**, but often the term Brownian motion is used instead of the term Wiener process.

Remark 1.2. *If the W and P processes were to start at a different initial time other than zero, say, at $t = t_0$, then the initial conditions would be changed to $W(t_0^+) = 0^+$ and $P(t_0^+) = 0^+$, respectively. There is nothing special about the zero initial conditions, just convenience and standardization*

The simplest and most useful view of these differential stochastic processes is to consider them defined as increments, i.e.,

$$dW(t) \equiv W(t + dt) - W(t) \tag{1.3}$$

and

$$dP(t) \equiv P(t + dt) - P(t), \tag{1.4}$$

for infinitesimal increments in time dt. The property that

$$\text{Var}[dW(t)] = \text{E}[(dW(t))^2] = dt \tag{1.5}$$

is motivation for the nondifferentiability of the $W(t)$ process since the limit of

$$\sqrt{\text{Var}[dW(t)]}/dt = \sqrt{\text{E}[(dW(t))^2]}/dt = \frac{1}{\sqrt{dt}} \to +\infty \tag{1.6}$$

as $dt \to 0^+$, i.e., the variance of the ratio of differentials $\text{Var}[dW(t)/dt] \to +\infty$ as $dt \to 0^+$. Hence, the differentiability of $W(t)$ is inconsistent with the failure of the variance of the quotient $dW(t)/dt$ in the limit $dt \to 0^+$. Equation (1.6) says that the root mean square (RMS) derivative becomes unbounded as $dt \to 0^+$. This is not a rigorous proof that $W(t)$ is a nonsmooth process, although $W(t)$ is a continuous process from (1.1). (For a proof that $W(t)$ is nondifferentiable, see Theorem 1.9.)

1.2   Wiener Process Basic Properties

The assumptions for the Wiener process, including that of being normally distributed, are defined by the following properties.

Properties 1.3. *The standard Wiener process* $W(t)$

- $W(t)$ *is a **continuous process**, since*

$$W(t^+) = W(t) = W(t^-), \quad t > 0.$$

- $W(t)$ *has **independent increments**, since the Wiener increments*

$$\Delta W(t_i) = W(t_i + \Delta t_i) - W(t_i)$$

*are mutually independent for all t_i on nonoverlapping time-intervals. The **nonover-lapping time-intervals** are defined such that $t_i \geq 0$, $t_{i+1} = t_i + \Delta t_i$, and any $\Delta t_i > 0$ for $0 = 1 : n$, so that*

$$t_i < t_{i+1} \text{ for } i = 0:n.$$

Note that $W(t_i) = W(0) + \sum_{j=0}^{i-1} \Delta W(t_j)$ depends on all preceding increments, recalling that $W(0) = 0$ with probability one at $t_0 = 0$, i.e.,

$$\text{Prob}[\Delta W(t_i) \leq w_i, \Delta W(t_j) \leq w_j] = \text{Prob}[\Delta W(t_i) \leq w_i] \cdot \text{Prob}[\Delta W(t_j) \leq w_j],$$

if $j \neq i$, such that there is no overlap in the time-intervals $[t_i, t_{i+1})$ and $[t_j, t_{j+1})$. Note that $\Delta W(t_i)$, as a forward increment, is independent of $W(t_i)$ (see Definition B.35 for independent random variables) and that $\Delta W(t_i) \equiv W(t_i + \Delta t_i) - W(t_i)$ is associated with the time-interval $[t_j, t_j + \Delta t_j)$, open on the right to be compatible with right-continuity of the Poisson process.

- $W(t)$ *is a **stationary process**, since the distribution of the increment $\Delta W(t) = W(t + \Delta t) - W(t)$, with $\Delta t > 0$, is independent of t.*

- $W(t)$ *is a **Markov process**, since*

$$\text{Prob}[W(t + \Delta t) = w \mid W(s), s \leq t] = \text{Prob}[W(t + \Delta t) = w | W(t)]$$

for any $t \geq 0$ and any $\Delta t \geq 0$. (It is helpful to note that $W(t)$ is synonymous with the increment $(W(t) - W(0))$.)

- $W(t)$ *is normally distributed with mean $\mu = 0$ and variance $\sigma^2 = t$, $t > 0$; i.e., the density of $W(t)$ is*

$$\phi_{W(t)}(w) = \phi_n(w; 0, t) = \frac{1}{\sqrt{2\pi t}} \exp\left(-\frac{w^2}{2t}\right) \qquad (1.7)$$

when $-\infty < w < +\infty$ and $t > 0$. (The actual distribution function for $W(t)$, $\Phi_{W(t)}(w)$, is given in (B.22).)

- $W(0) = 0$ **with probability one**, *since* $\phi_{W(0^+)}(w) = \delta(w)$ *from* (1.7), *i.e., in the limit as* $t \to 0^+$. *(See Exercise 22 on p. B73 in Section B.16.)*

Thus, the increments $\Delta[W(t+i\,\Delta t)] \equiv W(t+(i+1)\Delta t) - W(t+i\,\Delta t)$ for $i = 0, 1, \ldots$ are stationary, independent and identically distributed (IID) as a normal distribution given time-step Δt and $t \geq 0$, i.e.,

$$\phi_{\Delta W(t)}(w) = \phi_n(w; 0, \Delta t) = \frac{1}{\sqrt{2\pi\,\Delta t}} \exp\left(-\frac{w^2}{2\Delta t}\right) \tag{1.8}$$

when $-\infty < w < +\infty$ and $\Delta t > 0$. So the basic moments of the Wiener increments are

$$\mathrm{E}[\Delta W(t)] = 0, \quad \mathrm{Var}[\Delta W(t)] = \Delta t. \tag{1.9}$$

Similarly, the stationarity property of the $dW(t) = W(t + dt) - W(t)$ differential process, when $dt > 0$, has the same probability distribution as the process $W(dt)$ when $t > 0$. Thus, the distribution from (1.7) is normal with mean $\mu = 0$ and variance $\sigma^2 = dt$,

$$\phi_{dW(t)}(w) = \phi_n(w; 0, dt) = \frac{1}{\sqrt{2\pi\,dt}} \exp\left(-\frac{w^2}{2dt}\right) \tag{1.10}$$

when $-\infty < w < +\infty$ and $dt > 0$.

Theorem 1.4. *Covariance of $W(t)$.*
If $W(t)$ is a Wiener process, then

$$\mathrm{Cov}[W(t), W(s)] = \min[t, s]. \tag{1.11}$$

Proof. This theorem is a very elementary application of the independent increment and mean-zero properties of Wiener or diffusion processes, also demonstrating how applications of independent increments rely on the zero-mean property. The zero-mean property implies that $\mathrm{E}[W(t)] = 0 = \mathrm{E}[W(s)]$. First consider the case $s < t$ and write $W(t) = W(s) + (W(t) - W(s))$, i.e., as independent increments (see Definition B.35 for expectations of products of independent random variables) and note that the first increment is $W(s) - W(0) = W(s)$ on $[0, s)$ since $W(0) = 0$ and the second increment is on $[s, t)$; then

$$\begin{aligned}
\mathrm{Cov}[W(t), W(s)] &= \mathrm{E}[W(t)W(s)] = \mathrm{E}[W^2(s) + W(s)(W(t) - W(s))] \\
&= \mathrm{E}[W^2(s)] + \mathrm{E}[W(s)(W(t) - W(s))] \\
&= \mathrm{Var}[W(s)] + \mathrm{E}[W(s)]\mathrm{E}[(W(t) - W(s))] \\
&= s + 0 \cdot 0 = s,
\end{aligned}$$

using the linearity of the expectation operator (B.9), the definition of the variance (B.10) together with the separability of expectations (B.80) for independent increments $W(s)$ and $(W(t) - W(s))$, and finally note that $W(s)$ denotes the independent increment $W(s) - W(0)$ with variance s. See (B.22) and (1.7). In the case $t < s$: $\mathrm{Cov}[W(t), W(s)] = t$ by symmetry using the splitting $W(s) = W(t) + (W(s) - W(t))$. Combining both cases produces the conclusion $\mathrm{Cov}[W(t), W(s)] = \min[s, t]$, where the function $\min[s, t]$ denotes the minimum of s and t. $\square$

When computing diffusion sample paths, i.e., the trajectory of $W(t)$ in time t, it is necessary to break up the time domain, say, $[0, T]$, into small increments $\Delta T = T/N$, where N is the number of random samples that will be used, so that each corresponding Wiener increment $\Delta W(t_i)$ will be independent. Since $W(0) = 0$ **with probability one**, let $t_i = i \cdot \Delta T$ for $i = 0 : N$; then

$$W(t_{i+1}) = \sum_{j=0}^{i} \Delta W(t_j).$$

Using MATLAB, for instance, an integer state, say, 0, is selected with the MATLAB command

```
randn('state',0);
```

where **'state'** is a literal script argument specifying that this function call is to set the random state of the function **randn**. A row N-vector set of diffusion increments can be computed wholesale by the formula

```
DWv = sqrt(DT)*randn(1,N); % scaled normal RNG used for Wiener,
```

where **randn(N,1)** is the $N \times 1$ standard zero-mean, unit-variance normal random number generator (RNG) of MATLAB. The factor **sqrt(DT)** is the Wiener scaling for the square root of the variance (1.9). Then the simulated trajectory can be computed by

```
% percent sign marks a comment.
tv = 0:DT:T;  % time vector tv(1:N+1).
Wv = zeros(1,N+1); % pre-declare size for MATLAB efficiency.
for  i = 1:N  % could replace loop using cumsum function.
     Wv(i+1) = sum((DWv(1:i)));
end
```

assuming $Wv(1) = 0.0$ in the MATLAB shifted subscript base at one, rather than at zero. In the above code fragment, the i-loop and the sum-function could be efficiently replaced with the MATLAB cumsum-function, but then the code would not be as transparent to many readers.

Finally, the diffusion sample path can be plotted with

```
plot(tv,Wv,'k-'); % much more is needed to make a nice graph,
```

and results for four sample paths are displayed in Figure 1.1(a) using $N = 1000$, $T = 1.0$, and $k = 1:4$ **randn** states. The MATLAB program used to generate this part of the figure is given in Program C.7, called wiener06fig1.m in Online Appendix C.

In Figure 1.1(b), the variation of the fine structure of the sample path is displayed, with time-step size using subsets of the same random sample state. The sample paths in this case differ markedly since the sample subsets are quite different in quantity, being $N = 1000$, 100, and 10 random sample points for $\Delta t = 10^{-3}$, 10^{-2}, and 10^{-1}, respectively, so the different cumulative set of random points leads to quite different random trajectories. The code used to generate this part of the figure is given in Program C.8, called wiener06fig2.m in Online Appendix C.

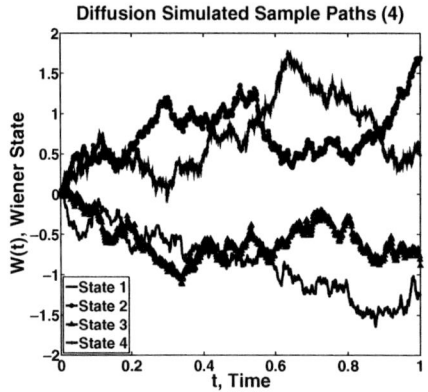
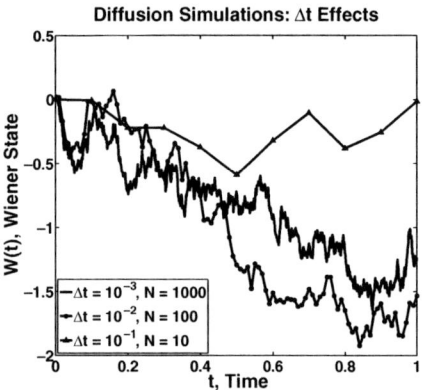

(a) Diffusion sample paths using four random states.

(b) Diffusion sample paths using three different time-steps.

Figure 1.1. (a) *Paths were simulated using MATLAB* [210] *with* $N = 1000$ *sample points, four* `randn` *states, and maximum time* $T = 1.0$. (b) *Paths were simulated using subsets of the same random state of* `randn` *used for the finer grids using* $N = 1000, 100, 10$.

1.3 More Wiener Process Moments

The expectations for the integer powers of the Wiener increment follow from the mean using the Wiener increment normal density (1.8). Only the even integer powers, $m = 2k$, need some calculation since the means will be zero for the odd integer powers due to the evenness of the density on $(-\infty, +\infty)$, i.e., $E[(\Delta W(t))^{2k+1}] = 0$ when $k = 0, 1, 2, \ldots,$

$$E\left[(\Delta W(t))^{m}\right] = E\left[(\Delta W(t))^{2k}\right] = \int_{-\infty}^{+\infty} \phi_n(w; 0, \Delta t) w^{2k} dw$$

$$= \frac{2}{\sqrt{2\pi \Delta t}} \int_{0}^{+\infty} \exp\left(-\frac{w^2}{2\Delta t}\right) w^{2k} dw$$

$$= \frac{(2\Delta t)^k}{\sqrt{\pi}} \int_{0}^{+\infty} \exp\left(-u\right) u^{k-1/2} du$$

$$= \frac{(2\Delta t)^k \Gamma(k + 1/2)}{\Gamma(1/2)} \tag{1.12}$$

for $k = 0, 1, 2, \ldots,$ where Γ is the gamma function [2] defined by

$$\Gamma(x) \equiv \int_{0}^{\infty} e^{-u} u^{x-1} du, \quad x > 0, \tag{1.13}$$

with initial condition $\Gamma(1) \equiv 1$ and special value $\Gamma(1/2) = \sqrt{\pi}$. The gamma function is like a generalized factorial function, due to the recursive form $\Gamma(x + 1) = x\Gamma(x)$ so that $\Gamma(x + 1) = x!$. The final formula (1.12) satisfies the recursion

$$g_{2k+2}(\Delta t) \equiv E[(\Delta W)^{2k+2}(t)] = (k + 1/2)(2\Delta t) g_{2k}(\Delta t).$$

Further, note that the final formula (1.12) holds for any integer m when the $\Delta W(t)$ is replaced by the absolute value, i.e.,

$$E[|\Delta W(t)|^m] = (2\Delta t)^{m/2}\Gamma((m+1)/2)/\Gamma(1/2). \tag{1.14}$$

The final formula (1.12) satisfies the recursion

$$g_{m+2}(\Delta t) \equiv E[(\Delta W)^{m+2}(t)] = (m+1)\Delta t g_m(\Delta t)$$

for $m = 0 : \infty$, starting from $g_0(\Delta t) = 1$ or $g_1(\Delta t) = \sqrt{2\Delta t/\pi}$. The results for the first few powers are summarized in Table 1.1. The function $(2k-1)!!$ is defined as follows.

Table 1.1. *Some expected moments (powers) of absolute value of the Wiener increments.*

m	$E[|\Delta W(t)|^m]$
0	1
1	$\sqrt{2\Delta t/\pi}$
2	Δt
3	$2\Delta t\sqrt{2\Delta t/\pi}$
4	$3(\Delta t)^2$
5	$8(\Delta t)^2\sqrt{2\Delta t/\pi}$
6	$15(\Delta t)^3$
$\vdots$	$\vdots$
$2k$	$(2k-1)!!(\Delta t)^k$
$2k+1$	$k!(2\Delta t)^k\sqrt{2\Delta t/\pi}$

Definition 1.5. *Double Factorial Function.*

$$(2k-1)!! = (2k-1)\cdot(2k-3)\ldots 1 \tag{1.15}$$

*denotes the **double factorial function**, given here for odd arguments. For example, $1!! = 1$, $3!! = 3$, and $5!! = 15$.*

For even arguments the double factorial function is proportional to the standard factorial function,

$$(2k)!! = 2^k k!.$$

Example 1.6. *These results can be applied to other expected moments, for example,*

$$\begin{aligned}
\text{Var}[(\Delta W)^2(t)] &= E[((\Delta W)^2(t) - \Delta t)^2]\\
&= E[(\Delta W)^4(t)] - 2\Delta t E[(\Delta W)^2(t)] + (\Delta t)^2 E[1]\\
&= 2(\Delta t)^2, \tag{1.16}
\end{aligned}$$

upon expanding the square and using the linear property of the expectation.

The moment calculation in (1.12) can be implemented directly by symbolic computation, for example, by Maple.

Example 1.7. *Wiener Moments by Maple.*

 # Maple functions for Wiener moments and Wiener deviation moments:

> restart: interface(showassumed = 0): assume($s2 > 0$):
> fnormal:= $(x, m, s2)-> \exp(-(x - m) * (x - m)/(2 * s2))/\text{sqrt}(2*\text{Pi}*s2)$;

$$\textit{fnormal} := (x, m, s2) \rightarrow \frac{e^{\left(1/2\frac{(x-m)^2}{s2}\right)}}{\sqrt{2\pi s2}}$$

> momentdw:= $(n, m, s2)->$ simplify(int(x^n*fnormal($x, m, s2$),
> $x = -\text{infinity..infinity}$));

$$\textit{momentdw} := (n, m, s2)) \rightarrow \textit{simplify}\left(\int_{-\infty}^{\infty} x^n \textit{fnormal}(x, m, s2)dx\right)$$

> momentdevdw:= $(n, m, s2)->$ simplify(int($(x - m)^n$ fnormal($x, m, s2$),
> $x = -\text{infinity..infinity}$));

$$\textit{momentdevdw} := (n, m, s2) \rightarrow \textit{simplify}\left(\int_{-\infty}^{\infty} (x - m)^n \textit{fnormal}(x, m, s2)dx\right)$$

 # Sample illustrations for moment functions:

> assume($dt > 0$): assume(sigma> 0):
> mom6dw:=collect(momentdw(6,mu*dt,sigma^2*dt),dt);

$$\textit{mom6dw} := \mu^6 dt^6 + 15\mu^4\sigma^2 dt^5 + 45\mu^2\sigma^4 + 15\sigma^6 dt^3$$

> mom6devdw:=momentdevdw(6,mu*dt,sigma^2*dt);

$$\textit{mom6dw} := 15dt^3$$

> mom5absdevdw:=momentabsdevdw(5,mu*dt,sigma^2*dt);

$$\textit{mom6dw} := \frac{8dt^{(5/2)}\sqrt{2}}{\sqrt{\pi}}$$

Remarks 1.8.

- *The results can also be applied to expected moments of Wiener differential process, $dW(t) = W(t + dt) - W(t)$, by replacing single-appearances of Δt by dt, i.e., $\Delta t \rightarrow dt$ is assumed, and neglecting terms of $O^2(\Delta t)$ as $\Delta t \rightarrow 0^+$ since they are treated as negligible compared to terms of $\text{ord}(\Delta t)$ as $\Delta t \rightarrow 0^+$.*

- *Sometimes, to keep the steps in a derivation simple, the infinitesimal dt will be treated as being an infinitesimally small object such that as $dt \rightarrow 0^+$, $(dt)^2 \ll 1$ or $(dt)^2 < \text{ord}(dt)$ and similarly for higher powers of dt. However, when there are no order Δt terms in the answer, then, as in (1.16), the proper leading order (by definition nonzero) would be of interest. Expected moments of $W(t)$ also follow by replacing Δt with t, except the higher powers of t would not be negligible compared to the first power, unless t were small.*

1.4　Wiener Process Nondifferentiability

Theorem 1.9. *Nondifferentiability of W(t).*
For any fixed $x > 0$ and $t > 0$,

$$\text{Prob}\left[\lim_{\Delta t \to 0^+}\left[\left|\frac{\Delta W(t)}{\Delta t}\right| > x\right]\right] = 1. \tag{1.17}$$

Proof. Let $x > 0$ be fixed, $t > 0$, $0 < \Delta t \ll 1$; then by interchanging the limit with the probability operations, since time is deterministic, and using the normal distribution of the increment $\Delta W(t) = W(t + \Delta t) - W(t)$ in (1.8),

$$\text{Prob}\left[\lim_{\Delta t \to 0^+}\left[\left|\frac{\Delta W(t)}{\Delta t}\right| > x\right]\right] = \lim_{\Delta t \to 0^+}\left[\text{Prob}\left[\left|\frac{\Delta W(t)}{\Delta t}\right| > x\right]\right]$$

$$= \lim_{\Delta t \to 0^+}\left[\text{Prob}\left[|\Delta W(t)| > x\Delta t\right]\right]$$

$$= \lim_{\Delta t \to 0^+}\left[\frac{2}{\sqrt{2\pi\,\Delta t}}\int_{x\Delta t}^\infty \exp\left(-\frac{w^2}{2\Delta t}\right)dw\right]$$

$$= \lim_{\Delta t \to 0^+}\left[\frac{2}{\sqrt{2\pi}}\int_{x\sqrt{\Delta t}}^\infty \exp\left(-\frac{v^2}{2}\right)dv\right]$$

$$= \frac{2}{\sqrt{2\pi}}\int_0^\infty \exp\left(-\frac{v^2}{2}\right)dv = 1$$

for any $x > 0$ and $t > 0$ fixed. Note that the error is

$$\frac{2}{\sqrt{2\pi}}\int_0^{x\sqrt{\Delta t}}\exp\left(-\frac{v^2}{2}\right)dv \le \frac{2}{\sqrt{2\pi}}\int_0^{x\sqrt{\Delta t}} 1\,dv = \frac{2}{\sqrt{2\pi}}x\sqrt{\Delta t} \ll 1,$$

since $\exp(-v^2/2) \le 1$. Further note that we can take x as large as we please, as long as it is fixed, so that $\Delta W(t)/\Delta t$ must be unbounded as $\Delta t \to 0^+$ **with probability one** for each t. Hence, the Wiener process $W(t)$ is nondifferentiable or nonsmooth with probability one for each t. (See also Mikosch [209, Section A3, p. 188], for a similar proof using less direct methods; see Steele [256, Section 5.2, p. 63] for more precise conditions.) □

1.5　Wiener Process Expectations Conditioned on the Past

Example 1.10. *Illustration of Independent Increments and Markov Properities for Wiener Process.*

- $\text{E}[W(t)|W(r), 0 \le r \le s] = W(\min[s, t])$.
  Note that the conditioning set $\{W(r), 0 \le r \le s\}$ denotes the past when $t > s \ge 0$, viewing $W(t)$ as the sum of two independent increments $(W(s) - W(0)) + (W(t) - W(s))$, noting that $W(0) = 0$. However, when $0 \le t \le s$, the increment $W(t) \equiv (W(t) - W(0))$ is a constant relative to the conditioning set, so the result depends

on the relation between t and s using the rule $\mathrm{E}[f(X)|X] = f(X)$ *given in Online Appendix* B *on p.* B27. *Hence,*

$$\mathrm{E}[W(t)|W(r), 0 \le r \le s] = \left\{ \begin{array}{l} W(t),\ 0 \le t \le s \\ \mathrm{E}[W(s) + (W(t) - W(s))|W(r),\ 0 \le r \le s], \\ \hspace{6cm} 0 \le s < t \end{array} \right\}$$

$$= \left\{ \begin{array}{l} W(t),\ 0 \le t \le s \\ \mathrm{E}[W(s)|W(r), 0 \le r \le s] + \mathrm{E}[(W(t) - W(s))], \\ \hspace{6cm} 0 \le s < t \end{array} \right\}$$

$$= \left\{ \begin{array}{l} W(t),\ 0 \le t \le s \\ W(s) + 0,\ 0 \le s < t \end{array} \right\} = \left\{ \begin{array}{l} W(t),\ 0 \le t \le s \\ W(s),\ 0 \le s < t \end{array} \right\}$$

$$= W(\min[s, t]),$$

where the independent increment property was used along with the zero-mean property of the increment, $\mathrm{E}[\Delta W(t)] = 0$, *and the completely conditioned rule that* $\mathrm{E}[f(X)|X] = f(X)$. *The function* $\min[s, t]$ *denotes the minimum of s and t. The linear property of the conditional expectation was also used.*

When $0 \le s < t$, *then the formula*

$$\mathrm{E}[W(t)|W(r), 0 \le r \le s] = W(s) \tag{1.18}$$

signifies that the average information conditioned on the past data, $\{W(r), r \in [0, s]\}$, *is given by the most recent past data* $W(s)$, *which may imply a significant reduction in uncertainty for the present data,* $W(t)$.

The form of the expectation result (1.18) *is the principal characteristic form for a* **martingale** $X(t)$,

$$\mathrm{E}[X(t)|X(r), 0 \le r \le s] = X(s), \tag{1.19}$$

where $X(t) = f(W(t))$, *for instance. The martingale is an abstract model of a fair game. (See the beginning preliminary chapter of Mikosch* [209] *for a clear description of martingales, but in an abstract presentation. Martingales are described in this book in Chapter 12 with full qualifications.)*

- $\mathrm{E}[W^2(t)|W(r), 0 \le r \le s] = W^2(\min[s, t]) + (t - s)H(t - s)$.
  Here $H(X)$ *is the Heaviside step function* (B.156). *This result is derived similarly to the prior result for the conditional mean, but much more algebra is required, although many of the small details of the prior derivation are omitted.*

$$
\mathrm{E}[W^2(t)|W(r), 0 \le r \le s] = \begin{cases} W^2(t),\ 0 \le t \le s \\ \mathrm{E}[(W(s)+(W(t)-W(s)))^2|W(r),\ 0 \le r \le s], \\ \hspace{4cm} 0 \le s < t \end{cases}
$$

$$
= \begin{cases} W^2(t),\ 0 \le t \le s \\ W^2(s)+2W(s)\mathrm{E}[(W(t)-W(s))]+\mathrm{E}[(W(t)-W(s))^2], \\ \hspace{5cm} 0 \le s < t \end{cases}
$$

$$
= \begin{cases} W^2(t),\ 0 \le t \le s \\ W^2(s)+2W(s)\cdot 0+(t-s),\ 0 \le s < t \end{cases}
$$

$$
= \begin{cases} W^2(t),\ 0 \le t \le s \\ W^2(s)+(t-s),\ 0 \le s < t \end{cases}
$$

$$
= W^2(\min[s,t])+(t-s)H(t-s).
$$

Here, the increment variance $\mathrm{Var}[\Delta W(t)] = \Delta t$ *has been used.*

The general technique for powers $W^m(t)$ *when* $s < t$ *with conditioning on* $W(s)$ *is to use the decomposition into independent increments* $W(t) = W(s) + (W(t) - W(s))$ *and then expand the power of m by the binomial expansion* (B.150)

$$
(W(s)+(W(t)-W(s)))^m = \sum_{k=0}^{m} \binom{m}{k} W^k(s)(W(t)-W(s))^{m-k},
$$

and then use independence of the increments and conditioning to calculate for each term

$$
\mathrm{E}\left[\binom{m}{k} W^k(s)(W(t)-W(s))^{m-k} \middle| W(r), 0 \le r \le s \right]
$$

$$
= \binom{m}{k} W^k(s)\mathrm{E}\left[(W(t)-W(s))^{m-k} \right],
$$

relying on Table 1.1 *for the remaining expectation.*

The term normal distribution is more often used in mathematics and statistics, while the term Gaussian distribution may be used more often in the sciences and in engineering.

1.6 Poisson Process Basic Properties

Since the Poisson process suffers from positive jumps of integer magnitude, the Poisson process is also discontinuous, which makes the differentiability problems of the Poisson process of secondary importance. For this reason, the Poisson process is also called a **counting process** or **point process**. Thus, the analytical problems are even more severe than for the Wiener process, since the singularities of the Poisson process arise at the zeroth order with the value of $P(t)$ jumping, while those of $W(t)$ arise at the first order derivative.

However, the jumps of the Poisson process have a modeling benefit over the Wiener process in that the Poisson process is useful for applications with disasters or crashes and those with bonanzas or rallies.

In summary, **Poisson process $P(t)$ is a discontinuous process** and the process is defined by the following properties.

Properties 1.11. *Simple Poisson Process $P(t)$.*

- $P(t)$ has **unit jumps**, *since if the value of $P(t)$ jumps at time $T_k > 0$, then*

$$P(T_k^+) = P(T_k^-) + 1,$$

where $P(T_k^+)$ denotes the limit from the right and $P(T_k^-)$ the limit from the left, so $P(t)$ is discontinuous, increasing, and has instantaneous jumps.

- $P(t)$ is **right-continuous**, *since*

$$P(t^+) = P(t) \geq P(t^-), \quad t > 0. \tag{1.20}$$

- $P(t)$ has **independent increments**, *since the Poisson increments*

$$\Delta P(t_i) \equiv P(t_i + \Delta t_i) - P(t_i)$$

are mutually independent for all t_i on nonoverlapping time-intervals defined such that $t_i \geq 0$, $t_{i+1} = t_i + \Delta t_i$, and any $\Delta t_i > 0$ for $0 = 1 : n$ so that

$$t_i < t_{i+1} \text{ for } i = 0 : n,$$

noting that $P(t_i) = P(0) + \sum_{j=0}^{i-1} \Delta P(t_j)$, depending on all preceding increments, recalling that $P(0) = 0$ with probability one at $t_0 = 0$, i.e.,

$$\mathrm{Prob}[\Delta P(t_i) \leq p_i, \Delta P(t_j) \leq p_j] = \mathrm{Prob}[\Delta P(t_i) \leq p_i] \cdot \mathrm{Prob}[\Delta P(t_j) \leq p_j],$$

if $j \neq i$, such that there is no overlap in the time-intervals $(t_i, t_{i+1}]$ and $(t_j, t_{j+1}]$. Note that $\Delta P(t_i)$ as a forward increment is independent (see Definition B.35 for expectations of products of independent random variables) of $P(t_i)$ and recall that $\Delta P(t_i) \equiv P(t_i + \Delta t_i) - P(t_i)$ is associated with the time-interval $[t_j, t_j + \Delta t_j)$, open on the right since the process $P(t_i)$ is right-continuous.

- $P(t)$ is a **stationary process**, *since the distribution of the increment $\Delta P(t) = P(t + \Delta t) - P(t)$ is independent of t.*

- $P(t)$ is a **Markov process**, *since*

$$\mathrm{Prob}[P(t + \Delta t) = k \,|\, P(s), s \leq t] = \mathrm{Prob}[P(t + \Delta t) = k \,|\, P(t)]$$

for any $t \geq 0$ and any $\Delta t > 0$. (It is helpful to note that $P(t)$ is synonymous with the increment $(P(t) - P(0))$.)

- $P(t)$ is **Poisson distributed** with mean $\mu = \lambda t$ and variance $\sigma^2 = \lambda t$, $t > 0$, i.e.,

$$\Phi_{P(t)}(k; \lambda t) = \text{Prob}[P(t) = k] \equiv p_k(\lambda t) = e^{-\lambda t}\frac{(\lambda t)^k}{k!}, \quad (1.21)$$

for integer values $k = 0, 1, 2, \ldots$ with constant $\lambda > 0$ and $t \geq 0$.

- $P(0^+) = 0^+$ **with probability one**, since from (1.21), $p_k(0^+) = \delta_{k,0}$, i.e., in the limit as $t \to 0^+$.

See also Çinlar [56] or Snyder and Miller [252] for a more essential list of assumptions.

Thus, for $P(t)$, the increments $\Delta[P(t + i\Delta t)] \equiv P(t + (i + 1)\Delta t) - P(t + i\Delta t)$ for $i = 0, 1, \ldots$ are IID given time-step $\Delta t > 0$ when $t \geq 0$.

By the stationarity property of the Poisson process, increment $\Delta P(t) = P(t + \Delta t) - P(t)$ has the same discrete distribution as $P(\Delta t)$ in (1.21) and so has the parameter $\lambda\Delta t$ instead of the λt in (B.50), i.e.,

$$\Phi_{\Delta P(t)}(k; \lambda\Delta t) = \text{Prob}[\Delta P(t) = k] = p_k(\lambda\Delta t) = e^{-\lambda\Delta t}\frac{(\lambda\Delta t)^k}{k!} \quad (1.22)$$

for $k = 0, 1, 2, \ldots, t \geq 0$, and $\Delta t \geq 0$.

Similarly, by the stationarity property of the differential, $dP(t) = P(t + dt) - P(t)$, the Poisson process has the same discrete distribution as $P(dt)$ in (1.21), except that $dP(t)$ and $P(dt)$ have the parameter λdt instead of the λt in (B.50) for $P(t)$. Thus $dP(t)$ has the distribution,

$$\Phi_{dP(t)}(k; \lambda dt) = \text{Prob}[dP(t) = k] = p_k(\lambda dt) = e^{-\lambda dt}\frac{(\lambda dt)^k}{k!} \quad (1.23)$$

for $k = 0, 1, 2, \ldots, t \geq 0$, and $dt \geq 0$. The distribution (1.23) might be considered as a limiting version of the more basic and proper incremental version in (1.22).

The simulation of the simple Poisson process $P(t)$ is usually based upon simulating the time between jumps, the interarrival time $T_{k+1} - T_k$, since the interarrival time can be shown to be exponentially distributed as sketched in Online Appendix B.

Lemma 1.12. *Exponential Distribution of Time Between Jumps.*
*Let $P(t)$ be a simple Poisson process with fixed jump frequency $\lambda > 0$ and let T_j denote the jth jump-time; then the distribution of the **interjump-time** $\Delta T_j \equiv T_{j+1} - T_j$ for $j = 0, 1, 2, \ldots$, defining $T_0 \equiv 0$, conditioned on T_j, is*

$$\Phi_{\Delta T_j}(\Delta t) = \text{Prob}[\Delta T_j \leq \Delta t \mid T_j] = 1 - e^{-\lambda\Delta t}. \quad (1.24)$$

Proof. The basic idea of this proof is that the probability of the time between jumps $\Delta T_j = T_{j+1} - T_j$ less than Δt, conditioned on the prior jump-time T_j, will be the same as the probability that there is at least one jump in the time-interval, which is the same as one minus the probability that there are no jumps in the time-interval, i.e.,

$$\text{Prob}[\Delta T_j \leq \Delta t \mid T_j] = 1 - \text{Prob}[\Delta T_j > \Delta t \mid T_j]$$
$$= 1 - \text{Prob}[\Delta P(T_j) = 0 \mid T_j].$$

However, by the stationarity property of the simple Poisson process $P(t)$ the probability of the difference depends not on the common time T_j but on the difference in time ΔT_j,

$$\begin{aligned} \text{Prob}[\Delta T_j \le \Delta t \mid T_j] &= 1 - \text{Prob}[P(\Delta t) - P(0) = 0] \\ &= 1 - \text{Prob}[P(\Delta t) = 0] = 1 - p_0(\lambda \Delta t) \\ &= 1 - e^{-\lambda \Delta t} = \Phi_e(\Delta t; 1/\lambda), \end{aligned}$$

where the fact that $P(0) = 0$ with probability one has been used, the Poisson distribution $p_k(\lambda \Delta t)$ is given in (1.22), and the exponential distribution $\Phi_e(t; \mu)$ is given in (B.40). $\quad\square$

Using MATLAB with the efficient and fundamental distribution transformation from uniform to exponential distribution (B.42), a uniformly distributed pseudo-random number generator can be used. These numbers can be generated wholesale, in vector form, for plotting or other applications, using given K samples and the Poisson parameter value lambda, by the code fragment

```
Uv = rand(1,K); % uniform RNG.
T = zeros(1,K+1); kv = zeros(1,K+1); % pre-declare.
for k = 1:K, kv(k+1) = k;
    T(k+1) = T(k) - log(Uv(k))/lambda;
end
plot(kv,T,'k-');
```

where log is the MATLAB natural logarithm notation. See the comments about (B.44) explaining why the proper term $\log(\text{Uv}(k))$ is used here rather than the less efficient term $\log(1 - \text{Uv}(k))$.

Since the natural time variable for the Poisson process is scaled as $\lambda * t$, four sample paths for $P(t)$ are illustrated in Figure 1.2(a) versus the dimensionless time $\lambda * t$. The variation with the jump-rate λ can be deduced since higher frequencies ($\lambda > 1$) compress the time axis and lower frequencies ($\lambda < 1$) expand the time axis. Note that the exponentially distributed interjump or interarrival times must be used for simulating $P(t)$ since the Poisson distribution is not useful in simulating the jump-times directly. The code used to generate this part of the figure is given in Program C.9, called poisson03fig2.m in Online Appendix C.

Figure 1.2(b) shows the corresponding sample paths for the Poisson process increment $\Delta P(t)$ when the time increments between jumps are sufficiently small so that the zero-one jump law, discussed more extensively in Theorem 1.19 in Section 1.7, applies and the time between jumps is uniformly distributed with asymptotic probability $\lambda \Delta t$ for the next jump and $(1 - \lambda \Delta t)$ for zero jumps, since

$$\text{Prob}[T_{k+1} - T_k \le \Delta t \mid T_k] = 1 - e^{-\lambda \Delta t} \sim \lambda \Delta t,$$

provided $\lambda \Delta t \ll 1$, i.e., small, taking $\Delta t = 0.05$ and $\lambda = 1.0$. The small time increment process can be numerically simulated by a standard uniform number generator such as rand from MATLAB and the **method of acceptance-rejection** [230, 97] such that the

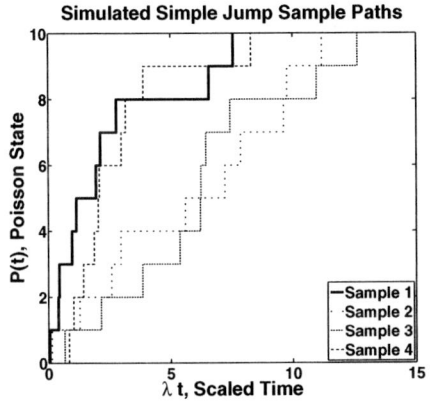

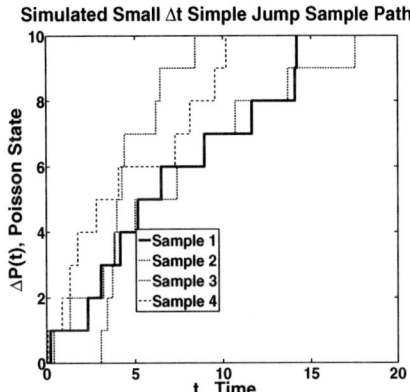

(a) Four Poisson jump $P(t)$ sample paths using exponentially distributed jump-times.

(b) Incremental Poisson jump $\Delta P(t)$ sample paths using the zero-one jump law for small time-steps.

Figure 1.2. (a) *Simulated sample paths for the simple Poisson process $P(t)$ versus the dimensionless time λt using four different MATLAB [210] random states for four different sample paths and using the exponential distribution of the time between jumps.* (b) *Similar illustration for the simple Poisson process increment simulations versus t with $\lambda = 1.0$ and $\Delta t = 0.05$, based upon the zero-one jump law implemented with uniform distribution paths and simulated using subsets of the same random state of* rand *used with a finer grid of 501 time-steps so the zero-one jump law was a good approximation.*

open interval $(0, 1)$ is partitioned into a centered interval of length $\lambda \Delta t$ and the complement of $(0, 1)$. When a uniformly generated point lands in the centered interval, then a jump is counted, while there is no jump if it lands in the complement. The centered interval, $[(1 - \lambda \Delta t)/2, (1 + \lambda \Delta t)/2]$, is used to avoid the bias of the open interval property of pseudo-random number generators, where the neighborhood of the end points is excluded by a very small amount that is the order of the **machine epsilon** (the smallest positive floating number that is significant when added to one, theoretically 2^{-53} in IEEE standard double precision arithmetic). A sufficiently large sample should approximately satisfy the zero-one jump law probabilities, since the rejection method is equivalent to the unit step function $U(X_u; (1-\lambda \Delta t)/2, (1+\lambda \Delta t)/2)$ applied to the uniform variate X_u approximately generated by rand, and the expectation is

$$
\begin{aligned}
E[U(X_u; (1 - \lambda \Delta t)/2, (1 + \lambda \Delta t)/2)] &= \int_0^1 U(u; (1 - \lambda \Delta t)/2, (1 + \lambda \Delta t)/2)du \\
&= \int_{(1-\lambda \Delta t)/2}^{(1+\lambda \Delta t)/2} du = \lambda \Delta t.
\end{aligned}
$$

The code used to generate this part of the figure is given in Program C.10, called poisson03fig3.m in Online Appendix C.

Theorem 1.13. *Covariance of* $P(t)$.
If $P(t)$ *is a Poisson process, then*

$$\text{Cov}[P(t), P(s)] = \lambda \min[t, s]. \tag{1.25}$$

Proof. This theorem is a very elementary application of the independent increment property of Poisson or Markov jump processes, also demonstrating how application of independent increments relies on the zero-mean properties. For the Poisson process, unlike the standardized diffusion process, the zero-mean property comes from using the Poisson deviation or centered Poisson processes $\widehat{P}(t) = P(t) - \lambda t$, where $\text{E}[P(t)] = \lambda t$, such that $\text{E}[\widehat{P}(t)] = 0 = \text{E}[\widehat{P}(s)]$. First consider the case $s < t$ and write

$$\widehat{P}(t) = \widehat{P}(s) + (\widehat{P}(t) - \widehat{P}(s)),$$

i.e., as independent increments, noting the time increment $\Delta t = t - s$, the first increment is $\widehat{P}(s) - \widehat{P}(0) = \widehat{P}(s)$, since $\widehat{P}(0) = 0$, and that subtracting the mean terms λt and λs preserves the independent increment property since functions of independent random variables are independent (B.80). Then

$$\begin{aligned}
\text{Cov}[P(t), P(s)] &= \text{E}[\widehat{P}(t)\widehat{P}(s)] = \text{E}[\widehat{P}^2(s) + \widehat{P}(s)(\widehat{P}(t) - \widehat{P}(s))] \\
&= \text{E}[\widehat{P}^2(s)] + \text{E}[\widehat{P}(s)(\widehat{P}(t) - \widehat{P}(s))] \\
&= \text{Var}[P(s)] + \text{E}[\widehat{P}(s)]\text{E}[\widehat{P}(t) - \widehat{P}(s)] \\
&= \lambda s + 0 \cdot 0 = \lambda s,
\end{aligned}$$

using the linearity of the expectation operator (B.9), the definition of the variance (B.10) together with the independence of the expectations (B.80) for independent increments $\widehat{P}(s)$ and $(\widehat{P}(t) - \widehat{P}(s))$, and finally note that $\widehat{P}(s)$, with $P(s)$, has variance λs (1.21). In the case $t < s$, we then have $\text{Cov}[P(t), P(s)] = \lambda t$ by symmetry, and both cases together produce the conclusion $\text{Cov}[P(t), P(s)] = \lambda \min[s, t]$. □

1.7 Poisson Process Moments

The expectations for the integer powers of the Poisson increment follow from the mean over the Poisson distribution (1.22) and summed by differentiation of the exponential series (B.53).

Lemma 1.14. *Poisson Sums by Differentiation.*

$$\text{E}[(\Delta P)^m(t)] = e^{-\lambda \Delta t} \sum_{k=0}^{\infty} \frac{(\lambda \Delta t)^k k^m}{k!} \tag{1.26a}$$

$$= \left[e^{-u} \left(u \frac{d}{du} \right)^m e^u \right]\Bigg|_{u=\lambda \Delta t} \tag{1.26b}$$

for $m = 0, 1, 2, \ldots$.

The result (1.26b) can be shown by induction from the definition (1.26). Either the direct summation form (1.26) or the differentiation form (1.26b) can be implemented by symbolic computation; for example, the summation definition form can be coded in Maple as follows.

Example 1.15. *Poisson Moment Summations by Maple.*

 # Maple functions for Poisson moments and Poisson deviation moments:

> fpoisson := $(k, u)- >$ exp$(-u) * u^k/k!$;

$$fpoisson := (k, u) \rightarrow \frac{e^{(-u)} u^k}{k!}$$

> momentdp := $(n, u)- >$ simplify(sum($'k^n$ fpoisson$(k, u)','k' = 0..$infinity));

$$momentdp := (n, u) \rightarrow simplify \left(\sum_{'k'=0}^{\infty} 'k^n fpoisson(k, u)' \right)$$

> momentdevdp := $(n, u)- >$ simplify(sum($'(k - u)^n$ fpoisson$(k, u)'$,

> $'k' = 0..$infinity));

$$momentdevdp := (n, u) \rightarrow simplify \left(\sum_{'k'=0}^{\infty} '(k - u)^n fpoisson(k, u)' \right)$$

 # Sample illustrations for fifth moment of both moment functions:

> mom5dp := momentdp(5,lambda $*dt$);

$$mom5dp := \lambda dt (1 + 15\lambda dt + 25\lambda^2 dt^2 + 10\lambda^3 dt^3 + \lambda^4 dt^4)$$

> mom5devdp := momentdevdp(5,lambda $*dt$);

$$mom5dp := 10\lambda^2 dt^2 + \lambda dt$$

The results for the first few powers are summarized in Table 1.2. The second column of this table can be quickly calculated by recursion, since if $u = \lambda \Delta t$ and $g_m(u) = $ E$[(\Delta P)^m(t)]$, then it can be shown that $g_{m+1}(u) = u \cdot (g_m(u) + g'_m(u))$. See Exercise 6 on p. 27 for the asymptotic form of E$[(\Delta P)^m(t)]$. The expectation of a general function, E$[f(\Delta P(t))]$, in terms of an infinite series of the finite differences of $f(0)$, which terminates if $f(\Delta P(t))$ is an integer power of $\Delta P(t)$, is the topic of Exercise 7 on p. 27.

These tabulated results can be applied to other expected moments, for example,

$$\text{Var}[\Delta P(t)] = \text{E}[(\Delta P(t) - \lambda \Delta t)^2]$$
$$= \text{E}[(\Delta P)^2(t)] - 2\lambda \Delta t \text{E}[\Delta P(t)] + (\lambda \Delta t)^2 \text{E}[1] = \lambda \Delta t,$$

upon expanding the square and using the linear property of the expectation. See the third column of Table 1.2. The results can also be applied to expected moments of a Poisson

Table 1.2. *Some expected moments (powers) of Poisson increments and their deviations.*

m	$E[(\Delta P)^m(t)]$	$E[(\Delta P(t) - \lambda \Delta t)^m]$
0	1	—
1	$\lambda \Delta t$	0
2	$\lambda \Delta t (1 + \lambda \Delta t)$	$\lambda \Delta t$
3	$\lambda \Delta t (1 + 3\lambda \Delta t + (\lambda \Delta t)^2)$	$\lambda \Delta t$
4	$\lambda \Delta t (1 + 7\lambda \Delta t + 6(\lambda \Delta t)^2 + (\lambda \Delta t)^3)$	$\lambda \Delta t (1 + 3\lambda \Delta t)$
5	$\lambda \Delta t (1 + 15\lambda \Delta t + 25(\lambda \Delta t)^2 + 10(\lambda \Delta t)^3 + (\lambda \Delta t)^4)$	$\lambda \Delta t (1 + 10\lambda \Delta t)$

differential process as an increment process, $dP(t) = P(t + dt) - P(t)$, by replacing Δt with dt and neglecting terms of $O^2(dt)$ since they are treated as negligible compared to terms of ord(dt), dt being infinitesimally small. Expected moments of $P(t)$ also follow by replacing Δt with t, except the higher powers of t would not be negligible compared to the first power, unless t were small.

1.8   Poisson Zero-One Jump Law

Theorem 1.16. *Zero-One Jump Law Order of Magnitude of Error for* $\Delta P(t)$.
As $\Delta t \to 0^+$ with $\lambda > 0$ and bounded, then

$$\text{Prob}[\Delta P(t) = 0] = 1 - \lambda \Delta t + O^2(\lambda \Delta t), \tag{1.27}$$

$$\text{Prob}[\Delta P(t) = 1] = \lambda \Delta t + O^2(\lambda \Delta t), \tag{1.28}$$

$$\text{Prob}[\Delta P(t) > 1] = O^2(\lambda \Delta t), \tag{1.29}$$

$$\text{Prob}[(\Delta P)^m(t) = \Delta P(t)] = 1 - \frac{1}{2}(\lambda \Delta t)^2 + O^3(\lambda \Delta t), \quad m \geq 2. \tag{1.30}$$

Proof. Taking the Poisson increment distribution (1.22) and expanding it asymptotically using primarily the exponential series expansion (B.53) for $\Delta t \ll 1$ yields

$$\text{Prob}[\Delta P(t) = 0] = e^{-\lambda \Delta t} = 1 - \lambda \Delta t + \frac{1}{2}(\lambda \Delta t)^2 + O^3(\lambda \Delta t),$$

$$\text{Prob}[\Delta P(t) = 1] = e^{-\lambda \Delta t} \lambda \Delta t = \lambda \Delta t - (\lambda \Delta t)^2 + O^3(\lambda \Delta t),$$

$$\text{Prob}[\Delta P(t) > 1] = 1 - \text{Prob}[\Delta P(t) = 0] - \text{Prob}[\Delta P(t) = 1]$$

$$= \frac{1}{2}(\lambda \Delta t)^2 + O^3(\lambda \Delta t).$$

Since $O^2(\lambda \Delta t) + O^3(\lambda \Delta t) = O^2(\lambda \Delta t)$, the first three equations are proved. The last equation (1.30) follows from the fact that $x^m = x$ is true for $m \geq 2$ only if $x = 0$ or $x = 1$, so

$$\text{Prob}[(\Delta P)^m(t) = \Delta P(t)] = \text{Prob}[\Delta P(t) = 0] + \text{Prob}[\Delta P(t) = 1]$$

$$= 1 - \text{Prob}[\Delta P(t) > 1] = 1 - \frac{1}{2}(\lambda \Delta t)^2 + O^3(\lambda \Delta t).$$

The significance of this result is that if $\lambda \Delta t$ is sufficiently small and terms of order $(\lambda \Delta t)^2$ can be neglected, then only jumps of zero or one are very likely, i.e., very probable.   □

Remarks 1.17.

- *In some other texts, the three small Poisson increment properties, equations (1.27), (1.28), (1.29), are used as an elementary definition of the simple Poisson process. Here, we have started with a higher-level definition to facilitate the use of the Poisson process in applications.*

- *Combining the asymptotic probability relations (1.28) for $\Delta P(t) = 1$ and (1.29) for $\Delta P(t) > 1$ leads to*

$$\text{Prob}[\Delta P(t) > 1] \ll \text{Prob}[\Delta P(t) = 1]$$

  *when $\lambda \Delta t \ll 1$. This asymptotic relationship characterizes the **orderliness** property of the Poisson process. (See Snyder and Miller [252].)*

With this result, the corresponding results for differential Poisson processes follow. First, we need a definition to specify that the square of a differential has been neglected.

Definition 1.18. *Equality to Precision-dt.*
Let $f(dt; x)$ and $g(x)$ be bounded functions for $dt \geq 0$ and parameter x. Write

$$f(dt; x) \stackrel{dt}{=} g(x)dt \tag{1.31}$$

if
$$f(dt; x) = g(x)dt + o(dt)$$

as $dt \to 0^+$ and fixed x.

Theorem 1.19. *Zero-One Jump Law for $dP(t)$.*
Let $dt > 0$ and let λ be positive and bounded. Then

$$\text{Prob}[dP(t) = 0] \stackrel{dt}{=} 1 - \lambda dt, \tag{1.32}$$

$$\text{Prob}[dP(t) = 1] \stackrel{dt}{=} \lambda dt, \tag{1.33}$$

$$\text{Prob}[dP(t) > 1] \stackrel{dt}{=} 0, \tag{1.34}$$

$$\text{Prob}[(dP)^2(t) = dP(t)] \stackrel{dt}{=} 1, \tag{1.35}$$

$$\text{Prob}[(dP)^m(t) = dP(t)] \stackrel{dt}{=} 1, m > 0. \tag{1.36}$$

Proof. The proof follows easily from the increment approximation Theorem 1.16 upon neglecting all terms $O^2(\lambda \Delta t)$. The last equation in precision-dt, (1.36), holds for the same reason that the prior equation (1.35) holds as long as $m > 0$. Note that $(dP)^m(t) = dP(t)$ is obviously valid for $dP(t) = 0$, but if $dP(t) \neq 0$, then division by $dP(t)$ is permissible, so $(dP)^{m-1}(t) = 1$ and we must have $dP(t) = 1$, one being the only real root in this real problem. It is reasonable to assume that $\lambda dt \leq 1$ to avoid nominally violating probability bounds $\text{Prob}[dP(t) = 0] \geq 0$ and $\text{Prob}[dP(t) = 1] \leq 1$, even though the error is hidden in the order symbols. The rules (1.32)–(1.36) will come in very handy for simplifying powers of $dP(t)$ in the Poisson jump calculus later in this text. □

This **zero-one jump law** immediately leads to the following corollary for Poisson differential distribution and expectations.

Corollary 1.20. *Zero-One Distribution and Expectation for dP(t).*

$$\Phi_{dP(t)}(k) = p_k(\lambda dt) \stackrel{\text{dt}}{=} (1 - \lambda dt)\delta_{k,0} + \lambda dt \delta_{k,1} \tag{1.37}$$

is a generalized representation of the differential Poisson distribution and

$$\text{E}[f(dP(t))] \stackrel{\text{dt}}{=} (1 - \lambda dt)f(0) + \lambda dt f(1) \tag{1.38}$$

is the expectation, provided $f(p)$ is a bounded and continuous function.

The Poisson zero-one jump law is a special case of a **Bernoulli distribution**, concerning Bernoulli trials that have only two outcomes, here with failure probability $p = 1 - \lambda dt$ for zero jump or success probability $1 - p = \lambda dt$ for one jump, provided λdt is small compared to unity.

1.9   Temporal, Nonstationary Poisson Process

Properties 1.21. *Temporal Poisson Process.*

- *For the **temporal or nonstationary Poisson process** $P(t)$ the jump-rate is time-dependent, $\lambda = \lambda(t)$, so that $P(t)$ is no longer simple or stationary but is nonstationary. First consider the differential process $dP(t)$, replacing the simple Poisson jump-count λdt by the time-dependent one,*

$$d\Lambda(t) \equiv \lambda(t)dt. \tag{1.39}$$

Letting $\Lambda(0) = 0$ initially, we then have

$$\Lambda(t) = \int_0^t \lambda(s)ds \tag{1.40}$$

with increment

$$\Delta\Lambda(t) \equiv \Lambda(t + \Delta t) - \Lambda(t) = \int_t^{t+\Delta t} \lambda(s)ds. \tag{1.41}$$

Thus, $\Delta\Lambda(t) \sim \lambda(t)\Delta t$ only when $\Delta t \ll 1$, i.e., is small.

- *The **temporal Poisson distribution** for the differential Poisson process $dP(t)$ remains **unchanged** from the fixed jump-rate Poisson, except for $\lambda = \lambda(t)$ and*

$$\Phi_{dP(t)}(k; \lambda(t)dt) = \text{Prob}[dP(t) = k]$$
$$= p_k(\lambda(t)dt) = e^{-\lambda(t)dt} \frac{(\lambda(t)dt)^k}{k!} \tag{1.42}$$

for $k = 0, 1, 2, \ldots$ with $t \geq 0$ and temporal parameter $\lambda(t) > 0$.

*However, the **Poisson distribution property** (1.21) of the Poisson process needs to be changed for the temporal increment process $\Delta P(t)$ (1.22) using the modified parameter $\Delta\Lambda(t)$,*

$$\Phi_{\Delta P(t)}(k; \Delta\Lambda(t)) = \text{Prob}[\Delta P(t) = k]$$
$$= p_k(\Delta\Lambda(t)) = e^{-\Delta\Lambda(t)} \frac{(\Delta\Lambda(t))^k}{k!} \tag{1.43}$$

*for $k = 0, 1, 2, \ldots$ with $t \geq 0$, $\Delta t \geq 0$, and temporal parameter $\Delta\Lambda(t)$. Thus, the temporal Poisson process is also a **time-inhomogeneous** process. The Poisson increment distribution is fundamental for the temporal Poisson process. Note that $\Lambda(t)$ will be nondecreasing if $\lambda(t) > 0$ and continuous.*

Finally, since the full temporal Poisson process $P(t)$ is the increment $P(t) - P(0) = P(t)$, then it has the distribution

$$\Phi_{P(t)}(k; \Lambda(t)) = \text{Prob}[P(t) = k]$$
$$= p_k(\Lambda(t)) = e^{-\Lambda(t)} \frac{(\Lambda(t))^k}{k!}, \tag{1.44}$$

inherited from (1.43).

- *The **nonstationary behavior** follows from the fact that the distribution of the increment (1.43) depends on t through the parameter $\Delta\Lambda(t)$ or, more simply, from the **Poisson increment expectation** given in (B.51) and Table 1.2 with $\Delta\Lambda(t)$ replacing the parameter $\lambda\Delta t$,*

$$\text{E}[\Delta P(t)] = \Delta\Lambda(t), \tag{1.45}$$

since it will be, in general, a function of time t. Thus,

$$\text{E}[P(t) - P(t_0)] = \Lambda(t) - \Lambda(t_0) = \int_{t_0}^{t} \lambda(s)ds.$$

*The **Poisson increment variance** must be the same as its expectation ((B.51)–(B.52)),*

$$\text{Var}[\Delta P(t)] = \Delta\Lambda(t). \tag{1.46}$$

However, treating the increment as an integral leads to another form:

$$\begin{aligned}
\text{Var}[\Delta P(t)] &= \text{Var}\left[\int_t^{t+\Delta t} dP(s)\right] \\
&= \text{E}\left[\left(\int_t^{t+\Delta t} dP(s) - \Delta\Lambda(t)\right)^2\right] \\
&= \text{E}\left[\left(\int_t^{t+\Delta t} (dP(s) - \lambda(s)ds)\right)^2\right] \\
&= \text{E}\left[\int_t^{t+\Delta t} (dP(s_1) - \lambda(s_1)ds_1)\int_t^{t+\Delta t} (dP(s_2) - \lambda(s_2)ds_2)\right] \\
&= \int_t^{t+\Delta t}\int_t^{t+\Delta t} \text{E}\left[(dP(s_1) - \lambda(s_1)ds_1)(dP(s_2) - \lambda(s_2)ds_2)\right] \\
&= \int_t^{t+\Delta t}\int_t^{t+\Delta t} \text{Cov}[dP(s_1), dP(s_2)].
\end{aligned} \tag{1.47}$$

Thus,

$$\text{Var}[\Delta P(t)] = \Delta\Lambda(t) = \int_t^{t+\Delta t} \lambda(s)ds,$$

since $dP(s_1)$ and $dP(s_2)$ are independent increments as differentials as long as $s_2 \neq s_1$. Hence, $\text{Cov}[dP(s_1), dP(s_2)] \neq 0$ only if $s_2 = s_1$ when it has the value $\text{Cov}[dP(s_1), dP(s_1)] = \text{Var}[dP(s_1)]$. Consequently,

$$\text{Cov}[dP(s_1), dP(s_2)] \overset{\text{gen}}{=} \lambda(s_1)\delta(s_1 - s_2)ds_1 ds_2 \tag{1.48}$$

for arbitrary Δt, so the inner integral of (1.47) will be

$$\int_t^{t+\Delta t} \text{Cov}[dP(s_1), dP(s_2)] = \lambda(s_1)ds_1, \tag{1.49}$$

and (1.47) yields the same answer as (1.46).

- *The **temporal Poisson differential process distribution for $dP(t)$** to precision-dt is*

$$\begin{aligned}
\Phi_{dP(t)}(k; d\Lambda(t)) &= \text{Prob}[dP(t) = k] \\
&= p_k(d\Lambda(t)) \overset{\text{dt}}{=} (1 - \lambda(t)dt)\delta_{k,0} + \lambda(t)dt\delta_{k,1}, \tag{1.50}
\end{aligned}$$

  which simply follows from (1.43) for sufficiently small Δt, and the corresponding simple process zero-one law result (1.37), if $\lambda(t) > 0$.

- *The **interjump-times for the nonstationary Poisson process are exponentially distributed.** The increasing property of $\Lambda(t)$ ($d\Lambda(t) > 0$) means that it can be used as a substitute "clock" in place of t, but for $\Lambda(t)$ to be a full range clock it is necessary that $\Lambda(t)$ be unbounded, i.e., $\Lambda(t) \to +\infty$ as $t \to +\infty$. Let T_j be the jth jump-time of the temporal $P(t)$ for $j \geq 1$ ($T_0 \equiv 0$ is the initial time) and $\Delta T_{j-1} \equiv T_j - T_{j-1}$ be the* *interjump-time (also called interarrival time) for $j \geq 1$, so $T_1 = \Delta T_0$. Slightly modifying the arguments for the exponential distribution of ΔT_j for the stationary $P(t)$ in (1.24), we see that the nonstationary distribution and its corresponding density conditioned on the most recent jump-time T_{j-1} are given by the following.*

Theorem 1.22. *Provided that $\Lambda(t) \to \infty$ as $t \to \infty$, the nonstationary distribution of the Poisson interjump-time ΔT_{j-1} for $j = 1, 2, \ldots$ interjump-times is*

$$\Phi_{\Delta T_{j-1}|T_{j-1}}(\Delta t) = 1 - \exp\left(-\int_{T_{j-1}}^{T_{j-1}+\Delta t} \lambda(t)dt\right) \tag{1.51}$$

with density

$$\phi_{\Delta T_{j-1}|T_{j-1}}(\Delta t) = \lambda(T_{j-1} + \Delta t) \exp\left(-\int_{T_{j-1}}^{T_{j-1}+\Delta t} \lambda(t)dt\right), \tag{1.52}$$

or alternatively, in terms of the jump times T_j for $j \geq 1$,

$$\phi_{T_j|T_{j-1}}(t) = \lambda(t) \exp\left(-\int_{T_{j-1}}^{t} \lambda(s)ds\right). \tag{1.53}$$

Proof. *The modified stationary proof is as follows:*

$$\begin{aligned}
\Phi_{\Delta T_{j-1}|T_{j-1}}(\Delta t) &\equiv \mathrm{Prob}[\Delta T_{j-1} \leq \Delta t \mid T_{j-1}] \\
&= 1 - \mathrm{Prob}[\Delta T_{j-1} > \Delta t \mid T_{j-1}] \\
&= 1 - \mathrm{Prob}[\Delta P(T_{j-1}) \equiv P(T_{j-1} + \Delta t) - P(T_{j-1}) = 0 \mid T_{j-1}] \\
&= 1 - p_0(\Lambda(T_{j-1} + \Delta t) - \Lambda(T_{j-1})) = 1 - p_0(\Delta\Lambda(T_{j-1})) \\
&= 1 - e^{-\Delta\Lambda(T_{j-1})} = 1 - \exp\left(-\int_{T_{j-1}}^{T_{j-1}+\Delta t} \lambda(t)dt\right) \\
&= \Phi_e(\Delta\Lambda(T_{j-1}); 1),
\end{aligned}$$

where $\Phi_e(\Delta\Lambda(T_{j-1}); 1)$ is the exponential distribution (B.40) in $\Delta\Lambda(T_{j-1})$ with $\Delta\Lambda(T_{j-1}) \equiv \Lambda(T_{j-1} + \Delta t) - \Lambda(T_{j-1})$ and mean $\mu = 1$, i.e., still exponentially distributed but the distribution depends on T_{j-1}. $\square$

Caution. *If $\Lambda(t)$ is finite, then $\Phi_{\Delta T_{j-1}|T_{j-1}}(\Delta t)$ as derived is **not a proper** probability distribution since $1 - \exp\left(-\int_{T_{j-1}}^{+\infty} \lambda(t)dt\right) < 1$ with $\Lambda(+\infty) < +\infty$).*

- *For more general properties and for extended information see Snyder and Miller [252] or Çinlar [56].*

1.10 Poisson Process Expectations Conditioned on the Past

Example 1.23. *Illustration of Independent Increments and Markov Properties for the Poisson Process.*

- $E[P(t)|P(r), 0 \le r \le s] = P(\min[s, t]) + \lambda(t - s)H(t - s).$

  Here $H(X)$ is the Heaviside step function (B.156). The techniques are similar to those for the Wiener process, except that there is no zero mean, but the mean increment is the same as the increment variance, i.e., $E[\Delta P(t)] = \lambda\Delta t = \mathrm{Var}[\Delta P(t)]$. Also, $P(0)$ is zero by definition with probability one:

$$E[P(t)|P(r), 0 \le r \le s] = \left\{ \begin{array}{l} P(t),\ 0 \le t \le s \\ E[(P(t) - P(s)) + (P(s) - P(0))|P(r), 0 \le r \le s], \\ \qquad\qquad\qquad\qquad\qquad\qquad 0 \le s < t \end{array} \right\}$$

$$= \left\{ \begin{array}{l} P(t),\ 0 \le t \le s \\ \lambda(t - s) + P(s),\ 0 \le s < t \end{array} \right\}$$

$$= P(\min[s, t]) + \lambda(t - s)H(t - s).$$

When $0 \le s < t$, then the above formula symmetrized using the Poisson deviation process, $(P(t) - \lambda t)$, having zero mean, with $H(t - s) = 1$ for $s < t$, has the form

$$E[P(t) - \lambda t|P(r), 0 \le r \le s] = P(s) - \lambda s. \qquad (1.54)$$

This signifies that for the deviation the average information conditioned on the past data, $\{P(r), r \in [0, s]\}$, is given by the most recent past deviation $P(s) - \lambda s$, which may imply a significant reduction in uncertainty for the present data, $P(t)$.

*The form of the result (1.54) is again the principal characteristic form for a **martingale**, as was (1.18) with $X(t) = f(P(t))$ or (1.18) for $W(t)$, i.e., an abstract model of a fair game. (See the beginning chapter of Mikosch [209] for a clear description of martingales, but in an elementary abstract presentation; martingales are described in Chapter 12 of our book.)*

- $E[P^2(t)|P(r), 0 \le r \le s] = P^2(\min[s, t]) + \lambda(t - s)(1 + 2P(s) + \lambda(t - s))H(t - s).$
  The derivation is similar to that for the conditional mean above:

$$E[P^2(t)|P(r), 0 \le r \le s] = \left\{ \begin{array}{l} P^2(t),\ 0 \le t \le s \\ E[((P(t) - P(s)) + (P(s) - P(0)))^2|P(r), 0 \le r \le s], \\ \qquad\qquad\qquad\qquad\qquad\qquad 0 \le s < t \end{array} \right\}$$

$$= \left\{ \begin{array}{l} P^2(t),\ 0 \le t \le s \\ E[(P(t) - P(s))^2] + 2P(s)E[(P(t) - P(s))] + P^2(s), \\ \qquad\qquad\qquad\qquad\qquad\qquad 0 \le s < t \end{array} \right\}$$

$$= \left\{ \begin{array}{l} P^2(t),\ 0 \le t \le s \\ \lambda(t - s)(1 + \lambda(t - s)) + 2P(s) \cdot \lambda(t - s) + P^2(s), \\ \qquad\qquad\qquad\qquad\qquad\qquad 0 \le s < t \end{array} \right\}$$

$$= P^2(\min[s, t]) + \lambda(t - s)(1 + 2P(s) + \lambda(t - s))H(t - s).$$

Table 1.2 has to be used for $E[(\Delta P)^2(s)]$ with $\Delta t = (t - s)$.

Similar to the techniques used previously for the Wiener process with conditioning on the past, the general technique for powers $P^m(t)$, when $s < t$ with conditioning on $P(s)$, is to use the decomposition into independent increments $P(t) = P(s) + (P(t) - P(s))$ and then expand the power of m by the binomial expansion (B.150) so that

$$(P(s) + (P(t) - P(s)))^m = \sum_{k=0}^{m} \binom{m}{k} P^k(s)(P(t) - P(s))^{m-k},$$

and then use independence of the increments and conditioning to calculate for each term

$$E\left[\binom{m}{k} P^k(s)(P(t) - P(s))^{m-k} \,\middle|\, P(r), 0 \leq r \leq s \right]$$

$$= \binom{m}{k} P^k(s)E\left[(P(t) - P(s))^{m-k}\right],$$

relying on Table 1.2 for the remaining expectations.

1.11 Exercises

1. Show formally that

$$\phi_{dW(t)}(w) \stackrel{dt}{=} \delta(w) + \frac{1}{2}dt\delta''(w), \tag{1.55}$$

i.e., has a **delta-density** in the generalized sense, by showing that

$$E[f(dW(t))] = \int_{\infty}^{+\infty} \phi_{dW(t)}(w)f(w)dw \stackrel{dt}{=} f(0) + \frac{1}{2}dtf''(0),$$

i.e., to precision-dt, neglecting terms o(dt). Also, show that the integral of the delta-density on the right-hand side of (1.55) has the same effect as the integral of the left-hand side. Assume that $f(w)$ is three times continuously differentiable and with $f(w)$ and its derivatives vanishing sufficiently at infinity.
(*Hint: Only a formal expansion of $f(w)$ should be needed here. The exponential properties of $\phi_{dW(t)}(w)$ ensure uniformity to allow expansion inside the integral, so that Laplace's or a higher order asymptotic method should not be needed.*)

2. Let $\{t_i : t_{i+1} = t_i + \Delta t_i, i = 0 : n, t_0 = 0; t_{n+1} = T\}$ be a variably spaced partition of the time-interval $[0, T]$ with $\Delta t_i > 0$. Show the following properties and justify them by giving a reason for every step, such as a property of the process or a property of expectations:

   (a) Let $G(t) = \mu_0 t + \sigma_0 W(t)$ and $\Delta G(t_i) \equiv G(t_i + \Delta t_i) - G(t_i)$ with μ and $\sigma_0 > 0$ constants; then show

   $$\text{Cov}[\Delta G(t_i), \Delta G(t_j)] = \sigma_0^2 \Delta t_i \, \delta_{i,j}$$

   for $i, j = 0 : n$, where $\delta_{i,j}$ is the Kronecker delta.

(b) Let $H(t) = v_0 P(t)$ and $\Delta H(t_i) \equiv H(t_i + \Delta t_i) - H(t_i)$ with $\lambda_0 > 0$ and $v_0 > -1$ constants; then show

$$\text{Cov}[\Delta H(t_i), \Delta H(t_j)] = v_0^2 \lambda_0 \Delta t_i \delta_{i,j}$$

for $i, j = 0 : n$.

(c) Let $\Delta W(t_i) \equiv W(t_i + \Delta t_i) - W(t_i)$, but $\Delta^\theta W(t_i) \equiv W(t_i + \theta \Delta t_i) - W(t_i)$ with $0 \le \theta \le 1$; then show

$$\text{Cov}[\Delta W(t_i), \Delta^\theta W(t_j)] = \theta \Delta t_i \, \delta_{i,j}$$

for $i, j = 0 : n$.

3. (a) Verify the $m = 3 : 4$ entries in Table 1.1 for $\text{E}[|\Delta W(t)|^m]$.

   (b) Verify the $m = 3 : 4$ entries in Table 1.2 for $\text{E}[(\Delta P(t))^m]$ and $\text{E}[(\Delta P(t) - \lambda \Delta t)^m]$.

4. (a) Show that when $0 \le s \le t$,

$$\text{E}[W^3(t)|W(r), 0 \le r \le s] = W^3(s) + 3(t - s)W(s),$$

justifying every step with a reason, such as a property of the process or a property of conditional expectations.

(b) Use this result to verify the martingale form (1.18),

$$\text{E}[W^3(t) - 3t W(t)|W(r), 0 \le r \le s] = W^3(s) - 3s W(s).$$

(Hint: The general technique is to seek the expectation of mth power in the separable form,

$$\text{E}[M_m(W(t), t)|W(r), 0 \le r \le s] = M_m(W(s), s),$$

where

$$M_m(W(t), t) = W^m(t) + \sum_{k=0}^{m-1} \alpha_k(t) W^k(t),$$

satisfied for the sequence of coefficient functions $\{\alpha_0(t), \ldots, \alpha_{m-1}(t)\}$ for the separable form, so that the conditional expectations of the lower order powers

$$\text{E}[W^k(t)|W(r), 0 \le r \le s]$$

can be recursively obtained in the order $k = 0 : m - 1$.)

5. (a) Show that when $0 \le s \le t$,

$$\text{E}[W^4(t)|W(r), 0 \le r \le s] = W^4(s) + 6(t - s)W^2(s) + 3(t - s)^2,$$

justifying every step with a reason, such as a property of the process or a property of conditional expectations.

(b) Use this result to verify the martingale form (1.18)

$$E[W^4(t) - 6tW^2(t) + 3t^2|W(r), 0 \le r \le s] = W^4(s) - 6sW^2(s) + 3s^2,$$

together with the form for similar conditional expectation of $W^2(t)$ or that for $W^2(t) - t$.

(See the Hint in Exercise 4.)

6. Show that

$$E[(\Delta P)^m(t)] = \lambda \Delta t (1 + O(\lambda \Delta t)) \qquad (1.56)$$

for $\lambda \Delta t \ll 1$, by induction for $m \ge 1$.

7. Show that for the Poisson increment process, $\Delta P(t)$, the expectation can be expanded as

$$E[f(\Delta P(t))] = \sum_{k=0}^{\infty} \frac{(\lambda \Delta t)^k}{k!} \Delta^k[f(0)],$$

assuming that $f(p)$ is a bounded function so that the sum converges. The kth order finite difference is defined inductively such that

$$\Delta^{k+1}[f(i)] \equiv \Delta[\Delta^k[f(i)]]$$

starting from $\Delta^0[f(i)] = f(i)$ and $\Delta^1[f(i)] = \Delta[f(i)] \equiv f(i+1) - f(i)$.
(Hint: Use the zero-step $I_0[f(i)] \equiv f(i)$ and one-step $I_1[f(i)] \equiv f(i+1)$ operators, so that $\Delta = I_1 - I_0$ and $\Delta^k = (I_1 - I_0)^k$, for which the binomial expansion can be used.)

8. Show that the temporal Poisson process increment distribution, $p_k(\Delta \Lambda(t))$, satisfies the differential-difference equation (DDE),

$$\frac{d}{dt} [p_k(\Delta \Lambda(t))] = \lambda(t) (p_k(\Delta \Lambda(t)) - p_{k-1}(\Delta \Lambda(t))), \qquad (1.57)$$

i.e., differential in t, but difference equation in k.

Show the following characteristic function (Fourier transform) formulas in the constant coefficient case (you need only assume that the imaginary unit $i \equiv \sqrt{-1}$ is a constant with $i^2 = -1$ when integrating for the expectation or that $\zeta = i \cdot z$ can be treated the same as a real variable):

(a) For the *Gaussian process* with time-linear drift, $G(t) = \mu_0 t + \sigma_0 W(t)$, where μ_0 and $\sigma_0 > 0$ are constants,

$$C[G](z) \equiv E[\exp(izG(t))] = \exp\left(iz\mu_0 t - z^2\sigma_0^2 t/2\right);$$

(b) for the Poisson process, $\nu_0 P$, with constant jump-rate $\lambda_0 > 0$ and constant jump-amplitude ν_0,

$$C[\nu_0 P](z) \equiv \mathrm{E}[\exp(iz\nu_0 P(t))] = \exp\left(\lambda_0 t \left(\exp(iz\nu_0) - 1\right)\right);$$

(c) and finally for the jump-diffusion process, assuming that $W(t)$ and $P(t)$ are independent processes,

$$C[X](z) \equiv \mathrm{E}[\exp(izX(t))] = \exp\left(iz\mu_0 t - z^2\sigma_0^2 t/2 + \lambda_0 t \left(\exp(iz\nu_0) - 1\right)\right).$$

9. (a) Show that when $0 \le s < t$ and the constant jump-rate is λ_0 (see the general result in Section 1.10, but verify independently this special result),

$$\mathrm{E}[P^2(t)|P(r), 0 \le r \le s] = P^2(s) + 2\lambda(t-s)P(s)$$
$$+\lambda_0(t-s)(1 + \lambda_0(t-s)),$$

justifying every step with a reason for its validity.

(b) Find the time polynomials $\alpha_1(t)$ and $\alpha_0(t)$ so that

$$MP_2(t) = P^2(t) + \alpha_1(t)P(t) + \alpha_0(t)$$

is a martingale. Assume $\alpha_k(0) = 0$ for $k = 0:1$.
(Note: The primary martingale property is that $\mathrm{E}[X(t)|X(r), 0 \le r \le s] = X(s)$ *for some process $X(t)$ and in this case $X(t) = f(P(t))$, but there are additional technical conditions for defining a martingale form. Also, by a simple form of the principle of separation of variables, if $f(t) = g(s)$ for arbitrary values of t and s, then $f(t) = C = g(s)$, where C is a constant.)*

10. (a) Show that when $0 \le s < t$,

$$\mathrm{E}[P^3(t)|P(r), 0 \le r \le s]$$
$$= P^3(s) + 3\lambda(t-s)P^2(s) + 3\lambda(t-s)(1 + \lambda(t-s))P(s)$$
$$+\lambda(t-s)(1 + 3\lambda(t-s) + \lambda^2(t-s)^2),$$

justifying every step with a reason, such as a property of the process or a property of conditional expectations.

(b) Use this result to verify the martingale form (1.18)

$$\mathrm{E}[P^3(t) - 3\lambda t P^2(t) - 3\lambda t(1 - \lambda t)P(t) - \lambda t(1 - 3\lambda t + \lambda^2 t^2)|P(r), 0 \le r \le s]$$
$$= P^3(s) - 3\lambda s P^2(s) - 3\lambda s(1 - \lambda s)P(s) - \lambda s(1 - 3\lambda s + \lambda^2 s^2).$$

(Hint: See the Hint in Exercise 4 in this section for $W^3(t)$ conditional expectation.)

Suggested References for Further Reading

- Arnold, 1974 [13]

- Çinlar, 1975 [56]

- Gard, 1988 [92]

- Jazwinski, 1970 [155]

- Karlin and Taylor, 1981 [163]

- Klebaner, 1998 [165]

- Mikosch, 1998 [209]

- Øksendal, 1998 [222]

- Schuss, 1980 [244]

- Snyder and Miller, 1991 [252]

- Steele, 2001 [256]

- Taylor and Karlin, 1998 [265]

- Tuckwell, 1995 [270]

Chapter 2

Stochastic Integration for Diffusions

My major aim in this was to find facts which would guarantee as much as possible the existence of atoms of definite finite size.
—Albert Einstein (1879–1955) in the first of four "Annus Mirabilis" papers in the *Annalen der Physik* concerning Brownian motion, 1905

Brownian motion, as described by Bachelier in 1900 and Einstein in 1905, was provided a rigorous mathematical definition by Wiener (1894–1964) in Wiener (1923, 1930) by proving the existence of an appropriate measure on a space of functions-of-time.
—Harry M. Markowitz in the foreword to [245]

Jump-diffusion stochastic differential equations (SDEs) with initial conditions are of the form

$$dX(t) = f(X(t), t)dt + g(X(t), t)dW(t) + h(X(t), t)dP(t), \quad X(0) = x_0, \qquad (2.1)$$

where the Poisson process $dP(t)$ supplies the jumps and the Wiener process $dW(t)$ supplies the diffusion. The initial value problem (2.1), unlike the ordinary differential equations (ODEs) with initial conditions, are symbolic equations. They are not fully defined until the method of integration for solving an SDE is specified, given the coefficient functions $\{f(x, t), h(x, t), g(x, t)\}$. More precisely, the SDE (2.1) is not fully specified until the methods of integration for the three types of integrals in the formal integral solution,

$$X(t) = x_0 + \int_0^t f(X(s), s)ds + \int_0^t g(X(s), s)dW(s) + \int_0^t h(X(s), s)dP(s), \qquad (2.2)$$

with respect to t, $W(t)$, and $P(t)$, respectively, have been defined. Until then, the stochastic integral equation (2.2) is as symbolic as the SDE in (2.1), since the evaluation of the second

31

and third integrals in (2.2) is very sensitive to the method of integration used due to the random and singular properties of $dW(t)$ and $dP(t)$. It will be necessary to reexamine the foundations for ordinary or Riemann integration to motivate the inclusion of integrands with randomness, nonsmoothness, and jump discontinuities contributed by the stochastic processes $W(t)$ and $P(t)$ to the state process $X(t)$. This re-examination of integration will also be useful for subsequent numerical approximations of the new definitions as well as provide a basis for new types of integrals that will arise.

In this chapter, the integrals of the second type in (2.2), i.e.,

$$\int_0^t g(X(s), s)dW(s),$$

where the integration is with respect to the diffusion process $W(t)$, will be treated primarily. However, the short treatment of ordinary integration will be sufficient for integrals of the first type, i.e.,

$$\int_0^t f(X(s), s)ds,$$

where the integration is with respect to the time t and the stochastic process $X(t)$ is only in the integrand. The third type of integral will be treated in the next chapter.

When considering higher approximations or other difficult behavior in the numerical solution of ODEs, it is often necessary to work with the corresponding integral equation. Similarly, the proper form for solving SDEs (which in general can be considered as a symbolic concept anyway) is the exact and numerical analysis of the corresponding stochastic integral equation.

Once the foundations for stochastic integrations have been made, as they would be for ordinary integration in a good calculus course, and the definition is illustrated for a few simple examples, then some simpler formal chain rules will be developed that will make calculations of integrals, where possible, much easier. This chapter on stochastic integration of diffusions and a similar one on jumps that follows, present the basis for the SDE models of this book. Although the level of analysis is much higher than would be expected for an applied text, it is important to have a good reference source when treating new types of problems that do not fit the current models or theories to facilitate the modification of the current theories.

2.1   Ordinary or Riemann Integration

The theory of ordinary or Riemann integration is quickly reviewed here as an intermediate step to build up to the treatment of stochastic integration. Let the ordinary integral be symbolically defined as

$$I[f](t) = \int_0^t f(s)ds, \tag{2.3}$$

where $f(t)$ is a continuous function on $0 \le t \le T$, but continuity is really more than what would be needed in general here. For general functions f, the integral interval $[0, t]$ is partitioned into $n + 1$ subintervals, $[t_i, t_{i+1}]$ of width $\Delta t_i \equiv t_{i+1} - t_i > 0$ for $i = 0 : n$, i.e., a grid of $n + 2$ points such that

$$0 = t_0 < t_1 < t_2 < \cdots < t_n < t_{n+1} = t. \tag{2.4}$$

On each subinterval an approximation point $t_i^* \equiv t_{i+\theta_i} \equiv t_i + \theta_i \Delta t_i$ is selected with $0 \leq \theta_i \leq 1$ provided that the θ_i's are chosen so that the t_i's are distinct as in (2.4), and the area on the subinterval is approximated by the simplest geometry, a rectangle of width Δt_i and height $f_i^* \equiv f_{i+\theta_i} \equiv f(t_{i+\theta_i})$, with area $f(t_{i+\theta_i}) \Delta t_i$. Next let the grid size be specified as $\delta t_n \equiv \max_{i=0:n} [\Delta t_i]$ such that $\delta t_n \rightarrow 0^+$ as $n \rightarrow \infty$ to ensure that all subintervals shrink to zero in the limit as $n \rightarrow \infty$. Finally, let

$$I_n^{(\theta)}[f](t) \equiv \sum_{i=0}^{n} f_{i+\theta_i} \Delta t_i \tag{2.5}$$

be the discrete approximation of the integral and define constructively the **Riemann integral** as

$$I[f](t) = \lim_{\substack{n \to \infty \\ \delta t_n \to 0}} \left[I_n^{(\theta)}[f](t) \right], \tag{2.6}$$

provided the limit exists. It is important to note that the limit is independent of $\theta_i, 0 \leq \theta_i \leq 1$.

Usually, only a constant value of θ_i is used in practice, so let $\theta_i = \theta$. Also, for simplicity, the grid partition will be assumed to be evenly spaced, so that $\Delta t_i = \Delta t$, with nodes starting at t_0 and successive nodes at $t_{i+1} = t_i + \Delta t$, but integrand approximation points at $t_{i+\theta} = t_i + \theta \Delta t$ for $i = 0 : n$. Also, $t_i = i * \Delta t$ for $i = 0 : (n+1)$. Since the step size is constant, we then have

$$\delta t_n = \Delta t = (t_{n+1} - t_0)/(n + 1) = t/(n + 1) \rightarrow 0^+,$$

as $n \rightarrow +\infty$, so the extra condition that $\delta t_n \rightarrow 0^+$ is not needed.

Fortunately, the limiting definition (2.6) does not have to be used much in ordinary calculus, but the **Riemann sum** (2.5) can be used for simply numerically approximating integrals. When $\theta = 0$ and $t_{i+\theta} = t_i$, the left-hand end point of the ith subinterval, the numerical forward integration rule is called the **left rectangular rule** or **Euler's explicit method** or tangent-line method for ODEs. When $\theta = 1$ and $t_{i+\theta} = t_{i+1}$, the right-hand end point of the ith subinterval, the numerical backward integration rule is called the **right rectangular rule** or implicit **backward Euler's method** for ODEs. When $\theta = 1/2$ and $t_{i+\theta} = (t_i + t_{i+1})/2$, the midpoint of the ith subinterval, the numerical integration rule is called the **midpoint rectangular rule**, which is more accurate by an order of magnitude in δt_n, provided $f(t)$ is sufficiently differentiable.

Since the process $W(t)$ is continuous with probability one, then integrals of composite functions $f(W(t), t)$ with respect to t can be defined by Riemann integration, i.e.,

$$\int_0^t f(W(s), s) ds = \lim_{n \to \infty} \left[\sum_{i=0}^{n} f(W(t_i), t_i) \Delta t_i \right], \tag{2.7}$$

choosing $\theta = 0$ here, although other values would be suitable. Similarly, when the integrand is for the composite process $X(t)$ with implied dependence on the diffusion $W(t)$ and also the jump process $P(t)$ through (2.2), the integral will be defined by Riemann integration, i.e.,

$$\int_0^t f(X(s), s)ds = \lim_{n \to \infty} \left[\sum_{i=0}^n f(X(t_i), t_i)\Delta t_i \right]. \tag{2.8}$$

The Poisson jump process, while discontinuous, is right-continuous with left limits, i.e., it is also a piecewise continuous step function, so it fits nicely in the framework of the use of forward integration, which is effectively a sequence of step-function approximations. However, the jumps are stochastic and not predictable, although once a jump is generated through simulation or observation, it will be known.

Sometimes, a deterministic integration is needed with respect to the position on the path $X(t)$. In this case, let the $f(s)ds$ in (2.3) be replaced by $f(X(s), s)dX(s)$, which could also come from the form $f(X(s), s)X'(s)ds$, provided the velocity $v(s) = X'(s)$ or $dX(s) = X'(s)ds$ exists; then this leads to the **Stieltjes integral**, or Riemann–Stieltjes integral, constructive definition,

$$\int_0^t f(X(s), s)dX(s) = \lim_{n \to \infty} \left[\sum_{i=0}^n f(X(t_{i+\theta}), t_{i+\theta})(X(t_{i+1}) - X(t_i)) \right], \tag{2.9}$$

provided $X(t)$ is continuous and has bounded variation [169], i.e.,

$$\sum_{i=0}^n |X(t_{i+1}) - X(t_i)| < B,$$

for some constant $B > 0$ for all partitions (2.4) of $[0, t]$, and provided $f(X(t), t)$ is continuous. (These conditions are stronger than needed, and Mikosch [209] gives weaker but more complicated conditions.) Another example is the Stieltjes form for the expectation in terms of the probability distribution $\Phi_X(x)$ in the random variable X,

$$E_X[f(X)] = \int_{-\infty}^{\infty} f(x)d\Phi_X(x),$$

sometimes used to permit the use of more general distributions than would be possible under the usual Riemann integration conditions. In the next section, the Stieltjes integration form will be modified for the stochastic integration relative to $W(t)$.

2.2 Stochastic Integration in $W(t)$: The Foundations

As in elementary calculus, the presentation starts with a fairly simple example. The integral that forms the basis for the formulation that follows is the stochastic Stieltjes integral,

$$I[W](t) = \int_0^t W(s)dW(s), \tag{2.10}$$

which has a stochastic correction for the simple deterministic calculus Stieltjes integral,

$$I^{(det)}[X](t) = \int_0^t X(s)\,dX(s) = \frac{1}{2}\int_0^t d(X^2)(s) = \frac{1}{2}\left(X^2(t) - X^2(0)\right). \qquad (2.11)$$

This follows from the ordinary calculus chain rule, $d(X^2)(s) = 2X(s)\,dX(s)$, for differentials, to form an exact differential.

However, in the case of the stochastic integral (2.10), $W(t)$ is a random process, is nowhere differentiable, and can be shown to have unbounded variation. Note that for even spacing $\delta t_n = \Delta t = (t - 0)/(n + 1)$ for $i = 0 : n$, so that the expected variation, from Table 1.1, is

$$\mathrm{E}\left[\sum_{i=0}^n |\Delta W_i|\right] = \sum_{i=0}^n \sqrt{2\Delta t/\pi} = (n+1)\sqrt{2t/(\pi(n+1))} = \sqrt{2t(n+1)/\pi} \to +\infty,$$

as $n \to +\infty$, so the variation must be unbounded since the expected variation must not exceed the supremum of the variation and the supremum must be unbounded as well. (See Mikosch [209] for another justification.)

In the first step in finding a constructive definition for the stochastic integral (2.10), following Itô [150], a left end point rectangular or forward integration rule ($\theta = 0$) is initially used to approximate the integral so that the independent increment property of $W(t)$ is preserved,

$$I_n^{(0)}[W](t) = \sum_{i=0}^n W(t_i)\Delta W(t_i) = \sum_{i=0}^n W_i \Delta W_i, \qquad (2.12)$$

with W_i independent of ΔW_i as intended, where the simplifying numerical notations $W_i \equiv W(t_i)$ and $\Delta W_i \equiv \Delta W(t_i) \equiv W(t_{i+1}) - W(t_i)$ have been used. The form (2.12) is not very useful for summing or approximating, but the following two general identities are.

Lemma 2.1. *Let $\{x_i | i = 0 : n + 1\}$ be **any** sequence of numbers, and let $\Delta x_i = x_{i+1} - x_i$ for $i = 0 : n$. Then*

$$\sum_{i=0}^n \Delta x_i = x_{n+1} - x_0, \qquad (2.13)$$

$$\sum_{i=0}^n x_i \Delta x_i = \frac{1}{2}\left(x_{n+1}^2 - x_0^2 - \sum_{i=0}^n (\Delta x_i)^2\right). \qquad (2.14)$$

Proof. The first identity (2.13) is trivial, since adding two successive increments cancels the common value of those increments, i.e.,

$$\Delta x_i + \Delta x_{i+1} = (x_{i+1} - x_i) + (x_{i+2} - x_{i+1}) = x_{i+2} - x_i.$$

Verifying the second, and important, identity is much easier by expanding the summand on the right-hand side of (2.14) to obtain the left-hand side rather than vice versa:

$$
\begin{aligned}
\frac{1}{2}\left(x_{n+1}^2 - x_0^2 - \sum_{i=0}^{n}(\Delta x_i)^2\right) &= \frac{1}{2}\left(x_{n+1}^2 - x_0^2 - \sum_{i=0}^{n}(x_{i+1} - x_i)^2\right)\\
&= \frac{1}{2}\left(x_{n+1}^2 - x_0^2 - \sum_{i=0}^{n}(x_{i+1}^2 - 2x_i x_{i+1} + x_i^2)\right)\\
&= \frac{1}{2}\left(x_{n+1}^2 - x_0^2 - \sum_{i=0}^{n}x_{i+1}^2 + 2\sum_{i=0}^{n}x_i x_{i+1} - \sum_{i=0}^{n}x_i^2\right)\\
&= \frac{1}{2}\left(x_{n+1}^2 - x_0^2 - \left(x_{n+1}^2 + \sum_{j=0}^{n}x_j^2 - x_0^2\right)\right.\\
&\qquad\left. + 2\sum_{i=0}^{n}x_i x_{i+1} - \sum_{i=0}^{n}x_i^2\right)\\
&= \sum_{i=0}^{n}x_i\,\Delta x_i,
\end{aligned}
\tag{2.15}
$$

where

$$
\sum_{i=0}^{n}x_{i+1}^2 = \sum_{j=1}^{n+1}x_j^2
$$

has been transformed by a **change of index** to combine with a similar sum.   □

    The benefit of the form (2.14) when used as $x_i = W_i$ is that the end points are explicitly given by $W_{n+1} = W(t)$ and $W_0 = 0$ with probability one, so the discrete approximation to the stochastic integral of $W(t)$ becomes

$$
I_n^{(0)}[W](t) = \frac{1}{2}\left(W^2(t) - \sum_{i=0}^{n}(\Delta W_i)^2\right).
\tag{2.16}
$$

Using Table 1.1 again, the expectation of $I_n^{(0)}[W](t)$ is

$$
E\left[I_n^{(0)}[W](t)\right] = \frac{1}{2}\left(t - \sum_{i=0}^{n}\Delta t_i\right) = \frac{1}{2}(t - t) = 0,
$$

returning to more general spacing Δt_i, where the (2.13) identity $\sum_{i=0}^{n}\Delta t_i = t_{n+1} - t_0 = t$ has also been used. This result suggests that a reasonable form for the stochastic integral (2.10) corresponds to ($\approx$)

$$
I[W](t) \approx \frac{1}{2}(W^2(t) - t),
\tag{2.17}
$$

where the term $(-\frac{1}{2}t)$ is the correction to the ordinary calculus or Riemann integration answer. However, since the proposed answer is not a true equality, another condition is appropriate for the stochastic nature of the problem, and that condition is the mean square limit or mean square convergence.

Definition 2.2. *Mean Square Limit or Convergence.*
*The random variable $I_n^{(0)}(t)$ **converges in the mean square** to the random variable $I(t)$ if*

$$\mathrm{E}\left[\left(I_n^{(0)}(t) - I(t)\right)^2\right] \to 0 \tag{2.18}$$

as $n \to \infty$, assuming that both random variables have bounded mean squares, i.e.,

$$\mathrm{E}\left[(I_n^{(0)})^2(t)\right] < \infty \ \text{ and } \ \mathrm{E}\left[I^2(t)\right] < \infty.$$

*If the limit (2.18) exists, then denote the **mean square limit** as*

$$I(t) = \overset{\mathrm{ms}}{\underset{n\to\infty}{\lim}} \left[I_n^{(0)}(t)\right].$$

As an abbreviation, sometimes $\overset{\mathrm{ims}}{=}$ will be used for $= \lim_{n\to\infty}^{\mathrm{ms}}$, where $\overset{\mathrm{ims}}{=}$ means "Itô mean square equals."

Some related **general stochastic convergence principles** follow.

Definition 2.3. *Convergence in Probability.*
*The random variable $I_n^{(0)}(t)$ **converges in probability** to the random variable $I(t)$ if for every $\epsilon > 0$,*

$$\mathrm{Prob}\left[\left|I_n^{(0)}(t) - I(t)\right| \geq \epsilon\right] \to 0 \tag{2.19}$$

*as $n \to \infty$. If the limit (2.19) exists, then denote the **limit in probability** as*

$$I(t) = \overset{\mathrm{prob}}{\underset{n\to\infty}{\lim}} [I_n^{(0)}(t)].$$

Definition 2.4. *Convergence in Mean.*
*The random variable $I_n^{(0)}(t)$ **converges in the mean** to the random variable $I(t)$ if for every $\epsilon > 0$,*

$$\mathrm{E}\left[\left|I_n^{(0)}(t) - I(t)\right|\right] \to 0 \tag{2.20}$$

*as $n \to \infty$. If the limit (2.20) exists, then denote the **limit in the mean** as*

$$I(t) = \overset{\mathrm{mean}}{\underset{n\to\infty}{\lim}} [I_n^{(0)}(t)].$$

Theorem 2.5. *Convergence in Mean Square $\Longrightarrow$ Convergence in Probability.*

$$I(t) = \overset{\mathrm{ms}}{\underset{n\to\infty}{\lim}} [I_n^{(0)}(t)] \quad \Longrightarrow \quad I(t) = \overset{\mathrm{prob}}{\underset{n\to\infty}{\lim}} [I_n^{(0)}(t)]. \tag{2.21}$$

Similarly,

Convergence in Mean Square $\implies$ Convergence in Mean.

$$I(t) = \overset{\text{ms}}{\underset{n\to\infty}{\lim}} [I_n^{(0)}(t)] \implies I(t) = \overset{\text{mean}}{\underset{n\to\infty}{\lim}} [I_n^{(0)}(t)]. \tag{2.22}$$

Proof. Let $\epsilon > 0$. Tacitly the mean square expectation of the limit $I(t)$ and the approximation are assumed as conditions for mean square convergence, which implies that $E[|I(t) - I_n^{(0)}(t)|^2] \to 0^+$ as $n \to \infty$. The theorem follows from the **Chebyshev inequality** (B.189) of Exercise 4 on p. B69 in Online Appendix B, which is written in a simplified but convenient form,

$$\text{Prob}[|X| \geq \epsilon] \leq E[|X|^2]/\epsilon^2, \tag{2.23}$$

where $\epsilon > 0$. Let $X = I(t) - I_n^{(0)}(t)$ and thus

$$E[|I(t) - I_n^{(0)}(t)|^2] \geq \epsilon^2 \text{Prob}[|I(t) - I_n^{(0)}(t)| \geq \epsilon].$$

Hence, as $n \to \infty$, $\text{Prob}[|I(t) - I_n^{(0)}(t)| \geq \epsilon] \to 0^+$ by being squeezed from above by the mean square deviation as it goes to zero, i.e., $I_n^{(0)}(t) \to I(t)$ in probability if $I_n^{(0)}(t) \to I(t)$ in the mean square.

Similarly, the Schwarz (Cauchy–Schwarz) inequality (B.190) of Exercise 5 on p. B69 in Online Appendix B, truncated to one variable,

$$E^2[X] \leq E[X^2],$$

can be used to show that convergence in the mean square implies convergence in the mean, i.e., $I_n^{(0)}(t) \to I(t)$ in the mean if $I_n^{(0)}(t) \to I(t)$ in the mean square. $\square$

The expectation of the proposed random variable answer is

$$E[I[W](t)] = E\left[\frac{1}{2}(W^2(t) - t)\right] = \frac{1}{2}(t - t) = 0,$$

the same as for the approximation.

To focus on the crucial term and to simplify the demonstration of the mean square limit, which is conjectured to be t, consider the following lemma.

Lemma 2.6. *Let*

$$J_n^{(0)}(t) \equiv \sum_{i=0}^{n} (\Delta W_i)^2; \tag{2.24}$$

then

$$t = \overset{\text{ms}}{\underset{n\to\infty}{\lim}} [J_n^{(0)}(t)]. \tag{2.25}$$

Proof. The mean t of $J_n^{(0)}(t)$ is absorbed into the summation by (2.13) with $x_i = t_i$; the square of the mean square argument leads to a double sum which is separated into diagonal parts ($j = i$) and off-diagonal parts ($j \neq i$), allowing the splitting of the expectations using the independent increment property, so

$$E\left[\left(J_n^{(0)}(t) - t\right)^2\right] = \text{Var}\left[J_n^{(0)}(t)\right]$$

$$= E\left[\left(\sum_{i=0}^n (\Delta W_i)^2 - t\right)^2\right]$$

$$= E\left[\left(\sum_{i=0}^n \left((\Delta W_i)^2 - \Delta t_i\right)\right)^2\right]$$

$$= E\left[\sum_{i=0}^n \left((\Delta W_i)^2 - \Delta t_i\right) \sum_{j=0}^n \left((\Delta W_j)^2 - \Delta t_j\right)\right]$$

$$= \sum_{i=0}^n E\left[\left((\Delta W_i)^2 - \Delta t_i\right)^2\right]$$

$$+ \sum_{i=0}^n E\left[(\Delta W_i)^2 - \Delta t_i\right] \sum_{\substack{j=0 \\ j \neq i}}^n E\left[(\Delta W_j)^2 - \Delta t_j\right]$$

$$= \sum_{i=0}^n \text{Var}\left[(\Delta W_i)^2\right] + 0 \cdot 0 = \sum_{i=0}^n \left(E\left[(\Delta W_i)^4\right] - E^2\left[(\Delta W_i)^2\right]\right)$$

$$= \sum_{i=0}^n \left(3(\Delta t_i)^2 - (\Delta t_i)^2\right) = 2\sum_{i=0}^n (\Delta t_i)^2,$$

the last two steps relying on the results of Table 1.1. Since $\Delta t_i \leq \delta t_n = \max_j[\Delta t_j]$, we then have

$$E\left[\left(J_n^{(0)}(t) - t\right)^2\right] = 2\sum_{i=0}^n (\Delta t_i)^2 \leq 2\delta t_n \sum_{i=0}^n \Delta t_i = 2t\delta t_n \to 0$$

as $n \to \infty$ showing that

$$t = \underset{n \to \infty}{\overset{ms}{\lim}} [J_n^{(0)}(t)].$$

Clearly both $J_n^{(0)}(t)$ and t have bounded mean squares for bounded t. Hence, $J_n^{(0)}(t) = I_n^{(0)}[dW](t)$, in our functional notation. □

Lemma 2.7.

$$\frac{1}{2}\left(W^2(t) - t\right) = \underset{n \to \infty}{\overset{ms}{\lim}} \left[I_n^{(0)}[W](t)\right], \tag{2.26}$$

where $t < \infty$ and

$$I_n^{(0)}[W](t) = \sum_{i=0}^n W_i \Delta W_i.$$

Proof. Note that

$$E[((W^2(t) - t)/2)^2] = E[W^4(t) - 2tW^2(t) + t^2]/4 = (3t^2 - 2t^2 + t^2)/4 = t^2/2,$$

again using the convenient Table 1.1, so $(W^2(t) - t)/2$ has a bounded mean square so long as t is bounded. Similarly, one can show that $I_n^{(0)}[W](t)$ has a bounded mean square. The mean square convergence of $I_n^{(0)}[W](t)$ is obvious since $J_n^{(0)}(t)$ converged in the mean square to t and $J_n^{(0)}(t)$ is the only term that depends on the grid variable n. In fact,

$$E\left[\left(I[W](t) - I_n^{(0)}[W](t)\right)^2\right] = \frac{1}{4}E\left[\left(t - J_n^{(0)}(t)\right)^2\right] \to 0^+,$$

as $n \to \infty$, converging for the same reason that $J_n^{(0)}(t)$ did in the mean square. This mean square relation follows due to the affine difference in forms $I_n^{(0)}[W](t) = (W^2(t) - J_n^{(0)}(t))/2$ in (2.16) with (2.24) and $I[W](t) \overset{\text{ims}}{=} (W^2(t) - t)/2$ in (2.17), no longer a proposed answer. In more general cases the decomposition of $I_n^{(0)}[W](t)$ will not be as simple as that between $I_n^{(0)}[W](t)$ and the part $J_n^{(0)}(t)$. $\square$

Definition 2.8. *Denote the **Itô mean square (IMS) limit stochastic integral** corresponding to the stochastic integral form*

$$I[g](t) = \int_{t_0}^t g(W(s), s)dW(s)$$

with associated forward integration (left rectangular rule or Euler's method) approximation

$$I_n^{(0)}[g](t) = \sum_{i=0}^n g(W(t_i), t_i)(W(t_{i+1}) - W(t_i))$$

by

$$I^{(\text{ims})}[g](t) = \overset{\text{ms}}{\underset{n\to\infty}{\lim}} \left[I_n^{(0)}[g](t)\right], \tag{2.27}$$

where $0 \le t_0 \le t$, assuming the integrand process $g(W(t), t)$ has a bounded mean integral of its square, i.e.,

$$E\left[\int_{t_0}^t g^2(W(s), s)ds\right] < \infty,$$

and the grid partitioning satisfies

$$0 \le t_0 < t_1 < \cdots < t_{n+1} = t \tag{2.28}$$

with

$$\delta t_n = \max_{i=0:n}[\Delta t_i \equiv t_{i+1} - t_i] \ll 1$$

as $n \to \infty$.

    Provided the Itô mean square limit (2.27) exists,

$$I[g](t) \overset{\text{ims}}{=} I^{(\text{ims})}[g](t). \tag{2.29}$$

In addition, the definition holds, since the independent increments property remains valid in a more general case, namely, if the function g depends on the past and present history of the Wiener process,

$$\mathcal{W}(t) = \{W(r), 0 \le r \le t\},$$

*i.e., $g = g(\mathcal{W}(t), t)$, in which case g is called **nonanticipatory** or **adapted** to the process set $\mathcal{W}(t)$.*

Remarks 2.9.

- *For most of what follows, general functions with dependence on $W(t)$ and t, i.e., $g(W(t), t)$, will be used in stochastic diffusion integrals, but the reader can easily extend results to functions of the type $g(\mathcal{W}(t), t)$ adapted to $\mathcal{W}(t)$.*

- *If the Itô mean square limit (2.27),*

$$I_n^{(0)}[g](t) \to I[g](t) = I^{(\text{ims})}[g](t)$$

  ***in the mean square** as $n \to \infty$, exists, then by Theorem 2.5*

$$I_n^{(0)}[g](t) \to I[g](t)$$

  ***in probability** as $n \to \infty$.*

- *In our notation, $I[g](t) = I^{(\text{ims})}[g](t)$ denotes the mean square limit of the Itô forward integration approximation $I_n^{(0)}[g](t)$ with $\theta = 0$ meaning that the integral g is evaluated at t_i on the ith step. The Itô forward integration approximations denote particular evaluations or approximations of the purely symbolic $I[g](t)$ integral representation which can have other evaluations using other integration rules with values of θ or using other rules relying on nonrectangular approximations.*

Thus, summarizing the results for the crucial simple example when $g(W(t), t) = W(t)$ is the following theorem.

Theorem 2.10. *Itô Fundamental Mean Square Stochastic Integrals.*

$$\int_0^t (dW)^2(s) \overset{\text{ims}}{=} t \tag{2.30}$$

and

$$\int_0^t W(s)dW(s) \overset{\text{ims}}{=} I^{(\text{ims})}[W](t) = \frac{1}{2}\left(W^2(t) - t\right). \tag{2.31}$$

Sketch of Proof. A heuristic justification is given here.

- In ordinary deterministic integral calculus, the symbol $\int_0^t (dx)^2(s)$ would be considered nonsense, but in Itô stochastic integration the symbol

$$\int_0^t (dW)^2(s) \overset{\text{ims}}{=} \overset{\text{ms}}{\underset{n\to\infty}{\lim}}\left[\sum_{i=0}^n (\Delta W)^2(t_i)\right] = t$$

makes perfect sense, since the Itô mean square (IMS) limit is well defined and leads to the Itô correction to the ordinary calculus rule for the differential of $x^2(t)$, i.e., $x(t)dx(t) = \frac{1}{2}d(x^2)(t)$.

- In fact, this leads to a corresponding symbolic **Itô mean square** $\overset{ims}{\underset{sym}{=}}$ version for differentials,

$$(dW)^2(t) \overset{dt}{\underset{ms}{=}} dt \tag{2.32}$$

and

$$W(t)dW(t) \overset{dt}{\underset{ms}{=}} \frac{1}{2}(d(W^2)(t) - dt). \tag{2.33}$$

Formally, we might rewrite (2.33) with the symbol $\overset{dt}{\underset{ms}{=}}$ for "equals in dt-precision mean square," or simply $\overset{dt}{=}$ for "dt-precision," denoting a commutative operation,

$$d(W^2)(t) \overset{dt}{=} 2W(t)dW(t) + dt. \tag{2.34}$$

Using the formal increment definition of the differential (1.3), $dW(t) \equiv W(t+dt) - W(t)$, or the alternate form $W(t + dt) = W(t) + dW(t)$, a quick calculation leads to

$$
\begin{aligned}
d(W^2)(t) &= W^2(t + dt) - W^2(t) - (W + dW)^2(t) - W^2(t) \\
&= (W^2 + 2WdW + (dW)^2 - W^2)(t) \\
&\overset{dt}{\underset{ms}{=}} 2WdW(t) + dt,
\end{aligned}
\tag{2.35}
$$

using a little bit of algebra and the symbolic fact that $(dW)^2 \overset{dt}{\underset{ms}{=}} dt$, formally justifying (2.34), demonstrating a fast technique that would be useful when fast answers are needed.

□

Remarks 2.11.

- *The Itô mean square result symbolized by $(dW)^2(t) \overset{dt}{\underset{ms}{=}} dt$ represents a remarkable paradox, since the differential $(dW)^2(t)$ is deterministic because dt is deterministic, but $dW(t)$ is stochastic or random.*

- *In the deterministic continuously differential case, the corresponding quadratic of a differential, $(dx)^2(t)$, would be negligible relative to terms of order dt. If the integral of such a term were considered, the limit of its finite difference approximation would* *be zero:*

$$\int_0^t (dx)^2(s) = \lim_{n \to \infty} \left[\sum_0^n (\Delta x_i)^2 \right]$$

$$= \lim_{n \to \infty} \left[\sum_0^n (x_{i+1} - x_i) \Delta x_i \right]$$

$$= \lim_{n \to \infty} \left[\sum_0^n x_{i+1} \Delta x_i \right] - \lim_{n \to \infty} \left[\sum_0^n x_i \Delta x_i \right]$$

$$= \lim_{n \to \infty} \left[I_n^{(1)}[x](t) \right] - \lim_{n \to \infty} \left[I_n^{(0)}[x](t) \right]$$

$$= I[x](t) - I[x](t) = 0,$$

since the regular integral of $\int_0^t x(s)dx(s)$ is independent in the limit of the particular approximation parameter used, whether $\theta = 1$ or $\theta = 0$ as in the above final lines.

- *See also Exercise 1, which is to demonstrate that the density, $\phi_{dW(t)}(w)$, for $dW(t)$ is the sum of two delta functions in the generalized sense that considerably constrains functions of $dW(t)$.*

- *Computational confirmation of the Itô fundamental mean square stochastic integrals is the subject of Exercise 3 for the $(dW)^2(t)$ integrand in (2.30) and Exercise 4 for the $(WdW)(t)$ integrand in (2.31). For example, Figure 2.1 is an illustration of the*

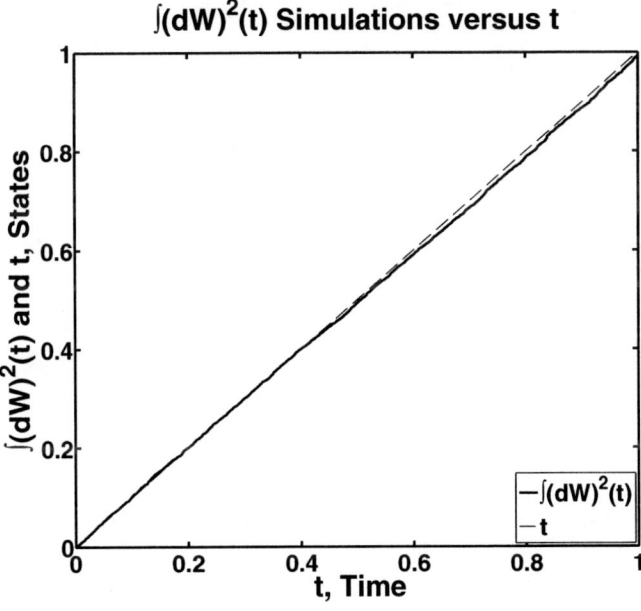

Figure 2.1. *Simulated sample path for the Itô forward integration approximating sum of $\int (dW)^2(t) \overset{\text{ims}}{=} t \simeq \sum_i (\Delta W_i)^2$ for $n = 10^4$ MATLAB* randn *sample size.*

computational confirmation of the Itô fundamental forward integration approximating sum

$$\int_0^t (dW)^2(s) \overset{\text{ims}}{=} t \simeq \sum_{i=0}^n (\Delta W_i)^2$$

with $n = 10^4$. The confirmation is remarkable considering it is a pointwise comparison of the approximating sum with the exact Itô answer t, and not a demonstration of convergence in the Itô mean square limit. The sample size has to be sufficiently large; otherwise the approximating sum tends away from the t answer due to the slope of the tangent line bias. This is also a feature of deterministic ODE applications of Euler's method.

- *The general code for simulating the stochastic diffusion integral with integrand $g(W(t), t)$ by the Itô forward integration approximation*

$$I[g](t) = \int_{t_0}^t g(W(s), s)ds \simeq \sum_{i=0}^n g_i \Delta W_i$$

in an abbreviated MATLAB fragment might be

```
%%%%%%%%%%%%%%%%%%%%%%%%%%%%%%%%%%%%%%%%%%%%%%%%%%%%%%%%%%%%%%%%%%%
function intdwdw0
% Example MATLAB code for integral of (dW)^2.
clc % clear variables;
t0 = 0.0; tf = 1.0;
n = 1.0e+4; nf = n + 1; % set time grid: (n+1) subintervals
dt = (tf-t0)/nf; %         and (n+2) points;
% replace these particular values according to the application;
t(1) = t0; % set initial time at i = 1 for MATLAB;
sqrtdt = sqrt(dt); % dW(i) noise time scale so E[dW] = 0;
kstate = 1; randn('state',kstate); % Set randn state
%     for repeatability;
dW = sqrtdt*randn(nf,1); % simulate (n+1)-dW(i)'s sample;
t = t0:dt:tf; % get time vector t;
W = zeros(1,nf+1); sumdw2 = zeros(1,nf+1); % pre-declare/initialize.
for i = 1:nf % simulate integral sample path.
    W(i+1) = W(i) + dW(i); % sum diffusion noise;
    sumdw2(i+1) = sumdw2(i) + (dW(i))^2; % sum whole integrand;
end
plot(t,sumdw2,'k-',t,t,'k--','LineWidth',2); % plot sum;
title('\int(dW)^2(t) Simulations versus t');
ylabel('\int(dW)^2(t) and t, States');
xlabel('t, Time');
legend('\int(dW)^2(t)','t',0);
% End Code
%%%%%%%%%%%%%%%%%%%%%%%%%%%%%%%%%%%%%%%%%%%%%%%%%%%%%%%%%%%%%%%%%%%
```

The complete code used to generate the figure is given in Program C.11, called intdwdw.m *in Online Appendix C.*

- *The form for the simulation of the Wiener increment process $\Delta W(t)$ by a standard normal distribution Z scaled by $\sqrt{\Delta t}$ in the above code fragment is based upon* *the following change of variables (or change of measure) result, showing that both $\Delta W(t)$ and $\sqrt{\Delta t}\, Z$ have the same distribution.*

Theorem 2.12. *Wiener Simulations by Standard Normal.*
Let Z be a random variable with a standard normal distribution, $\Phi_Z(z) = \Phi_n(z; 0, 1)$. Then

$$\Phi_{\Delta W(t)}(w) = \Phi_{\sqrt{\Delta t}\, Z}(w), \qquad (2.36)$$

where $\Delta t > 0$.

Proof. *From properties of the normal distribution,*

$$\Phi_Z(z) = \mathrm{Prob}[Z \le z] = \frac{1}{\sqrt{2\pi}} \int_{-\infty}^{z} e^{-y^2/2} dy$$

and

$$\begin{aligned}
\Phi_{\Delta W(t)}(w) &= \mathrm{Prob}[\Delta W(t) \le w] = \frac{1}{\sqrt{2\pi\,\Delta t}} \int_{-\infty}^{w} e^{-v^2/(2\Delta t)} dv \\
&= \frac{1}{\sqrt{2\pi}} \int_{-\infty}^{w/\sqrt{\Delta t}} e^{-y^2/2} dy = \mathrm{Prob}\left[z \le w/\sqrt{\Delta t}\right] \\
&= \mathrm{Prob}\left[\sqrt{\Delta t}\, Z \le w\right] = \Phi_{\sqrt{\Delta t}\, Z}(w),
\end{aligned}$$

since $\mathrm{Prob}[aZ \le w] = \mathrm{Prob}[Z \le w/a]$ provided $a > 0$. □

- *See also the full version of this MATLAB code in Program C.11 of Online Appendix C used for the actual typeset figure.*

- *See also Figure 4.1 in Chapter 4, which illustrates the application to $g(W(t), t) = \exp(W(t) - t/2)$ that yields an exact differential in the Itô mean square sense.*

- *Computational simulation is another way to get fast answers when they are needed.*

However, the Itô stochastic integration of exact differentials is easy as the following theorem shows.

Theorem 2.13. *Fundamental Theorem of Itô Stochastic Diffusion Calculus.*
Let $g(w)$ be continuous and $G(w)$ be continuously differentiable. Then

(a)

$$d\left[\int_0^t g(W(s))dW(s)\right] \overset{\mathrm{ims}}{=} g(W(t))dW(t) \qquad (2.37)$$

and

(b)

$$\int_0^t dG(W(s)) \stackrel{\text{ims}}{=} G(W(t)) - G(0) \tag{2.38}$$

for $0 \le t$.

Proof. Part (a) of the fundamental theorem benefits from the Itô forward integration approximation and continuity of g, but mostly from the continuity of W. Consider the increment version for sufficiently small increments Δt,

$$\Delta \left[\int_0^t g(W(s))dW(s) \right] = \left(\int_0^{t+\Delta t} - \int_0^t \right) g(W(s))dW(s)$$
$$= \int_t^{t+\Delta t} g(W(s))dW(s)$$
$$\simeq g(W(t))\Delta W(t)$$
$$\to g(W(t))dW(t),$$

as $\Delta t \to 0^+$, using the continuity of both g and W.

For part (b) of the fundamental theorem, using the Itô stochastic integration Definition 2.8, we have

$$\int_0^t dG(W(s)) \stackrel{\text{ims}}{=} \stackrel{\text{ms}}{\lim_{n \to \infty}} \left[\sum_{i=0}^n (G(W(t_{i+1})) - G(W(t_i))) \right]$$
$$= \stackrel{\text{ms}}{\lim_{n \to \infty}} \left[\sum_{i=0}^n (\Delta G(W(t_i))) \right] = \stackrel{\text{ms}}{\lim_{n \to \infty}} [G(W(t_{n+1})) - G(W(t_0))]$$
$$= \stackrel{\text{ms}}{\lim_{n \to \infty}} [G(W(t)) - G(0)] = G(W(t)) - G(0),$$

upon using the facts that $t_0 = 0$ and $t_{n+1} = t$, and for any sum over all increments it is the total increment from (2.13) of Lemma 2.1. Assuming that $G(W(t))$ is bounded on $[0, t]$ should be all that is needed. Thus, for exact derivatives, Itô stochastic integration and ordinary deterministic or Riemann integration agree. See Kolmogorov and Fomin [169] or Protter [232] for the importance of bounded variation as well, although these details are beyond the scope of this book.   ☐

Remarks 2.14.

- *Part* (a) *relates the integral to the differential formulation, and part* (b) *is useful since it is one of the main ways of finding stochastic integrals which are not often found in closed form. Usually, part* (b) *is used to reduce a more complicated stochastic integral to a closed form plus a simpler, perhaps Riemann, integral.*

- *Note that in the proof of part* (a), *there is a difference in the exact **increment of an integral** and its approximate increment for small Δt. Using a more general form in* *some process $X(t)$ for the integral, the exact increment has the form*

$$\Delta[I[G]](t) \equiv I[G](t + \Delta t) - I[G](t) = \int_t^{t+\Delta t} G(X(s), s)dX(s)$$

that holds for arbitrary Δt as long as the integral can be defined, while the approximate integral has the form

$$\Delta[I[G]](t) \simeq G(X(t), t)\Delta X(t)$$

for sufficiently small Δt. So which form is used in an application depends on the application and the size of the time increment Δt.

When dealing with Itô stochastic integrals more information may be needed, because of more general functions of Markov stochastic processes such as $g(W(t))$, $g(W(t), t)$, or $g(X(t), t)$, where $X(t)$ may itself be a stochastic process that is a functional of $W(t)$ and also $P(t)$. In particular, more assumptions or theorems beyond the scope of this applied book may be be needed to demonstrate the mean square convergence of the stochastic integrals. Typically, the usual assumptions [13, 166, 209, 232] require that the integrand function, say, $Y(t) = g(X(t), t)$, has a piecewise-constant, right-continuous approximation that is compatible with the Itô forward summation approximation and that permits satisfaction of the mean square limit criterion. Such assumptions are unnecessary when there is an explicit function of $W(t)$ since, as will be seen, the mean square limit property can be verified directly. However, when a general function is considered with little information, then this extra piecewise constant assumption will be necessary.

Assumption 2.15. *Piecewise-Constant Approximations in the Itô Sense (i-PWCA) for General Mean Square Limits.*

1. *Let $Z(t)$ be a **piecewise-constant, right-continuous** stochastic process such that*

$$Z_n(s) = \{\zeta_i : \tau_i \le s < \tau_{i+1}; \text{ for } i = 0 : n\}, \tag{2.39}$$

*where the times τ_i belong to a partition of $[0, t]$ such that $\tau_0 = 0$ and $\tau_{n+1} = t$, so $Z_n(t) = \zeta_{n+1}$ if needed but does not contribute to the integral. The ζ_i are a sequence of discrete stochastic processes depending on the past Wiener processes $\mathcal{W}_i = \{W(s) \mid 0 \le s \le \tau_i\}$, i.e., **adapted to $\mathcal{W}_i$ for** $i = 0 : n + 1$, **in the Itô sense**. Let $\mathcal{W}$ be the set of all $\mathcal{W}_i$.*

2. *Let $Y(t)$ be a stochastic process depending on $\mathcal{W}$ where $Y(s)$ can be approximated in the Itô sense by the piecewise-constant, right-continuous stochastic process $Z_n(s)$ for $0 \le s \le t$ such that*

$$\lim_{n\to\infty} E\left[\int_0^t (Y(s) - Z_n(s))^2 \, ds\right] \to 0 \tag{2.40}$$

as $n \to +\infty$.

Remark 2.16. *An example of an approximation by the Itô piecewise-constant approximation (i-PWCA) is the function on $[t_0, t]$,*

$$\mathcal{G}_n(s) = \{g(W_i, t_i) : t_i \leq s < t_{i+1}; \text{ for } i = 0 : n\}, \tag{2.41}$$

where $g(w, t)$ is a continuous function of (w, t), $W_i = W(t_i)$, and the set

$$\{t_i : t_{i+1} = t_i + \Delta t_i, \ \Delta t_i > 0, \text{ for } i = 0 : n; \ t_{n+1} = t\} \tag{2.42}$$

is the time partition of $[t_0, t]$.

Theorem 2.17. *Mean of Itô Stochastic Integral.*

$$\mathrm{E}\left[\int_{t_0}^{t} g(W(s), s) dW(s)\right] \overset{\text{ims}}{=} 0, \ \ 0 \leq t_0 \leq t, \tag{2.43}$$

assuming the mean square integrability condition

$$\mathrm{E}\left[\int_{t_0}^{t} g^2(W(s), s) ds\right] < \infty \tag{2.44}$$

and the i-PWCA mean square limits Assumption 2.15 for $Y(t) = g(W(t), t)$.

Proof. Only heuristic justification will be given here to keep this presentation simple. For more elaborate justification using sequences of approximate step function sums, consult the works of Arnold [13], Schuss [244], Øksendal [222], Mikosch [209], or Steele [256].

- Using the Itô mean square limit (2.27), we then have the formal finite sum approximation using partition (2.42),

$$\int_{t_0}^{t} g(W(s), s) dW(s) \simeq \sum_{i=0}^{n} g(W(t_i), t_i)(W(t_{i+1}) - W(t_i)) = \sum_{i=0}^{n} g_i \Delta W_i,$$

where $g_i = g(W(t_i), t_i)$ from $\mathcal{G}_n(s)$ (2.41) and $\Delta W_i = W(t_{i+1}) - W(t_i)$. Since the right-hand side sum is finite, the operations of expectation and summation can be interchanged, so

$$\mathrm{E}\left[\int_{t_0}^{t} g(W(s), s) dW(s)\right] \simeq \sum_{i=0}^{n} \mathrm{E}[g_i \Delta W_i] = \sum_{i=0}^{n} \mathrm{E}[g_i] \mathrm{E}[\Delta W_i]$$

$$= \sum_{i=0}^{n} \mathrm{E}[g_i] \cdot 0 = 0,$$

the last line using the independent increments and zero-mean properties.

Note that the forward integration rule of Itô is not used, and thus the mean-zero result of (2.43) will not be true in general. (See Exercise 10 on p. 61 for a θ-rule counterexample.)

- The final justification requires justifying the interchange of the expectation operator, a Riemann integral, and the mean square limit operator. The underlying integrability assumption can be rewritten using Itô's forward integration choice, leading to the approximation

$$\mathrm{E}\left[\int_{t_0}^t g^2(W(s), s)ds\right] = \int_{t_0}^t \mathrm{E}\left[g^2(W(s), s)\right] ds \simeq \sum_{i=0}^n \mathrm{E}[g_i^2]\Delta t_i.$$

- This approximation can be compared with the expected absolute value of the original Itô approximated sum of interest followed by a one-component Schwarz's inequality $\left(\overset{csi}{\leq}\right)$, to put it into a usable quadratic form and rearrange it into independent increments $\left(\overset{ind}{\underset{inc}{=}}\right)$,

$$\mathrm{E}\left[\left|\int_{t_0}^t g(W(s), s)dW(s)\right|\right] \simeq \mathrm{E}\left[\left|\sum_{i=0}^n g_i \Delta W_i\right|\right]$$

$$\overset{csi}{\leq} \sqrt{\mathrm{E}\left[\sum_{i=0}^n g_i \Delta W_i \cdot \sum_{j=0}^n g_j \Delta W_j\right]}$$

$$= \sqrt{\mathrm{E}\left[\sum_{i=0}^n g_i^2 (\Delta W_i)^2 + \sum_{i=0}^n g_i \Delta W_i \left(\sum_{j=0}^{i-1} + \sum_{j=i+1}^n\right) g_j \Delta W_j\right]}$$

$$\overset{ind}{\underset{inc}{=}} \left[\sum_{i=0}^n \mathrm{E}\left[g_i^2\right] \mathrm{E}\left[(\Delta W_i)^2\right]\right.$$

$$\left. + \sum_{i=0}^n \left(\sum_{j=0}^{i-1} \mathrm{E}\left[g_i g_j \Delta W_j\right] \mathrm{E}[\Delta W_i] + \sum_{j=i+1}^n \mathrm{E}\left[g_i g_j \Delta W_i\right] \mathrm{E}[\Delta W_j]\right)\right]^{0.5}$$

$$= \sqrt{\sum_{i=0}^n \mathrm{E}\left[g_i^2\right] \Delta t_i + 0},$$

where the zero mean and Δt_i variance properties of ΔW_i were used in the last step. The **expectation Schwarz (Cauchy–Schwarz) inequality**

$$\mathrm{E}[|XY|] \leq \sqrt{\mathrm{E}[X^2] \cdot \mathrm{E}[Y^2]} \tag{2.45}$$

has been used with $X = \sum_{i=0}^n g_i \Delta W_i$ and $Y = 1$ to relate the magnitude of the sum to the square root of the sum of squares. Hence, in the mean square sense as $n \to +\infty$, we formally have that the expected absolute value of the stochastic diffusion integral is majorized by the square root of the integral of the expected square of the integrand,

$$\mathrm{E}\left[\left|\int_{t_0}^t g(W(s), s)dW(s)\right|\right] \leq \sqrt{\int_{t_0}^t \mathrm{E}\left[g^2(W(s), s)\right] ds}. \tag{2.46}$$

It has been assumed that the sums are bounded on the bounded interval $[t_0, t]$, so that, in the absence of stochasticity, we can expect uniform convergence of the sums and that the operations of expectation and mean square limit can be interchanged.

- Note that this mean-zero (2.43) for the Itô stochastic integral result depends heavily on the Itô forward or left end point integration choice, and, as will be seen later, the mean-zero result will not hold for other rectangular integration rule choices.

- Under similar conditions, a quadratic or "ims-covariance" version of this theorem holds for interchanging expectation and mean square limit.

☐

Theorem 2.18. *Itô Covariance of the Stochastic Integral.*

$$E\left[\int_{t_0}^t f(W(s), s)dW(s) \int_{t_0}^t g(W(r), r)dW(r)\right]$$
$$\stackrel{\text{ims}}{=} \int_{t_0}^t E[f(W(s), s)g(W(s), s)]ds \tag{2.47}$$

for $0 \leq t_0 \leq t$, assuming that $f(W(t), t)$ and $g(W(t), t)$ satisfy the mean square integrability condition (2.44) and the i-PWCA mean square limits Assumption 2.15 for $Y(t) = g(W(t), t)$.

Proof. Again, heuristic justifications are presented here. Replacing the expectation of the Itô integral product with that of the corresponding product of finite sum approximations leads to

$$J_2(t) = E\left[\int_{t_0}^t f(W(s), s)dW(s)\int_{t_0}^t g(W(r), r)dW(r)\right] \simeq \sum_{i=0}^n \sum_{j=0}^n E[f_i \Delta W_i g_j \Delta W_j],$$

but the independent increments are intermingled in the sums and the argument of the expectation of $f_i \Delta W_i g_j \Delta W_j$. However, if $j < i$, then the increment ΔW_i will be independent of f_i, g_j, and ΔW_j, while if $j > i$, then ΔW_j will be independent of f_i, g_j, and ΔW_i, and for $i = j$ the usual independent increment form is obtained. Thus, taking these independence properties to split the double sum three ways and using independent increment properties leads to

$$J_2(t) \simeq \sum_{i=0}^n E[f_i g_i]E[(\Delta W_i)^2] + \sum_{i=0}^n \sum_{j=0}^{i-1} E[f_i g_j \Delta W_j]E[\Delta W_i]$$
$$+ \sum_{i=0}^n \sum_{j=i+1}^n E[f_i g_j \Delta W_i]E[\Delta W_j]$$
$$= \sum_{i=0}^n E[f_i g_i]\Delta t_i$$
$$\stackrel{\text{ims}}{\longrightarrow} \int_{t_0}^t E[f(W(s), s)g(W(s), s)]ds,$$

giving the desired conclusion, except for replacing the approximately equals ($\simeq$) by the mean square limit as $n \to \infty$

Replacing the function f by g leads to the immediate corollary for the ims-variance of the Itô stochastic integral in the following. ☐

Corollary 2.19. *Itô Variance of the Stochastic Integral.*

$$\mathrm{E}\left[\left(\int_{t_0}^{t} g(W(s), s) dW(s)\right)^2\right] \overset{\mathrm{ims}}{=} \int_{t_0}^{t} \mathrm{E}\left[g^2(W(s), s)\right] ds \qquad (2.48)$$

for $0 \leq t_0 \leq t$, assuming that $g(W(t), t)$ satisfies the mean square integrability condition (2.44).

Result (2.48) is also called **Itô isometry or martingale isometry**.

Theorem 2.20. *Itô Stochastic Integral Simple Rules.*
Let g, g_1, and g_2 satisfy the mean square integrability condition (2.44) on $0 \leq t_0 \leq t$ and the i-PWCA mean square limits Assumption 2.15, while letting c_1 and c_2 be constants.
These are the following simple rules:

- *Operator linearity:*

$$\int_{t_0}^{t} [c_1 g_1(W(s), s) + c_2 g_2(W(s), s)] dW(s)$$

$$\overset{\mathrm{ims}}{=} c_1 \int_{t_0}^{t} g_1(W(s), s) dW(s) + c_2 \int_{t_0}^{t} g_2(W(s), s) dW(s).$$

- *Additivity over subintervals:*

$$\int_{t_0}^{t} g(W(s), s) dW(s) \overset{\mathrm{ims}}{=} \int_{t_0}^{r} g(W(s), s) dW(s) + \int_{r}^{t} g(W(s), s) dW(s)$$

for $0 \leq t_0 \leq r \leq t$.

- *Continuity of sample paths for*

$$I[g](t) = \int_{t_0}^{t} g(W(s), s) dW(s)$$

  with probability one.

Proof. The first two are clearly true by examining the forward integration approximation. For the last item note that

$$\Delta I[g](t) = I[g](t + \Delta t) - I[g](t) = \int_{t}^{t+\Delta t} g(W(s), s) dW(s) \overset{\mathrm{ims}}{=} g(W(t), t) \Delta W(t) \to 0$$

with probability one as $\Delta t \to 0^+$. □

For later use in formal stochastic calculations, it will be helpful to know how to handle powers of $dW(t)$ greater than square powers. The critical problem is to know when to truncate a differential expansion, such as that for $\exp(dW(t))$, at a power of $dW(t)$ beyond which the higher powers are zero in the sense of the Itô mean square limit. For example, $\exp(dW(t))$ can be formally expanded by Taylor series as

$$\exp(dW(t)) = 1 + dW(t) + (dW)^2(t)/2! + (dW)^3(t)/3! + (dW)^4(t)/4! + \cdots$$

and it turns out we can justify stopping at the quadratic term for the mean square limit. The consequence will be the famous Itô stochastic chain rule discussed for jump-diffusions in Chapter 4 and will lead to more rapid calculations. The main purpose of the current chapter is to set up the foundational justification for this chain rule.

Lemma 2.21. *Powers of $dW(t)$.*
Let the integer $m \geq 3$, then there exists the power rule:

$$\int_0^t (dW)^m(s) \overset{\text{ims}}{=} 0 \tag{2.49}$$

or in symbolic differential notation

$$(dW)^m(t) \overset{\text{dt}}{\underset{\text{ms}}{=}} 0. \tag{2.50}$$

Proof. Let $m \geq 3$ and

$$I[(dW)^{m-1}](t) = I(t; m) \equiv \int_0^t (dW)^m(s) \simeq I_n^{(0)}(t; m) = \sum_{i=0}^n (\Delta W_i)^m. \tag{2.51}$$

The expectation of the Itô approximate sum $I_n^{(0)}(t; m)$ yields different formulae for odd values, $m = 2k - 1$ for $k \geq 2$,

$$\mathrm{E}[I_n^{(0)}(t; 2k - 1)] = \sum_{i=0}^n \mathrm{E}\left[(\Delta W_i)^{2k-1}\right] = 0,$$

while for even values, $m = 2k$ for $k \geq 2$,

$$\mathrm{E}[I_n^{(0)}(t; 2k)] = \sum_{i=0}^n \mathrm{E}\left[(\Delta W_i)^{2k}\right] = (2k - 1)!! \sum_{i=0}^n (\Delta t_i)^k$$

$$\leq (2k - 1)!! \, t \, (\delta t_n)^{k-1} \to 0,$$

as $n \to \infty$, where $(2k - 1)!!$ is the double factorial function (1.15). For odd or even values of m and $m \geq 3$, then the results suggest that the Itô mean square value is given by

$$I(t; m) \overset{\text{ims}}{=} I^{(\text{ims})}(t; m) \equiv \lim_{n \to \infty} [I_n^{(0)}(t; m)] = 0.$$

The justification requires confirmation of mean square convergence,

$$\lim_{n \to \infty} \mathrm{E}\left[(I_n^{(0)}(t; m) - I^{(\text{ims})}(t; m))^2\right] = \lim_{n \to \infty} \mathrm{E}\left[(I_n^{(0)})^2(t; m)\right].$$

For odd values, with $m = 2k - 1$, by separating out the diagonal part of the quadratic to separate the independent increments, then

$$\mathrm{E}\left[(I_n^{(0)})^2(t; 2k - 1)\right] = \sum_{i=0}^n \mathrm{E}\left[(\Delta W_i)^{2(2k-1)} + \sum_{j \neq i}(\Delta W_i)^{2k-1}(\Delta W_j)^{2k-1}\right]$$

$$= (4k - 3)!! \sum_{i=0}^n (\Delta t_i)^{2k-1}$$

$$\leq (4k - 3)!! \, t \, (\delta t_n)^{2k-2} \to 0,$$

as $n \to \infty$. We see that off-diagonal odd power terms do not contribute. Here $(4k-3)!!$ is the double factorial function (1.15). For even values, $m = 2k$, the off-diagonal terms contribute since they are products of even powers of increments in i and j, so upon completing the double sum over $j \neq i$ and subtracting the completed amount from the single sum, we get

$$
\begin{aligned}
\mathrm{E}\left[(I_n^{(0)})^2(t; 2k)\right] &= \sum_{i=0}^{n} \mathrm{E}\left[(\Delta W_i)^{4k} + \sum_{j \neq i}(\Delta W_i)^{2k}(\Delta W_j)^{2k}\right] \\
&= ((4k-1)!! - ((2k-1)!!)^2) \sum_{i=0}^{n}(\Delta t_i)^{2k} \\
&\quad + ((2k-1)!!)^2 \sum_{i=0}^{n}(\Delta t_i)^k \sum_{j=0}^{n}(\Delta t_j)^k \\
&\leq (4k-1)!! \, t(\delta t_n)^{2k-1} + ((2k-1)!!)^2 t(\delta t_n)^{2k-2}(t - \delta t_n) \to 0,
\end{aligned}
$$

as $n \to \infty$. Thus, denoting the conclusion symbolically, $(dW)^m(t) \overset{dt}{\underset{ms}{=}} 0$, provided $m \geq 3$ to an accuracy with error $o(dt)$. $\square$

Another differential product whose Itô mean square limit will be useful is $dt \, dW(t)$ since it arises in the expansions of functions of stochastic differentials:

Lemma 2.22. *Differential Product $dt \, dW(t)$.*

$$
\int_0^t ds \, dW(s) \overset{ims}{=} 0 \tag{2.52}
$$

or in symbolic notation

$$
dt \, dW(t) \overset{dt}{\underset{ms}{=}} 0. \tag{2.53}
$$

Proof. Let

$$
I[dt](t) = \int_0^t ds \, dW(s) \simeq I_n^{(0)}[dt](t) \equiv \sum_{i=0}^{n} \Delta t_i \Delta W_i \tag{2.54}
$$

with some abuse of the notation by replacing functional argument g by dt. The expectation of the sum $I_n^{(0)}[dt](t)$ yields

$$
\mathrm{E}[I_n^{(0)}[dt](t)] = \sum_{i=0}^{n} \mathrm{E}[\Delta t_i \Delta W_i] = 0.
$$

The result suggests that the Itô mean square value is given by

$$
I[dt](t; m) \overset{ims}{=} \lim_{n \to \infty} [I_n^{(0)}[dt](t; m)] = 0.
$$

The justification requires confirmation of mean square convergence by separating out the diagonal part of the quadratic to separate the independent increments,

$$
\mathrm{E}\left[(I_n^{(0)})^2[dt](t)\right] = \sum_{i=0}^{n} \mathrm{E}\left[(\Delta t_i)^2(\Delta W_i)^2 + \sum_{j \neq i} \Delta t_i \Delta t_j \Delta W_i \Delta W_j\right]
$$
$$
= \sum_{i=0}^{n}(\Delta t_i)^3 \leq t(\delta t_n)^2 \to 0,
$$

as $n \to \infty$, and thus off-diagonal terms do not contribute. Thus, $dt\ dW(t) \overset{dt}{\underset{ms}{=}} 0$ to an accuracy with error $o(dt)$.   □

Remarks 2.23.

- *Of the Itô differentiable forms that have zero limit in the mean square, $dt\,dW(t)$ is one of the most marginal to approximate due to the randomness of $dW(t)$, even though we know $\mathrm{E}[dt\,dW(t)] = 0$ and $\mathrm{E}[|dW(t)|] = \sqrt{2\Delta t/\pi}$ from convenient Table 1.1. Hence, the justification of $\int_0^t ds\,dW(s) \overset{ims}{=} 0$ by showing the mean square limit is especially important. Note that for even spacing of time increments, the RMS of the bound of the mean square approximation above is $\sqrt{t(\delta t_n)^2} = t\sqrt{t}/(n+1) \to 0$ as $n \to \infty$. However, see Exercise 2 for a more cutting-edge example.*

- *See Exercise 5 for how to computationally confirm Lemma 2.22.*

The mean square limits to an accuracy with error $o(dt)$ are summarized in Table 2.1.

Table 2.1. *Some Itô stochastic diffusion differentials with an accuracy with error $o(dt)$ as $dt \to 0^+$.*

Differential Diffusion Form	Itô Mean Square Limit
$dW(t)$	$dW(t)$
dt	dt
$dt\,dW(t)$	0
$(dW)^2(t)$	dt
$(dW)^m(t)$	$0,\ m \geq 3$
$(dt)^\alpha(dW)^m(t)$	$0,\ \alpha > 0,\ m \geq 1$

The more general form,

$$
(dt)^p(dW)^q(t) \overset{dt}{\underset{ms}{=}} \delta_{2p+q,0} + dW(t)\delta_{2p+q,1} + dt\,\delta_{2p+q,2}, \tag{2.55}
$$

when p and q are nonnegative integers, is left as Exercise 1 on p. 57.

Remark 2.24. *In using Table 2.1, the differential entries are just symbols of the underlying integral basis, and care should be taken when applying them to find the mean square* *representation of differentials, especially when they appear in multiplicative combinations. For instance, one might be tempted to replace $(dW)^4(t)$ by $(dW)^2(t)(dW)^2(t)$, then replace those terms with $(dW)^2(t) \overset{\text{dt}}{\underset{\text{ms}}{=}} dt$ to get to $(dt)^2 \overset{\text{dt}}{\underset{\text{ms}}{=}} 0$, which is the correct but crudely found answer for $(dW)^4(t)$. Note that for finite increments, $\mathrm{E}[(\Delta W_i)^4] = 3(\Delta t_i)^2$, while $\mathrm{E}^2[(\Delta W_i)^2] = (\Delta t_i)^2$, differing by a factor of 3.*

2.3   Stratonovich and Other Stochastic Integration Rules

In this section, other definitions of stochastic integration rules, other than Itô's choice of the forward left end point rule, are explored for the purpose of comparison and understanding Itô's choice. This comparison will be illustrated by the simple stochastic integral of $W(t)$.

Let the integration θ-rule approximation point be

$$t_{i+\theta} \equiv t_i + \theta \Delta t_i, \tag{2.56}$$

where $0 \leq \theta \leq 1$, so the Itô rule is when $\theta = 0$ with $\Delta t_i \equiv t_{i+1} - t_i$. Let the interval of integration be $[0, t]$ with partition (2.4). Let the approximate integrand be $W_{i+\theta} \equiv W(t_{i+\theta})$. The technique of splitting terms into independent increments is similar to that for Itô's rule, except that there are extra independent increments,

$$\Delta^\theta W_i \equiv W_{i+\theta} - W_i \tag{2.57}$$

and its complement

$$\Delta^\theta_c W_i \equiv \Delta W_i - \Delta^\theta W_i = W_{i+1} - W_{i+\theta} \tag{2.58}$$

for intermediate approximation points when $\theta > 0$, such that $\Delta^\theta W_i + \Delta^\theta_c W_i = \Delta W_i$. We also reuse (2.14) of the reduction lemma, Lemma 2.1, for the Itô case in the more general case here:

$$
\begin{aligned}
I[W](t) = \int_0^t W(s)dW(s) &\simeq I_n^{(\theta)}[W](t) \equiv \sum_{i=0}^n W_{i+\theta} \Delta W_i \\
&= \sum_{i=0}^n (W_i + \Delta^\theta W_i)(\Delta^\theta W_i + \Delta^\theta_c W_i) \\
&= \sum_{i=0}^n \left(W_i \Delta W_i + (\Delta^\theta W_i)^2 + \Delta^\theta W_i \Delta^\theta_c W_i \right) \\
&= \frac{1}{2} \left(W_{n+1}^2 - \sum_{i=0}^n (\Delta W_i)^2 \right) + \sum_{i=0}^n (\Delta^\theta W_i)^2 + \sum_{i=0}^n \Delta^\theta W_i \Delta^\theta_c W_i.
\end{aligned}
$$

Since $W_{n+1} = W(t)$ with this $[0, t]$ partition and the mean square limit of $\sum_{i=0}^n (\Delta W_i)^2$ has been shown to be t, similarly, the mean square limit of $\sum_{i=0}^n (\Delta^\theta W_i)^2$ will be the expected value θt, and the last sum will not contribute to the mean being the product of independent increments, and the mean square limit corresponding to the Itô Lemma 2.7 can be stated as follows.

Lemma 2.25.

$$\int_0^t W(s)dW(s) \overset{\theta}{\underset{\text{ms}}{=}} I^{(\theta)}[W](t) = \frac{1}{2}W^2(t) - \left(\frac{1}{2} - \theta\right)t \tag{2.59}$$

$$= \overset{\text{ms}}{\underset{n\to\infty}{\lim}} \left[I_n^{(\theta)}[W](t)\right].$$

Proof. The mean square limit justifications are quite lengthy and somewhat tangent to our goals here, so only the general end result is given, and the details are left to the reader:

$$\text{E}\left[\left(I_n^{(\theta)}[W](t) - I^{(\theta)}[W](t)\right)^2\right] = 2\left|\frac{1}{2} - \theta\right| \sum_{i=0}^n (\Delta t_i)^2$$

$$\leq 2\left|\frac{1}{2} - \theta\right| t\delta t_n \to 0,$$

where $\delta t_n = \max_{i=0:n}[\Delta t_i] \to 0^+$ as $n \to \infty$. □

Remark 2.26. *Stratonovich and Other Stochastic Integration Rules.*
*The mean square limit is exact for any n in the case $\theta = 1/2$, where $t_{i+0.5} = (t_i + t_{i+1})/2$ is the midpoint of $[t_i, t_{i+1}]$ and the integration rule is called the midpoint rule or **Stratonovich** **stochastic integration** [260]. For Stratonovich integration,*

$$\int_0^t W(s)dW(s) \overset{\theta}{\underset{\text{ms}}{=}} I^{(0,5)}[W](t) = W^2(t)/2,$$

which is the deterministic integral answer, containing no correction as in the case of Itô's rule. This deterministic property might offer some benefit in some applications, but at the expense of more complicated overlapping dependence of increments in time.

Lemma 2.27.

$$\text{E}\left[I^{(\theta)}[W](t)\right] = \text{E}\left[\frac{1}{2}W^2(t) - \left(\frac{1}{2} - \theta\right)t\right] = \theta t. \tag{2.60}$$

Proof. The result is immediate since $\text{E}[W^2(t)] = t$ from Table 1.1 when $n = 2$ with $|\Delta W|^2(t)$ replaced by $W^2(t)$ and Δt by t. □

Remarks 2.28.

- *When $\theta \neq 0$, the useful Itô **expectation-integration interchange property**,*

$$\text{E}\left[\int_0^t f(W(s), s)dW(s)\right] \overset{\text{ims}}{=} \int_0^t \text{E}[f(W(s), s)]\text{E}[dW(s)] = 0,$$

is no longer valid as implied by (2.43). This is a quite nice concrete property, but for abstract analysis it is more crucial since it means, with appropriate qualification on $f(W(t), t)$, that the Itô integral is a martingale.

- *Decades ago, there was a larger controversy as to*

  whether Itô or Stratonovich stochastic integration

  should be used. The question sometimes centered around what was more appropriate for the application at hand (see, for instance, Turelli [271] for a discussion involving biological applications), but the benefits of Itô's choice of forward integration facilitating the use of independent increments of the processes, and the fact that many Stratonovich properties were derived by Itô stochastic calculus, have made the Itô calculus dominant.

- *Perhaps the current, almost exclusive, dominance of the Itô calculus is the predominant use of and tremendous interest in stochastic models of mathematical finance, as suggested by a reviewer.*

2.4   Conclusion

In this chapter, the foundations have been laid for the integrals of the second type in the integrated SDE (2.2), i.e., using the stochastic diffusion integral of Itô of Definition 2.8 extended to the more general case.

Definition 2.29. *Stochastic Diffusion Integration.*

$$\int_0^t g(X(s), s)dW(s) \overset{\text{ims}}{=} \lim_{n \to \infty} \left[\sum_{i=0}^n g(X(t_i), t_i)dW(t_i) \right], \tag{2.61}$$

where $X(t)$ in the integrand function g has an implied dependence on the diffusion process $W(t)$ but also depends on the jump process $P(t)$. The integrand process $g(X(t), t)$ is also assumed to have a bounded mean square,

$$E\left[\int_0^t g^2(X(s), s)ds \right] < \infty,$$

and to satisfy the i-PWCA mean square limits Assumption 2.15 for $Y(t) = g(X(t), t)$.

However, as previously explained, the Poisson jump process fits within the framework of Itô stochastic integration since it is piecewise continuous. The stochastic diffusion integration rule (2.61) has been motivated and illustrated by a number of examples using functions and powers of the diffusion process $W(t)$.

2.5   Exercises

In all computational exercises, Mathematica, MATLAB, Maple, or other programming may be used where appropriate, but both figures and codes should be submitted for evaluation.

1. Justify the general form (2.55) by mean square convergence,

$$(dt)^p(dW)^q(t) \overset{dt}{\underset{ms}{=}} \delta_{2p+q,0} + dW(t)\delta_{2p+q,1} + dt\, \delta_{2p+q,2},$$

   when p and q are nonnegative integers.

(Note: It may be assumed that the cases $2p + q = 0 : 2$ are well known, so we need to show mean square convergence results for $2p + q \geq 3$ in general.)

2. Show the limit in the mean square for

$$I\left[(dt)^\alpha\right](t) \equiv \int_0^t (ds)^\alpha dW(s) \overset{\text{ims}}{=} 0,$$

provided $\alpha > 0$ and is real.
(Hint: See Lemma 2.22 for the case $\alpha = 1$.)

3. Computationally confirm the mean square limit for Itô's most fundamental stochastic integral given as

$$\int_0^t (dW)^2(s) \overset{\text{ims}}{=} t$$

by demonstrating that the Itô forward integration approximating sum

$$I_n^{(0)}[dW](t) = \sum_{i=0}^n (\Delta W_i)^2$$

gives a close approximation to t for sufficiently large n. Apply a modification of the algorithm of Program C.7 in Appendix C, used in generating Figure 1.1, to the approximation $I_n^{(0)}[dW](t)$, submitting your modification of the code. Use $n_1 = 1000$ and $n_2 = 10,000$ sample sizes, plotting the $I_n^{(0)}[dW](t)$ with the limit t versus t for $t \in [0, 2]$. Plot separately the errors for each n between the approximation sum and the exact Itô mean square (IMS) answer. Also report the standard deviation (std in MATLAB) of the errors for each n. Characterize the convergence on the average by assuming that the standard deviation satisfies the simple rule $std_j \simeq C/n_j^\beta$ as $n_j \to \infty$ for $j = 1 : 2$, and find the average convergence rate β from the two sample step sizes n_j for $j = 1 : 2$.
(Caution: In this problem and the next two, you are not asked to verify the mean square limit but to verify that the forward approximation comes close in this example.)

4. Computationally confirm the mean square limit for Itô's other very fundamental stochastic integral given as

$$\int_0^t W(s)dW(s) \overset{\text{ims}}{=} I^{(\text{ims})}[W](t) = \frac{1}{2}\left(W^2(t) - t\right)$$

by demonstrating that the Itô forward integration approximating sum

$$I_n^{(0)}[W](t) = \sum_{i=0}^n W_i \Delta W_i$$

gives a close approximation to $(W^2(t) - t)/2$ for sufficiently large n. Apply a modification of the algorithm of Program C.7 in Appendix C, used in generating Figure 1.1, to the approximation $I_n^{(0)}[W](t)$. Use $n_{1,k} = 100$ and $n_{2,k} = 10,000$ sample sizes and for $k = 1 : 4$ different states or seeds. Plot the approximation $I_n^{(0)}[W](t)$ with $(W^2(t) - t)/2$ and the error $E_n[W](t) = I_n^{(0)}[W](t) - (W^2(t) - t)/2$ versus t for $t \in [0, 2]$. Plot separately the errors for each n between the approximation sum and the exact Itô mean square answer. Also report the standard deviation (std in MAT-LAB) of the errors for each nj, k for $j = 1 : 2$ sample sizes and $k = 1 : 4$ states. From these values compute the common rate β_k for both $j = 1 : 2$ sizes and for each fixed state $k = 1 : 4$, assuming $\mathrm{std}_{j,k} = C_k/n_{j,k}^{\beta_k}$ for each $j = 1 : 2$ and $k = 1 : 4$, finally computing the average β_k over $k = 1 : 4$. Does the larger value of n make Itô's stochastic integration model more convincing than the smaller value?

5. Computationally confirm the mean square limit for another of Itô's more obvious fundamental stochastic integrals,

$$\int_0^t ds\, dW(s) \overset{\text{ims}}{=} I^{(\text{ims})}[dt](t) = 0$$

by demonstrating that the Itô forward integration approximating sum

$$I_n^{(0)}[dt](t) = \sum_{i=0}^{n} \Delta t_i \Delta W_i$$

gives a close approximation to 0 for sufficiently large n. Apply a modification of the algorithm of Program C.7 in Appendix C, used in generating Figure 1.1, to the approximation $I_n^{(0)}[dt](t)$. Use $n_1 = 1000$ and $n_2 = 10{,}000$ sample sizes, plotting the common value of the approximation and error $I_n^{(0)}[dt](t) = E_n[dt](t)$ and the noise $W(t)$ for $t \in [0, 2]$. Plot separately the errors for each n between the approximation sum and the exact IMS answer. Also report the standard deviation (std in MATLAB) of the errors for each nj, k for $j = 1 : 2$ sample sizes and $k = 1 : 4$ states. From these values compute the common rate β_k for both $j = 1 : 2$ sizes and for each fixed state $k = 1 : 4$, assuming $\mathrm{std}_{j,k} = C_k/n_{j,k}^{\beta_k}$ for each $j = 1 : 2$ and $k = 1 : 4$, finally computing the average β_k over $k = 1 : 4$. Does the larger value of n make Itô's stochastic integration model more convincing than the smaller value?

6. Computationally check the Itô mean square limit for convergence of the Itô approximating sum of the stochastic integral of $(dW)^2(t)$ to the limit t by directly computing the K-sample mean square

$$S_{i,n}^{(K)} = \frac{1}{K} \sum_{k=1}^{K} \left(\sum_{j=1}^{i} \left(\left(\Delta W_j^{(k)} \right)^2 - \Delta t_j \right) \right)^2,$$

where the identity $t = t_{n+1} = \sum_{i=0}^{n} \Delta t_i$ has been used to merge t into the approximating sum. Select $K = 5$ random states or seeds, $n = 10^m$ for $m = 2 : 5$ sample sizes, constant $\Delta t_i = \Delta t$, $i = n$ and $t = 1$, as an example. Plot $\log_{10}(S_{n,n}^{(K)})$ versus $m = \log_{10}(n)$. What rate of convergence is suggested by this graph?
(*Hint: If* $\Delta t = 10^m$ *and* $S \sim C \cdot (\Delta t)^a$, *then* $\log_{10}(S) \sim a \cdot m + \log_{10}(C)$. *In MATLAB, for instance, recall that* `randn('state',k)` ; *sets the k normal random number state.*)

7. Show that the non-Itô approximate backward integration rule $(\theta = 1)$ for the stochastic integral

$$\int_{t_0}^{t} W(s)dW(s) \simeq I_n^{(1)}(t) = \sum_{i=0}^{n} W_{i+1}\Delta W_i$$

differs from the Itô rule $(\theta = 0)$ by a deterministic factor of t in the mean square limit, i.e.,

$$I_n^{(1)}(t) - I_n^{(0)}(t) \xrightarrow{\text{ims}} t.$$

(Hint: The mean square limit is not needed if the approximate integral is related to the Itô integral for $(dW)^2(t)$.)

8. Show that the non-Itô approximate trapezoidal integration rule, a variant of the Stratonovich integral, for the stochastic integral

$$\int_{t_0}^{t} W(s)dW(s) \simeq I_n^{(trap)}(t) = \frac{1}{2}\sum_{i=0}^{n}(W_i + W_{i+1})\Delta W_i$$

differs from the Itô rule $(\theta = 0)$ by a deterministic factor of $t/2$ in the mean square limit, i.e.,

$$I_n^{(trap)}(t) - I_n^{(0)}(t) \xrightarrow{\text{ims}} \frac{1}{2}t.$$

(Hint: The mean square limit is not needed if the approximate integral is related to the one for $(dW)^2(t)$.)

9. Demonstrate that the **trapezoidal rule** leads to Stratonovich or regular calculus by approximating the stochastic integral example

$$\int_{t_0}^{t} W^2(s)dW(s)$$

with

$$I_n^{(0)}(t) = \frac{1}{2}\sum_{i=0}^{n}(W_i^2 + W_{i+1}^2)\Delta W_i.$$

In particular, show that

$$I_n^{(0)}(t) = \frac{1}{3}(W^3(t_{n+1}) - W^3(t_0)) + \frac{1}{6}\sum_{i=0}^{n}(\Delta W_i)^3$$

by forming convenient powers of independent increments. Formally, justify that the mean square limit is just the first term using elementary mean square properties for the powers of increments $(\Delta W)^p(t_i)$. You are not required to rigorously show mean square convergence, unless you want to show it.

(Note: In numerical integration of deterministic integrands, both the midpoint rectangular rule and the trapezoidal rule yield the same order of error estimate when the integrand is sufficiently continuous.)

10. Formally show that the θ-rule expansion (no mean square convergence justification requested) leads to

$$\mathrm{E}\left[\int_0^t g(W(s))dW(s)\right] \underset{\mathrm{ms}}{\overset{\theta}{=}} \mathrm{E}\left[I^{(\theta)}[g(W)](t)\right] = \theta \int_0^t \mathrm{E}\left[g'(W(s))\right]ds,$$

where $0 \le \theta \le 1$. Assume that the basic θ-rule approximation for the stochastic integral is

$$\int_0^t g(W(s))dW(s) \simeq I_n^{(\theta)}[g(W)](t) \equiv \sum_{i=0}^n g_{i+\theta}\Delta W_i,$$

where g has a bounded mean square expectation (2.44) with bounded derivatives of all orders, $g_{i+\theta} = g(W_{i+\theta}) = g(W(t_{i+\theta})) = g(W_i + \Delta^\theta W_i)$ for $t_{i+\theta} = t_i + \theta\Delta t_i$ from (2.56). Also assume that g satisfies the second order Taylor approximation with third order error,

$$g(w_0 + \Delta W) = g(w_0) + g'(w_0)\Delta W + \frac{1}{2}g''(w_0)(\Delta W)^2 + (\Delta W)^3 \mathrm{O}(1),$$

sufficiently uniform with respect to the density $\phi_{\Delta W(t)}(w)$ on $(-\infty, +\infty)$ to allow termwise expectations, provided you can show that $\mathrm{E}[(\Delta^\theta W_i)^m] = \mathrm{O}^2(\theta\Delta t_i)$ for $m \ge 3$ and sufficiently small Δt_i. See also the θ-decomposition (2.57)–(2.58) of ΔW_i.

(Note: Thus, this demonstrates that in the Itô sense, Theorem 2.17 is generally limited to $\theta = 0$.)

Suggested References for Further Reading

- Arnold, 1974 [13]

- Gard, 1988 [92]

- Itô, 1951 [150]

- Karlin and Taylor, 1981 [163]

- Kloeden and Platen, 1999 [166]

- Kolmogorov and Fomin, 1970 [169]

- Mikosch, 1998 [209]

- Øksendal, 1998 [222]

- Protter, 1990 [232]

- Schuss, 1980 [244]

- Taylor and Karlin, 1998 [265]

Chapter 3

Stochastic Integration for Jumps

It is common experience that markets are discontinuous: from time to time they 'jump,' usually downwards.
—Paul Wilmott, 1998 [283]

However, the Black–Scholes solution is not vaild, even in the continuous limit, when the stock-price dynamics cannot be represented by a stochastic process with a continuous sample path.
—Robert C. Merton, 1976 [202]

In particular the Poisson process seems not to be known to many students trained in Black–Scholes theory!
—Rama Cont and Peter Tankov, 2004 [60]

A unique feature of this chapter is the greater emphasis on the importance of the lack of continuity that leads to deviations from the chain rule of regular calculus, namely, the discontinuity of Poisson jumps in time and the nonsmooth behavior of Wiener processes. The Poisson jump processes are given in terms of special right-continuous step and impulse functions. Unless otherwise stated, a fixed jump-rate λ is assumed. The Poisson jump calculus is also formulated in terms of finite difference algebraic recursions.

3.1 Stochastic Integration in $P(t)$: The Foundations

In this chapter, foundations will be laid for the integrals of the third type in the integrated SDE (2.2), i.e., using the notion of the Itô stochastic integral of Definition 2.8 on p. 40 by extending it to the jump case.

Definition 3.1. *Poisson Jump Stochastic Integration.*

$$\int_0^t h(X(s), s)dP(s) \stackrel{\text{ims}}{=} \lim_{n \to \infty}^{\text{ms}} \left[\sum_{i=0}^n h(X(t_i), t_i)\Delta P(t_i) \right], \tag{3.1}$$

where $X(t)$ in the integrand function h has an implied dependence on the diffusion process
$W(t)$ but also depends on the jump process $P(t)$. The integrand process $h(X(t),t)$ is also
*assumed to have a **bounded mean integral of squares**,*

$$\mathrm{E}\left[\int_0^t h^2(X(s),s)ds\right] < \infty, \tag{3.2}$$

and to satisfy the Itô piecewise constant approximation (i-PWCA) of the mean square limits
Assumption 2.15 on p. 47 for $Y(t) = h(X(t),t)$, with the usual grid partition specifications
on $[0,t]$.

For most problems encountered in practice, there will not be a need for this elaborate,
yet fundamental, mean square definition. The definition may be needed as a reference for
unusual applications with stochastic jumps.

For instance, if an exact differential in $P(t)$ can be formed, then as with stochastic
diffusion integration, i.e., when the variable of integration is the random diffusion process
$W(t)$, there will be no need for mean square justification. Since much of the work of
stochastic integration was performed in the previous chapter, with some very general results,
it will be possible to move more quickly through this chapter.

Theorem 3.2. *Fundamental Theorem of Poisson Jump Calculus.*
Let $h(p)$ be continuous and $\mathcal{H}(p)$ be continuously differentiable. Then

(a)
$$d\left(\int_0^t h(P(s))dP(s)\right) \overset{\text{ims}}{=} h(P(t))dP(t) \tag{3.3}$$

and

(b)
$$\int_0^t d\mathcal{H}(P(s)) \overset{\text{ims}}{=} \mathcal{H}(P(t)) - \mathcal{H}(0), \ 0 \le t. \tag{3.4}$$

Proof. The proof is almost the same as for the analogous result (2.37), (2.38), except for
the change in names from $W(t)$ to $P(t)$ and that the issue of unbounded variation need not
be considered.

However, the right-continuity property of $P(t)$ is essential to account for a jump at t
for part (a). Consider the increment version for sufficiently small increments Δt,

$$\Delta\left(\int_0^t h(P(s))dP(s)\right) = \left(\int_0^{t+\Delta t} - \int_0^t\right) h(P(s))dP(s)$$
$$= \int_t^{t+\Delta t} h(P(s))dP(s)$$
$$\simeq h(P(t))\Delta P(t) = h(P(t))(P(t+\Delta t) - P(t))$$
$$\to h(P(t))dP(t)$$

as $\Delta t \to 0^+$, using the increment definition, subinterval additivity (see (3.23) later in this
chapter), the continuity h, and piecewise continuity of P, such that any last-minute jump is
captured in $\Delta P(t)$ or $dP(t)$.

See the proof of the diffusion part (b) in (2.38) for the jump part (b) in (3.4). $\square$

First, consider the most basic jump-integral, the integral of $P(t)$ with respect to $P(t)$, namely,

$$I[P](t) = \int_0^t P(s)dP(s),$$

which will be evaluated directly through precision-dt calculus and indirectly by showing that the defining mean square limit is satisfied.

Theorem 3.3. *Jump-Integral of $\int PdP$.*

$$I[P](t) = \int_0^t P(s)dP(s) \overset{\text{ims}}{=} I^{(ims)}[P](t) \equiv \frac{1}{2}(P(P-1))(t) \qquad (3.5)$$

*is the **mean square limit integral**,*

$$I^{(ims)}[P](t) \overset{\text{ims}}{=} \underset{n\to\infty}{\overset{\text{ms}}{\lim}} \left[I_n^{(0)}[P](t) \right], \qquad (3.6)$$

where the forward integration approximation is

$$I_n^{(0)}[P](t) = \sum_{i=0}^n P(t_i)\Delta P(t_i). \qquad (3.7)$$

Proof. Starting with the Poisson increment and the square $P^2(t)$, as in the diffusion case, since $d(x^2) = 2xdx$ in smooth deterministic calculus, we have

$$\Delta(P^2) \equiv P^2(t+\Delta t) - P^2(t) = \left((P+\Delta P)^2 - P^2\right)(t)$$
$$= \left(2P\Delta P + (\Delta P)^2\right)(t).$$

Taking the limit $\Delta t \to 0^+$, replacing ΔP by dP, and using the zero-one jump law (1.35) to let $(dP)^2 \overset{\text{dt}}{=} dP$ **with probability one (w.p.o.)** upon neglect of smaller order terms leads to

$$d(P^2)(t) \overset{\text{dt}}{=} (2PdP + dP)(t)$$

in probability. Solving for the integrand-differential while forming an exact differential yields in probability

$$(PdP)(t) \overset{\text{dt}}{=} \frac{1}{2}d\left(P^2 - P\right)(t).$$

Therefore, integration by the fundamental theorem of stochastic jump integration (3.3) leads to

$$\int_0^t (PdP)(s) \overset{\text{ims}}{=} \frac{1}{2}\int_0^t (d\left(P^2 - P\right)(t))(s) = \frac{1}{2}\left(P^2 - P\right)(t) = I^{(ims)}[P](t),$$

where the initial Poisson condition $P(0) = 0$ **w.p.o.** has been used to eliminate the initial value of the integral. That takes care of the first part of the proof, but the technique is general enough for other powers.

For the second part, the forward integration approximation can be simplified by the useful finite difference identity (2.14),

$$I_n^{(0)}[P](t) = \sum_{i=0}^{n} P_i \Delta P_i = \frac{1}{2}\left(P^2(t) - \sum_{i=0}^{n}(\Delta P_i)^2\right)$$

for the partition

$$0 = t_0 < t_1 < \cdots < t_{n+1} = t,$$

and by using the fact (2.13) that

$$P(t) = P_{n+1} = \sum_{i=0}^{n} \Delta P_i,$$

we see that the difference between the approximation and the limit reduces to

$$I_n^{(0)}[P](t) - I^{(\text{ims})}[P](t) = \frac{1}{2}\sum_{i=0}^{n}\left(\Delta P_i - (\Delta P_i)^2\right).$$

The mean square again is reduced by splitting the sums due to the square into independent increments before termwise passing the mean over the sums,

$$\mathrm{E}\left[\left(I_n^{(0)}[P](t) - I^{(\text{ims})}[P](t)\right)^2\right] = \frac{1}{4}\mathrm{E}\left[\left(\sum_{i=0}^{n}\left(\Delta P_i - (\Delta P_i)^2\right)\right)^2\right]$$

$$= \frac{1}{4}\sum_{i=0}^{n}\mathrm{E}\left[\left(\Delta P_i - (\Delta P_i)^2\right)^2\right]$$

$$+ \frac{1}{4}\sum_{i=0}^{n}\sum_{j\neq i}\mathrm{E}\left[\left(\Delta P_i - (\Delta P_i)^2\right)\cdot\left(\Delta P_j - (\Delta P_j)^2\right)\right]$$

$$= \frac{1}{4}\sum_{i=0}^{n}\mathrm{E}\left[(\Delta P_i)^2 - 2(\Delta P_i)^3 + (\Delta P_i)^4\right]$$

$$+ \frac{1}{4}\sum_{i=0}^{n}\mathrm{E}\left[\Delta P_i - (\Delta P_i)^2\right]\sum_{j\neq i}\mathrm{E}\left[\Delta P_j - (\Delta P_j)^2\right]$$

$$= \frac{1}{4}\sum_{i=0}^{n}\left(\lambda\Delta t_i(1 + \lambda\Delta t_i) - 2\lambda\Delta t_i(1 + 3\lambda\Delta t_i + (\lambda\Delta t_i)^2)\right.$$

$$\left. + \lambda\Delta t_i(1 + 7\lambda\Delta t_i + 6(\lambda\Delta t_i)^2 + (\lambda\Delta t_i)^3)\right)$$

$$+ \frac{1}{4}\sum_{i=0}^{n}\left(\lambda\Delta t_i - \lambda\Delta t_i(1 + \lambda\Delta t_i)\right)\sum_{j\neq i}\left(\lambda\Delta t_j - \lambda\Delta t_j(1 + \lambda\Delta t_j)\right)$$

$$\leq \frac{1}{4}\left(\sum_{i=0}^{n}(\lambda\Delta t_i)^2(2 + 4\lambda\Delta t_i) + \sum_{i=0}^{n}(\lambda\Delta t_i)^2\sum_{j=0}^{n}(\lambda\Delta t_j)^2\right)$$

$$\leq \frac{1}{4}\left(\lambda t(2\lambda\delta t_n + 4(\lambda\delta t_n)^2) + (\lambda t^2)(\lambda\delta t_n)^2\right) \longrightarrow 0$$

as $n \to \infty$ and bounded t. For the evaluation of the expectations of powers of Poisson increments, the convenient Table 1.2 has been frequently used. Therefore, the mean square limit has been proven. $\square$

Remarks 3.4.

- *The main result (3.5),*

$$\int_0^t P(s)dP(s) \overset{\text{ims}}{=} \frac{1}{2}(P(P-1))(t),$$

*for this basic integral has an interesting mathematical interpretation. Since $P(t)$ is integer valued, the answer is the Pythagorean $(P(t)-1)$th **triangular number** given by the successive sum of $n = P(t) - 1$ integers,*

$$S_n^{(1)} = \sum_{k=0}^{n} k = n(n+1)/2. \tag{3.8}$$

*The interpretation is not a coincidence, since when $P(t)$ jumps instantaneously by one unit and adds it to its count, $dP(t)$ jumps by one only momentarily so that the integral in (3.5) serves as a **triangular number counter**. The forward integration approximation serves to keep the count short of the last jump, e.g., the forward approximation is zero when $P(t) = 1$.*

- *The derivation of (3.8) by finite differences gives useful techniques for calculating and interpreting other Poisson jump-integrals. The basic lemma for the difference inversion ("discrete integration") is given by the following lemma.*

Lemma 3.5. *If*

$$\Delta[a_n] = \Delta[b_n]$$

for two sequences and any integer n, then

$$a_n = b_n + C,$$

where C is an arbitrary constant.

- *The proof is obvious since a constant sequence is the only sequence of elements that produces zero difference.*

- *Since $\Delta[S_n^{(1)}] = S_{n+1}^{(1)} - S_n^{(1)} = (n+1)$, $\Delta[n] = 1$, and $\Delta[n^2] = 2n+1 = 2n + \Delta[n]$ or $n = \frac{1}{2}\Delta[n^2 - n]$, then $\Delta[S_n^{(1)}] = \Delta[(n^2 - n)/2 + n]$ and $S_n^{(1)} = n(n+1)/2$, upon elimination of the constant of discrete integration by the initial condition $S_0^{(1)} = 0$. This proves the first triangular number sum (3.8) by finite differences using Lemma 3.5.*

The first few Poisson power integrals are listed with an accuracy with error $o(dt)$ in Table 3.1.

Table 3.1. *Some stochastic jump-integrals of powers with an accuracy with error* $o(dt)$ *as* $dt \rightarrow 0^+$.

m	precision-dt: $\int_0^t (P^m dP)(s)$
0	$P(t)$
1	$(P(P-1))(t)/2$
2	$(P(P-1)(2P-1))(t)/6$
3	$(P^2(P-1)^2)(t)/4$

Remarks 3.6.

- *The proofs of the formulas for* $m = 2$ *and* $m = 3$ *are left for the reader as in Exercise 1 on p. 77.*

- *The integral results of Table 3.1 are all in the form of generalized or super-triangular numbers of order m when* $n = P(t) - 1$.

Definition 3.7. *The **supertriangular numbers** of order m for the first* $n + 1$ *nonnegative integers are defined as*

$$S_n^{(m)} = \sum_{k=0}^{n} k^m$$

for integers $m \geq 0$ *and* $n > 0$.

The summation form of a pure Poisson integral is generalized in the following theorem.

Theorem 3.8. *Pure Poisson Integral as Sum Form.*
Let $h(p)$ *be a continuous function and let the process* $h(P(t))$ *have a bounded mean integral of squares (3.2). Then,*

$$\int_0^t h(P(s))dP(s) \stackrel{\text{ims}}{=} \sum_{k=0}^{P(t)-1} h(k) \tag{3.9}$$

with the usual summation convention for irregular forms that

$$\sum_{k=0}^{-1} h(k) \equiv 0 \tag{3.10}$$

for the case that $P(t) = 0$.

Proof. It is necessary only to confirm that both sides of (3.9) satisfy the same differential. The tools used will be the fundamental theorem of stochastic calculus (3.3) and the idea of the zero-one jump power law (1.36). By the fundamental theorem, the differential of the left-hand side of (3.9) is

$$d\left(\int_0^t h(P(s))dP(s)\right) \overset{\text{dt}}{=} h(P(t))dP(t) \ .$$

Then, by using the incremental definition of the differential for the right-hand side of (3.9),

$$d\left(\sum_{k=0}^{P(t)-1} h(k)\right) = \sum_{k=0}^{P(t)+dP(t)-1} h(k) - \sum_{k=0}^{P(t)-1} h(k) \overset{\text{dt}}{=} h(P(t))dP(t),$$

where the last step is due to the zero-one jump law since the difference in the two sums in the first line is zero if $dP(t) = 0$; else there is only one extra term in the first sum in the alternate case $dP(t) = 1$. Also $dP(t) = 1$ is used in the argument of h. Hence, the differentials of both sides of (3.9) are the same. The final result then follows for these reasons:

1. Both sides satisfy the same initial condition.

2. The vanishing of the jump-integral in the limit is

$$\lim_{t\to 0^+} \int_0^t h(P(s))dP(s) = 0.$$

3. The vanishing of the Poisson sum in the limit is

$$\lim_{t\to 0^+} \sum_{k=0}^{P(t)-1} h(k) = \sum_{k=0}^{-1} h(k) \equiv 0.$$

4. $P(0^+) = 0$.

5. The irregular summation convention (3.10) is valid.

The argument is analogous to that of mathematical induction, since we have shown that both sides of (3.9) satisfy the same initial condition and the same changes and so lead to the same result hypothesized in the theorem. $\quad\square$

Remarks 3.9.

- *Note that in this theorem the sum is over all $P(t)$ jump-amplitudes for $k+1 = 1 : P(t)$ jumps but that the jump-amplitude h is evaluated at the prejump value $h(k)$ for $k = 0 : P(t) - 1$ by the definition of the Poisson jump with amplitude determined by the function h. This jump-amplitude evaluation is consistent with the Itô forward integral approximation,*

$$\Delta \int_0^t h(P(s))dP(s) \simeq h(P(t))\Delta P(t)$$

for a single, sufficiently small time-step Δt, picking the prior value of h at $P(t)$ in the case $\Delta P(t) > 0$; although it is not that obvious for the simple jump-amplitude dependence $h(P(t))$, the choice of the prejump value is also a consequence of the right-continuity property of the Poisson process (1.20).

Corollary 3.10.

$$\int_{t=0}^{t} P^m(s)dP(s) \stackrel{\text{ims}}{=} S_{P(t)-1}^{(m)} = \sum_{k=0}^{P(t)-1} k^m \tag{3.11}$$

for $m \geq 0$ and the irregular summation convention (3.10) *is applicable.*

Remark 3.11. *A simple consistency check on* (3.11) *is to verify the simplest case when $m = 0$, and the integral of $(P^m dP)(t) = dP(t)$ on $[0, t]$ must be $P(t)$ by the fundamental theorem. The right-hand side of* (3.11)*, with $k^m = 1$, is*

$$\sum_{k=0}^{P(t)-1} 1 = (P(t) - 1 + 1) \cdot 1 = P(t).$$

Theorem 3.12. *General Poisson Stochastic Integral.*
Let $h(x, t)$ be a continuous function and let the process $h(X(t), t)$ have a bounded mean integral of squares (3.2) *and satisfy the i-PWCA mean square limits Assumption 2.15 for $Y(t) = h(X(t), t)$. Then,*

$$\int_0^t h(X(s), s)dP(s) \stackrel{\text{ims}}{=} \sum_{k=1}^{P(t)} h(X(T_k^-), T_k^-), \tag{3.12}$$

where T_k is the kth jump of Poisson process $P(t)$.

Proof. Here, we rely explicitly on both the Itô forward integration rule ($\theta = 0$) and the right-continuity property of $P(t)$. It is sufficient to examine the processes $P(t)$, $\Delta P(t)$, and $h(X(t), t)$ in the very neighborhood of the kth jump at time T_k, such that Δt is small enough so we can exclude the prior jump at T_{k-1} and the next jump at T_{k+1} with $T_{k-1} < t < T_{k+1}$. After all, the Poisson process is a rare event process. Thus, the Poisson process has the simple, right-continuous form

$$P(t) = \begin{cases} k-1, & T_{k-1} < t \leq T_k^- \\ k, & T_k = T_k^+ \leq t < T_{k+1} \end{cases},$$

where $1 \leq k \leq P(t)$. However, the increment $\Delta P(t_i) = P(t_i + \Delta t) - P(t_i)$ is a function of both t_i and Δt for $i = 1 : n$, but we are interested in the limit as $\Delta t \to 0^+$ with t_i fixed in (T_{k-1}, T_{k+1}), so there are three cases, both t_i and $t_i + \Delta t$ to the left of T_k, T_k between t_i and $t_i + \Delta t$, and both on the right of T_k, i.e.,

$h(X(t_i), t_i)\Delta P(t_i)$

$$= \begin{cases} 0, & T_{k-1} < t_i < t_i + \Delta t \leq T_k^- \\ h(X(t_i), t_i), & T_{k-1} < t_i \leq T_k^- < T_k = T_k^+ \leq t_i + \Delta t < T_{k+1} \\ 0, & T_k = T_k^+ \leq t_i < t_i + \Delta t < T_{k+1} \end{cases}$$

$$\to \begin{cases} 0, & T_{k-1} < t_i < T_k^- \\ h(X(T_k^-), T_k^-), & T_{k-1} < t_i = T_k^- \\ 0, & T_k = T_k^+ \leq t_i < T_{k+1} \end{cases},$$

as $\Delta t \to 0^+$ with t_i fixed in (T_{k-1}, T_{k+1}), and this is valid for $1 \leq k \leq P(t)$. Thus, the Itô approximate sum is

$$\int_0^t h(X(s), s)dP(s) \simeq \sum_{i=0}^n h(X(t_i), t_i)\Delta P(t_i)$$

$$\to \sum_{k=1}^{P(t)} h(X(T_k^-), T_k^-),$$

as $n \to +\infty$ and $\delta t_n = \max_j[\Delta t_j] \to 0^+$, since for large n the $\Delta P(t_i)$ will be mostly zero and only the time-intervals that straddle a jump T_k^- will be selected. The state process, different from the simple jump Poisson process, will in general undergo continuous changes between jumps of $P(t)$, but the right-continuity causes the immediate prejump value of the jump-amplitude at T_k^- to be chosen for each jump-time T_k. $\quad \Box$

Remark 3.13. *Obviously, if $h(X(t), t) = 1$, then $\sum_{k=1}^{P(t)} 1 = P(t)$. Another simple consistency check on (3.12) is to verify the case when $h(X(t), t) = P(t)$ and the integral of $(PdP)(t)$ on $[0, t]$ must be $(P(P-1))(t)/2$ by (3.5). The right-hand side of (3.11), with $h(X(t), t) = P(t)$, $P(T_k^-) = k - 1$, is*

$$\sum_{k=1}^{P(t)} P(T_k^-) = \sum_{k=1}^{P(t)} (k-1) = P(t)(P(t) - 1)/2,$$

using the standard triangular number summation. Hence, (3.12) is consistent with (3.9).

Definition 3.14. *Jump Function $[X](t)$.*
The jump value of the state X at the prejump-time T_k^- is defined as

$$[X](T_k) \equiv X(T_k^+) - X(T_k^-) \tag{3.13}$$

*when the kth jump is at time T_k. For finite discontinuities, the jump function includes all the change of the function, the **zeroth change or discrete derivative** of the state $X(t)$.*

Example 3.15. *Let*

$$Y(t) = \int_0^t h(X(s), s)dP(s)$$

and

$$\Delta Y(t) = \int_t^{t+\Delta t} h(X(s), s)dP(s) \simeq h(X(t), t)\Delta P(t)$$

for $0 < \Delta t \ll 1$, so

$$[Y](t) \equiv Y(t^+) - Y(t^-) = \int_{t^-}^{t^+} h(X(s), s)dP(s) = h(X(t^-), t^-)dP(t), \tag{3.14}$$

since $dP(t) = dP(t^-)$ with both being one when $t = T_k^-$ or $t^- = T_k^-$ but otherwise zero when $T_{k-1} < t < T_k^-$ or $T_{k-1} < t^- < T_k^-$.

In the non-Itô integration approximation, $0 < \theta \leq 1$,

$$\Delta \int_0^t h(P(s))dP(s) \simeq h(P(t + \theta\Delta t))\Delta P(t),$$

so if the last jump is T_k and the next one is T_{k+1}, such that $T_k < t < T_{k+1} < t + \Delta t$, i.e., within the single time-step, then $P(t) = k$ and we see that the jump-amplitude is $h(k)$ if the jump is late, $t + \theta\Delta t < T_{k+1} < t + \Delta t$, since $P(t + \theta\Delta t) = k$, but we see that the amplitude $h(k+1)$ if the jump is early, $t < T_{k+1} < t + \theta\Delta t$, since $P(t + \theta\Delta t) = k + 1$. Thus, the Itô formulation has much less complexity and is more straightforward to implement.

Some other jump differential products whose mean square limits will be useful are $dt\, dP(t)$ and $dP(t)\, dW(t)$, since they arise in the expansions of functions of stochastic differentials.

Lemma 3.16. *Differential Products $dt\, dP(t)$ and $dP(t)\, dW(t)$.*

$$\int_0^t ds\, dP(s) \overset{\text{ims}}{=} 0, \tag{3.15}$$

or in symbolic notation

$$dt\, dP(t) \overset{\text{dt}}{=} 0, \tag{3.16}$$

and

$$\int_0^t dP(s)\, dW(s) \overset{\text{ims}}{=} 0, \tag{3.17}$$

or in symbolic notation

$$dP(t)\, dW(t) \overset{\text{dt}}{=} 0, \tag{3.18}$$

where $W(t)$ and $P(t)$ are independent random variables.

Proof. The proofs are similar to the proof for $dt\, dW(t)$, with a minor change in argument due to the nonzero incremental mean

$$\mathrm{E}[\Delta P(t_i)] = \lambda \Delta t_i.$$

Let

$$I[dt](t) = \int_0^t ds\, dP(s) \simeq I_n[dt](t) \equiv \sum_{i=0}^n \Delta t_i \Delta P(t_i). \tag{3.19}$$

The expectation of the sum $I_n[dt](t)$ yields

$$\mathrm{E}[I_n[dt](t)] = \sum_{i=0}^n \mathrm{E}\left[\Delta t_i \Delta P(t_i)\right] = \sum_{i=0}^n \lambda(\Delta t_i)^2$$
$$\leq \lambda t \delta t_n \to 0^+,$$

as $n \to +\infty$. The result suggests that the Itô mean square value is given by

$$I[dt](t) \overset{ims}{=} \overset{ms}{\underset{n \to \infty}{\lim}} \, I_n[dt](t) = 0.$$

This can be verified in the mean square limit by showing that the mean square limit is zero, while the splitting into independent increments is employed,

$$\mathrm{E}\left[\left(\sum_{i=0}^{n} \Delta t_i \Delta P_i - 0\right)^2\right] = \sum_{i=0}^{n} \left((\Delta t_i)^2 \mathrm{E}[(\Delta P_i)^2] + \sum_{j \neq i} \Delta t_i \Delta t_j \mathrm{E}[\Delta P_i] \mathrm{E}[\Delta P_i]\right)$$

$$= \sum_{i=0}^{n} \left(\lambda(\Delta t_i)^3 (1 + \lambda \Delta t_i) + \sum_{j \neq i} \lambda^2 (\Delta t_i \Delta t_j)^2\right)$$

$$= \mathrm{O}^2(\delta t_n) \to 0,$$

as $n \to +\infty$. So,

$$dt \, dP(t) \overset{dt}{=} 0.$$

The cross product of differentials $dP(t)dW(t)$ works out similarly, except here we have the benefit of independence of processes as well as independence of respective process increments. Let

$$J(t) = \int_0^t dP(s) \, dW(s) \simeq J_n(t) \equiv \sum_{i=0}^{n} \Delta P(t_i) \Delta W(t_i). \tag{3.20}$$

The expectation of the sum $J_n(t)$ yields

$$\mathrm{E}[J_n(t)] = \sum_{i=0}^{n} \mathrm{E}[\Delta P(t_i) \Delta W(t_i)] = \sum_{i=0}^{n} \lambda(\Delta t_i) \cdot 0 = 0.$$

This result suggests that the Itô mean square value is given by

$$J(t) \overset{ims}{=} \overset{ms}{\underset{n \to \infty}{\lim}} \, [J_n(t)] = 0,$$

so that it is intuitively clear that the mean square limit will also behave like the cases $dt \, dW(t)$ and $dt \, dP(t)$, but the verification of the mean square limit is still needed and is left as Exercise 3 for the reader.   □

Theorem 3.17. *Mean Square Limit Form of the Zero-One Law.*
Let m be a nonnegative integer and $\mathrm{E}[dP(t)] = \lambda(t)dt$ *with bounded maximum,* $\lambda^* = \max_t[\lambda(t)]$*. Then*

$$\int_0^t (dP)^m(s) \overset{ims}{=} P(t), \tag{3.21}$$

or in symbolic notation

$$(dP)^m(t) \overset{dt}{=} dP(t). \tag{3.22}$$

Proof. The mean square limit proof is left for the reader as Exercise 4.   □

Table 3.2 summarizes the Itô mean square limits to an accuracy with error $o(dt)$ in the case of the Poisson jump process.

Table 3.2. *Some Itô stochastic jump differentials with an accuracy with error $o(dt)$ as $dt \to 0^+$.*

Differential Jump Form	Itô Mean Square Limit
$dP(t)$	$dP(t)$
dt	dt
$dt\, dP(t)$	0
$(dP)^m(t)$	$dP(t), m \geq 1$
$dP(t)\, dW(t)$	0
$(dt)^k (dP)^m(t)$	$0, k \geq 1,\ m \geq 1$
$(dt)^k (dP)^m(t)(dW)^n(t)$	$0,\ k \geq 1,\ m \geq 1,\ n \geq 1$

Remarks 3.18.

- *In Table 3.2, the differential entries are just symbols of the underlying integral basis, and care should be taken when applying them to find the mean square representation of differentials, especially when they appear in multiplicative combinations.*

- *The mean square limit justification of the power rule $(dP)^m(t) \overset{dt}{=} dP(t)$ is left as Exercise 4, along with Exercise 3 previously mentioned for $dP(t)dW(t)$.*

3.2   Stochastic Jump Integration Rules and Expectations

Theorem 3.19. *Itô Stochastic Jump-Integral Simple Rules.*
Let h, h_1, and h_2 satisfy the mean square integrability condition (2.44) on $0 \leq t_0 \leq t$ while letting $X(t)$ be a Markov process, and let c_1 and c_2 be constants.

- *Operator linearity:*

$$\int_{t_0}^t [c_1 h_1(X(s), s) + c_2 h_2(X(s), s)]dP(s)$$

$$\overset{\text{ims}}{=} c_1 \int_{t_0}^t h_1(X(s), s)dP(s) + c_2 \int_{t_0}^t h_2(X(s), s)dP(s).$$

- *Additivity over subintervals:*

$$\int_{t_0}^t h(X(s), s)dP(s) \overset{\text{ims}}{=} \int_{t_0}^r h(X(s), s)dP(s) + \int_r^t h(X(s), s)dP(s) \quad (3.23)$$

for $0 \leq t_0 \leq r \leq t$.

Proof. That these are clearly true can be seen by examining the forward integration approximation. $\square$

Poisson jump processes may seem easier in terms of differentials, but they can lead to more difficulties when more complicated integral properties are considered.

Theorem 3.20. *Some Mean Stochastic Jump-Integrals.*
Let $h(X(t), t)$ satisfy the mean square integrability condition on $0 \leq t_0 \leq t$ and let $X(t)$ be a Markov process,

$$\mathrm{E}\left[\int_{t_0}^{t} h^2(X(s), s)ds\right] < \infty, \tag{3.24}$$

and the i-PWCA mean square limits Assumption 2.15 for $Y(t) = h(X(t), t)$, where $\mathrm{E}[dP(t)] = \lambda(t)dt$. Then

1. $\mathrm{E}[\int h(X(s), s)dP(s)]$:

$$\mathrm{E}\left[\int_{t_0}^{t} h(X(s), s)dP(s)\right] \overset{\text{ims}}{=} \int_{t_0}^{t} \mathrm{E}[h(X(s), s)]\lambda(s)ds. \tag{3.25}$$

2. $\mathrm{E}[\int h(X(s), s)d\widehat{P}(s)]$: *Letting*

$$d\widehat{P}(t) \equiv dP(t) - \lambda(t)dt \tag{3.26}$$

*be the simple **mean-zero Poisson process**,*

$$\mathrm{E}\left[\int_{t_0}^{t} h(X(s), s)d\widehat{P}(s)\right] \overset{\text{ims}}{=} 0. \tag{3.27}$$

3. $\mathrm{E}[|\int h(X(s), s)dP(s)|]$: *Estimate*

$$\mathrm{E}\left[\left|\int_{t_0}^{t} h(X(s), s)dP(s)\right|\right] \leq \int_{t_0}^{t} \mathrm{E}[|h(X(s), s)|]\lambda(s)ds, \tag{3.28}$$

where the inequality is in the mean square sense.

4. $\mathrm{E}[\int h_1(X(s), s)d\widehat{P}(s) \int h_2(X(r), r)d\widehat{P}(r)]$: *Let $h_1(X(t), t)$ and $h_2(X(t), t)$ satisfy the same mean square integrability condition (2.44) as $h(X(t), t)$ on $0 \leq t_0 \leq t$; then the Itô covariance for jump stochastic integrals is*

$$\mathrm{E}\left[\int_{t_0}^{t} h_1(X(s), s)d\widehat{P}(s) \int_{t_0}^{t} h_2(X(r), r)d\widehat{P}(r)\right]$$
$$\overset{\text{ims}}{=} \int_{t_0}^{t} \mathrm{E}[h_1(X(s), s)h_2(X(s), s)]\lambda(s)ds. \tag{3.29}$$

5. $\mathrm{E}[(\int h(X(s), s)d\widehat{P}(s))^2]$: *The Itô variance for jump stochastic integrals is given by*

$$\mathrm{E}\left[\left(\int_{t_0}^{t} h(X(s), s)d\widehat{P}(s)\right)^2\right] \overset{\text{ims}}{=} \int_{t_0}^{t} \mathrm{E}[h^2(X(s), s)]\lambda(s)ds. \tag{3.30}$$

Sketch of Proof. Only fast heuristic or formal justification will be given here to keep this presentation simple, since many of the techniques were given earlier for diffusion $W(t)$ and our interests are in applications.

1. Using the Itô mean square limit (2.27), we have the formal finite sum approximation using partition (2.28) with $h_i = h(X(t_i), t_i)$ for the expectation

$$
\mathrm{E}\left[\int_{t_0}^{t} h(X(s), s)dP(s)\right] \simeq \sum_{i=0}^{n} \mathrm{E}[h_i \Delta P_i] = \sum_{i=0}^{n} \mathrm{E}[h_i]\mathrm{E}[\Delta P_i]
$$

$$
= \sum_{i=0}^{n} \mathrm{E}[h_i]\lambda_i \Delta t_i,
$$

the last line using the independent increments and mean properties. Hence (3.25) is formally justified.

2. The form (3.27) follows immediately by combining both sides of the mean square equation in part (1).

3. Again using the forward integration approximation, but with the triangular inequality, the expectation of the absolute value of the stochastic jump-integral formally follows as

$$
\mathrm{E}\left[\left|\int_{t_0}^{t} h(X(s), s)dP(s)\right|\right] \simeq \mathrm{E}\left[\left|\sum_{i=0}^{n} h_i \Delta P_i\right|\right] \le \sum_{i=0}^{n} \mathrm{E}[|h_i|\Delta P_i]
$$

$$
= \sum_{i=0}^{n} \mathrm{E}[|h_i|]\mathrm{E}[\Delta P_i] = \sum_{i=0}^{n} \mathrm{E}[|h_i|]\lambda_i \Delta t_i
$$

$$
\xrightarrow{\text{ims}} \int_{t_0}^{t} \mathrm{E}[|h(X(s), s)|]\lambda(s)ds,
$$

as $n \to +\infty$, using the mean square limit in the last step to get the desired limiting estimate.

4. Due to the mean zero property (3.27) of the stochastic jump-integral with respect to the mean zero process $d\widehat{P}(t)$ (3.26), the Itô forward integration approximation to the covariance of the stochastic jump-integral follows. However, the use of the mean zero process is critical; otherwise, the independent increment property is not very helpful. As in the $W(t)$ diffusion case, the approximate finite difference double sum is split up into three parts—diagonal $(j = i)$, lower diagonal $(j < i)$, and upper diagonal $(j > i)$.

$$
\mathrm{E}\left[\int_{t_0}^{t} h_1(X(s), s)d\widehat{P}(s) \int_{t_0}^{t} h_2(X(r), r)d\widehat{P}(r)\right]
$$

$$
\simeq \sum_{i=0}^{n} \sum_{j=0}^{n} \mathrm{E}[h_{1,i}\Delta\widehat{P}_i h_{2,i}\Delta\widehat{P}_j]
$$

$$
\simeq \sum_{i=0}^{n} \mathrm{E}[h_{1,i}h_{2,i}]\mathrm{E}[(\Delta\widehat{P}_i)^2] + \sum_{i=0}^{n} \sum_{j=0}^{i-1} \mathrm{E}[h_{1,i}h_{2,j}\Delta\widehat{P}_j]\mathrm{E}[\Delta\widehat{P}_i]
$$

$$
+ \sum_{i=0}^{n} \sum_{j=i+1}^{n} \mathrm{E}[h_{1,i}h_{2,j}\Delta\widehat{P}_i]\mathrm{E}[\Delta\widehat{P}_j]
$$

$$
= \sum_{i=0}^{n} \mathrm{E}[h_{1,i}h_{2,i}]\lambda_i \Delta t_i
$$

$$
\xrightarrow{\text{ims}} \int_{t_0}^{t} \mathrm{E}[h_1(X(s), s)h_2(X(s), s)]\lambda(s)ds,
$$

giving the desired conclusion, except for replacing the approximately equals ($\simeq$) by the mean square limit as $n \to \infty$.

5. The Itô variance stochastic jump-integral follows immediately from part (4) for the Itô covariance stochastic jump-integral by replacing the functions h_1 and h_2 by h. This result (3.30) is also called *Itô isometry* or *martingale isometry* since $\widehat{P}(t)$ is a martingale.   □

3.3 Conclusion

In this chapter, the foundations have been laid for the integrals of the third type in the integrated SDE (2.2), i.e., using the stochastic jump-integral of Itô of Definition 2.8 extended to the more general case and defined in Definition 3.1 at the beginning of this chapter:

$$\int_0^t h(X(s), s)dP(s) \stackrel{\text{ims}}{=} \lim_{n \to \infty}^{\text{ms}} \left[\sum_{i=0}^n h(X(t_i), t_i)dP(t_i) \right]$$

$$= \sum_{k=1}^{P(t)} h(X(T_k^-), T_k^-), \tag{3.31}$$

where $X(t)$ in the integrand function h has an implied dependence on the simple Poisson jump process $P(t)$ but also depends on the diffusion process $W(t)$. The integrand process $h(X(t), t)$ is also assumed to have a bounded mean integral of squares (3.2),

$$\mathrm{E} \left[\int_0^t h^2(X(s), s)ds \right] < \infty,$$

with the usual grid partition specifications on $[0, t]$. However, as previously explained, the Poisson jump process fits within the framework of Itô stochastic integration since it is piecewise continuous. The stochastic jump integration rule (3.31) has been motivated and illustrated by a number of examples using functions and powers of the jump process $P(t)$.

3.4 Exercises

1. Show that the power rules for stochastic integration for Poisson noise can be written as the recursion

$$\int_0^t P^m(s)dP(s) = \frac{1}{m+1} \left(P^{m+1}(t) - \sum_{k=2}^{m+1} \binom{m+1}{k} \int_0^t P^{m+1-k}(s)dP(s) \right)$$

using the jump form of the stochastic chain rule and the binomial theorem.

   (a) Illustrate the application of the formulae for $P(t)$ to confirm the results for $m = 0:3$ in Table 3.1.

   (b) Alternatively, show the general result for $m \geq 1$.

2. Show that the partial sums of the **geometric series** can be summed as

$$S_n(x) \equiv \sum_{k=0}^{n} x^k = T_n(x) \equiv \left\{ \begin{array}{ll} \frac{1 - x^{n+1}}{1 - x}, & x \neq 1 \\ n + 1, & x = 1 \end{array} \right\} \tag{3.32}$$

for integers $n \geq 0$ by showing that the difference of the defined summation, $\Delta S_n(x)$, and the difference of the summed answer, $\Delta T_n(x)$, to the far right are the same and that the discrete initial conditions are the same at $n = 0$.

3. Show the mean square limit for the product of $dP(t)$ and $dW(t)$ in (3.17)–(3.18) by proving that

$$\text{Var} \left[\sum_{i=0}^{n} \Delta P_i \Delta W_i \right] \rightarrow 0, \tag{3.33}$$

as $n \rightarrow +\infty$ and $\delta t_n \rightarrow 0^+$.

4. Show the mean square limit for the Poisson differential power $(dP)^m(t)$ version of the zero-one jump law in Theorem 3.17 by showing the following.

   (a) Let $M_m(\Delta \Lambda_j) = \text{E}[(\Delta P_j)^m]$ be the mth power of the jth Poisson increment for $\Delta \Lambda_j = M_1(\Delta \Lambda_j)$ and bounded maximum jump-rate $\lambda^* = \max_t[\lambda(t)]$, with nonnegative integers m and j. Then $M_m(u)$ satisfies the recursion relation

$$M_{m+1}(u) = u \cdot (M_m(u) + M'_m(u)). \tag{3.34}$$

   (b) Let $M_m(u) = u + K_m(u)u^2$. Then $K_m(u) \geq 0$, $K_m(u) = O(1)$, and $K'_m(u) = O(1)$, both as $u \rightarrow 0^+$.

   (c) Finally,

$$\text{E} \left[\left(\sum_{i=0}^{n} \left((\Delta P_i)^m - \Delta P_i \right) \right)^2 \right] \rightarrow 0, \tag{3.35}$$

as $n \rightarrow +\infty$ and the mesh $\delta t_n \rightarrow 0^+$ for $m \geq 1$. Hence,

$$(dP)^m(t) \stackrel{\text{dt}}{=} dP(t),$$

the symbolic version of the mean square limit form of the zero-one law.

5. Show that

$$\int_0^t e^{aP(s)} dP(s) = \left\{ \begin{array}{ll} \frac{e^{aP(t)} - 1}{e^a - 1}, & e^a \neq 1 \text{ or } a \neq 0 \\ P(t), & e^a = 1 \text{ or } a = 0 \end{array} \right\} \tag{3.36}$$

for real constant a, in two ways, showing that they give the same answers:

   (a) Using the Poisson sum form $\sum_{k=0}^{P(t)-1} h(k)$ of Theorem 3.8 and the geometric series partial sum results in (3.32) of this Exercise section, and

   (b) using the zero-one jump law and the fundamental theorem of jump calculus 3.4(b) applied to $d \exp(aP(t))$ to evaluate the integral.

Suggested References for Further Reading

- Çinlar, 1975 [56]

- Protter, 1990 [232]

- Snyder and Miller, 1991 [252]

- Tuckwell, 1995 [270]

Chapter 4

Stochastic Calculus for Jump-Diffusions: Elementary SDEs

Stochastic differential equations and Itô's lemma
are the key mathematical tools used in the analysis
of continuous-time models of economic processes.
—Robert C. Merton, 1992 [203]

This chapter could have been entitled
"A beginner's guide to stochastic calculus with jumps."
—Rama Cont and Peter Tankov, 2004 [60]

In Chapter 2 for diffusions and Chapter 3 for jumps, the foundations of Itô stochastic jump-diffusion integrals have been given. In Table 2.1 of Chapter 2, the mean square differential forms for diffusions, i.e., powers of $dW(t)$ and dt, were summarized, such that higher order differential forms are zero symbolically in the Itô mean square sense to dt-precision, for example,

$$(dW)^3(t) \stackrel{dt}{=} 0.$$

In Table 3.2 of Chapter 3, the mean square differential forms for Poisson jumps, i.e., powers of $dP(t)$ and dt, were summarized. Different from diffusion differential forms, the powers of $(dP)(t)$ are generally nonzero except when multiplied by a positive power of dt, but they have the zero-one jump law property that $(dP)^m(t) \stackrel{dt}{\underset{zol}{=}} dP(t)$ for integers $m > 0$.

Similar rules apply in the algebra of deterministic differentials and in constructing deterministic models; e.g., terms with the factor $(dt)^2$ are neglected compared to terms with just the factor dt in both deterministic and stochastic differential models. For stochastic differentials, the nondifferentiability of $W(t)$ and the jump discontinuities of $P(t)$ produce notable exceptions from deterministic differential rules.

For the mean square limits of more general functions and their approximations where there is insufficient information for a proof, the mean square integrability assumption and the i-PWCA mean square limits (2.44) Assumption 2.15 will be assumed to be satisfied. This is applicable to both diffusion and jump-integrals, and this will be an underlying assumption throughout this chapter. However, the primary focus of this chapter will be faster, more efficient formal stochastic calculations.

4.1   Diffusion Process Calculus Rules

The most basic rule (2.32) for diffusions in the Itô mean square sense is

$$(dW)^2(t) \overset{\text{dt}}{=} dt \tag{4.1}$$

symbolically, while the higher order differential forms are zero in the Itô mean square sense, beginning with

$$(dW)^3(t) \overset{\text{dt}}{=} 0, \quad dt\,dW(t) \overset{\text{dt}}{=} 0, \quad \text{and} \quad (dt)^2 \overset{\text{dt}}{=} 0,$$

using summary Table 2.1.

Another basic rule or principle is the use of increments both for increments themselves in single steps of Δt,

$$\begin{aligned}
\Delta G(W(t), t) &\equiv G(W(t + \Delta t), t + \Delta t) - G(W(t), t) \\
&= G(W(t) + \Delta W(t), t + \Delta t) - G(W(t), t),
\end{aligned} \tag{4.2}$$

with functions of the form $G(w(t), t)$ and $\Delta W(t) \equiv W(t + \Delta t) - W(t)$, as well as for differentials as increments,

$$\begin{aligned}
dG(W(t), t) &\equiv G(W(t + dt), t + dt) - G(W(t), t) \\
&= G(W(t) + dW(t), t + dt) - G(W(t), t)
\end{aligned} \tag{4.3}$$

with $dW(t) \equiv W(t + dt) - W(t)$.

The increment (4.2) and differential (4.3) rules can be used, with the rest of Table 2.1, to develop a fast and efficient procedure for deriving stochastic formulas. When there are problems, it is best to go back and check the result by more precise Itô stochastic integral procedures.

4.1.1   Functions of Diffusions Alone, $G(W(t))$

Some simple calculus-like examples are given below as an introduction. Although we could just as well work with differentials at the start, we will start with the increments at t and then get the differential form in the limit as $\Delta t \to 0^+$, then later switch to starting with the differential forms as increment forms in dt.

Examples 4.1.

• *Cubic integral:*

$$\Delta \left[W^3 \right] (t) = (W + \Delta W)^3(t) - W^3(t) = \left(3W^2 \Delta W + 3W(\Delta W)^2 + (\Delta W)^3 \right)(t)$$

using the cubic expansion. As $\Delta t \to 0$, $(\Delta W)^2(t) \to (dW)^2(t) \overset{\mathrm{dt}}{=} dt$ and $(\Delta W)^3(t) \to (dW)^3(t) \overset{\mathrm{dt}}{=} 0$, so the corresponding differential form is

$$d\left[W^3\right](t) \overset{\mathrm{dt}}{=} \left(3W^2 dW + 3W dt\right)(t).$$

The first term is the deterministic differential, since $d(w^3) = 3w^2 dw$, but with an Itô stochastic correction $3W(t)dt$. Solving for $W^2(t)dW(t)$, the Itô integral of the square of $W(t)$ yields

$$\int_{t_0}^t W^2(s)dW(s) \overset{\mathrm{ims}}{=} \frac{1}{3}\left(W^3(t) - W^3(t_0)\right) - \int_{t_0}^t W(s)ds.$$

The Itô integral of $W^2(t)$ is reduced to an exact stochastic-Riemann integral of $d(W^3)(t)$, and the Itô correction is reduced to the Riemann integral of $W(t)$ and looks simple, but it cannot be Itô integrated exactly and must be numerically simulated if needed.

- *General integer power integral:*
  By using the full binomial theorem (B.150),

$$\Delta\left[W^{m+1}\right](t) = (W + \Delta W)^{m+1}(t) - W^{m+1}(t)$$

$$= \sum_{i=0}^{m}\binom{m+1}{i} W^i(t)\Delta W^{m+1-i}(t),$$

where the passage to the limit as $\Delta t \to 0$ and the Itô mean square limit leading to the integral form

$$\int_0^t W^m(s)dW(s),$$

have been left as Exercise 5 on p. 123 in Section 4.5.

- *Exponential integral:*
  Using laws of exponents and the first few terms of the exponential expansion (B.53), going directly to the formal differential form and skipping the more general increment form to expedite applied stochastic calculations,

$$d\left[e^W\right](t) = \left(e^{W+dW} - e^W\right)(t) = \left(e^W\left(e^{dW} - 1\right)\right)(t)$$

$$\overset{\mathrm{dt}}{=} \left(e^W\left(dW + \frac{1}{2}(dW)^2\right)\right)(t),$$

neglecting differential forms that are zero in the Itô mean square limit, such as $dW^3(t) \overset{\mathrm{dt}}{=} 0$, $dt dW(t) \overset{\mathrm{dt}}{=} 0$, $(dt)^2 \overset{\mathrm{dt}}{=} 0$, and higher powers with this zero mean square limit property.

Using the basic mean square limit differential form (4.1), $(dW)^2(t) \overset{\mathrm{dt}}{=} dt$, so

$$d\left[e^W\right](t) \overset{\mathrm{dt}}{=} \left(e^W\left(dW + \frac{1}{2}dt\right)\right)(t). \tag{4.4}$$

This is almost like the deterministic differential, $d(e^w) = e^w dw$, but here with an Itô stochastic correction $e^{W(t)} dt/2$. Solving for $e^{W(t)} dW(t)$, the Itô integral of the exponential of $W(t)$ yields the implicit integration

$$\int_{t_0}^t e^{W(s)} dW(s) \overset{\text{ims}}{=} e^{W(t)} - e^{W(t_0)} - \frac{1}{2} \int_{t_0}^t e^{W(s)} ds. \tag{4.5}$$

As with the integral of $W^2(t)$, the Itô integral of $e^{W(t)}$ cannot be Itô integrated exactly and must be numerically simulated if needed. The simulations are presented later in Figure 4.2 on p. 93 in Subsection 4.1.3 for the Itô partial sums form for the stochastic exponential,

$$S_{i+1} = \sum_{j=0}^i \exp(W_i) \Delta W_i$$

for $t = t_{i+1} = (i+1)\Delta t$ for $t_0 = 0$ evenly spaced using $\Delta t_i = \Delta t$, where

$$W_{i+1} = \sum_{j=0}^i \Delta W_j.$$

The error is

$$E_{i+1} = S_{i+1} - R_{i+1}$$

between the partial sums S_{i+1} and the difference approximation to the right-hand side

$$R_{i+1} = \exp(W_{i+1}) - 1 - \frac{1}{2} \sum_{j=0}^i \exp(W_j) \Delta t$$

of (4.5); note that $t_0 = 0$, so $\exp(W(t_0)) = 1$. Remember that the cumulative noise W_i must always be approximated by sums of simulated independent increments ΔW_j for $j = 0 : i - 1$.

In the differential (4.4) of the pure exponential, there is a clue to an exact differential in the Itô mean square sense, since the factor $(dW + dt/2)$ suggests subtracting $t/2$ from $W(t)$. In fact,

$$d\left[e^{W(t)-t/2}\right] \overset{\text{dt}}{=} e^{W(t)-t/2} dW(t). \tag{4.6}$$

So

$$\int_0^t e^{W(s)-s/2} dW(s) \overset{\text{ims}}{=} e^{W(t)-t/2} - 1. \tag{4.7}$$

In the forthcoming Subsection 4.1.3 on p. 88, a method for systematically finding general exact integrals is presented, provided they exist. The simulations are presented later in Figure 4.1 on p. 92 in Subsection 4.1.3 for the Itô partial sums form $S_{i+1} = \sum_{j=0}^i \exp(W_i - t_i/2) \Delta W_i$ and the error between the partial sums and the difference approximation of (4.7).

More general rules can be derived by the same techniques.

Rule 4.2. *Chain Rule for $G(W(t))$.*
Let $G(w)$ be twice continuously differentiable. Then the differential form of the Itô stochastic chain rule for $G(W(t))$ is

$$dG(W(t)) \overset{\text{dt}}{=} G'(W(t))dW(t) + \frac{1}{2}G''(W(t))dt, \qquad (4.8)$$

*corresponding to the integral form of the Itô stochastic **chain rule for $G(W(t))$**,*

$$G(W(t)) \overset{\text{ims}}{=} G(W(t_0)) + \int_{t_0}^{t} G'(W(s))dW(s) + \frac{1}{2}\int_{t_0}^{t} G''(W(s))ds \qquad (4.9)$$

for $0 \le t_0 \le t$.

Sketch of Proof. Assuming $G(w)$ is twice continuously differentiable in the argument w, we then see that $G(W(t))$ has the differential

$$\begin{aligned} dG(W(t)) &= G(W(t) + dW(t)) - G(W(t)) \\ &\overset{\text{dt}}{=} G'(W(t))dW(t) + \frac{1}{2}G''(W(t))(dW)^2(t). \end{aligned}$$

Taking the Itô mean square limit, neglecting error terms that are zero in the mean square limit, such as $dW^3(t)$, $dtdW(t)$, and $(dt)^2$, then using $(dW)^2(t) \overset{\text{dt}}{=} dt$ yields the differential form (4.8) of the Itô stochastic **chain rule for $G(W(t))$**. The last term in the second derivative is the Itô stochastic correction to the deterministic chain rule. Immediately, we have the Itô stochastic integral form (4.9), which provides substantial meaning to the symbolic differential form.   □

Rewriting (4.9) yields the **fundamental theorem of calculus** according to the Itô version [150] as follows.

Corollary 4.3. *Itô's Fundamental Theorem of Calculus for Stochastic Diffusions.*
Let $G(w)$ be twice continuously differentiable. Then

$$\int_{t_0}^{t} G'(W(s))dW(s) \overset{\text{ims}}{=} G(W(t)) - G(W(t_0)) - \frac{1}{2}\int_{t_0}^{t} G''(W(s))ds. \qquad (4.10)$$

Remark 4.4. *Recall the more elementary integral of a differential form of the fundamental theorem of stochastic diffusion calculus in (2.38), which in fact leads to the exact part of the Itô version, using G in (2.38),*

$$\int_{t_0}^{t} dG(W(s)) \overset{\text{ims}}{=} G(W(t)) - G(W(t_0)).$$

4.1.2   Functions of Diffusions and Time, $G(W(t), t)$

Rule 4.5. *Chain Rule for $G(W(t), t)$ (Itô's Lemma or Formula).*
*Let $G(w, t)$ be twice continuously differentiable in w and once continuously differentiable in t. Then the differential Itô stochastic **chain rule for** $G(W(t), t)$ is*

$$dG(W(t), t) \stackrel{\text{dt}}{=} \left(G_t + \frac{1}{2} G_{ww} \right) (W(t), t)dt + G_w(W(t), t)dW(t), \qquad (4.11)$$

*corresponding to the integral form of the Itô stochastic **chain rule for** $G(W(t), t)$,*

$$G(W(t), t) \stackrel{\text{ims}}{=} G(W(t_0), t_0) + \int_{t_0}^{t} G_w(W(s), s)dW(s)$$

$$+ \int_{t_0}^{t} \left(G_t + \frac{1}{2} G_{ww} \right) (W(s), s)ds \qquad (4.12)$$

for $0 \leq t_0 \leq t$.

Sketch of Proof. Assuming $G(w, t)$ is twice continuously differentiable in the argument w and once continuously differentiable in t, by using a mean square order modification of the Taylor approximation in (B.181), $G(W(t), t)$ has the differential

$$dG(W(t), t) = G(W(t) + dW(t), t + dt) - G(W(t), t)$$

$$\stackrel{\text{dt}}{=} G_t(W(t), t)dt + G_w(W(t), t)dW(t) + \frac{1}{2}G_{ww}(W(t), t)(dW)^2(t),$$

where the partial derivatives are denoted with subscripts, i.e.,

$$G_w(w, t) = \frac{\partial G}{\partial w}(w, t), \quad G_t(w, t) = \frac{\partial G}{\partial t}(w, t), \quad G_{ww}(w, t) = \frac{\partial^2 G}{\partial w^2}(w, t).$$

Taking the Itô mean square limit with $(dW)^2(t) \stackrel{\text{dt}}{=} dt$ and neglecting the higher order differential forms that are zero in the Itô mean square sense, such as $dW^3(t)$, $dtdW(t)$, and $(dt)^2$, yields (4.11), which is called the Itô stochastic **chain rule for** $G(W(t), t)$. Again, the last term in the second derivative is the Itô stochastic correction to the deterministic chain rule. Translating the symbolic differential form into the substantial Itô stochastic integral form gives (4.12). $\square$

Remarks 4.6. *Functions, Values, and Partial Derivatives.*

- *For readers without much PDE background, we note that there are certain concepts that are important and there are subtle differences in the function and its values $G(w, t)$. This is particularly true when there are two or more independent variables, such as the $w = W(t)$ and t in $G(W(t), t)$. This does not arise when there is just one independent variable, such as x in $y = f(x)$. Another complication is that the $W(t)$ is a nondifferentiable function, so we never form its derivative but only compute its differential $dW(t)$, and that is best done formally by the increment form of the differential.*

- *The symbol G denotes a function specified by a set of rules for its calculation, while G(w, t) is the value of that function with its first argument evaluated at w and with the second argument at time t. Similarly, G(W(t), t) is the value of G specified at the random variable W(t) at time t in place of the realized or dummy variable w. Further, X(t) = G(W(t), t) is the path of the state in time and is nondifferentiable along with W(t), i.e., X(t) is a composite function in time through both arguments of G, implicitly through W(t) and explicitly through the second argument t.*

- *Using limits of Newton's quotient for derivatives, the partial derivatives of G(w, t) are defined, also giving several alternate notations, at (w, t) as*

$$G_w(w, t) = \frac{\partial G}{\partial w}(w, t) = \left(\frac{\partial G}{\partial w}\right)\bigg|_{t \atop \text{fixed}} (w, t) = \lim_{\Delta w \to 0} \frac{G(w + \Delta w, t) - G(w, t)}{\Delta w}$$

and

$$G_t(w, t) = \frac{\partial G}{\partial t}(w, t) = \left(\frac{\partial G}{\partial t}\right)\bigg|_{w \atop \text{fixed}} (w, t) = \lim_{\Delta t \to 0} \frac{G(w, t + \Delta t) - G(w, t)}{\Delta t},$$

provided the limits exist. Hence, partial derivatives with one of the variables fixed are based on the definition of ordinary derivatives.

- *The partial derivatives G_w and G_t are defined as rules based upon the target function rule G. For the topics here, when the first argument is a random variable w = W(t),*

$$\frac{\partial G}{\partial w}(W(t), t)$$

*is just G_w evaluated at the first variable w = W(t) after differentiation. We would **never** write $G_{W(t)}$ due to the nondifferentiable properties of W(t). The partial derivative is calculated first, and then it is evaluated. For example, $G_w(1, 1)$ can be computed if we know G_w and it has a unique value at (1, 1), but $(G(1, 1))_w = 0$ since G(1, 1) has a fixed, constant value, presumably unique, at (1, 1). The order of partial differentiation and partial derivative function evaluation are very important.*

- *Another, more relevant, example illustrating the difference in the differential is multiplying by dt to avoid obtaining the singular derivative of W(t), i.e.,*

$$dG(W(t), t) \overset{\text{dt}}{=} \left(G_t dt + G_w dW(t) + \frac{1}{2} G_{ww} dt\right)(W(t), t),$$

contains the partial derivative of the function G with respect to t evaluated at (W(t), t),

$$\frac{\partial G}{\partial t}(W(t), t)dt,$$

rather than the partial derivative with respect to t written as the derivative of the value G(W(t), t),

$$\frac{\partial G(W(t), t)}{\partial t}dt,$$

which makes no sense since it would involve the derivative of the nondifferentiable W(t) in t with probability one. (Recall Theorem 1.9 on p. 9.)

Corollary 4.7. *Let $g(W(t), t)$ satisfy the conditions of Definition 2.8 for an Itô stochastic integral and be once continuously differentiable in w. Let $G(w, t)$ be the antiderivative of $g(w, t)$ with respect to w, i.e., $G_w(w, t) = g(w, t)$, and let $G(w, t)$ be twice continuously differentiable in w, but only once in t. Then,*

$$\int_{t_0}^{t} g(W(s), s) dW(s) \overset{\text{ims}}{=} G(W(t), t) - G(W(t_0), t_0)$$

$$- \int_{t_0}^{t} (G_t + 0.5 * g_w)(W(s), s) ds \qquad (4.13)$$

for $0 \le t_0 \le t$.

Proof. The proof follows directly from (4.12) by rearranging terms, since $G_w = g$ and $G_{ww} = g_w$. □

Remark 4.8. *Thus, the Itô stochastic diffusion integral of $g(W(t), t)$ can be reduced to an exact integral $G(W(t), t) - G(W(t_0), t_0)$ with respect to w less a quasi-deterministic Riemann integral over the diffusion shifted drift function $(G_t + 0.5 * g_w)(W(t), t)$. Thus, if the partial differential equation $(G_t + 0.5 * g_w)(w, t) = 0$ is valid with $g_w(w, t) = G_{ww}(w, t)$, then the integral of $g(W(t), t)$ is equal to the exactly integrated part $G(W(t), t) - G(W(t_0), t_0)$ in the Itô mean square sense. This idea can be the basis for constructing exact stochastic diffusion integrals.*

Example 4.9. *Merton's Analysis of the Black–Scholes Option Pricing Model.*

At this point in the text, a good application in finance is the survey of Merton's analysis [201]; [203, Chapter 8] of the Black–Scholes [34] financial options pricing model in Section 10.2. This survey follows the tone of this book, although Merton's model has several state dimensions—the bond, the stock, and the option. While multidimension SDEs are covered in the next chapter, this treatment will serve as motivation for the next chapter, which contains details not in Merton's paper.

4.1.3 Itô Stochastic Natural Exponential Construction

From the differential of $\exp(W(t))$ in (4.4) it is seen that the stochastic exponential is not like the deterministic natural exponential, where the derivative is proportional to the original function, e.g., the natural exponential e^x in the natural base e has the differential property

$$d\left(e^x\right) = e^x dx,$$

returning the original function times dx, and has the inverse relationship to the natural logarithm

$$e^{\ln(x)} = x$$

for $x > 0$, whereas when $b > 0$ and in particular $b \neq e$ for the base b, then

$$d\left(b^x\right) = d\left(e^{x \ln(b)}\right) = b^x \ln(b) dx,$$

returning an additional factor $\ln(b)$.

For more generality, consider the deterministic model

$$d\left(e^{ax}\right) = ae^{ax}dx,$$

where the parameter a is a nonzero constant. The corresponding stochastic model is the process $X(t) = G(W(t), t)$ such that

$$dX(t) = dG(W(t), t) \stackrel{\mathrm{dt}}{=} aG(W(t), t)dW(t) = aX(t)dW(t). \qquad (4.14)$$

The explicit t dependence is needed to avoid correction factors in dt. Applying the appropriate stochastic chain rule (4.11) to illustrate a technique for inverting the chain rule to get the desired model in terms of the composite function G, we have

$$aG(W(t), t)dW(t) \stackrel{\mathrm{dt}}{=} dG(W(t), t)$$
$$\stackrel{\mathrm{dt}}{=} \left(G_t(W(t), t) + \frac{1}{2}G_{ww}(W(t), t) \right) dt + G_w(W(t), t)dW(t).$$

Since the differentials, $dW(t)$ and dt, can be independently varied in this equation, the coefficients of $dW(t)$ and dt can be separately set equal to their values on both sides of the equation (dropping the arguments of G for simplicity):

$$G_w = aG \quad \text{and} \quad G_t + \frac{1}{2}G_{ww} = 0. \qquad (4.15)$$

The solution of the first partial differential equation (PDE), $G_w = aG$, in (4.15), being effectively an ODE with t held fixed, is

$$G(w, t) = A(t)e^{aw}, \qquad (4.16)$$

since $d(e^{-aw})/dw = -ae^{-aw}$ (differentiation is allowable for a regular continuous, i.e., nonstochastic, function) so

$$d\left(e^{-aw}G\right)_w = e^{-aw}\left(G_w - aG\right) = 0,$$

which shows that (4.16) satisfies the first PDE by substitution, $e^{-aw} \neq 0$. Here, $A(t)$ is a **function of integration** since the differential equation is only in w and t is arbitrary, although held fixed in the equation. Given a differentiable function $F(w, t)$, the notation $F_w(w, t) = 0$ is shorthand for the partial deriviative

$$\left(\frac{\partial F}{\partial w} \right)_{\substack{t \\ \text{fixed}}} (w, t) = 0.$$

This means that $F(w, t) = A(t)$ for some function A of t, since t is held fixed in the partial differentiation with respect to w.

Upon substituting this current functional form into the second PDE, $G_t + 0.5G_{ww} = 0$, using

$$\left(A(t)e^{aw}\right)_t = e^{aw}(A(t))_t = A'(t)e^{aw},$$
$$\left(A(t)e^{aw}\right)_{ww} = A(t)(e^{aw})_{ww} = a^2A(t)e^{aw},$$

then

$$A'(t)e^{aw} + \frac{a^2}{2}A(t)e^{aw} = 0.$$

Canceling out the common nonzero factor e^{aw},

$$A'(t) + \frac{a^2}{2}A(t) = 0, \tag{4.17}$$

and solving for the function of integration yields

$$A(t) = Ce^{-a^2t/2}, \tag{4.18}$$

where C is a **genuine constant of integration**.

Remark 4.10. *Note that an ultimate test of a solution of a differential equation solution is the **substitution test**, i.e., substituting the solution back into the equation and verifying that the equation and any conditions are satisfied.*

For (4.18), substitution into the ODE (4.17) leads to

$$A'(t) + \frac{a^2}{2}A(t) = Ce^{-a^2t/2} \cdot \left(-\frac{a^2}{2} + \frac{a^2}{2}\right) = 0.$$

By reassembling the parts of the solution, we obtain the Itô general stochastic form of the natural exponential (exponential in the natural base e),

$$X(t) = G(W(t), t) = Ce^{aW(t) - a^2t/2}, \tag{4.19}$$

systematically deriving what previously was a guess in (4.6). The extra exponential term $(-a^2t/2)$ is the special Itô correction that forces the simple linear growth model $dX(t) = aX(t)dt$ for the exponential growth in the diffusion $W(t)$.

Since $W(0^+) = 0$ with probability one, $X(0^+) = G(0, 0^+) = C$, with probability one, is the initial value of the state $X(t)$, while a is a rate of growth. The basic moments of the state trajectory can be calculated by using the density $\phi_{W(t)}(w)$ for $W(t)$ in (1.7).

Some of the details are given to illustrate the use of the **completing the square** technique when computing exponential moments with respect to normal distributions. An illustration of the completing the square technique is presented for the expectation of an exponential whose exponent is linear (or affine) in $W(t)$, i.e., $\exp(a(t)W(t) + b(t))$.

Lemma 4.11. *Completing the Square for* $\mathrm{E}[K(t)\exp(a(t)W(t) + b(t))]$.
Let $a(t) \neq 0$, $b(t)$, and $K(t) \neq 0$ be bounded deterministic functions of t. Then

$$\mathrm{E}\left[K(t)e^{a(t)W(t) + b(t)}\right] = K(t)e^{a^2(t)t/2 + b(t)}. \tag{4.20}$$

Proof. Since the Wiener process density,

$$\phi_{W(t)}(w) = \frac{1}{\sqrt{2\pi t}}e^{-w^2/(2t)},$$

$-\infty < w < +\infty$, from (1.7), is essentially a function of the sampled dummy variable w, and t is only a parameter that we can hold fixed during the integration, the deterministic functions of time are treated as constants. By the laws of exponents, the exponent of the density and the exponent of the argument of the expectation with the dummy variable substitution $W(t) = w$ are added together to obtain a complete square of all w terms,

$$-w^2/(2t) + a(t)w + b(t) = -(w - a(t)t)^2/(2t) + a^2(t)t/2 + b(t).$$

Thus,

$$
\begin{aligned}
\mathrm{E}\left[K(t)e^{a(t)W(t)+b(t)}\right] &= K(t)\frac{1}{\sqrt{2\pi t}}\int_{-\infty}^{+\infty} e^{-(w-a(t)t)^2/(2t)+a^2(t)t/2+b(t)}dw \\
&= K(t)e^{a^2(t)t/2+b(t)}\frac{1}{\sqrt{2\pi t}}\int_{-\infty}^{+\infty} e^{-v^2/(2t)}dv \\
&= K(t)e^{a^2(t)t/2+b(t)}\mathrm{E}[1], \\
&= K(t)e^{a^2(t)t/2+b(t)},
\end{aligned}
$$

where the fixed part of the integral with exponent $(a^2(t)t/2 + b(t))$ has been separated out and the change of variables $v = w - a(t)t$ with $dv = dw$, t being fixed, in the integral has been used to transform the completed square part of the expectation integral into conservation of probability $\mathrm{E}[1] = 1$ for the standard Wiener process. □

The mean state $X(t)$ using Lemma 4.11 is

$$\mathrm{E}\left[Ce^{aW(t)-a^2t/2}\right] = C = X(0^+), \qquad (4.21)$$

so the mean trajectory is a constant at the initial level $X(0^+)$. However, the state variance, again using Lemma 4.11 but with $a(t)$ replaced by $2a$ following application of the variance-expectation identity (B.186), $\mathrm{Var}[X] = \mathrm{E}[X^2]-\mathrm{E}^2[X]$, to use the expectation result (4.21), is

$$
\begin{aligned}
\mathrm{Var}\left[Ce^{aW(t)-a^2t/2}\right] &= \mathrm{E}\left[\left(Ce^{aW(t)-a^2t/2}\right)^2\right] - \mathrm{E}^2\left[Ce^{aW(t)-a^2t/2}\right] \\
&= C^2\mathrm{E}\left[e^{2aW(t)-a^2t}\right] - C^2 \\
&= C^2\left(e^{a^2t} - 1\right).
\end{aligned}
$$

Examining the standard deviation, or square root of the variance,

$$\sigma_{X(t)} = \sqrt{\mathrm{Var}[X(t)]} = C\sqrt{e^{a^2t} - 1} \sim Ce^{a^2t/2}$$

as $t \to \infty$, it is seen that the RMS of stochastic fluctuations grows exponentially with exponent $a^2t/2$ starting initially at $\sigma_{X(0^+)} = 0^+$.

Figure 4.1 is an illustration of the simulation of the integral of this natural exponential in the special case

$$I[g](t) = \int_0^t g(W(s), s)dW(s) = \int_0^t e^{W(s)-s/2}dW(s) \overset{\mathrm{ims}}{=} e^{W(t)-t/2} - 1, \qquad (4.22)$$

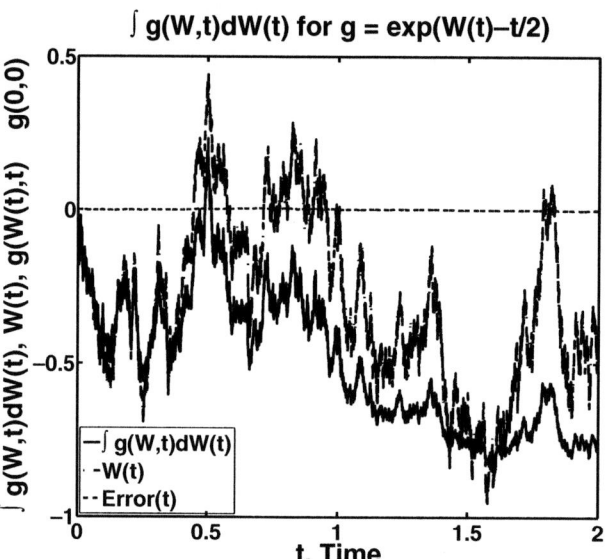

Figure 4.1. *Example of a simulated Itô discrete approximation to the stochastic diffusion integral* $I_n[g](t_{i+1}) = \sum_{j=0}^{i} g_j \Delta W_j$ *for* $i = 0 : n$, *using MATLAB* randn *with sample size* $n = 10{,}000$ *on* $0 \leq t \leq 2.0$. *Presented are the simulated Itô partial sums* S_{i+1}, *the simulated noise* W_{i+1}, *and the error* E_{i+1} *relative to the exact integral,* $I^{(\mathrm{ims})}[g](t_{i+1}) \stackrel{\mathrm{ims}}{=} \exp(W_{i+1} - t_{i+1}/2) - 1$, *in the Itô mean square sense.*

i.e., when $a = 1 = C$. Also plotted is the diffusion process $W(t)$ for comparison and the error,

$$E_{i+1} = S_{i+1} - I_{i+1},$$

between the simulation of the integral by Itô finite difference partial sums,

$$S_{i+1} = \sum_{j=0}^{i} g_j \Delta W_j,$$

and the simulation of the exact mean square integral value in (4.22)

$$I_{i+1} = g_{i+1} - 1$$

for $i = 0 : n$, where the integrand is

$$g_i = \exp(W_i - t_i/2)$$

with $W_i = \sum_{j=0}^{i-1} \Delta W_j$ and $t_i = i * \Delta t$ for $i = 0 : n + 1$. Observe that the integral initially tracks the W_i simulated noise but eventually diverges from it. Also, the error slowly degrades as time t_i becomes long (not shown), in this case for $n = 10{,}000$ (note that this is an approximate sample size since random sample size is $n + 1 = 10{,}001$ random increments) and $t = 2.0$. The MATLAB code for the exactly integrable $g(W(t), t)$ in the Itô mean square diffusion integral sense is given in Program C.12, called `intgwtdw.m` in Online Appendix C.

In Figure 4.2, the chain rule formulation of the Itô diffusion integral of the simple exponential $g(W(t), t) = \exp(W(t))$ of Example 4.1 is compared to the Itô partial sums $S_{i+1} = \sum_{j=0}^{i} g_j \Delta W_j$. Unlike the stochastic natural exponential $\exp(W(t) - t/2)$, the simple exponential is not exactly integrable in the Itô mean square sense since the stochastic chain rule introduces a quasi-deterministic, regular-type integral for the diffusion term

$$-0.5 G_w(w, t) = -0.5 g(w, t) = -0.5 \exp(w).$$

The partially integrated chain rule form is thus

$$I_{i+1} = \exp(W_i) - 1 - 0.5 * \sum_{j=0}^{i} \exp(W_j) \Delta t \tag{4.23}$$

with $G_t(w, t) = 0$. In the figure the error between the two approximations of the integral $E_{i+1} = S_{i+1} - I_{i+1}$ and the underlying diffusive noise is $W(t)$. The error is very small for a sample size of $n = 10,000$. The integration significantly dampens the fluctuations in the original noise $W(t)$. The MATLAB code for this figure is given in Program C.13, called `intgxtdw.m` in Online Appendix C.

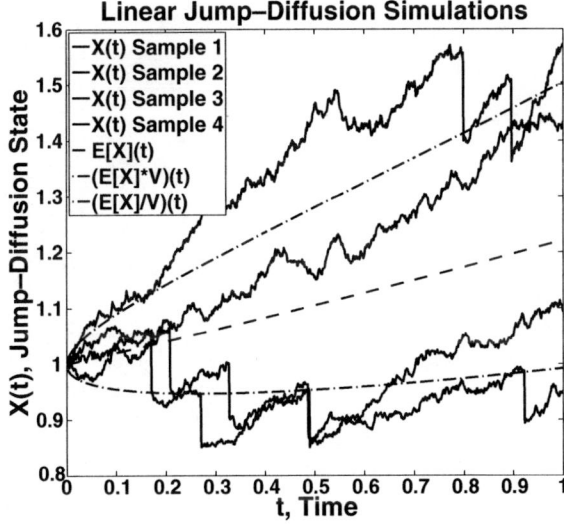

Figure 4.2. *Example of a simulated Itô discrete approximation to the stochastic diffusion integral $I_n[g](t_{i+1}) = \sum_{j=0}^{i} g_j \Delta W_j$ for $i = 0 : n$, using MATLAB `randn` with sample size $n + 1 = 10,001$ on $0 \leq t \leq 2.0$. Presented are the simulated Itô partial sums S_{i+1}, the simulated noise W_{i+1}, and the error E_{i+1} relative to the stochastic chain rule partially integrated form I_{i+1} given in (4.23).*

4.1.4   Transformations of Linear Diffusion SDEs

Consider the diffusion SDE, linear in the state process $X(t)$, with time-dependent coefficients

$$dX(t) = X(t) \left(\mu(t)dt + \sigma(t)dW(t) \right), \qquad (4.24)$$

where the initial condition is $X(t_0) = x_0 > 0$ with probability one, $\mu(t)$ is called the **drift** or deterministic coefficient and $\sigma(t)$ is called the **volatility** or standard deviation of the diffusion term. The **diffusion coefficient** is usually defined as $\mathcal{D} = \sigma^2(t)/2$, so $\sigma(t) = \sqrt{2\mathcal{D}}$. The linear form of (4.24) is sometimes called the **multiplicative noise** case, the state $X(t)$ multiplies the stochastic terms, and the word **noise** refers to the randomness or stochastic properties here. In the deterministic case, transforming the state variable into its logarithm makes the right-hand side independent of the transformed state variable, so let

$$Y(t) = F(X(t)) \equiv \ln(X(t)).$$

Since we have F depending on $X(t)$ rather than $W(t)$, we go back to the basic treatment of the change as an increment and expand the increment to second order,

$$
\begin{aligned}
dY(t) &= \log(X(t) + dX(t)) - \log(X(t)) \\
&\overset{\mathrm{dt}}{=} \frac{1}{X(t)} dX(t) - \frac{1}{2X^2(t)} (dX)^2(t) \\
&\overset{\mathrm{dt}}{=} (\mu(t)dt + \sigma(t)dW(t)) - 0.5\sigma^2(t)(dW)^2(t) \\
&\overset{\mathrm{dt}}{=} (\mu(t) - 0.5\sigma^2(t))dt + \sigma(t)dW(t), \qquad (4.25)
\end{aligned}
$$

where we again used $(dW)^2(t) \overset{\mathrm{dt}}{=} dt$ and dropped terms zero in the mean square. Use has been made of the following partial derivatives:

$$F_t(X(t)) \equiv 0, \quad F_x(X(t)) = 1/X(t), \quad F_{xx}(X(t)) = -1/X^2(t).$$

The final line in (4.25) is also called **additive noise** since it just adds to the state value and can be immediately integrated, as opposed to the multiplicative noise in the original SDE in (4.24). In the above derivation, the Itô stochastic correction on the drift $\mu(t)$ is the negative of the **diffusion coefficient** $\sigma^2(t)/2$. The final right-hand side (4.25) defines a differential simple Gaussian process (B.24) with infinitesimal mean $(\mu(t) - 0.5\sigma^2(t))dt$ and infinitesimal variance of $\sigma^2(t)dt$. The **infinitesimal mean** here is defined as

$$\mathrm{E}[dY(t)] \qquad (4.26)$$

and the **infinitesimal variance** is defined as

$$\mathrm{Var}[dY(t)], \qquad (4.27)$$

in each case neglecting orders smaller than ord(dt). An alternate method of deriving (4.25) is to use the Itô stochastic chain rule for $G(W(t))$, but with $W(t)$ replaced by $X(t)$, subsequently expanding the differentials $dX(t)$ and $(dX)^2(t)$, then replacing them by the SDE in (4.24) and neglecting any terms that are zero in the mean square.

Since the right-hand side of (4.25) does not depend on the state $Y(t)$, we can immediately integrate for $Y(t)$ given the coefficient functions leading to

$$Y(t) = y_0 + \int_{t_0}^t (\mu(s) - 0.5\sigma^2(s))ds + \int_{t_0}^t \sigma(s)dW(s), \qquad (4.28)$$

where $y_0 = \ln(x_0)$. Exponentiation leads to the formal solution for the original state,

$$X(t) = x_0 \exp\left(\int_{t_0}^t (\mu(s) - 0.5\sigma^2(s))ds + \int_{t_0}^t \sigma(s)dW(s) \right). \qquad (4.29)$$

Linear Diffusion SDEs with Constant Coefficients

If the SDE has **constant coefficients** $\mu(t) = \mu_0$ and $\sigma(t) = \sigma_0$, while letting $t_0 = 0$, then the solution is simpler:

$$X(t) = x_0 \exp\left((\mu_0 - 0.5\sigma_0^2)t + \sigma_0 W(t) \right). \qquad (4.30)$$

Note that if $X(0^+) = x_0$ is initially positive as declared, then the solution $X(t)$ will never become negative by the property of the exponential for real arguments, and the transformation $Y(t) = \ln(X(t))$ is proper with $X(t) > 0$. The state $X(t)$ positivity feature is very important in biological and financial applications. Aside from time-dependence, this is just a shift by the drift of the exponent, as in the Itô stochastic exponential in (4.19).

In the **additive noise** case, borrowing the exponent form in (4.25), the relation between the new and old values of Y is computed by adding the noise

$$Y(t + \Delta t) = Y(t) + (\mu_0 - 0.5\sigma_0^2)\Delta t + \sigma_0 \Delta W(t) \qquad (4.31)$$

or recursively in the time-step Δt_i from t_i to t_{i+1} and then summing the recursion,

$$\begin{aligned} Y_{i+1} &= Y_i + (\mu_0 - 0.5\sigma_0^2)\Delta t_i + \sigma_0 \Delta W_i \\ &= y_0 + \sum_{j=0}^i \left((\mu_0 - 0.5\sigma_0^2)\Delta t_j + \sigma_0 \Delta W_j \right). \end{aligned}$$

So taking the expectation,

$$\begin{aligned} \mathrm{E}[Y_{i+1}] &= y_0 + \sum_{j=0}^i \left(\mu_0 - 0.5\sigma_0^2 \right) \Delta t_j \\ &= y_0 + (\mu_0 - 0.5\sigma_0^2)\sum_{j=0}^i \Delta t_j. \end{aligned} \qquad (4.32)$$

This result should be compared to the corresponding deterministic additive or arithmetic recursion with constant a,

$$z_{i+1} = z_i + a \implies z_{i+1} = z_0 + (i + 1) \cdot a,$$

so the corresponding additive parameter form of the mean (4.32) is

$$E[Y_{i+1}] = y_0 + (i + 1)(\mu_0 - 0.5\sigma_0^2)\overline{\Delta t_i}^{(am)}, \tag{4.33}$$

where

$$\overline{\Delta t_i}^{(am)} = \frac{1}{i + 1} \sum_{j=0}^{i} \Delta t_j$$

is the **arithmetic mean** of the first $(i + 1)$ time-steps Δt_j for $j = 0 : i$.

As the **multiplicative noise** property can be seen by rewriting (4.30) as a single step,

$$X(t + \Delta t) = X(t) \exp\left((\mu_0 - 0.5\sigma_0^2)\Delta t + \sigma_0 \Delta W(t)\right), \tag{4.34}$$

so the new noise exponential contribution from $\Delta W(t)$ multiplies the current value of the solution $X(t)$ to produce the new value $X(t + \Delta t)$. The corresponding recursive form in the time-step Δt_i from t_i to t_{i+1}, followed by a summing of the recursion, yields

$$
\begin{aligned}
X_{i+1} &= X_i \exp\left((\mu_0 - 0.5\sigma_0^2)\Delta t_i + \sigma_0 \Delta W_i\right) \\
&= x_0 \exp\left(\sum_{j=0}^{i} \left((\mu_0 - 0.5\sigma_0^2)\Delta t_j + \sigma_0 \Delta W_j\right)\right) \\
&= x_0 \prod_{j=0}^{i} \exp\left((\mu_0 - 0.5\sigma_0^2)\Delta t_j + \sigma_0 \Delta W_j\right),
\end{aligned}
$$

using the laws of exponents to turn the exponential of a sum into a product of exponentials. Thus, taking the expectation and using the completing the squares Lemma 4.11,

$$E[X_{i+1}] = x_0 \prod_{j=0}^{i} \exp\left(\mu_0 \Delta t_j\right). \tag{4.35}$$

This result should be compared to the corresponding deterministic multiplicative recursion or geometric progression with constant r,

$$x_{i+1} = rx_i = x_0 r^{i+1},$$

so the corresponding multiplicative parameter form of the mean (4.35) is

$$E[X_{i+1}] = x_0 \left(\overline{\xi_i}^{(gm)}\right)^{i+1},$$

where

$$\overline{\xi_i}^{(gm)} = \left(\prod_{j=0}^{i} e^{\mu_0 \Delta t_j}\right)^{\frac{1}{i+1}}$$

is the **geometric mean** of the first $(i + 1)$ growth steps $\xi = e^{\mu_0 \Delta t_j}$ for $j = 0 : i$.

Applications include stochastic population growth, where $X(t)$ is the population size, $\mu(t)$ is an intrinsic growth rate (rate of growth in the absence of stochastic or other effects in the environment), and the $\sigma(t)X(t)dW(t)$ denotes the stochastic effects. The term $\sigma(t)X(t)dW(t)$ is called **demographic stochasticity** [271], since it looks like a stochastic perturbation from $\mu(t)$. Similarly, perturbations of nonlinear saturation terms are called **environmental stochasticity**. In biology, multiplicative or geometric noise is also called **density independent noise**, since $dX(t)/X(t)$ is independent of $X(t)$. See also Chapter 11 on **biological applications**.

Another application is financial engineering, where $X(t)$ is the investment return, $\mu(t)$ is the mean appreciation rate, and $\sigma(t)$ is the investment volatility. In stochastic finance, the process $X(t)$ is called **geometric Brownian motion (GBM)** due to the linear scaling on the right-hand side for the $dX(t)$ and, in particular, due to the stochastic noise being multiplied by the state process $X(t)$, i.e., the multiplicative noise. In finance, one of the earliest stochastic stock models was from the thesis of Bachelier [16], in which additive noise was used, but this work did not attract much attention until Black and Scholes [34], Merton [203], and others began using multiplicative noise stock and options models. Multiplicative models are more appropriate in finance as well as in biology, since random effects are more likely to compound rather than add. See also Chapter 10 on **financial engineering applications**.

For the constant coefficient case of the linear stochastic diffusion SDE, the solution can be shown to have a lognormal distribution.

Theorem 4.12. *The Solution of the Constant Coefficient, Linear Stochastic Diffusion SDE is LogNormally Distributed.*
Let $X(t)$ satisfy

$$dX(t) = X(t)\left(\mu_0 dt + \sigma_0 dW(t)\right), \tag{4.36}$$

$X(0) = x_0 > 0$ *with probability one, where μ_0 and $\sigma_0 > 0$ are constants. Then, the distribution of $X(t)$,*

$$\Phi_{X(t)}(x) = \Phi_n\left(\ln(x); \ln(x_0)\mu_n(t), (\sigma_n)^2(t)\right), \tag{4.37}$$

where Φ_n is the general normal distribution defined in (B.18),

$$\mu_n(t) = \ln(x_0) + \left(\mu_0 - 0.5\sigma_0^2\right)t,$$

and

$$(\sigma_n)^2(t) = \sigma_0^2 t.$$

Proof. Using the **probability inversion** Lemma B.19, the distribution for the solution $X(t)$ in (4.30) can be derived by reducing the distribution for $X(t)$ to the known one for the Wiener process $W(t)$ by inverting $X(t)$ in favor of $W(t)$. It is important here that $x_0 > 0$, $\sigma_0 > 0$, and that the natural logarithm $\ln(x)$ is an increasing function to preserve the direction of an inequality.

$$
\begin{aligned}
\Phi_{X(t)}(x) &= \mathrm{Prob}[X(t) \leq x] \\
&= \mathrm{Prob}\left[x_0 \exp\left((\mu_0 - 0.5\sigma_0^2)t + \sigma_0 W(t)\right) \leq x\right] \\
&= \mathrm{Prob}\left[\left(\mu_0 - 0.5\sigma_0^2\right)t + \sigma_0 W(t) \leq \ln(x/x_0)\right] \\
&= \mathrm{Prob}\left[W(t) \leq \left(\ln(x/x_0) - (\mu_0 - 0.5\sigma_0^2)t\right)/\sigma_0\right] \\
&= \Phi_{W(t)}\left(\left(\ln(x/x_0) - (\mu_0 - 0.5\sigma_0^2)t\right)/\sigma_0; 0, t\right) \\
&= \Phi_n\left(\ln(x); \ln(x_0) + \left(\mu_0 - 0.5\sigma_0^2\right)t, \sigma_0^2 t\right).
\end{aligned}
$$

The last step follows from the conversion identity from the standard Wiener distribution $\Phi_{W(t)}$ in (B.22) to the general normal distribution Φ_n, given for Φ_n in Exercise 9 on p. B70. Thus, the probability distribution of the solution $X(t)$ is the general **lognormal distribution** of Online Subsection B.1.6, where the exponent has the normal distribution mean

$$
\mu_n(t) = \ln(x_0) + \left(\mu_0 - 0.5\sigma_0^2\right)t
$$

and normal variance

$$
(\sigma_n)^2(t) = \sigma_0^2 t;
$$

i.e., the logarithm of the solution $X(t)$ has a general normal distribution, where the lognormal moment formulas are given in Properties B.20.   ☐

The probability density of $X(t)$ is found using the regular calculus chain rule by differentiating the distribution to yield

$$
\phi_{X(t)}(x) = x^{-1}\phi_n\left(\ln(x); \mu_n(t), (\sigma_n)^2(t)\right). \tag{4.38}
$$

Although the differentiation of the $\ln(x)$ distribution argument leads to an algebraic pole in $\phi_{X(t)}(x)$, $\phi_{X(t)}(0^+) \equiv 0$, which is in fact the limit as $x \to 0^+$. The leading part of the exponentially small normal distribution term $\exp(-\ln^2(x)/(2\sigma_0^2 t))$ dominates the simple algebraic pole $1/x = \exp(-\ln(x))$ as $x \to 0^+$ with the larger logarithmic exponent in magnitude.

4.1.5   Functions of General Diffusion States and Time: $F(X(t), t)$

The derivation for the special chain rule for the linear SDE logarithm transformation suggests that a more general chain rule for $F(X(t), t)$ will be needed.

Rule 4.13. *Chain Rule for Diffusion $F(X(t), t)$.*
Let $Y(t) = F(X(t), t)$, such that function $F(w, t)$ is twice continuously differentiable in x and once in t. Let the $X(t)$ process satisfy the diffusion SDE,

$$
dX(t) = f(X(t), t)dt + g(X(t), t)dW(t), \tag{4.39}
$$

$X(0) = x_0$ with probability one, while $f(X(t), t)$ and $g(X(t), t)$ satisfy the mean square integrability conditions (2.44) with the $W(t)$ argument replaced by the $X(t)$ arguments of

f and g. Then

$$dY(t) = dF(X(t), t)$$
$$\overset{dt}{=} \left(F_t + f F_x + \frac{1}{2} g^2 F_{xx} \right) (X(t), t) dt + (g F_x) (X(t), t) dW(t), \quad (4.40)$$

where wholesale arguments have been used for the coefficient functions multiplying dt and dW(t), respectively.

Sketch of Proof. Formally, using the increment form of the differential,

$$dY(t) = Y(t + dt) - Y(t)$$
$$= F(X(t + dt), t + dt) - F(X(t), t)$$
$$= F(X(t) + dX(t), t + dt) - F(X(t), t).$$

Next, mean square approximations are used with their implied precision-dt,

$$dY(t) \overset{dt}{=} F_t(X(t), t) dt + F_x(X(t), t) dX(t) + \frac{1}{2} F_{xx}(X(t), t)(dX)^2(t)$$

$$\overset{dt}{=} F_t(X(t), t) dt + F_x(X(t), t)(f(X(t), t) dt + g(X(t), t) dW(t))$$
$$+ \frac{1}{2} F_{xx}(X(t), t) g^2(X(t), t) dt$$

$$\overset{dt}{=} \left(F_t(X(t), t) + (f F_x)(X(t), t) + \frac{1}{2} \left(g^2 F_{xx} \right)(X(t), t) \right) dt$$
$$+ (g F_x)(X(t), t) dW(t),$$

where the diffusion SDE (4.39) has been substituted for $dX(t)$ and its square, the latter being truncated by the basic diffusion rule $(dW)^2(t) \overset{dt}{=} dt$ and other rules to neglect zero terms in the mean square, such as $(dW)^3(t)$, $dt dW(t)$, and $(dt)^2$, from the useful Table 2.1. □

4.2 Poisson Jump Process Calculus Rules

The Poisson process is quite different from the continuous diffusion process, primarily because of the discontinuity property of the Poisson process and the property that multiple jumps are highly unlikely during small increments in time Δt.

4.2.1 Jump Calculus Rule for $h(dP(t))$

Thus, the most basic rule is the *zero-one law (ZOL) for jumps* (1.36), in precision-dt compact differential form,

$$(dP)^m(t) \overset{dt}{\underset{zol}{=}} dP(t), \quad (4.41)$$

provided the integer $m \geq 1$, the case $m = 0$ being trivial. An immediate generalization of this law is the following corollary.

Corollary 4.14. *Zero-One Jump Law for* $h(dP(t))$.

$$h(dP(t)) \underset{zol}{\overset{dt}{=}} h(1)dP(t) + h(0)(1 - dP(t)), \tag{4.42}$$

with probability one, provided the function $h(p)$ *is right-continuous, such that values* $h(0)$ *and* $h(1)$ *exist and are bounded.*

Proof. This follows by simple substitution of the *zero-one jump law,*

$$h(dP(t)) \underset{zol}{\overset{dt}{=}} \begin{cases} h(1), & dP(t) = 1 \\ h(0), & dP(t) = 0 \end{cases} \underset{zol}{\overset{dt}{=}} h(1)dP(t) + h(0)(1 - dP(t)), \tag{4.43}$$

$dP(t) = 0$, or $dP(t) = 1$ with probability one to precision-dt. $\square$

Formally, the differential $dP(t)$ can be treated as a condition to test whether there has been a jump. This form (4.42) of the zero-one law suggests another extension of the **jump function definitions** (B.178), (B.179). For example, recall in (B.185) for a jump at t_1,

$$[F](X(t_1), t_1) = F(X(t_1^+), t_1^+) - F(X(t_1^-), t_1^-).$$

Definition 4.15. *Jump Function* $[h](dP(t))$.

$$[h](dP(t)) \underset{zol}{\overset{dt}{=}} h(dP(t)) - h(0) \tag{4.44}$$

to precision-dt, provided $h(p)$ is right-continuous, such that values $h(0)$ and $h(dP(t))$ exist and are bounded.

With this definition, version (4.42) of the **zero-one law** can immediately be written.

Corollary 4.16. *Zero-One Jump Law for* $h(dP(t))$ *with Jump Function.*

$$h(dP(t)) \underset{zol}{\overset{dt}{=}} h(0) + [h](dP(t)) \tag{4.45}$$

in terms of the jump function $[h](dP(t))$. Alternatively, the jump function is written as

$$[h](dP(t)) \underset{zol}{\overset{dt}{=}} (h(1) - h(0))dP(t). \tag{4.46}$$

4.2.2   Jump Calculus Rule for $\mathcal{H}(P(t), t)$

Equations (4.45), (4.46) are a primitive differential chain rule for functions of only the Poisson differential $dP(t)$. However, more complex rules will be needed, for instance, a chain rule for a combination of a simple Poisson jump process in $P(t)$ and a deterministic process with explicit dependence on t.

Rule 4.17. *Chain Rule for* $\mathcal{H}(P(t), t)$.
Let $\mathcal{H}(p, t)$ *be once continuously differentiable in t and right-continuous in p.*

$$d\mathcal{H}(P(t), t) \underset{zol}{\overset{dt}{=}} \mathcal{H}_t(P(t), t)dt + [\mathcal{H}](P(t), t), \qquad (4.47)$$

where

$$[\mathcal{H}](P(t), t) \underset{zol}{\overset{dt}{=}} (\mathcal{H}(P(t) + 1, t) - \mathcal{H}(P(t), t))\, dP(t) \qquad (4.48)$$

is the corresponding jump function definition for functions of $P(t)$ *and t.*

Sketch of Proof. Proceeding formally with differential precision-dt, the differential definition as an increment yields

$$d\mathcal{H}(P(t), t) = \mathcal{H}(P(t + dt), t + dt) - \mathcal{H}(P(t), t)$$
$$= \mathcal{H}(P(t) + dP(t), t + dt) - \mathcal{H}(P(t), t).$$

Next, using the zero-one jump law (4.42) for $h(dP(t))$ on

$$\mathcal{H}(P(t) + dP(t), t + dt)$$

for fixed $(P(t), t)$ to take $dP(t)$ out of its first argument, and then expanding the second argument dt to two terms up to $\mathcal{H}_t$,

$$d\mathcal{H}(P(t), t) \underset{zol}{\overset{dt}{=}} \mathcal{H}(P(t) + 1, t + dt)dP(t)$$
$$+ \mathcal{H}(P(t) + 0, t + dt)(1 - dP(t)) - \mathcal{H}(P(t), t)$$
$$\underset{zol}{\overset{dt}{=}} (\mathcal{H}(P(t), t) + \mathcal{H}_t(P(t), t)dt)(1 - dP(t))$$
$$+ (\mathcal{H}(P(t) + 1, t) + \mathcal{H}_t(P(t) + 1, t)dt)dP(t) - \mathcal{H}(P(t), t)$$
$$\underset{zol}{\overset{dt}{=}} \mathcal{H}_t(P(t), t)dt + (\mathcal{H}(P(t) + 1, t) - \mathcal{H}(P(t), t))dP(t)$$
$$\underset{zol}{\overset{dt}{=}} \mathcal{H}_t(P(t), t)dt + [\mathcal{H}](P(t), t),$$

the last line due to using the jump function definition (4.48). Also used was the bilinear differential form

$$dt\, dP(t) \underset{zol}{\overset{dt}{=}} 0,$$

which is mainly responsible for the elimination of combined continuous and jump changes.

The precision-dt jump differential Table 3.2 was used to eliminate terms smaller than precision-dt terms in the mean square sense. The dt factor $\mathcal{H}_t(p, t)$ is the partial derivative of $\mathcal{H}$ with respect to t while p is held fixed. Note that the jump function is defined for all t so that if there is no Poisson jump, then the jump function is identically zero since $dP(t) = 0$, the zero jump case. $\quad\square$

Remarks 4.18.

- *The bilinear differential form $dt\, dP(t) \overset{dt}{\underset{zol}{=}} 0$ is consistent with the fact that the Poisson process has jump discontinuities, and thus jumps must be instantaneous. Consequently, continuous changes and jump changes can be computed independently, since there are zero continuous changes at each jump instant.*

- *This leads to the alternate form of Rule 4.17.*

Rule 4.19. *Alternate Chain Rule for $\mathcal{H}(P(t), t)$.*
Let $\mathcal{H}(p, t)$ be once continuously differentiable in t and right-continuous in p.

$$d\mathcal{H}(P(t), t) \overset{dt}{\underset{zol}{=}} d_{(\text{cont})}\mathcal{H}(P(t), t) + d_{(\text{jump})}\mathcal{H}(P(t), t), \qquad (4.49)$$

where

$$d_{(\text{cont})}\mathcal{H}(P(t), t) \equiv \mathcal{H}_t(P(t), t)dt \qquad (4.50)$$

and

$$d_{(\text{jump})}\mathcal{H}(P(t), t) \equiv [\mathcal{H}](P(t), t). \qquad (4.51)$$

Example 4.20. *Stochastic Jump Power.*
Let $a \neq 0$ and $b > 0$. Using the stochastic jump chain rule (4.48) in differential form, we then have

$$d\left[b^{aP(t)+ct}\right] \overset{dt}{\underset{zol}{=}} c\ln(b)b^{aP(t)+ct}dt + \left(b^{a(P(t)+1)+ct} - b^{aP(t)+ct}\right)dP(t)$$

$$= b^{aP(t)+ct}\left(c\ln(b)dt + \left(b^a - 1\right)dP(t)\right),$$

where the calculus rule, $d(b^{ct}) = d(e^{c\ln(b)t}) = c\ln(b)b^{ct}$, for an arbitrary positive power base b with an exponential rule has been used.

The corresponding jump-integral derived from this formula is

$$\int_0^t b^{aP(s)+cs}dP(s) \overset{dt}{=} \frac{1}{b^a - 1}\left(\left(b^{aP(t)+ct} - 1\right) - c\ln(b)\int_0^t b^{aP(s)+cs}ds\right),$$

provided $b^a \neq 1$. This integral formula simplifies if $b = e$ and $c = 0$ to

$$\int_0^t \exp(aP(s))dP(s) \overset{dt}{=} (\exp(aP(t)) - 1)/(\exp(a) - 1)$$

but still is different from the deterministic version,

$$\int_0^t \exp(as)ds = (\exp(at) - 1)/a.$$

4.2.3   Jump Calculus Rule with General State $Y(t) = F(X(t), t)$

The chain rule for $F(P(t), t)$ is still too simple, so a chain rule for more general jump processes $X(t)$, such as for $F(X(t), t)$, is needed. First, a definition of a jump function for general transformations is needed.

Definition 4.21. $[Y](t)$ *for General* $Y(t) = F(X(t), t)$.
Let the process $Y(t) = F(X(t), t)$ *be a continuous transformation of the process* $X(t)$ *with jump function* $[X](t)$ *at t. Then the jump function in* $Y(t)$ *is defined as*

$$[Y](t) = [F](X(t), t) = F(X(t) + [X](t), t) - F(X(t), t). \qquad (4.52)$$

Lemma 4.22. $[Y](t)$ *for* $Y(t) = F(X(t), t)$ *with* $[X](t) = h(X(t), t)dP(t)$.
Let the process $Y(t) = F(X(t), t)$ *be a continuous transformation of the process* $X(t)$ *with jump function*

$$[X](t)h(X(t), t)dP(t)$$

at t; then

$$[Y](t) = [F](X(t), t) = (F(X(t) + h(X(t), t), t) - F(X(t), t)) \, dP(t). \quad (4.53)$$

Proof. This follows from the zero-one jump law (4.42) for $h(dP(t))$ upon substitution of the jump of $[X](t) = h(X(t), t)dP(t)$ into the definition (4.52), so that

$$\begin{aligned}
[Y](t) &\equiv F(X(t) + [X](t), t) - F(X(t), t) \\
&= F(X(t) + h(X(t), t)dP(t), t) - F(X(t), t) \\
&= (F(X(t) + h(X(t), t), t) - F(X(t), t)) \, dP(t). \quad \square
\end{aligned}$$

Rule 4.23. *Chain Rule for Jump in* $Y(t) = F(X(t), t)$.
Let $Y(t) = F(X(t), t)$, *such that the function* $F(x, t)$ *is continuously differentiable once in* x *and once in t. Let the* $X(t)$ *process satisfy the jump SDE,*

$$dX(t) = f(X(t), t)dt + h(X(t), t)dP(t), \qquad (4.54)$$

$X(0) = x_0$ *with probability one, while* $f(X(t), t)$ *and* $h(X(t), t)$ *satisfy the mean square integrability conditions* (2.44) *with the* $W(t)$ *argument replaced by the* $X(t)$ *arguments of* f *and* h. *In* (4.54), *the jump in* $X(t)$ *is* $[X](T_k^-) \equiv X(T_k^+) - X(T_k^-) = h(X(T_k^-), T_k^-)$ *for each kth jump-time* T_k *of* $P(t)$. *Then*

$$\begin{aligned}
dY(t) &= dF(X(t), t) \\
&\stackrel{dt}{\underset{zol}{=}} (F_t + f F_x) (X(t), t)dt + [F](X(t), t), \qquad (4.55)
\end{aligned}$$

where wholesale arguments have been used for the coefficient functions multiplying dt and dP(t), respectively, and where the jump in $Y(t) = F(X(t), t)$ *is given in* (4.53) *of Lemma 4.22.*

Sketch of Proof. Formally, a sketch of the proof uses the increment form of the differential

$$\begin{aligned}
dY(t) &= Y(t+dt) - Y(t) \\
&= F(X(t+dt), t+dt) - F(X(t), t) \\
&= F(X(t) + dX(t), t+dt) - F(X(t), t).
\end{aligned}$$

Next, as for (4.47), (4.49) of the two prior rules, the instantaneous jump changes (terms in $dP(t)$ only, such that $[X](t) = h(X(t), t)dP(t)$ are treated separately from the continuous and smooth deterministic changes (terms in dt only, such that $dX^{(\text{det})}(t) = f(X(t), t)dt$), then the mean square approximations are used with their implied precision-dt,

$$\begin{aligned}
dY(t) &\overset{dt}{\underset{zol}{=}} F_t(X(t), t)dt + F_x(X(t), t)f(X(t), t)dt \\
&\quad + (F(X(t) + [X](t), t) - F(X(t), t)) \\
&\overset{dt}{\underset{zol}{=}} (F_t + f F_x)(X(t), t)dt + (F(X(t) + h(X(t), t)dP(t), t) - F(X(t), t)) \\
&\overset{dt}{\underset{zol}{=}} (F_t + f F_x)(X(t), t)dt + (F(X(t) + h(X(t), t), t) - F(X(t), t))\, dP(t),
\end{aligned}$$

where the zero-one jump law (4.46) has been used to take the $dP(t)$ out of the argument of F and let it multiply the jump change in F in the last line of the above equation. Note that the jump change has been defined, so that if there is no Poisson jump, then the jump function is zero.   □

4.2.4 Transformations of Linear Jump with Drift SDEs

Consider the jump SDE, linear in the state process $X(t)$, with time-dependent coefficients,

$$dX(t) = X(t)\left(\mu(t)dt + \nu(t)dP(t)\right), \tag{4.56}$$

where here the initial condition is $X(t_0) = x_0 > 0$ with probability one, $\mu(t)$ is called the **drift** or deterministic coefficient, and $\nu(t)$ is called the **jump-amplitude** coefficient of the Poisson jump term. The jump in state is $[X](T_k) = \nu(T_k^-)$ for each jump of $P(t)$, i.e., $[P](T_k) = 1$ for each k. Assume that the rate coefficients $\mu(t)$ and $\nu(t)$ are bounded, while $\nu(t) > -1$. In the deterministic and linear diffusion cases, transforming the state variable to its logarithm makes the right-hand side independent of the transformed state variable, so let

$$Y(t) = F(X(t)) \equiv \ln(X(t)).$$

The most recent jump chain rule (4.55), (4.53) is applicable in this case with

$$f(X(t), t) = X(t)\mu(t)$$

and

$$h(X(t), t) = X(t)\nu(t),$$

although the increment form of $dF(X(t))$ can be directly expanded to get the same result. Since only the first partial derivative and the jump function of F are needed, while F does not depend on t, then

$$F_x(X(t)) = 1/X(t), \quad F_t(X(t)) \equiv 0,$$

and from (4.53)

$$[F](X(t)) \underset{zol}{\overset{dt}{=}} (\ln(X(t) + X(t)v(t)) - \ln(X(t)))\, dP(t) = \ln(1 + v(t))dP(t), \quad (4.57)$$

where the logarithm subtraction rule $\ln(A) - \ln(B) = \ln(A/B)$, provided $A > 0$ and $B > 0$, has been used to cancel out the linear state dependence in the jump term. Note that the jump-amplitude becomes singular as $v(t) \to (-1)^+$, approaching a massive disaster in the state. Thus,

$$dY(t) = dF(X(t)) = F_x(X(t))X(t)\mu(t)dt + [F](X(t))$$
$$\underset{zol}{\overset{dt}{=}} \mu(t)dt + \ln(1 + v(t))dP(t). \quad (4.58)$$

The infinitesimal mean of $Y(t)$, assuming the jump-rate is time-dependent $\mathrm{E}[dP(t)] = \lambda(t)dt$ too, is

$$\mathrm{E}[dY(t)] = (\mu(t) + \lambda(t)\ln(1 + v(t)))\, dt \quad (4.59)$$

and the **infinitesimal variance** is

$$\mathrm{Var}[dY(t)] \overset{dt}{=} \lambda(t)\ln^2(1 + v(t))dt, \quad (4.60)$$

noting that the jump-amplitude has a power effect between the infinitesimal expectation and the variance unlike the Poisson infinitesimal property that $\mathrm{Var}[dP(t)] = \mathrm{E}[dP(t)]$.

Since the final right-hand side of (4.58) does not depend on the state $Y(t)$, we can easily integrate for $Y(t)$ explicitly, leading to

$$Y(t) = Y(t_0) + \int_{t_0}^t \mu(s)ds + \int_{t_0}^t \ln(1 + v(s))dP(s). \quad (4.61)$$

Exponentiation leads to the formal solution for the original state,

$$X(t) = X(t_0) \exp\left(\int_{t_0}^t \mu(s)ds + \int_{t_0}^t \ln(1 + v(s))dP(s) \right). \quad (4.62)$$

Linear Jump SDEs with Constant Coefficients

If the SDE has **constant coefficients** $\mu(t) = \mu_0$, $v(t) = v_0$, and $\lambda(t) = \lambda_0$, then the solution is simpler:

$$X(t) \overset{ims}{=} X(t_0) \exp\left(\mu_0(t - t_0) + \ln(1 + v_0)(P(t) - P(t_0)) \right)$$
$$= X(t_0) \exp\left(\mu_0(t - t_0) \right)(1 + v_0)^{(P(t) - P(t_0))}, \quad (4.63)$$

where, in the last line, the exponential-logarithm inverse relation, $\exp(a\ln(b)) = b^a$, has been used to move the Poisson term out of the exponential.

In this pure jump with drift process, the moments are computed using the Poisson distribution (1.21) coupled with the stationary property that the distribution depends only on the increment,

$$\mathrm{Prob}[P(t) - P(t_0) = k] = \mathrm{Prob}[P(t - t_0) = k] = p_k(\lambda_0(t - t_0))$$
$$= e^{-\lambda_0(t-t_0)} \frac{(\lambda_0(t - t_0))^k}{k!}.$$

Thus, the calculation of the mean of the process in (4.63) is

$$
\begin{aligned}
\mathrm{E}[X(t)] &= x_0 e^{\mu_0(t-t_0)} e^{-\lambda_0(t-t_0)} \sum_{k=0} \frac{(\lambda_0(t-t_0))^k}{k!} (1+v_0)^k \\
&= x_0 e^{\mu_0(t-t_0) - \lambda_0(t-t_0)} e^{\lambda_0(t-t_0)(1+v_0)} \\
&= x_0 e^{(\mu_0 + \lambda_0 v_0)(t-t_0)},
\end{aligned}
$$

growing in time if $\mu_0 + \lambda_0 v_0 > 0$ but decaying if $\mu_0 + \lambda_0 v_0 < 0$. Note that $\lambda_0 > 0$, but both μ_0 and v_0 can be of any sign. The corresponding calculation of the variance of $X(t)$ is

$$
\begin{aligned}
\mathrm{Var}[X(t)] &= \mathrm{E}[X^2(t)] - \mathrm{E}^2[X(t)] \\
&= x_0^2 e^{2\mu_0(t-t_0)} e^{-\lambda_0(t-t_0)} \sum_{k=0} \frac{(\lambda_0(t-t_0))^k}{k!} (1+v_0)^{2k} - \mathrm{E}^2[X(t)] \\
&= x_0^2 e^{2\mu_0(t-t_0) - \lambda_0(t-t_0)} e^{\lambda_0(t-t_0)(1+v_0)^2} - x_0^2 e^{2(\mu_0+\lambda_0 v_0)(t-t_0)} \\
&= x_0^2 e^{2(\mu_0+\lambda_0 v_0)(t-t_0)} \left(e^{\lambda_0 v_0^2(t-t_0)} - 1 \right) \\
&= \mathrm{E}^2[X(t)] \left(e^{\lambda_0 v_0^2(t-t_0)} - 1 \right),
\end{aligned}
$$

so the growth or decay is proportional to the mean squared, but amplified asymptotically by the growing term $\exp(\lambda_0 v_0^2)(t-t_0)$. For the distribution, see Subsection 4.3.3 for the linear jump-diffusion SDE case.

Applications include stochastic population growth where $X(t)$ is the population size, such that the population grows exponentially at intrinsic growth rate $\mu(t)$ in the absence of stochastic disasters, but suffers from a random linear disaster if the jump-amplitude rate $-1 < v(t) < 0$ or from a random linear bonanza if $v(t) > 0$. See also Ryan and Hanson [241] or Chapter 11 on **biological applications**.

4.3   Jump-Diffusion Rules and SDEs

Wiener diffusion and simple Poisson jump processes provide an introduction to elementary SDEs in continuous time for the **simple jump-diffusion** state process $X(t)$,

$$
dX(t) = f(X(t), t)dt + g(X(t), t)dW(t) + h(X(t), t)dP(t), \tag{4.64}
$$

where $X(0) = x_0$, with a set of continuous coefficient functions $\{f, g, h\}$, possibly nonlinear in the state $X(t)$. However, in the process of introducing the component Markov processes, too many rules have been accumulated, and in this section most of these rules will be combined into one rule or a few rules.

4.3.1   Jump-Diffusion Conditional Infinitesimal Moments

The conditional infinitesimal moments for the state process are useful for application modeling and are given by

$$
\mathrm{E}[dX(t)|X(t) = x] = (f(x, t) + \lambda(t)h(x, t))dt \tag{4.65}
$$

and

$$\text{Var}[dX(t)|X(t) = x] = \left(g^2(x, t) + \lambda(t)h^2(x, t)\right) dt, \tag{4.66}$$

using (1.1), (1.2), (4.64) and assuming that the Poisson process is independent of the Wiener process.

The jump in the state at jumps T_k in the Poisson process, i.e., $[P](T_k) = 1$, is not an infinitesimal moment but serves as a simple property of the SDE and is given by

$$[X](T_k) \equiv X(T_k^+) - X(T_k^-) = h(X(T_k^-), T_k^-) \tag{4.67}$$

or

$$[X](t) = h(X(t), t)dP(t), \tag{4.68}$$

under the assumptions that the jumps are instantaneous so there are no time-continuous changes and that in the interval $(t, t + dt]$ there is time for only one jump, if any, of the Poisson term by the zero-one jump rule (1.35). Note that no $dP(t)$ appears in (4.67) since a jump is assumed at $t = T_k$. The jump-amplitude evaluation (4.67) at the prejump-time value T_k^- follows from the Itô forward integration approximation and the right-continuity of $P(t)$, as discussed in the previous chapter, and also means that the jump-amplitude depends only on the immediate prejump value of h but not on the postjump value, which in a sense is in the future.

The infinitesimal moment and jump properties are very useful for modeling approximations of real applications, by providing a basis for estimating the coefficient functions f, g, and h, as well as some of the process parameters, at least in the first approximation, through comparison to the empirical values of the basic probability corresponding to the stochastic integral equation.

4.3.2 Stochastic Jump-Diffusion Chain Rule

The corresponding stochastic chain rule for calculating the differential of a composite process $F(X(t), t)$ begins by interpreting the differential as an infinitesimal increment and recognizing that since the Poisson jumps are instantaneous there is no time for continuous changes. Thus, a critical concept in deriving the chain rule is that the continuous changes and jump changes can be calculated independently.

The state process is decomposed into continuous changes,

$$d_{(\text{cont})}X(t) = f(X(t), t)dt + g(X(t), t)dW(t), \tag{4.69}$$

and discontinuous or jump changes,

$$d_{(\text{jump})}X(t) = [X](t) = h(X(t), t)dP(t), \tag{4.70}$$

such that

$$dX(t) = d_{(\text{cont})}X(t) + d_{(\text{jump})}X(t). \tag{4.71}$$

Another critical concept is the transformation of the conditioning for the jump. The differential Poisson $dP(t)$ serves as the conditioning for the existence of a jump. This jump conditioning follows from the probability distribution for the differential Poisson process (1.23), which behaves asymptotically for small λdt as the zero-one jump law,

$$\Phi_{dP(t)}(k; \lambda dt) = \text{Prob}[dP(t) = k] = \begin{Bmatrix} 1 - \lambda dt, & k = 0 \\ \lambda dt, & k = 1 \\ 0, & k > 1 \end{Bmatrix} + \text{O}^2(\lambda dt), \qquad (4.72)$$

so that $dP(t)$ behaves as an **indicator function of the jump-counter** k with neglected error $\text{O}^2(dt) = \text{o}(dt)$, i.e., $dP(t) = 0$ with asymptotic probability $(1 - \lambda dt)$ if there is no jump and $dP(t) = 1$ with asymptotic probability (λdt) if there is a jump, while multiple jumps are likely to be negligible.

Thus, the change of a composite function of the state process $X(t)$, $dF(X(t), t)$, can be decomposed into the sum of continuous and discontinuous changes.

The function $F(x, t)$ is assumed to be at least twice continuously differentiable in x and once in t. Due to the nonsmoothness (1.6), a two-term Taylor approximation from continuous calculus yields, with subscripts denoting partial derivatives, the continuous change

$$d_{(\text{cont})}F(X(t), t) \simeq F_t(X(t), t)dt + F_x(X(t), t)d_{(\text{cont})}X(t)$$
$$+ \frac{1}{2}F_{xx}(X(t), t)(d_{(\text{cont})}X(t))^2,$$

which would be the chain rule for the compound function $F(X(t), t)$ of a deterministic function $X(t)$ with the nonsmooth property in (1.6). The discontinuous change follows from the transformation of the jump in $X(t)$ at time t given in the previous section to the jump in the composite function $Y(t) = F(X(t), t)$,

$$d_{(\text{jump})}F(X(t), t) = (F(X(t) + h(X(t), t), t) - F(X(t), t)) \, dP(t),$$

using the jump

$$[X](t) = h(X(t), t)dP(t)$$

and the continuity of F in t, such that when there is a jump at time T_k in $dP(t)$, the jump in F is evaluated at the prejump-time T_k^-; else the discontinuous contribution is zero. Combining the continuous and discontinuous process changes while neglecting nonzero terms of $\text{o}(dt)$ in the mean square limit sense yields

$$dF(X(t), t) = F(X(t) + dX(t), t + dt) - F(X(t), t)$$
$$= F_t(X(t), t)dt + F_x(X(t), t) \cdot (f(X(t), t)dt + g(X(t), t)dW(t))$$
$$+ \frac{1}{2}F_{xx}(X(t), t) \cdot g^2(X(t), t)dt \qquad (4.73)$$
$$+ (F(X(t) + h(X(t), t), t) - F(X(t), t))dP(t).$$

Rewriting (4.73) slightly leads to the final statement of the Itô stochastic chain rule for jump-diffusions with simple Poisson jumps.

Rule 4.24. *Jump-Diffusion Chain Rule or Itô's Lemma with Jumps.*
Let $F(x, t)$ be twice continuously differentiable in x and once in t.

$$dF(X(t), t) = \left(F_t + f F_x + \frac{1}{2} g^2 F_{xx} \right) (X(t), t) dt + (g F_x)(X(t), t) dW(t)$$
$$+ (F(X(t) + h(X(t), t), t) - F(X(t), t)) dP(t). \tag{4.74}$$

Here, to summarize, it is assumed that the Wiener process is **independent** of the Poisson processes and that the quadratic differential Wiener process $(dW)^2(t)$ can be replaced by its mean square limiting value which is dt within precision-dt. Thus, the part of the O(dt) change in dF due to the Wiener process requires a second derivative beyond the regular calculus first derivative Taylor approximation, and thus the nonsmooth Wiener property plays a strong role compared to its stochastic or random property. The second derivative term is a diffusion term and hence the Wiener process is called a diffusion process. However, the motivations for stochastic diffusions and physical diffusions are quite different, but they both lead to diffusion equations. The jump term uses the **zero-one jump indicator property** of $dP(t)$, so

$$[F](X(t), t) = F(X(t) + [X](t), t) - F(X(t), t)$$
$$= F(X(t) + h(X(t), t) dP(t), t) - F(X(t), t)$$
$$= (F(X(t) + h(X(t), t), t) - F(X(t), t)) dP(t)$$

to pass the jump differential $dP(t)$ from the state argument of $F(x, t)$ to a multiplying factor of the potential jump difference $F(x + h(x, t), t) - F(x, t)$. If there is a jump at $t = T_k$, then $dP(t)$ produces a change in the arguments $(X(t), t)$ of both F and h to $(x, t) = (X(T_k^-), T_k^-)$. However, if F and h are continuous in the explicit t-arguments, then $(x, t) = (X(T_k^-), T_k)$ can be used.

Remark 4.25. *Several authors use artificial arguments like $(X(t^-), t^-)$ when treating Markov jump process SDEs, or their corresponding integral equations, due to using an incomplete Poisson or related process model.*

4.3.3 Linear Jump-Diffusion SDEs

Let the linear jump-diffusion SDEs be combined into a single SDE,

$$dX(t) = X(t) (\mu(t) dt + \sigma(t) dW(t) + \nu(t) dP(t)), \tag{4.75}$$

$X(t_0) = x_0 > 0$ with probability one (this is for specificity, but only $x_0 \neq 0$ is sufficient), where the set of coefficients $\{\mu(t), \sigma(t), \nu(t), \lambda(t)\}$ is assumed to be bounded and integrable, with $\nu(t) > -1$ (otherwise, positivity of $X(t)$ cannot be maintained) and $\sigma(t) > 0$ (for consistency with the interpretation of $\sigma(t)$ as a standard deviation coefficient of the process). The logarithmic transformation of the state process $Y(t) = \ln(X(t))$ removes the state from the right-hand side of the SDE using the jump-diffusion chain rule (4.74) and the first two logarithmic derivatives, so

$$dY(t) = (\mu(t) - \sigma^2(t)/2) dt + \sigma(t) dW(t) + \ln(1 + \nu(t)) dP(t). \tag{4.76}$$

SDE (4.76) is a linear combination of the deterministic, diffusion, and jump processes with deterministic time-dependent coefficients and so can be immediately but formally integrated to yield

$$Y(t) = y_0 + \int_{t_0}^{t} \left((\mu(s) - \sigma^2(s)/2)ds + \sigma(s)dW(s) + \ln(1 + \nu(s))dP(s) \right), \quad (4.77)$$

where $y_0 = \ln(x_0)$, recalling that it has been assumed that $x_0 > 0$. Inverting logarithmic state $Y(t)$ back to the original state

$$X(t) = \exp(Y(t))$$

leads to

$$X(t) = x_0 \exp\left(\int_{t_0}^{t} \left((\mu(s) - \sigma^2(s)/2)ds + \sigma(s)dW(s) + \ln(1 + \nu(s))dP(s) \right) \right). \quad (4.78)$$

Linear Jump-Diffusion SDEs with Constant Coefficients

For the special case of constant rate coefficients, $\mu(t) = \mu_0$, $\sigma(t) = \sigma_0$, $\nu(t) = \nu_0$, and $\lambda(t) = \lambda_0$, also setting $t_0 = 0$, leads to the SDE

$$dX(t) = X(t) \left(\mu_0 dt + \sigma_0 dW(t) + \nu_0 dP(t) \right), \quad (4.79)$$

$X(t_0) = x_0 > 0$ with probability one with solution,

$$X(t) = x_0 \exp\left((\mu_0 - \sigma_0^2/2)t + \sigma_0 W(t) + \ln(1 + \nu_0)P(t) \right)$$
$$= x_0 (1 + \nu_0)^{P(t)} \exp\left((\mu_0 - \sigma_0^2/2)t + \sigma_0 W(t) \right), \quad (4.80)$$

applying the logarithm-exponential inverse property.

Using the density $\phi_{W(t)}(w)$ for the diffusion $W(t)$ in (1.7) and the discrete distribution $\Phi_{P(t)}(k) = p_k(\lambda_0 t)$ for the jump process $P(t)$, together with the pairwise independence of the two processes, the state expectation can be found directly as

$$E[X(t)] = x_0 e^{(\mu_0 - \sigma_0^2/2)t} e^{-\lambda_0 t} \sum_{k=0}^{\infty} \frac{(\lambda_0 t)^k}{k!} (1 + \nu_0)^k \frac{1}{\sqrt{2\pi t}} \int_{-\infty}^{+\infty} e^{-w^2/(2t)} e^{\sigma_0 w} dw$$

$$= x_0 e^{(\mu_0 - \sigma_0^2/2)t} e^{-\lambda_0 t} e^{\lambda_0 t(1 + \nu_0)} e^{\sigma_0^2 t/2}$$

$$= x_0 e^{(\mu_0 + \lambda_0 \nu_0)t}, \quad (4.81)$$

where the exponential series and **completing the square** technique have been used. It is interesting to note that the conditional infinitesimal expectation relative to the $X(t)$ for this constant coefficient case is

$$E[dX(t)|X(t)]/X(t) = (\mu_0 + \lambda_0 \nu_0)dt,$$

provided that the given condition value $X(t) \neq 0$, which means that if the above infinitesimal expected result is interpreted as implying the expected rate, then the state expectation in (4.81) is the same result as for the equivalent deterministic process. Note that the above equation is equivalent to $E[dX(t)|X(t) = x]/x = (\mu_0 + \lambda_0 \nu_0)dt$ with $x \neq 0$, but it is unnecessary to introduce the extra realized value x for $X(t)$, and later it will be seen that this extra introduction would be awkward in nested conditional expectations for stochastic dynamic programming in Chapter 6. For more on this **quasi-deterministic equivalence** for linear stochastic processes, see Hanson and Ryan [115].

Using similar applications of the same techniques, the state variance is computed to be

$$
\begin{aligned}
\mathrm{Var}[X(t)] &= \mathrm{E}\left[(X(t) - \mathrm{E}[X(t)])^2\right] = \mathrm{E}\left[X^2(t)\right] - \mathrm{E}^2[X(t)] \\
&= x_0^2 e^{2(\mu_0 - \sigma_0^2/2)t}\left(\mathrm{E}\left[e^{2\sigma_0 W(t)}(1 + \nu_0)^{2P(t)}\right] - \mathrm{E}^2\left[e^{\sigma_0 W(t)}(1 + \nu_0)^{P(t)}\right]\right) \\
&= x_0^2 e^{2(\mu_0 - \sigma_0^2/2)t}\left(e^{2\sigma_0^2 t}e^{\lambda_0 t((1+\nu_0)^2 - 1)} - e^{\sigma_0^2 t}e^{2\lambda_0 \nu_0 t}\right) \\
&= x_0^2 e^{2(\mu_0 + \lambda_0 \nu_0)t}\left(e^{(\sigma_0^2 + \lambda_0 \nu_0^2)t} - 1\right) \\
&= \mathrm{E}^2[X(t)]\left(e^{(\sigma_0^2 + \lambda_0 \nu_0^2)t} - 1\right).
\end{aligned}
\tag{4.82}
$$

The conditional infinitesimal variance relative to the square of the state, in this constant coefficient case, is

$$
\mathrm{Var}[dX(t)|X(t)]/X^2(t) = (\sigma_0^2 + \lambda_0 \nu_0^2)dt,
$$

provided $X(t) \neq 0$, which in turn is the time integral of the exponent, $(\sigma_0^2 + \lambda_0 \nu_0^2)t$, in the last line of (4.82) and since this exponent must be positive ($\lambda_0 > 0$), ensuring exponential amplification in time relative to the expectation exponential with exponent $((\mu_0 + \lambda_0 \nu_0)t)$, which could be of any sign. The usual measure of the relative changes of a random variable is called the **coefficient of variation**, which here is

$$
\mathrm{CV}[X(t)] \equiv \frac{\sqrt{\mathrm{Var}[X(t)]}}{\mathrm{E}[X(t)]} = \sqrt{e^{(\sigma_0^2 + \lambda_0 \nu_0^2)t} - 1},
\tag{4.83}
$$

provided $X(t) \neq 0$, which grows exponentially with time t. The $\mathrm{CV}[X(t)]$ is often used in the sciences to represent results, due to its dimensionless form. The dimensionless form makes it easier to pick out general trends or properties, especially if the $\mathrm{CV}[X(t)]$ can be distilled into something very simple.

The probability density for the solution $X(t)$ in (4.80) in the case of the constant coefficient, linear jump-diffusion SDE can be found by application of the **law of total probability** (B.92) and the probability inversion principle in Lemma B.19. Thus, assuming $x_0 > 0$ and $\sigma_0 > 0$,

$$
\begin{aligned}
\Phi_{X(t)}(x) &\equiv \text{Prob}[X(t) \le x] \\
&= \sum_{k=0}^{\infty} \text{Prob}\left[x_0 e^{(\mu_0 - 0.5\sigma_0^2)t + \sigma_0 W(t)}(1 + \nu_0)^{P(t)} \le x \,\Big|\, P(t) = k \right] \text{Prob}\left[P(t) = k \right] \\
&= \sum_{k=0}^{\infty} p_k(\lambda_0 t) \,\text{Prob}\left[x_0 e^{(\mu_0 - 0.5\sigma_0^2)t + \sigma_0 W(t)}(1 + \nu_0)^{k} \le x \right] \\
&= \sum_{k=0}^{\infty} p_k(\lambda_0 t) \,\text{Prob}\left[W(t) \le \left(\ln(x/x_0) - (\mu_0 - 0.5\sigma_0^2)t - k\ln(1 + \nu_0) \right)/\sigma_0 \right] \\
&= \sum_{k=0}^{\infty} p_k(\lambda_0 t) \,\Phi_{W(t)}\left(\left(\ln(x/x_0) - (\mu_0 - 0.5\sigma_0^2)t - k\ln(1 + \nu_0) \right)/\sigma_0 \right) \\
&= \sum_{k=0}^{\infty} p_k(\lambda_0 t) \,\Phi_n\left(\left(\ln(x/x_0) - (\mu_0 - 0.5\sigma_0^2)t - k\ln(1 + \nu_0) \right)/\sigma_0; 0, t \right) \\
&= \sum_{k=0}^{\infty} p_k(\lambda_0 t) \,\Phi_n\left(\ln(x); \ln(x_0) + (\mu_0 - 0.5\sigma_0^2)t + k\ln(1 + \nu_0),\, \sigma_0^2 t \right),
\end{aligned}
$$

where $\Phi_{W(t)}$ is the distribution of $W(t)$ (B.22) given in terms of the normal distribution Φ_n (B.18). The last step again follows from the conversion identity from standard to general normal distribution, given in Exercise 9 on p. B70. Thus, we have just proved the following jump-diffusion probability distribution theorem for the linear constant coefficient SDE by elementary probability principles.

Theorem 4.26. *Jump-Diffusion Probability Distribution for Linear Constant-Coefficient SDE.*
Let $X(t)$ formally satisfy the scalar, linear, constant coefficient SDE (4.79) with initial condition $X(0) = x_0 > 0$. Then for each value of the jump-counter k, the distribution is a sequence of distributions,

$$
\Phi_{X(t)}(x) = \sum_{k=0}^{\infty} p_k(\lambda_0 t)\Phi_{X(t)}^{(k)}(x),
$$

where each term of the sequence has the form

$$
\Phi_{X(t)}^{(k)}(x) = \Phi_n\left(\ln(x); \mu_n^{(k)}(t), \sigma_0^2 t \right),
$$

i.e., is a lognormal distribution (B.30) with normal mean

$$
\mu_n^{(k)}(t) = \ln(x_0) + \left(\mu_0 - 0.5\sigma_0^2 \right)t + k\ln(1 + \nu_0)
$$

and normal variance

$$
(\sigma_n)^2(t) = \sigma_0^2 t.
$$

For each k the logarithm of the solution $X(t)$ has a general normal distribution, where the lognormal moment formulas are given in Properties B.20. The probability density of $X(t)$ is found by chain-rule differentiating the distribution to yield

$$\phi_{X(t)}(x) = \sum_{k=0}^{\infty} p_k\,(\lambda_0 t)\,x^{-1}\phi_n\left(\ln(x);\,\mu_n^{(k)}(t),\,(\sigma_n)^2(t)\right) \tag{4.84}$$

for $x > 0$ such that $\phi_{X(t)}(0) \equiv \phi_{X(t)}(0^+) = 0$.

Remarks 4.27.

- *The fact $\phi_{X(t)}(0) \equiv \phi_{X(t)}(0^+) = 0$ is true because for the limit as $x \to 0^+$, the exponentially small normal distribution term dominates the simple algebraic pole from $1/x$.*

- *For each k, the normal mean is shifted by an amount $\ln(1 + \nu_0)$ and is weighted by the Poisson jump counting probability $p_k(\lambda_0 t) = \exp(-\lambda_0 t)(\lambda_0 t)^k/k!$, so the contributions decay like those of the exponential series.*

Solution Simulations for Linear Jump-Diffusion SDEs with Constant Coefficients

Upon merging and modifying the simulation algorithms for small time increments in Figure 1.1, and using the cumulative sum of normal random generated Wiener increment approximations, together with the cumulative sum of uniform random generated Poisson increment approximations with acceptance-rejection technique [230, 97] to model the zero-one jump law, we show simulations of the linear jump-diffusion process with constant parameters solution (4.80) in Figure 4.3. The basic simulation is performed on the approximate exponent increment

$$\Delta Y_i \simeq (\mu_0 - \sigma_0^2/2)\Delta t + \sigma_0 \Delta W_i + \ln(1 + \nu_0)\Delta P_i, \tag{4.85}$$

corresponding to SDE (4.85), where $\Delta t = 0.001$ for this MATLAB-generated figure,

$$\Delta W_i \simeq DW(i), \Delta t = Dt, \quad \text{where} \quad DW = \texttt{sqrt}(Dt)*\texttt{randn}(N, 1),$$

and

$$\Delta P_i \simeq DP(i) = U(DU(i); ul, ur), \quad \text{where} \quad DU = \texttt{rand}(N, 1)$$

for $i = 1 : N$ with $X(0) = x_0$, $U(u; ul, ur)$ is the unit step function on the centered interval $[ul = (1 - \lambda_0 \Delta t)/2, ur = (1 + \lambda_0 \Delta t)/2]$, approximating the zero-one jump law through the acceptance-rejection method [230, 97]. The sample path of the state exponent, YS, starting from a zero initial condition $YS(1) = 0$ rather than $\ln(x_0)$, for $i = 1 : N$, is approximated by

$$YS(i + 1) = YS(i) + (\mu_0 - \sigma_0^2/2)*Dt + \sigma_0*DW(i) + \texttt{log}(1 + \nu_0)*DP(i)$$

with $t(i + 1) = i * Dt$. The sample path of the desired state, XS, is approximated by

$$X(t(i + 1)) \simeq XS(i + 1) = x_0 * \exp(YS(i + 1)).$$

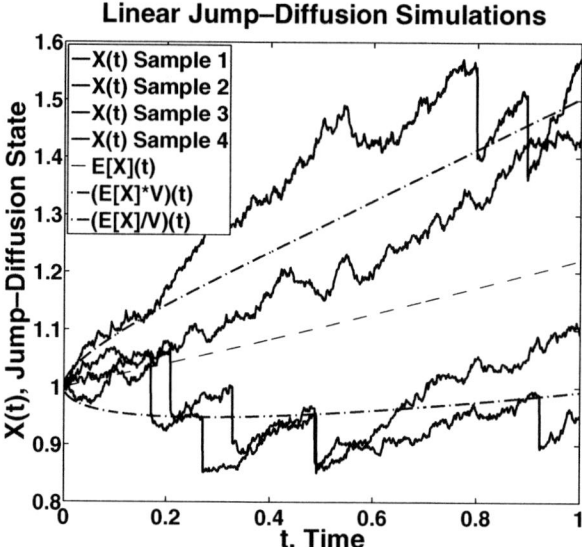

Figure 4.3. *Four linear jump-diffusion sample paths for constant coefficients are simulated using MATLAB [210] with $N = 1000$ sample points, maximum time $T = 1.0$, and four* randn *and four* rand *states. Parameter values are $\mu_0 = 0.5$, $\sigma_0 = 0.10$, $\nu_0 = -0.10$, $\lambda_0 = 3.0$, and $x_0 = 1.0$. In addition to the four simulated states, the expected state $\mathrm{E}[X(t)]$ and two deviation measures $\mathrm{E}[X(t)] * V(t)$ and $\mathrm{E}[X(t)]/V(t)$, are displayed where the factor $V(t)$ is based on the standard deviation of the state exponent $Y(t)$.*

The mean trajectory, XM, is given by

$$\mathrm{E}[X(t(i+1))] \simeq XM(i+1) = x_0 * \exp((\mu_0 + \lambda_0 * \nu_0)t(i+1))$$

is also displayed in the figure along with the upper XT exponential standard deviation estimate

$$\mathrm{E}[X(t(i+1))] * V(i+1) \simeq XT(i+1) = XM(i+1) * V(i+1)$$

and lower XB exponential standard deviation estimate

$$\mathrm{E}[X(t(i+1))]/V(i+1) \simeq XB(i+1) = XM(i+1)/V(i+1),$$

where the factor

$$V(i+1) = \exp(\sqrt{\mathrm{Var}[Y(t(i+1))]}) = \exp\left(\sqrt{(\sigma_0^2 + \lambda_0 * \log^2(1+\nu_0))t(i+1)}\right)$$

is the exponential of the standard deviation of the exponent process $Y(t)$ in discrete form. Alternatively, one plus or minus the coefficient of variation formula (4.83) could be used to form a deviation factor, but the factor $V(i+1)$ above is more appropriate since it corresponds better to the finite difference simulation approximation. Although the jump-amplitude is only a 10% decrement, the jumps are very noticeable in the figure, while both the jump and diffusion component processes result in excesses beyond the indicated upper and lower standard deviation estimates. The estimates correspond to rough confidence intervals and not bounds. See Program C.14, called $\mathtt{linjumpdiff03fig1.m}$ in the Online Appendix C, for the MATLAB code used to produce this figure.

The same code, in the case of constant parameters, can be used for the pure-diffusion model in the example (4.24) by setting $\nu_0 = 0$ for the diffusion as shown in Figure 4.4, or for the pure jump model in the example (4.56) by setting $\sigma_0 = 0$ for the jump process as shown in Figure 4.5.

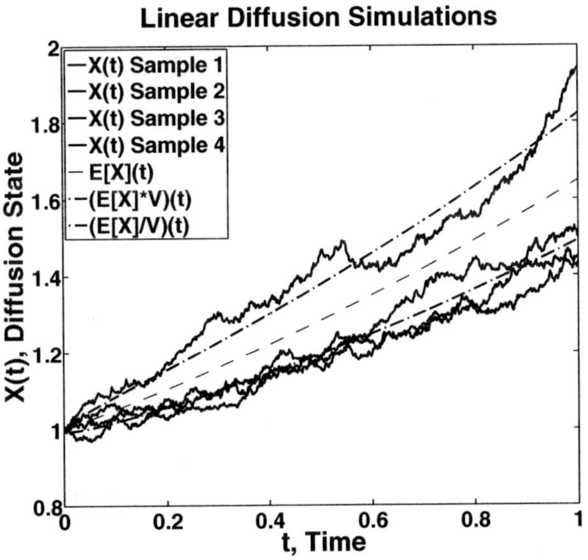

Figure 4.4. *Four linear pure-diffusion sample paths for constant coefficients are simulated using MATLAB [210] with $N = 1000$ sample points, maximum time $T = 1.0$, and four* $\mathtt{randn}$ *states. Parameter values are $\mu_0 = 0.5$, $\sigma_0 = 0.10$, $\nu_0 = 0.0$, and $x_0 = 1.0$. In addition to the four simulated states, the expected state $\mathrm{E}[X(t)]$ and two deviation measures $\mathrm{E}[X(t)] * V(t)$ and $\mathrm{E}[X(t)]/V(t)$ are displayed, where the factor $V(t)$ is based on the standard deviation of the state exponent $Y(t)$.*

Remarks 4.28.

- *Simulation caution: Note that the constant coefficient closed-form solution (4.80) is not used directly, i.e.,*

$$X_i = X_0(1 + \nu_0)^{P_i} \exp\left((\mu_0 - \sigma_0^2/2)t_i + \sigma_0 W_i\right)$$

for $i = 0 : (n+1)$, where $t_{n+1} = T$ is the final time, by directly simulating the random variables P_i and W_i, since they are not independent of either earlier or later values P_j and W_j for $j \neq i$. So such a simulation would be incorrectly approximated. However, simulating the increment set $\{\Delta P_i, \Delta W_i\}$ for $i = 0 : n$ would be an appropriate use

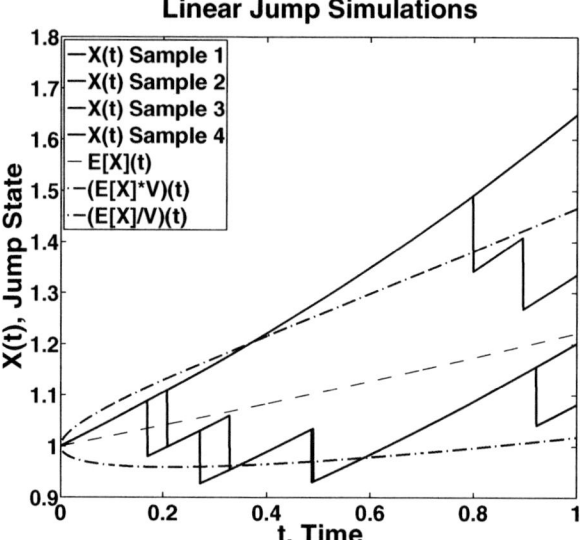

Figure 4.5. *Four linear pure jump with drift sample paths for constant coefficients are simulated using MATLAB [210] with $N = 1000$ sample points, maximum time $T = 1.0$, and four* randn *states. Parameter values are $\mu_0 = 0.5$, $\sigma_0 = 0.0$, $\nu_0 = -0.10$, and $x_0 = 1.0$. In addition to the four simulated states, the expected state $\mathrm{E}[X(t)]$ and two deviation measures $\mathrm{E}[X(t)] * V(t)$ and $\mathrm{E}[X(t)]/V(t)$ are displayed, where the factor $V(t)$ is based on the standard deviation of the state exponent $Y(t)$.*

of the approximate independence property of the pseudo-random number generators of ΔP_i and ΔW_i, i.e.,

$$X_{i+1} = X_i(1 + \nu_0)^{\Delta P_i} \exp\left((\mu_0 - \sigma_0^2/2)\Delta t_i + \sigma_0\Delta W_i\right)$$

for $i = 0 : n$, and we note that

$$\Delta Y_i = (\mu_0 - \sigma_0^2/2)\Delta t_i + \sigma_0\Delta W_i + \ln(1 + \nu_0)\Delta P_i$$

and that $\exp(\ln(1 + \nu_0)\Delta P_i) = (1 + \nu_0)^{\Delta P_i}$ using the exponential-logarithm inverse relationship. Considering finite precision arithmetic, this would be similar to using

$$W_{i+1} = \sum_{j=0}^{i} \Delta W_j \quad and \quad P_{i+1} = \sum_{j=0}^{i} \Delta P_j$$

for $(i + 1) = 1 : (n + 1)$.

*It is **important to build simulations in independent increments**.*

- *Other SDE codes can be found in the literature. Maple codes for jump-diffusions, along with higher approximations, can be found in the paper of Cyganowski, Grüne, and Kloeden [65]. In the numerical SDE tutorial review, D. Higham lists some very readable MATLAB codes modeled on techniques from the superb MATLAB guide of D. Higham and N. Higham. Both Maple and MATLAB codes for diffusion SDEs for finance, along with higher order approximations, can be found in D. Higham and Kloeden [144]. See also the recent probability and SDEs book of Cyganowski, Kloeden, and Ombach [67] for more on Maple codes for diffusions. For diffusion SDE codes in Mathematica see the computational finance-oriented book of Stojanovic [259]. Higher order, but older, diffusion SDE codes are found in the computational book of Kloeden, Platen, and Schurz [167] and are also used for the illustrations in the more theoretical treatise of Kloeden and Platen [166]. However, these codes are in Turbo-Pascal, a language not often used now.*

- *More computational SDE methods will be discussed in the compact Section 9.1. This section is a good introduction to SDE simulations for readers and instructors interested in exploring the topic further. Since the Itô forward integration uses the Euler or the tangent line method for purely deterministic processes, and Euler's method is perhaps the crudest numerical method for differential equations, higher order numerical methods are important when accuracy is important. Some sample codes are given in Section 9.1 and in Online Appendix C. See also [65, 144, 67, 140]. However, Euler's method is the most genuine application of Itô's stochastic integration for Markov processes in continuous time, although the simulation sample size should be large for reasonable representation of the stochastic processes.*

Linear Jump-Diffusion SDEs with Time-Dependent Coefficients

While linear constant coefficient SDEs often occur in applications such as elementary finance, time dependence of market parameters can also play an important role. For this reason, our attention returns to the time-dependent coefficients of the linear SDE solution (4.78) and the expected state trajectory. However, the procedure is more complex than in the simple constant coefficient case, since the expectations of exponentials of integrals are required. In the following two lemmas and related corollaries, we first consider the pure-diffusion case and then the pure jump case.

Lemma 4.29. *Expectation of* $\exp(\int \sigma \, dW(s))$.
Let $\sigma(t)$ be square integrable on $[t_0, t]$. Then

$$\mathrm{E}\left[\exp\left(\int_{t_0}^{t} \sigma(s)dW(s)\right)\right] \overset{\text{ims}}{=} \exp\left(\frac{1}{2}\int_{t_0}^{t} \sigma^2(s)ds\right). \tag{4.86}$$

Sketch of Proof. To keep the justification reasonably brief and maintain the usefulness as an integration technique, the stochastic diffusion integral will first be formally decomposed into a forward Itô sum, averaged, and then recomposed back into a deterministic integral. The justification of each step will be indicated in shorthand on the sign of the relation, but the more rigorous Itô limits will be omitted. Let $t_i = t_0 + i * \Delta t$ for $i = 0 : n+1$ be a proper partition of $[t_0, t]$ with $\Delta t = (t - t_0)/(n + 1)$, $\Delta W_i = W(t_{i+1}) - W(t_i)$, and $\sigma_i = \sigma(t_i)$ for $i = 0 : n$.

$$\mathrm{E}\left[\exp\left(\int_{t_0}^t \sigma(s)dW(s)\right)\right] \overset{ims}{\simeq} \mathrm{E}\left[\exp\left(\sum_{i=0}^n \sigma_i \Delta W_i\right)\right]$$

$$\overset{law}{\underset{exp}{=}} \mathrm{E}\left[\prod_{i=0}^n \exp(\sigma_i \Delta W_i)\right] \overset{ind}{\underset{inc}{=}} \prod_{i=0}^n \mathrm{E}_{\Delta W_i}[\exp(\sigma_i \Delta W_i)]$$

$$\overset{norm}{\underset{dist}{=}} \prod_{i=0}^n \int_{-\infty}^{+\infty} \frac{\exp\left(-\frac{w^2}{2\Delta t} + \sigma_i w\right)}{\sqrt{2\pi\Delta t}} dw \overset{comp}{\underset{sq}{=}} \prod_{i=0}^n \exp\left(\sigma_i^2 \Delta t/2\right)$$

$$\overset{ims}{\simeq} \exp\left(\frac{1}{2}\int_{t_0}^t \sigma^2(s)ds\right). \quad \square$$

Lemma 4.30. *Expectation of* $\exp(\int \ln(1 + v)dP(s))$.
Let $\lambda(t)v(t)$ *be integrable on* $[t_0, t]$. *Then*

$$\mathrm{E}\left[\exp\left(\int_{t_0}^t \ln(1 + v(s))dP(s)\right)\right] \overset{ims}{=} \exp\left(\int_{t_0}^t \lambda(s)v(s)ds\right). \tag{4.87}$$

Sketch of Proof. Again, to keep the justification reasonably brief and maintain the usefulness as an integration technique, the stochastic diffusion integral will first be formally decomposed into a forward Itô sum, averaged and then recomposed back into a deterministic integral. The justification of each step will be indicated in shorthand on the sign of the relation, but the more rigorous Itô limits will be omitted. Again, let $t_i = t_0 + i * \Delta t$ for $i = 0 : n+1$ be a proper partition of $[t_0, t]$ with $\Delta t = (t-t_0)/(n+1)$, $\Delta P_i = P(t_{i+1})-P(t_i)$, $\lambda_i = \lambda(t_i)$, and $v_i = v(t_i)$ for $i = 0 : n$.

$$\mathrm{E}\left[\exp\left(\int_{t_0}^t \ln(1 + v(s))dP(s)\right)\right] \overset{ims}{\simeq} \mathrm{E}\left[\exp\left(\sum_{i=0}^n \ln(1 + v_i)\Delta P_i\right)\right]$$

$$\overset{law}{\underset{exp}{=}} \mathrm{E}\left[\prod_{i=0}^n \exp(\ln(1 + v_i)\Delta P_i)\right] \overset{ind}{\underset{inc}{=}} \prod_{i=0}^n \mathrm{E}_{\Delta P_i}[\exp(\ln(1 + v_i)\Delta P_i)]$$

$$\overset{pois}{\underset{dist}{=}} \prod_{i=0}^n \sum_{k=0}^\infty e^{-\lambda_i \Delta t} \frac{\lambda_i \Delta t}{k!}(1 + v_i)^k \overset{exp}{\underset{sum}{=}} \prod_{i=0}^n e^{-\lambda_i \Delta t + \lambda_i \Delta t(1+v_i)}$$

$$\overset{ims}{\simeq} \exp\left(\int_{t_0}^t \lambda(s)v(s)ds\right). \quad \square$$

Using diffusion and jump Lemmas 4.29 and 4.30, the expectation of the state trajectory $X(t)$ (4.78) for the linear SDE with time-dependent coefficients (4.75) can be readily calculated, as follows.

Theorem 4.31. *Expectation of* $X(t)$ *in the Linear Jump-Diffusion SDE with Time-Dependent Coefficients Case.*

Let $\mu(t)$, $\sigma^2(t)$, and $\lambda(t)\nu(t)$ be integrable on $[t_0, t]$. Then

$$\mathrm{E}[X(t)] = \mathrm{E}\left[x_0 \exp\left(\int_{t_0}^{t} \left((\mu(s) - \sigma^2(s)/2)ds + \sigma(s)dW(s) + \ln(1 + \nu(s))dP(s)\right)\right)\right]$$

$$\stackrel{\mathrm{ims}}{=} x_0 \exp\left(\int_{t_0}^{t} (\mu(s) + \lambda(s)\nu(s))\, ds\right). \tag{4.88}$$

Proof. The proof is left as an algebraic exercise for the reader, using Lemmas 4.29 and 4.30. □

For the corresponding variance $\mathrm{Var}[X(t)]$, see Exercise 16 on p. 126 of this chapter. Note that the expectation and the variance results for the time-dependent case easily reduce to the linear SDE, constant coefficients results given in (4.81) for the expectation and (4.82) for the variance when the coefficients $\mu(t)$, $V(t)$, and $\sigma(t)$ become constants.

4.3.4 SDE Models Exactly Transformable to Purely Time-Varying Coefficients

In this section, a catalogue of exactly transformable jump-diffusion SDE models is given. First the notational correlations are listed for ease of interpreting the list of models and their transformations, where conditions are applicable.

SDE Models and Their Transformations:

- **Original SDE (4.64):**
  $$dX(t) = f(X(t), t)dt + g(X(t), t)dW(t) + h(X(t), t)dP(t).$$
- **Transformed process:** $Y(t) = F(X(t), t)$.
- **Transformed SDE:** $dY(t) = (F_t + F_x f + \frac{1}{2}F_{xx}g^2)dt + F_x g\, dW(t) + [F]dP(t)$.
- **Target explicit SDE:** $dY(t) = C_1(t)dt + C_2(t)dW(t) + C_3(t)dP(t)$.
- **Original target coefficient equations:**
  $$F_t + F_x f + \tfrac{1}{2}F_{xx}g^2 = C_1(t);$$
  $$F_x g = C_2(t);$$
  $$[F] \equiv F(x + h(x, t), t) - F(x, t) = C_3(t).$$
- **Original coefficients:**
  $$f(x, t) = (C_1(t) - F_t(x, t) - \tfrac{1}{2}F_{xx}(x, t)C_2^2(t)/F_x^2(x, t))/F_x(x, t);$$
  $$g(x, t) = C_2(t)/F_x(x, t);$$
  $$h(x, t) = -x + F^{-1}(F(x, t) + C_3(t)).$$

See Table 4.1 for examples.

Table 4.1. *Example transforms listing original coefficients in terms of target and transform coefficients.*

Transform $Y \to$ $F(x,t)$	Plant Coefficient $f(x,t)$	Gaussian Coefficient $g(x,t)$	Poisson Coefficient $h(x,t)$
x	$C_1(t)$	$C_2(t)$	$C_3(t)$
$a(t)x + b(t)$	$\frac{C_1(t)-a'(t)x-b'(t)}{a(t)}$	$\frac{C_2(t)}{a(t)}$	$\frac{C_3(t)}{a(t)}$
$a(t)x^2$	$\frac{C_1(t)-a'(t)x^2-\frac{C_2^2(t)}{4a(t)x^2}}{2a(t)x}$	$\frac{C_2(t)}{2a(t)x}$	$-x \pm \sqrt{x^2 + \frac{C_3(t)}{a(t)}}$
$\frac{a(t)}{x+b(t)}$	$\frac{C_2^2}{a^2}(x+b)^3 - \frac{C_1}{a}(x+b)^2 + \frac{a'}{a}(x+b) - b'$	$-\frac{C_2}{a}(x+b)^2$	$-\frac{C_3(x+b)^2}{C_3(x+b)+a}$
$(t)\frac{x+c(t)}{x+b(t)}$ $\{b \neq c\}$	$\frac{1}{b-c}\left(\frac{C_2^2(x+b)^3}{a^2(b-c)} + \frac{C_1}{a}(x+b)^2 + b'(x+c) - c'(x+b) - \frac{a'}{a}(x+b)(x+c) \right)$	$\frac{C_2}{a(b-c)}(x+b)^2$	$-\frac{C_3(x+b)^2}{C_3(x+b)-a(b-c)}$
$a(t)e^{b(t)x}$	$-\left(\frac{a'}{ab} + \frac{b'x}{b}\right) + \frac{C_1}{ab}e^{-bx} - \frac{1}{2}\frac{C_2^2}{a^2b}e^{-2bx}$	$\frac{C_2}{ab}e^{-bx}$	$\frac{1}{b}\ln\left(\frac{C_3}{a}e^{-bx} + 1\right)$
$a(t)\ln(x) + h(t)$	$\left(\frac{C_1}{a} + \frac{C_2^2}{2a^2} - \frac{a'}{a}\ln(x) - \frac{b'}{a}\right)x$	$\frac{C_2}{a}x$	$\left(e^{C_3/a} - 1\right)x$

In their theoretical and numerical treatise on SDEs, Kloeden and Platen [166, Section 4.4] list many exact solutions for diffusion SDEs. They also give a comprehensive treatment of convergence and stability of numerical approximations to solutions of SDEs that are well beyond this text.

4.4   Poisson Noise Is White Noise Too!

Noise can be rapid fluctuations or disturbances, so stochastic processes are sometimes called noise as well. Another typical feature of noise is that it contains many frequencies, so such noise can also be called **colored noise**. If the noise contains all frequencies, then it is called **white noise**, in analogy with white light, which contains all colors of the light spectrum.

There are two principal kinds of white noise in stochastic processes, **Gaussian white noise**, which occurs if the noise is normally distributed and **Poisson white noise**, which occurs if the noise is Poisson distributed. The whiteness of the noise relies heavily on the independent increment property.

However, many authors use the term white noise without qualification to refer to Gaussian white noise, perhaps because of unfamiliarity with the other main Markov process in continuous time that also is white noise, the Poisson process. An exception is Arnold [13], who treats mainly Gaussian white noise but properly mentions that Poisson noise is also white noise.

Consider the Gaussian case first. It is necessary to look at the covariance of the Wiener increments at different times relative to the time increment,

$$C_{\Delta W(t)}(h) \equiv \text{Cov}[\Delta W(t)/\Delta t, \Delta W(t+h)/\Delta t]$$
$$= \text{E}[\Delta W(t)\Delta W(t+h)]/(\Delta t)^2, \tag{4.89}$$

where $\Delta W(t) \equiv W(t+\Delta t) - W(t)$, the time increment $\Delta t > 0$ as usual, but $h \neq 0$. The covariance in (4.89) is also related to the correlation coefficient (B.140) between $\Delta W(t)$ and $\Delta W(t+h)$ (note that the reciprocal $1/\Delta t$ scales out of the correlation coefficient) by

$$\rho_{\Delta W(t)}(h) = \Delta t \cdot C_{\Delta W(t)}(h), \tag{4.90}$$

using $\sqrt{\text{Var}[\Delta W(t)]} = \sqrt{\text{Var}[\Delta W(t+h)]} = \sqrt{\Delta t}$. Since $W(t)$ is not differentiable, the finite difference approximation $\Delta W(t)/\Delta t$ is used in place of its rate or velocity, so we can eventually let $\Delta t \to 0^+$. Using the independent increments property and the zero-mean property $\text{E}[\Delta W(t)] = 0$, separating $\Delta W(t)$ and $\Delta W(t+h)$ into independent and common increments, we get

$$C_{\Delta W(t)}(h) = \frac{1}{(\Delta t)^2} \begin{cases} 0, & h \leq -\Delta t \\ \Delta t + h, & -\Delta t \leq h \leq 0 \\ \Delta t - h, & 0 \leq h \leq \Delta t \\ 0, & h \geq +\Delta t \end{cases}$$
$$= \frac{(\Delta t - |h|)}{(\Delta t)^2} U(h; -\Delta t, +\Delta t), \tag{4.91}$$

where $U(x; a, b)$ is the unit step function on $[a, b]$ and is used to give $C_{\Delta W(t)}(h)$ a more compact form.

Next, we seek the limiting generalized behavior of $C_{\Delta W(t)}(h)$ as $\Delta t \to 0^+$ by considering the integral of a sufficiently well-behaved "test" function, $F(h)$, and by using the step function representation in (4.91),

$$\int_{-\infty}^{+\infty} C_{\Delta W(t)}(h)F(h)dh = \int_{-\Delta t}^{+\Delta t} \frac{(\Delta t - |h|)}{(\Delta t)^2} F(h)dh$$
$$= \int_{-1}^{+1} (1 - |u|)F(u\Delta t)du$$
$$\to F(0) \int_{-1}^{+1} (1 - |u|)du = F(0), \tag{4.92}$$

where the change of variables $h = u\Delta t$ moved all Δt's into the argument of f, and subsequently an expansion retained the leading term and neglected errors of order Δt. Thus, we have the generalized form corresponding to the covariance of differential $dW(t)$,

$$C_{dW(t)}(h) = \delta(h), \tag{4.93}$$

where $\delta(h)$ is the Dirac delta function. Thus, Gaussian noise is also called **delta-correlated** and delta-correlation is closely connected with the notion of white noise, but note that the actual correlation coefficient (4.90) goes to zero as $\Delta t \to 0^+$.

Finally, to examine the frequency spectrum of $C_{dW(t)}(h)$, consider the power density spectrum using the Fourier transform

$$\mathcal{F}[C_{dW(t)}](k) = \frac{1}{\pi} \int_{-\infty}^{+\infty} e^{-ikh} C_{dW(t)}(h) dh$$

$$\stackrel{\text{gen}}{=} \frac{1}{\pi} \int_{-\infty}^{+\infty} e^{-ikh} \delta(h) dh = \frac{1}{\pi}, \tag{4.94}$$

which is certainly constant, so $C_{dW(t)}(h)$ has a flat frequency spectrum and thus represents an approximation to Gaussian white noise, $dW(t)$ being normally distributed.

Similarly, for the simple Poisson process, which we consider in the zero-mean form,

$$\Delta \widehat{P}(t) \equiv \Delta P(t) - \lambda_0 \Delta t,$$

where $\lambda_0 > 0$ is a constant jump-rate; then the covariance of the time-separated finite difference velocities is

$$\begin{aligned} C_{\Delta P(t)}(h) &\equiv \text{Cov}[\Delta \widehat{P}(t)/\Delta t, \Delta \widehat{P}(t+h)/\Delta t] \\ &= \text{E}[\Delta \widehat{P}(t) \Delta \widehat{P}(t+h)]/(\Delta t)^2 \\ &= \frac{\lambda_0(\Delta t - |h|)}{(\Delta t)^2} U(h; -\Delta t, +\Delta t) \\ &\stackrel{\text{gen}}{\longrightarrow} \lambda_0 \delta(h) \stackrel{\text{gen}}{=} C_{dP(t)}(h), \end{aligned} \tag{4.95}$$

taking a similar limit as $\Delta t \to 0^+$ as with $C_{\Delta W(t)}(h)$ above. Hence, Poisson noise is also **delta-correlated**. For the Poisson increment process, recalling $\text{Var}[\Delta P(t)] = \text{Var}[\Delta \widehat{P}(t)] = \lambda_0 \Delta t$, the corresponding correlation coefficient is

$$\rho_{\Delta P(t)}(h) = \Delta t \cdot C_{\Delta P(t)}(h)/\lambda_0. \tag{4.96}$$

Finally, taking the Fourier transform of $C_{dP(t)}(h)$,

$$\mathcal{F}[C_{dP(t)}](k) = \frac{1}{\pi} \int_{-\infty}^{+\infty} e^{-ikh} C_{dP(t)}(h) dh$$

$$\stackrel{\text{gen}}{=} \lambda_0 \frac{1}{\pi} \int_{-\infty}^{+\infty} e^{-ikh} \delta(h) dh = \lambda_0 \frac{1}{\pi}, \tag{4.97}$$

which is also a constant so that **Poisson noise is also white noise**.

4.5 Exercises

1. Derive the Itô stochastic integral formulas for

$$\int_0^t \cos(aW(s)) dW(s) \quad \text{and} \quad \int_0^t \sin(aW(s)) dW(s),$$

where $a \neq 0$ and is a real constant. Also, derive the results when $a = 0$.

2. Find $X(t)$ if

$$\int_0^t X(s)dP(s) = b^{P(t)}\ln(aP(t)+c),$$

where $a > 0$, $b > 0$, and $c > 0$ are real constants.

3. Derive the following, using stochastic calculus:

   (a) $\int_0^t \sin(\pi P(s))dP(s) \overset{dt}{=} -\frac{1}{2}\sin(\pi P(t))$.

   (b) $\int_0^t \cos(\pi P(s))dP(s) \overset{dt}{=} \frac{1}{2}(1 - \cos(\pi P(t)))$.

   (*Hint: You may need some elementary trigonometric identities.*)

4. Consider the simple linear jump-diffusion SDE,

$$dX(t) = (\mu_d dt + \sigma_d dW(t) + J dP(t))X(t),$$

where the $\{\mu_d, \sigma_d, \mu_j, \sigma_j, \lambda_0\}$ are constants and λ_0 is the Poisson jump rate, while μ_j is the mean of the jump-amplitude J and σ_j^2 is the jump-amplitude variance. The diffusion process $W(t)$ is independent of the jump process $P(t)$ and the jump-amplitude J independent of $P(t)$ conditioned on a jump of $P(t)$. Show that the conditional infinitesimal mean is given by

$$E[dX(t)|X(t) = x] = (\mu_d + \lambda_0\mu_j)xdt$$

and the conditional infinitesimal variance is given by

$$\text{Var}[dX(t)|X(t) = x] \overset{dt}{=} (\sigma_d^2 + \lambda_0(\sigma_j^2 + \mu_j^2))x^2dt,$$

explaining why equality in dt-precision (see Chapter 1 for a definition) is required in the latter but not in the former conditional moment.

5. Show that the **power rules** for (Itô) stochastic integration for Wiener noise can be written as the recursion

$$\int_0^t W^m(s)dW(s) = \frac{1}{m+1}W^{m+1}(t) - \frac{m}{2}\int_0^t W^{m-1}(s)ds. \qquad (4.98)$$

   (a) Illustrate the application of the formula to find the results for the cases $m = 2$ and $m = 3$.

   (b) Alternatively, using the (Itô) stochastic chain rule and mathematical induction, show the general result.

6. Solve the following (Itô) diffusion SDE for $X(t)$, $E[X(t)]$, and $\text{Var}[X(t)]$:

$$dX(t) = \left(a\sqrt{X(t)} + b^2/4\right)dt + b\sqrt{X(t)}dW(t),$$

where a and b are real constants, and $X(0) = x_0 > 0$, with probability one.
(*Hint: Seek a transformation $Y(t) = f(X(t))$ for some f such that $Y(t)$ satisfies a constant coefficient SDE.*)

7. Solve the following (Itô) diffusion SDE for $X(t)$, $E[X(t)]$, and $Var[X(t)]$:

$$dX(t) = \left(aX^2(t) + b^2X^3(t)\right)dt + bX^2(t)dW(t),$$

where a and b are real constants, and $X(0) = x_0 > 0$, with probability one. (*Hint: Seek a transformation* $Y(t) = f(X(t))$ *for some* f *such that* $Y(t)$ *satisfies a constant coefficient SDE.*)

8. Solve the following diffusion SDE for $X(t)$ and $E[X(t)]$:

$$dX(t) = \left(aX^{3/4}(t) + \frac{3}{8}b^2X^{1/2}(t)\right)dt + bX^{3/4}(t)dW(t),$$

where a and b are real constants, and $X(0) = x_0 > 0$, with probability one. (*Hint: Find a power transformation to convert the SDE to a constant coefficient SDE.*)

9. Solve the following (Itô) jump SDE for $X(t)$, $E[X(t)]$, and $Var[X(t)]$:

$$dX(t) = -aX^2(t)dt - \frac{cX^2(t)}{1 + cX(t)}dP(t),$$

where $E[P(t)] = \lambda_0 t$ and $X(0) = x_0 > 0$, with probability one, while $a > 0$, $b > 0$, and $\lambda_0 > 0$ are constants. The answer may be left as a sum over the Poisson distribution.

(*Hint: Seek a transformation* $Y(t) = f(X(t))$ *for some* f *such that* $Y(t)$ *satisfies a constant coefficient SDE.*)

10. Solve the following Poisson jump SDE for $X(t)$ and $E[X(t)]$:

$$dX(t) = a\sqrt{X(t)}dt + b\left(b + 2\sqrt{X(t)}\right)dP(t),$$

where $E[P(t)] = \lambda_0 t$ and $X(0) = x_0 > 0$, with probability one, while λ_0, a, and b are real constants. (*Hint: Find a power transformation to convert the SDE to a constant coefficient SDE.*)

11. Show that the (Itô) jump-diffusion SDE for $X(t)$,

$$dX(t) = f(X(t))dt + bX^r(t)dW(t) + h(X(t))dP(t),$$

can be transformed by $Y(t) = F(X(t))$ to a **constant coefficient SDE**, where $E[P(t)] = \lambda_0 t$ and $X(0) = x_0 > 0$, with probability one, while λ_0, b, and $r \neq 0$ are real constants. In a proper answer, the power forms of $f(X(t))$ and $h(X(t))$ must be derived from the constant coefficient SDE condition.

12. **Martingales:** A **martingale** in continuous time satisfies the essential property that

$$E[M(t)|M(s)] = M(s) \tag{4.99}$$

for all $0 \leq s < t$ provided its absolute value has finite expectation, $E[|M(t)|] < \infty$ for all $t \geq 0$ plus some other technical properties. (See Mikosch [209], for instance.)

Driftless Log-Linear Process $\Longrightarrow$ Martingale?
Show directly that

$$M_1(t) = \ln(X(t)) - \mathrm{E}[\ln(X(t))] \tag{4.100}$$

is a martingale using that $Y(t) = \ln(X(t))$ symbolically satisfies the solution to a general linear SDE transformed to state-independent, time-dependent form (4.76).

Remark 4.32. *This type of problem is applicable to many financial problems where the return on a linear financial asset $X(t)$ is transformed into a log-return form $\ln(X(t))$, forming an SDE with state-independent coefficients, so the driftless deviation $M_1(t)$ form in (4.100) is a martingale, a log-martingale. However, readers must be aware of all the assumptions involved. See the next exercise.*

13. **Driftless $\Longrightarrow$ Martingale?**
    Prove the following theorem, explicitly justifying every step where an underlined theorem assumption or expectation property is needed.

    Theorem 4.33. *If the **Markov** process $X(t)$ is **driftless** (i.e., $\mathrm{E}[X(t)] = 0$) and has **independent increments** (along with the boundedness and technical condition cited with (4.99)), then $X(t)$ is a martingale.*

    Remark 4.34. *Readers must be aware of the direction of the implication. For example, Hull [148, p. 507] states, "A martingale is a zero-drift stochastic process," while Baxter and Rennie [22, p. 79] state, "X is a martingale $\Longleftrightarrow$ X is driftless ($\mu_t \equiv 0$)," yet all the assumptions are not apparent. For example, the state independence of the SDE for the log-return is in the background.*

14. **Exponential-martingale counterexample to driftless martingale requirement.**

    (a) Derive the nonrandom function $\beta(t)$ that makes

    $$M_2(t) = \beta(t)X(t)$$

    a martingale if $X(t)$ symbolically satisfies the linear SDE (4.75).

    (b) Show that $M_2(t)$ is not driftless, i.e.,

    $$\mathrm{E}[M_2(t)] \neq 0,$$

    in the absence of trivial initial conditions, i.e., $x_0 \neq 0$.

    Remark 4.35. *This is a counterexample showing that if $M(t)$ is a martingale, then it in not necessarily a driftless process.*

15. **General exponential-expectation interchange formula for linear jump-diffusions.**
    Formally show that

$$E\left[\exp\left(\int_0^t d\ln(X)(s)\right)\right] = \exp\left(E\left[\int_0^t \left(\frac{dX}{X}\right)(s)\right]\right) \qquad (4.101)$$

    if $X(t)$ is a linear jump-diffusion process (4.75), verifying that both sides of (4.101)
    are equivalent. Assume all integrals of process coefficients are bounded.

16. For the solution $X(t)$ (4.78) of the linear SDE (4.75) with time-dependent coefficients,
    assuming all integrals of process coefficients are bounded,

    (a) calculate the expectation of the quadratic of the exponential of the diffusion
        integral in Lemma 4.29 by transforming the results of the lemma or by using
        the same techniques as in the lemma;

    (b) calculate the expectation of the quadratic of the exponential of the jump-integral
        in Lemma 4.30 by transforming the results of the lemma;

    (c) use the result of the first two parts of this exercise and the expectation theorem,
        Theorem 4.31 to prove the next theorem.

    Theorem 4.36. *Variance of $X(t)$ in the Linear SDE with Time-Dependent
    Coefficients Case.*

    Let $\mu(t)$, $\sigma^2(t)$, and $\lambda(t)\nu^j(t)$ for $j = 1 : 2$ be integrable on $[t_0, t]$. Then

$$\mathrm{Var}[X(t)] \overset{\text{ims}}{=} x_0^2 \exp\left(2\int_{t_0}^t (\mu(s) + \lambda(s)\nu(s))\,ds\right)$$
$$\cdot\left(\exp\left(\int_{t_0}^t \left(\sigma^2(s) + \lambda(s)\nu^2(s)\right) ds\right) - 1\right) \qquad (4.102)$$

    for the state trajectory $X(t)$ given in (4.78).

Suggested References for Further Reading

- Cyganowski, Grüne, and Kloeden, 2002 [65]

- Cyganowski, Kloeden, and Ombach, 2002 [67]

- Gard, 1988 [92]

- Glasserman, 2003 [97]

- D. Higham and Kloeden, 2002 [144] and 2005 [145]

- Jazwinski, 1970 [155]

- Karlin and Taylor, 1981 [163]

- Klebaner, 1998 [165]

- Kloeden and Platen, 1992 [166]

- Kloeden, Platen, and Schurz, 1994 [167]

- Mikosch, 1998 [209]

- Øksendal, 1998 [222]

- Schuss, 1980 [244]

- Shreve, 2004 [248]

- Snyder and Miller, 1991 [252]

- Taylor and Karlin, 1998 [265]

- Tuckwell, 1995 [270]

- Wonham, 1970 [286]

Stochastic Calculus for General Markov SDEs: Space-Time Poisson, State-Dependent Noise, and Multidimensions

*Not everything that counts can be counted,
and not everything that can be counted counts.*
—Albert Einstein (1879–1955)

*The only reason for time is so that everything doesn't happen
at once.*
—Albert Einstein at `http://www.brainyquote.com/`
`quotes/authors/a/albert_einstein.html`

*Time is nature's way of keeping everything from happening at once.
Space is what prevents everything from happening to me.*
—attributed to John Archibald Wheeler
`http://en.wikiquote.org/wiki/Time`

What about stochastic effects?
—Don Ludwig, University of British Columbia, printed on
his tee shirt to save having to ask it at each seminar

*We are born by accident into a purely random universe.
Our lives are determined by entirely fortuitous combinations
of genes. Whatever happens happens by chance. The
concepts of cause and effect are fallacies. There is only
seeming causes leading to apparent effects. Since nothing
truly follows from anything else, we swim each day through
seas of chaos, and nothing is predictable, not even the events
of the very next instant.*

Do you believe that?

*If you do, I pity you, because yours must be a bleak and
terrifying and comfortless life.*
—Robert Silverberg in *The Stochastic Man*, 1975

This chapter completes the generalization of Markov noise in continuous time for this book, by including space-time Poisson noise, state-dependent SDEs, and multidimensional SDEs.

5.1   Space-Time Poisson Process

Space-time Poisson processes are also called general compound Poisson processes, marked Poisson point processes, and Poisson noise with randomly distributed jump-amplitudes, conditioned on a Poisson jump in time. The marked adjective refers to the marks which are the underlying stochastic process for the Poisson jump-amplitude or the space component of the space-time Poisson process, whereas the jump amplitudes of the simple Poisson process are deterministic or fixed with unit magnitude. The space-time Poisson process is a generalization of the Poisson process. The space-time Poisson process formulation helps in understanding the mechanism for applying it to more general jump applications and generalization of the chain rules of stochastic calculus.

Properties 5.1.

- **Space-time Poisson differential process:** *The basic space-time or mark-time Poisson differential process denoted as*

$$d\Pi(t) = \int_Q h(t, q)\mathcal{P}(\mathbf{dt}, \mathbf{dq}) \tag{5.1}$$

*on the **Poisson mark space** Q can be defined using the **Poisson random measure** $\mathcal{P}(\mathbf{dt}, \mathbf{dq})$, which is shorthand measure notation for the measure-set equivalence $\mathcal{P}(\mathbf{dt}, \mathbf{dq}) = \mathcal{P}((t, t + dt], (q, q + dq])$. The jump-amplitude $h(t, q)$ is assumed to be continuous and bounded in its arguments.*

- **Poisson mark Q:** *The space Poisson mark Q is the underlying IID random variable for the mark-dependent jump-amplitude coefficient denoted by $h(t, Q) = 1$, i.e., the space part of the space-time Poisson process. The realized variable $Q = q$ is used in expectations or conditional expectations, as well as in definition of the type (5.1).*

- **Time-integrated, space-time Poisson process:**

$$\Pi(t) = \int_0^t \int_Q h(t, q)\mathcal{P}(\mathbf{dt}, \mathbf{dq})dt. \tag{5.2}$$

- **Unit jumps:** *However, if the jumps have unit amplitudes, $h(t, Q) \equiv 1$, then the space time process in (5.1) must be the same result as the simple differential Poisson process $dP(t; Q)$ modified with a mark parameter argument to allow for generating mark realizations, and we must have the equivalence*

$$\int_Q \mathcal{P}(\mathbf{dt}, \mathbf{dq}) \equiv dP(t; Q), \tag{5.3}$$

*giving the jump number count on $(t, t + dt]$. Integrating both sides of (5.3) on $[0, t]$
gives the jump-count up to time t,*

$$\int_0^t \int_Q \mathcal{P}(\mathbf{dt}, \mathbf{dq}) = \int_0^t dP(s; Q) = P(t; Q). \tag{5.4}$$

*Further, in terms of Poisson random measure $\mathcal{P}(\mathbf{dt}, \{1\})$ on the fixed set $Q = \{1\}$,
purely the number of jumps in $(t, t + dt]$ is obtained,*

$$\int_Q \mathcal{P}(\mathbf{dt}, \mathbf{dq}) = \mathcal{P}(\mathbf{dt}, \{1\}) = P(\mathbf{dt}) = dP(t; 1) \equiv dP(t),$$

and the marks are irrelevant.

- **Purely time-dependent jumps:** *If $h(t, Q) = h_1(t)$, then*

$$\int_Q h_1(t)\mathcal{P}(\mathbf{dt}, \mathbf{dq}) \equiv h_1(t)dP(t; Q). \tag{5.5}$$

- **Compound Poisson process form:** *An alternate form of the space-time Poisson
process (5.2) that many may find more comprehensible is the marked generalization
of the* **simple Poisson process** $P(t; Q)$, **with IID random mark generation**, *that is,
the counting sum called the* **compound Poisson process** *or* **marked point process**,

$$\Pi(t) = \sum_{k=1}^{P(t;Q)} h(T_k^-, Q_k), \tag{5.6}$$

*where $h(T_k^-, Q_k)$ is the kth jump-amplitude, T_k^- is the prejump value of the kth random
jump-time, Q_k is the corresponding random jump-amplitude mark realization, and
for the special case that $P(t; Q)$ is zero the following reverse-sum convention is used,*

$$\sum_{k=1}^{0} h(T_k^-, Q_k) \equiv 0 \tag{5.7}$$

for any h. The corresponding differential process has the expectation

$$\mathrm{E}[dP(t; Q)] = \lambda(t)dt,$$

*although it is possible that the jump-rate is mark-dependent (see [223], for example)
so that*

$$\mathrm{E}[dP(t; Q)] = \mathrm{E}_Q[\lambda(t; Q)]dt.$$

*However, it will be assumed here that the jump-rate is mark-independent to avoid
complexities with iterated expectations later.*

• **Zero-one law compound Poisson differential process form:** *Given the Poisson compound process form in (5.6), the corresponding* **zero-one jump law** *for the compound Poisson differential process is*

$$d\Pi(t) = h(t, Q)dP(t; Q) \tag{5.8}$$

such that the jump in $\Pi(t)$ at $t = T_k$ is given by

$$[\Pi](T_k) \equiv \Pi(T_k^+) - \Pi(T_k^-) = h(T_k^-, Q_k). \tag{5.9}$$

For consistency with the Poisson random measure and compound Poisson process forms, it is necessary that

$$\int_0^t h(s, Q)dP(s; Q) = \int_0^t \int_Q h(s, q)\mathcal{P}(ds, dq) = \sum_{k=1}^{P(t; Q)} h(T_k^-, Q_k),$$

so

$$\int_0^t dP(s; Q) = \int_0^t \int_Q \mathcal{P}(ds, dq) = P(t; Q)$$

and

$$dP(t; Q) = \int_Q \mathcal{P}(dt, dq).$$

Note that the selection of the random marks depends on the existence of the Poisson jumps and that the mechanism is embedded in $dP(t; Q)$ in the formulation of this book.

• *In the* **Poisson random measure notation** $\mathcal{P}(\mathbf{dt}, \mathbf{dq})$, *the arguments* **dt** *and* **dq** *are semiclosed subintervals when these arguments are expanded,*

$$\mathcal{P}(\mathbf{dt}, \mathbf{dq}) = \mathcal{P}((t, t + dt], (q, q + dq]).$$

These subintervals are closed on the left and open on the right due to the definition of the increment, leaving no overlap between differential increments and corresponding to the simple Poisson right continuity property that

$$\Delta P(t; Q) \to P(t^+; Q) - P(t; Q) \text{ as } \Delta t \to 0^+,$$

so we can write $\Delta P(t; Q) = P((t, t + \Delta t]; Q)$ and $dP(t; Q) = P((t, t + dt]; Q)$. When $t_n = t$ and $t_{i+1} = t_i + \Delta t_i$, the covering set of intervals is $\{[t_i, t_i + \Delta t_i) \text{ for } i = 0:n\}$ plus t. If the marks Q are continuously distributed, then closed subintervals can also be used in the q argument. For the one-dimensional mark space Q, Q can be a finite interval such as $Q = [a, b]$ or an infinite interval such as $Q = (-\infty, +\infty)$. Also, these subintervals are convenient in partitioning continuous intervals since they avoid overlap at the nodes.

- $\mathcal{P}$ has **independent increments** on nonoverlapping intervals in time t and marks q, i.e., $\mathcal{P}_{i,k} = \mathcal{P}((t_i, t_i + \Delta t_i], (q_k, q_k + \Delta q_k])$ is independent of $\mathcal{P}_{j,\ell} = \mathcal{P}((t_j, t_j + \Delta t_j], (q_\ell, q_\ell + \Delta q_\ell])$, provided that the time-interval $(t_j, t_j + \Delta t_j]$ has no overlap with $(t_i, t_i + \Delta t_i]$ and the mark interval $(q_k, q_k + \Delta q_k]$ has no overlap with $(q_\ell, q_\ell + \Delta q_\ell]$. Recall that $\Delta P(t_i; Q) \equiv P(t_i + \Delta t_i; Q) - P(t_i; Q)$ is associated with the time-interval $(t_i, t_i + \Delta t_j]$, open on the left since the process $P(t_i; Q)$ has been subtracted to form the increment.

- The **expectation of** $\mathcal{P}(\mathbf{dt}, \mathbf{dq})$ is

$$\mathrm{E}[\mathcal{P}(\mathbf{dt}, \mathbf{dq})] = \Phi_Q(\mathbf{dq})\lambda(t)dt \overset{\mathrm{gen}}{=} \phi_Q(q)dq\lambda(t)dt, \tag{5.10}$$

where, in detail,

$$\begin{aligned}\Phi_Q(\mathbf{dq}) &= \Phi_Q((q, q + dq]) = \Phi_Q(q + dq) - \Phi_Q(q) \\ &= \mathrm{Prob}[Q \le q + dq] - \mathrm{Prob}[Q \le q] = \mathrm{Prob}[q < Q \le q + dq] \\ &\overset{\mathrm{gen}}{=} \phi_Q(q)dq\end{aligned}$$

is the probability distribution measure of the Poisson amplitude mark in measure-theoretic notation corresponding to the mark distribution function $\Phi_Q(q)$ and where $\mathbf{dq}$ is shorthand for the arguments $(q, q + dq]$, just as the $\mathbf{dt}$ in $\mathcal{P}(\mathbf{dt}, \mathbf{dq})$ is shorthand for $(t, t + dt]$. The corresponding mark density will be equal to $\phi_Q(q)$ if Q is continuously distributed with continuously differentiable distribution function and also if the mark density is equal to $\phi_Q(q)$ in the generalized sense (symbol $\overset{\mathrm{gen}}{=}$), for instance, if Q is discretely distributed. Generalized densities will be assumed for almost all distributions encountered in applications. It is also assumed that Φ_Q is a proper distribution,

$$\int_Q \Phi_Q(\mathbf{dq}) = \int_Q \phi_Q(q)dq = 1.$$

- **Poisson random measure** $\mathcal{P}(\Delta \mathbf{t}_i, \Delta \mathbf{q}_j)$ is **Poisson distributed**, i.e.,

$$\mathrm{Prob}[\mathcal{P}(\Delta \mathbf{t}_i, \Delta \mathbf{q}_j) = k] = e^{-\overline{\mathcal{P}}_{i,j}} \left(\overline{\mathcal{P}}_{i,j}\right)^k / k!, \tag{5.11}$$

where

$$\overline{\mathcal{P}}_{i,j} = \mathrm{E}[\mathcal{P}(\Delta \mathbf{t}_i, \Delta \mathbf{q}_j)] = \Phi_Q(\Delta \mathbf{q}_j) \int_{\Delta \mathbf{t}_i} \lambda(t)dt = \Phi_Q(\Delta \mathbf{q}_j)\Lambda(\Delta \mathbf{t}_i)$$

for sets $\Delta \mathbf{t}_i \equiv [t_i, t_i + \Delta t_i)$ in time and $\Delta \mathbf{q}_j \equiv [q_j, q_j + \Delta q_j)$ in marks.

Thus, as Δt_i and Δq_j approach 0^+, they can be replaced by dt and dq, respectively, so

$$\mathrm{Prob}[\mathcal{P}(\mathbf{dt}, \mathbf{dq}) = k] = e^{-\overline{\mathcal{P}}} \left(\overline{\mathcal{P}}\right)^k / k!, \tag{5.12}$$

where

$$\overline{\mathcal{P}} = \mathrm{E}[\mathcal{P}(\mathbf{dt}, \mathbf{dq})] = \phi_Q(q)dq\lambda(t)dt,$$

so by the zero-one jump law,

$$\mathrm{Prob}[\mathcal{P}(\mathbf{dt}, \mathbf{dq}) = k] \overset{dt}{\underset{zol}{=}} (1 - \overline{\mathcal{P}})\delta_{k,0} + \mathcal{P}\delta_{k,1}.$$

- *The **expectation** of* $dP(t; Q) = \int_Q \mathcal{P}(\mathbf{dt}, \mathbf{dq})$ *is*

$$\mathrm{E}\left[\int_Q \mathcal{P}(\mathbf{dt}, \mathbf{dq})\right] = \lambda(t)dt \int_Q \phi_Q(q)dq = \lambda(t)dt = \mathrm{E}[dP(t; Q)], \quad (5.13)$$

corresponding to the earlier Poisson equivalence (5.3) and using the above proper distribution property. Similarly,

$$\mathrm{E}\left[\int_0^t \int_Q \mathcal{P}(\mathbf{ds}, \mathbf{dq})\right] = \mathrm{E}[P(t; Q)] = \int_0^t \lambda(s)ds = \Lambda(t).$$

- *The **variance** of* $\int_Q \mathcal{P}(\mathbf{dt}, \mathbf{dq}) \equiv dP(t; Q)$ *is by definition*

$$\mathrm{Var}\left[\int_Q \mathcal{P}(\mathbf{dt}, \mathbf{dq})\right] = \mathrm{Var}[dP(t; Q)] = \lambda(t)dt. \quad (5.14)$$

Since

$$\mathrm{Var}\left[\int_Q \mathcal{P}(\mathbf{dt}, \mathbf{dq})\right] = \int_Q \int_Q \mathrm{Cov}[\mathcal{P}(\mathbf{dt}, \mathbf{dq_1}), \mathcal{P}(\mathbf{dt}, \mathbf{dq_2})],$$

then

$$\mathrm{Cov}[\mathcal{P}(\mathbf{dt}, \mathbf{dq_1}), \mathcal{P}(\mathbf{dt}, \mathbf{dq_2})] \overset{\mathrm{gen}}{=} \lambda(t)dt\phi_Q(q_1)\delta(q_1 - q_2)dq_1dq_2, \quad (5.15)$$

analogous to (1.48) for $\mathrm{Cov}[dP(s_1), dP(s_2)]$. *Similarly, since*

$$\mathrm{Var}\left[\int_t^{t+\Delta t} \int_Q \mathcal{P}(\mathbf{ds}, \mathbf{dq})\right] = \mathrm{Var}[\Delta P(t; Q)] = \Delta\Lambda(t)$$

and

$$\mathrm{Var}\left[\int_t^{t+\Delta t} \int_Q \mathcal{P}(\mathbf{ds}, \mathbf{dq})\right] = \int_t^{t+\Delta t} \int_t^{t+\Delta t} \int_Q \int_Q \mathrm{Cov}[\mathcal{P}(\mathbf{ds_1}, \mathbf{dq_1}), \mathcal{P}(\mathbf{ds_2}, \mathbf{dq_2})],$$

then

$$\mathrm{Cov}[\mathcal{P}(\mathbf{ds_1}, \mathbf{dq_1}), \mathcal{P}(\mathbf{ds_2}, \mathbf{dq_2})] \overset{\mathrm{gen}}{=} \lambda(s_1)\delta(s_2 - s_1)ds_1ds_2$$
$$\cdot \phi_Q(q_1)\delta(q_1 - q_2)dq_1dq_2, \quad (5.16)$$

embodying the independent increment properties in both time and mark arguments of the space-time or mark-time Poisson process in differential form.

- *It is assumed that jump-amplitude function h has **finite second order moments**, i.e.,*

$$\int_Q |h(t,q)|^2 \phi_Q(q) dq < \infty \qquad (5.17)$$

for all $t \geq 0$ and, in particular,

$$\int_0^t \int_Q |h(s,q)|^2 \phi_Q(q) dq \lambda(s) ds < \infty. \qquad (5.18)$$

- *From Theorem 3.12 and (3.12) on p. 70, a generalization of the standard compound Poisson process is obtained,*

$$\int_0^t \int_Q h(s,q) \mathcal{P}(\mathbf{ds},\mathbf{dq}) = \sum_{k=1}^{P(t;Q)} h(T_k^-, Q_k), \qquad (5.19)$$

i.e., the jump-amplitude counting version of the space-time integral, where T_k is the kth jump-time of a Poisson process $P(t; Q)$ and provided comparable assumptions are satisfied. This is also consistent for the infinitesimal counting sum form in (5.6), and the convention (5.7) applies for (5.19). This form is a special case of the filtered compound Poisson process considered in Snyder and Miller [252, Chapter 5]. The form (5.19) is somewhat awkward due to the presence of three random variables, $P(t; Q)$, T_k, and Q_k, requiring multiple iterated expectations.

- *For a **compound Poisson process with time-independent jump-amplitude**, $h(t,q) = h_2(q)$ (the simplest case being $h(t,q) = q$), we then have*

$$\Pi_2(t) = \int_0^t \int_Q h_2(q) \mathcal{P}(\mathbf{ds},\mathbf{dq}) = \int_Q h_2(q) \mathcal{P}([0,t),\mathbf{dq}) = \sum_{k=1}^{P(t;Q)} h_2(Q_k), \qquad (5.20)$$

where the sum is zero when $P(t; Q) = 0$, and the jump-amplitudes $h_2(Q_k)$ form a set of IID random variables independent of the jump-times of the Poisson process $P(t; Q)$; see [56] and Snyder and Miller [252, Chapter 4]. The mean can be computed by double iterated expectations, since the jump-rate is mark-independent,

$$E[\Pi_2(t)] = E_{P(t;Q)}\left[\sum_{k=1}^{P(t;Q)} E_Q[h_2(Q_k)|P(t;Q)]\right]$$

$$= E_{P(t;Q)}\left[P(t;Q)E_Q[h_2(Q)]\right] = E_Q[h_2(Q)]\Lambda(t),$$

where the IID property and more have been used, e.g., $\Lambda(t) = \int_0^t \lambda(s) ds$.

Similarly, the variance is calculated, letting $\bar{h}_2 \equiv \mathrm{E}_Q[h_2(Q)]$,

$$
\begin{aligned}
\mathrm{Var}[\Pi_2(t)] &= \mathrm{E}\left[\left(\sum_{k=1}^{P(t;Q)} h_2(Q_k) - \bar{h}_2\Lambda(t)\right)^2\right] \\
&= \mathrm{E}\left[\left(\sum_{k=1}^{P(t;Q)} \left(h_2(Q_k) - \bar{h}_2\right) + \bar{h}_2(P(t;Q) - \Lambda(t))\right)^2\right] \\
&= \mathrm{E}_{P(t;Q)}\left[\sum_{k_1=1}^{P(t;Q)}\sum_{k_2=1}^{P(t;Q)} \mathrm{E}_Q\left[\left(h_2(Q_{k_1}) - \bar{h}_2\right)\left(h_2(Q_{k_2}) - \bar{h}_2\right)\right]\right. \\
&\qquad \left. + 2\bar{h}_2(P(t;Q) - \Lambda(t))\sum_{k=1}^{P(t;Q)} \mathrm{E}_Q\left[h_2(Q_k) - \bar{h}_2\right] + \bar{h}_2^2(P(t;Q) - \Lambda(t))^2\right] \\
&= \mathrm{E}_{P(t;Q)}\left[P(t;Q)\mathrm{Var}_Q[h_2(Q)] + 2\bar{h}_2(P(t;Q) - \Lambda(t))P(t;Q)\cdot 0 \right. \\
&\qquad \left. + \bar{h}_2^2(P(t;Q) - \Lambda(t))^2\right] \\
&= \left(\mathrm{Var}_Q[h_2(Q)] + \bar{h}_2^2\right)\Lambda(t) = \mathrm{E}_Q\left[h_2^2(Q)\right]\Lambda(t),
\end{aligned}
$$

using the IID property, separation into mean-zero forms, and the variance-expectation identity (B.186).

- *For **compound Poisson process with both time- and mark-dependence**, $h(t, q)$ and $\lambda(t; q)$, we then have*

$$
\Pi(t) = \int_0^t\int_Q h(s, q)\mathcal{P}(\mathbf{ds}, \mathbf{dq}) = \sum_{k=1}^{P(t;Q)} h(T_k^-, Q_k); \tag{5.21}
$$

however, the iterated expectations technique is not very useful for the compound Poisson form, due to the additional dependence introduced by the jump-time T_k and the jump-rate $\lambda(t; q)$, but the Poisson random measure form is more flexible:

$$
\begin{aligned}
\mathrm{E}[\Pi(t)] &= \mathrm{E}\left[\int_0^t\int_Q h(s, q)\mathcal{P}(\mathbf{ds}, \mathbf{dq})\right] = \int_0^t\int_Q \lambda(s, q)h(s, q)\phi_Q(q)dq\,ds \\
&= \int_0^t \mathrm{E}_Q[\lambda(s, Q)h(s, Q)]ds.
\end{aligned}
$$

- *Consider the generalization of **mean square limits** to include mark space integrals. For ease of integration in mean square limits, let the **mean-zero Poisson** random measure be denoted by*

$$
\widetilde{\mathcal{P}}(\mathbf{dt}, \mathbf{dq}) \equiv \mathcal{P}(\mathbf{dt}, \mathbf{dq}) - \mathrm{E}[\mathcal{P}(\mathbf{dt}, \mathbf{dq})] = \mathcal{P}(\mathbf{dt}, \mathbf{dq}) - \phi_Q(q)dq\lambda(t)dt \tag{5.22}
$$

and let the corresponding space-time integral be

$$
\widetilde{I} \equiv \int_Q h(t, q)\widetilde{\mathcal{P}}(\mathbf{dt}, \mathbf{dq}). \tag{5.23}
$$

Let $\mathcal{T}_n = \{t_i | t_{i+1} = t_i + \Delta t_i$ for $i = 0 : n, t_0 = 0, t_{n+1} = t, \max_i[\Delta t_i] \to 0$ as $n \to +\infty\}$ be a proper partition of $[0, t]$. Let $\mathcal{Q}_m = \{\Delta \mathcal{Q}_j$ for $j = 1 : m | \cup_{j=1}^m \Delta \mathcal{Q}_j = \mathcal{Q}\}$ be a proper partition of the mark space $\mathcal{Q}$, noting that it is implicit that the subsets $\Delta \mathcal{Q}_j$ are disjoint. Let $h(t, q)$ be a continuous function in time and marks. Let the corresponding partially discrete approximation

$$\widetilde{I}_{m,n} \equiv \sum_{i=0}^{n} \sum_{j=1}^{m} h(t_i, q_j^*) \int_{\mathcal{Q}_j} \widetilde{\mathcal{P}}([t_i, t_i + \Delta T), dq_j) \qquad (5.24)$$

for some $q_j^* \in \Delta \mathcal{Q}_j$. Note that if $\mathcal{Q}$ is a finite interval $[a, b]$, then $\mathcal{Q}_j = [q_j, q_j + \Delta q]$ using even spacing with $q_1 = a, q_{m+1} = b$, and $\Delta q = (b-a)/m$. Then $\widetilde{I}_{m,n}$ converges in the mean square limit to $\widetilde{I}$ if

$$\mathrm{E}[(\widetilde{I} - \widetilde{I}_{m,n})^2] \to 0 \qquad (5.25)$$

as m and $n \to +\infty$.

For more advanced and abstract treatments of the Poisson random measure, see Gihman and Skorohod [95, Part 2, Chapter 2], Snyder and Miller [252, Chapters 4 and 5], Cont and Tankov [60], and Øksendal and Sulem [223], or see Chapter 12.

Theorem 5.2. *Basic Infinitesimal Moments of the Space-Time Poisson Process.*

$$\mathrm{E}[d\Pi(t)] = \lambda(t)dt \int_{\mathcal{Q}} h(t, q)\phi_Q(q)dq \equiv \lambda(t)dt\mathrm{E}_Q[h(t, Q)] \equiv \lambda(t)dt\overline{h}(t) \qquad (5.26)$$

and

$$\mathrm{Var}[d\Pi(t)] = \lambda(t)dt \int_{\mathcal{Q}} h^2(t, q)\phi_Q(q)dq = \lambda(t)dt\mathrm{E}_Q[h^2(t; Q)] \equiv \lambda(t)dt\overline{h^2}(t). \qquad (5.27)$$

Proof. The jump-amplitude function $h(t, Q)$ is independently distributed, through the mark process Q, from the underlying Poisson counting process here, except that this jump in space is conditional on the occurrence of the jump-time or -count of the underlying Poisson process. However, the function $h(t, q)$ is deterministic since it depends on the realization q in the space-time Poisson definition, rather than the random variable Q. The infinitesimal mean (5.26) is straightforward:

$$\mathrm{E}[d\Pi(t)] = \mathrm{E}\left[\int_{\mathcal{Q}} h(t, q)\mathcal{P}(\mathbf{dt}, \mathbf{dq})\right] = \int_{\mathcal{Q}} h(t, q)\mathrm{E}[\mathcal{P}(\mathbf{dt}, \mathbf{dq})]$$

$$= \lambda(t)dt \int_{\mathcal{Q}} h(t, q)\phi_Q(q)dq = \lambda(t)dt\mathrm{E}_Q[h(t, Q)] \equiv \lambda(t)dt\overline{h}(t);$$

note that the expectation operator applied to the mark integral can be moved to apply just to the Poisson random measure $\mathcal{P}(\mathbf{dt}, \mathbf{dq})$.

However, the result for the variance in (5.27) is not so obvious, but the covariance formula for two Poisson random measures with differing mark variables $\mathrm{Cov}[\mathcal{P}(\mathbf{dt}, \mathbf{dq_1})$,

$\mathcal{P}(\textbf{dt}, \textbf{dq}_2)]$ in (5.15) will be made useful by converting it to the mean-zero Poisson random measure $\widetilde{\mathcal{P}}(\textbf{dt}, \textbf{dq})$ in (5.22),

$$
\begin{aligned}
\text{Var}[d\Pi(t)] &= \text{E}\left[\left(\int_{\mathcal{Q}} h(t,q)\mathcal{P}(\textbf{dt}, \textbf{dq}) - \overline{h}(t)\lambda(t)dt\right)^2\right] \\
&= \text{E}\left[\left(\int_{\mathcal{Q}} \left(h(t,q)\mathcal{P}(\textbf{dt}, \textbf{dq}) - h(t,q)\phi_Q(q)\lambda(t)dt\right)\right)^2\right] \\
&= \text{E}\left[\left(\int_{\mathcal{Q}} h(t,q)\widetilde{\mathcal{P}}(\textbf{dt}, \textbf{dq})\right)^2\right] \\
&= \text{E}\left[\int_{\mathcal{Q}} h(t,q_1)\int_{\mathcal{Q}} h(t,q_2)\widetilde{\mathcal{P}}(\textbf{dt}, \textbf{dq}_1)\widetilde{\mathcal{P}}(\textbf{dt}, \textbf{dq}_1)\right] \\
&= \int_{\mathcal{Q}} h(t,q_1)\int_{\mathcal{Q}} h(t,q_2)\text{Cov}\left[\widetilde{\mathcal{P}}(\textbf{dt}, \textbf{dq}_1), \widetilde{\mathcal{P}}(\textbf{dt}, \textbf{dq}_1)\right] \\
&= \lambda(t)dt\int_{\mathcal{Q}} h^2(t,q_1)\phi_Q(q_1)dq_1 = \lambda(t)dt\text{E}_Q\left[h^2(t,Q)\right] \equiv \lambda(t)dt\overline{h^2}(t). \quad \square
\end{aligned}
$$

Examples 5.3.

- *Uniformly distributed jump-amplitudes:*
  As an example of a continuous distribution, consider the uniform density for the jump-amplitude mark Q given by

$$\phi_Q(q) = \frac{1}{b-a}U(q; a, b), \ a < b, \tag{5.28}$$

  where $U(q; a, b) = \mathbf{1}_{q\in[a,b]}$ is the step or indicator function for the interval $[a, b]$, i.e., $U(q; a, b)$ is one when $a \leq q \leq b$ and zero otherwise. The first few moments are

$$
\begin{aligned}
\text{E}_Q[1] &= \frac{1}{b-a}\int_a^b dq = 1, \\
\text{E}_Q[Q] &= \frac{1}{b-a}\int_a^b qdq = \frac{b+a}{2}, \\
\text{Var}_Q[Q] &= \frac{1}{b-a}\int_a^b \left(q - \frac{b+a}{2}\right)^2 dq = \frac{(b-a)^2}{12}.
\end{aligned}
$$

  In the case of the log-uniform amplitude, letting $Q = \ln(1 + h(Q))$ be the mark-amplitude relationship using the log-transformation form from the linear SDE problem (4.76), we then have

$$h(Q) = e^Q - 1,$$

  and the expected jump-amplitude is

$$\text{E}_Q[h(Q)] = \frac{1}{b-a}\int_a^b (e^q - 1)dq = \frac{e^b - e^a}{b-a} - 1.$$

- *Poisson distributed jump-amplitudes:*
  As an example of a discrete distribution of jump-amplitudes, consider

$$\Phi_Q(k) = p_k(u) = e^{-u}\frac{u^k}{k!}$$

for $k = 0 : \infty$. Thus, the jump process is a Poisson–Poisson process or a Poisson–mark Poisson process. The mean and variance are

$$\mathrm{E}_Q[Q] = u,$$
$$\mathrm{Var}_Q[Q] = u.$$

Remark 5.4. *For the general discrete distribution,*

$$\Phi_Q(k) = p_k, \quad \sum_{k=0}^{\infty} p_k = 1,$$

the comparable continuous form is

$$\Phi_Q(q) \stackrel{\text{gen}}{=} \sum_{k=0}^{\infty} H_R(q - k)p_k = \sum_{k=0}^{\lfloor q \rfloor} p_k,$$

where $H_R(q)$ is again the right-continuous Heaviside step function and $\lfloor q \rfloor$ is the maximum integer not exceeding q. The corresponding generalized density is

$$\phi_Q(q) \stackrel{\text{gen}}{=} \sum_{k=0}^{\infty} \delta_R(q - k)p_k.$$

The reader should verify that this density yields the correct expectation and variance forms.

5.2 State-Dependent Generalization of Jump-Diffusion SDEs

5.2.1 State-Dependent Generalization for Space-Time Poisson Processes

The space-time Poisson process is generalized to include state dependence with $X(t)$ in both the jump-amplitude and the Poisson measure, such that

$$d\Pi(t; X(t), t) = \int_Q h(X(t), t, q)\mathcal{P}(\mathbf{dt}, \mathbf{dq}; X(t), t) \tag{5.29}$$

on the Poisson mark space Q with Poisson random measure $\mathcal{P}(\mathbf{dt}, \mathbf{dq}; X(t), t)$, which helps to describe the space-time Poisson mechanism and related calculus. The space-time state-dependent Poisson mark, $Q = q$, is again the underlying random variable for the state-dependent and mark-dependent jump-amplitude coefficient $h(x, t, q)$. The double time t arguments of $d\Pi$, dP, and $\mathcal{P}$ are not considered redundant for applications, since the first time t or time set dt is the usual Poisson jump process implicit time-dependence, while the second to the right of the semicolon denotes explicit or parametric time-dependence paired with explicit state dependence that is known in advance and is appropriate for the application model.

Alternatively, the Poisson zero-one law form may be used, i.e.,

$$d\Pi(t; X(t), t) \overset{dt}{\underset{zol}{=}} h(X(t), t, Q)dP(t; Q, X(t), t) \tag{5.30}$$

with the jump of $\Pi(t; X(t), t)$ being

$$[\Pi](T_k) = h(X(T_k^-), T_k^-, Q_k)$$

at jump-time T_k and jump-mark Q_k. The multitude of random variables in this sum means that expectations or other Poisson integrals will be very difficult to calculate even by conditional expectation iterations.

Definition 5.5. *The **conditional expectation** of $\mathcal{P}(\mathbf{dt}, \mathbf{dq}; X(t), t)$ is*

$$E[\mathcal{P}(\mathbf{dt}, \mathbf{dq}; X(t), t)|X(t) = x] = \phi_Q(q; x, t)dq\lambda(t; x, t)dt, \tag{5.31}$$

where $\phi_Q(q; x, t)dq$ is the probability density of the now state-dependent Poisson amplitude mark and the jump-rate $\lambda(t; x, t)$ now has state-time dependence. In this notation, the relationship to the simple counting process is given by

$$\int_Q \mathcal{P}(\mathbf{dt}, \mathbf{dq}; X(t), t) = dP(t; Q, X(t), t).$$

Hence, when $h(x, t, q) = \tilde{h}(x, t)$, i.e., independent of the mark q, the space-time Poisson is the simple jump process with mark-independent amplitude,

$$d\Pi(t; X(t), t) = \tilde{h}(X(t), t)dP(t; Q, X(t), t),$$

but with nonunit jumps in general. Effectively the same form is obtained when there is a single discrete mark, e.g., $\phi_Q(q) = \delta(q - 1)$, so $h(x, t, q) = h(x, t, 1)$ always.

Theorem 5.6. *Basic Conditional Infinitesimal Moments of the State-Dependent Poisson Process.*

$$E[d\Pi(t; X(t), t)|X(t) = x] = \int_Q h(x, t, q)\phi_Q(q; x, t)dq\lambda(t; x, t)dt$$

$$\equiv E_Q[h(x, t, Q)]\lambda(t; x, t)dt \tag{5.32}$$

and

$$\text{Var}[d\Pi(t; X(t), t)|X(t) = x] = \int_Q h^2(x, t, q)\phi_Q(q; x, t)dq\lambda(t; x, t)dt$$

$$\equiv E_Q[h^2(x, t; Q)]\lambda(t; x, t)dt. \tag{5.33}$$

Proof. The justification is the same justification as for (5.26)–(5.27). It is assumed that the jump-amplitude $h(x, t, Q)$ is independently distributed due to Q from the underlying Poisson counting process here, except that this jump in space is conditional on the occurrence of the jump-time of the underlying Poisson process. $\quad\square$

5.2.2 State-Dependent Jump-Diffusion SDEs

The general, scalar SDE takes the form

$$dX(t) = f(X(t), t)dt + g(X(t), t)dW(t) + \int_{\mathcal{Q}} h(X(t), t, q)\mathcal{P}(\mathbf{dt}, \mathbf{dq}; X(t), t) \tag{5.34}$$

$$\stackrel{\mathrm{dt}}{=} f(X(t), t)dt + g(X(t), t)dW(t) + h(X(t), t, Q)dP(t; Q, X(t), t)$$

for the state process $X(t)$ with a set of continuous coefficient functions $\{f, g, h\}$. However, the SDE model is just a useful symbolic model for many applied situations, but the more basic model relies on specifying the method of integration. So

$$
\begin{aligned}
X(t) = X(t_0) + \int_{t_0}^t & (f(X(s), s)ds + g(X(s), s)dW(s) \\
& + h(X(t), s, Q)dP(s; Q, X(s), s))
\end{aligned}
\tag{5.35}
$$

$$\stackrel{\mathrm{ims}}{=} X(t_0) + \stackrel{\mathrm{ms}}{\lim_{n \to \infty}} \left[\sum_{i=0}^n \left(f_i \Delta t_i + g_i \Delta W_i + \sum_{k=P_i+1}^{P_i + \Delta P_i} h_{i,k} \right) \right],$$

where $f_i = f(X_i, t_i)$, $g_i = g(X_i, t_i)$, $h_{i,k} = h(X_i, T_k, Q_k)$, $\Delta t_i = t_{i+1} - t_i$, $\Delta P_i = \Delta P(t_i; Q, X_i, t_i)$, and $\Delta W_i = \Delta W(t_i)$. Here, T_k is the kth jump-time and $\{Q, Q_k\}$ are the corresponding random marks.

The **conditional infinitesimal moments for the state process** are

$$\mathrm{E}[dX(t)|X(t) = x] = f(x, t)dt + \overline{h}(x, t)\lambda(t; x, t)dt, \tag{5.36}$$

$$\overline{h}(x, t)\lambda(t; x, t)dt \equiv \mathrm{E}_Q[h(x, t, Q)]\lambda(t; x, t)dt, \tag{5.37}$$

and

$$\mathrm{Var}[dX(t)|X(t) = x] = g^2(x, t)dt + \overline{h^2}(x, t)\lambda(t; x, t)dt, \tag{5.38}$$

$$\overline{h^2}(x, t)\lambda(t; x, t)dt \equiv \mathrm{E}_Q[h^2(x, t, Q)]\lambda(t; x, t)dt \tag{5.39}$$

using (1.1), (5.32), (5.33), (5.34) and assuming that the Poisson process is independent of the Wiener process. The jump in the state at jump-time T_k in the underlying Poisson process is

$$[X](T_k) \equiv X(T_k^+) - X(T_k^-) = h(X(T_k^-), T_k^-, Q_k) \tag{5.40}$$

for $k = 1, 2, \ldots$, now depending on the kth mark Q_k at the prejump-time T_k^- at the kth jump.

Rule 5.7. *Stochastic Chain Rule for State-Dependent SDEs.*
The **stochastic chain rule** *for a sufficiently differentiable function $Y(t) = F(X(t), t)$ has the form*

$$dY(t) = dF(X(t), t) \stackrel{\text{sym}}{=} F(X(t) + dX(t), t + dt) - F(X(t), t)$$
$$= d_{(\text{cont})} F(X(t), t) + d_{(\text{jump})} F(X(t), t)$$
$$\stackrel{\text{dt}}{=} F_t(X(t), t)dt + F_x(X(t), t)(f(X(t), t)dt + g(X(t), t)dW(t))$$
$$+ \frac{1}{2} F_{xx}(X(t), t)g^2(X(t), t)dt \tag{5.41}$$
$$+ \int_Q (F(X(t) + h(X(t), t, q), t) - F(X(t), t))\mathcal{P}(\mathbf{dt}, \mathbf{dq}; X(t), t)$$

to precision-dt. It is sufficient that F be twice continuously differentiable in x and once in t.

5.2.3   Linear State-Dependent SDEs

Let the state-dependent jump-diffusion process satisfy an SDE linear in the state process $X(t)$ with time-dependent rate coefficients

$$dX(t) \stackrel{\text{dt}}{=} X(t) \left(\mu_d(t)dt + \sigma_d(t)dW(t) + \nu(t, Q)dP(t; Q) \right) \tag{5.42}$$

for $t > t_0$ with $X(t_0) = X_0$ and $\mathrm{E}[dP(t; Q)] = \lambda(t)dt$, where $\mu_d(t)$ denotes the mean and $\sigma_d^2(t)$ denotes the variance of the diffusion process, while Q_k denotes the kth mark and T_k denotes the kth time of the jump process.

Again, using the log-transformation $Y(t) = \ln(X(t))$ and the stochastic chain rule (5.41),

$$dY(t) \stackrel{\text{dt}}{=} (\mu_d(t) - \sigma_d^2(t)/2)dt + \sigma_d(t)dW(t) + \ln(1 + \nu(t, Q))dP(t; Q) \tag{5.43}$$

with immediate integrals

$$Y(t) = \ln(x_0) + \int_{t_0}^{t} dY(s) \tag{5.44}$$

and

$$X(t) = x_0 \exp\left(\int_{t_0}^{t} dY(s) \right), \tag{5.45}$$

or in recursive form,

$$X(t + \Delta t) = X(t) \exp\left(\int_{t}^{t+\Delta t} dY(s) \right). \tag{5.46}$$

Linear Mark-Jump-Diffusion Simulation Forms

For simulations, a small time-step, $\Delta t_i \ll 1$, approximation of the recursive form (5.46) would be more useful with $X_i = X(t_i)$, $\mu_i = \mu_d(t_i)$, $\sigma_i = \sigma_d(t_i)$, $\Delta W_i = \Delta W(t_i)$, $\Delta P_i = \Delta P(t_i; Q)$, and the convenient jump-amplitude coefficient approximaton, $v(t, Q) \simeq v_0(Q) \equiv \exp(Q) - 1$, i.e.,

$$X_{i+1} \simeq X_i \exp\left((\mu_i - \sigma_i^2/2)\Delta t_i + \sigma_i \Delta W_i\right)(1 + v_0(Q))^{\Delta P_i} \tag{5.47}$$

for $i = 1 : N$ time-steps, where a zero-one jump law approximation has been used.

For the diffusion part, it has been shown that

$$\mathrm{E}\left[e^{\sigma_i \Delta W_i}\right] = e^{\sigma_i^2 \Delta t_i/2},$$

using the completing the square technique. In addition, there is the following lemma for the jump part of (5.47).

Lemma 5.8. *Jump Term Expectation.*

$$\mathrm{E}\left[(1 + v_0(Q))^{\Delta P_i}\right] = e^{\lambda_i \Delta t_i \mathrm{E}[v_0(Q)]}, \tag{5.48}$$

where $\mathrm{E}[\Delta P_i] = \lambda_i \Delta t_i$ *and* $v_0(Q) = \exp(Q) - 1$.

Proof. Using given forms, iterated expectations, the Poisson distribution, and the IID property of the marks Q_k, we then have

$$\mathrm{E}\left[(1 + v_0(Q))^{\Delta P_i}\right] = \mathrm{E}\left[e^{Q\Delta P_i}\right]$$

$$= e^{-\lambda_i \Delta t_i} \sum_{k=0}^{\infty} (\lambda_i \Delta t_i)^k \mathrm{E}_Q\left[e^{kQ}\right]$$

$$= e^{-\lambda_i \Delta t_i} \sum_{k=0}^{\infty} (\lambda_i \Delta t_i)^k \left(\mathrm{E}_Q\left[e^Q\right]\right)^k$$

$$= e^{-\lambda_i \Delta t_i} e^{\lambda_i \Delta t_i \mathrm{E}_Q\left[e^Q\right]}$$

$$= e^{\lambda_i \Delta t_i \mathrm{E}_Q[v_0(Q)]}. \quad \square$$

An immediate consequence of this result is the following corollary.

Corollary 5.9. *Discrete State Expectations.*

$$\mathrm{E}[X_{i+1}|X_i] \simeq X_i \exp((\mu_i + \lambda_i \mathrm{E}_Q[v_0(Q)])\Delta t_i) \tag{5.49}$$

and

$$\mathrm{E}[X_{i+1}] \simeq x_0 \exp\left(\sum_{j=0}^{i} (\mu_j + \lambda_j \mathrm{E}_Q[v_0(Q)])\Delta t_j\right). \tag{5.50}$$

Further, as Δt_i and $\delta t_n \to 0^+$, the continuous form of the expectation follows and is given later in Corollary 5.13 on p. 146 using other justification.

Example 5.10. *Linear, Time-Independent, Constant-Rate Coefficient Case.*
In the linear, time-independent, constant-rate coefficient case with $\mu_d(t) = \mu_0$, $\sigma_d(t) = \sigma_0$, $\lambda(t) = \lambda_0$, and $\nu(t, Q) = \nu_0(Q) = e^Q - 1$,

$$X(t) = x_0 \exp\left((\mu_0 - \sigma_0^2/2)(t - t_0) + \sigma_0(W(t) - W(t_0)) + \sum_{k=1}^{P(t;Q)-P(t_0;Q)} \nu_0 Q_k\right), \quad (5.51)$$

where the Poisson counting sum form is now more manageable since the marks do not depend on the prejump-times T_k^-.

Using the independence of the three underlying stochastic processes, $(W(t) - W(t_0))$, $(P(t; Q) - P(t_0; Q))$, and Q_i, as well as the stationarity of the first two and the law of exponents law to separate exponents, leads to partial reduction of the expected state process:

$$\mathrm{E}[X(t)] = x_0 e^{(\mu_0 - \sigma_0^2/2)(t-t_0)} \cdot \mathrm{E}_W\left[e^{\sigma_0 W(t-t_0)}\right] \cdot \sum_{k=0}^{\infty} \mathrm{E}[P(t - t_0; Q) = k] \mathrm{E}\left[e^{\sum_{i=1}^{k} Q_i}\right]$$

$$= x_0 e^{(\mu_0 - \sigma_0^2/2)(t-t_0)} \int_{-\infty}^{+\infty} \frac{e^{-w^2/(2(t-t_0))}}{\sqrt{2\pi(t - t_0)}} e^{\sigma_0 w} dw$$

$$\cdot e^{-\lambda_0(t-t_0)} \sum_{k=0}^{\infty} \frac{(\lambda_0(t - t_0))^k}{k!} \prod_{i=1}^{k} \mathrm{E}_Q\left[e^Q\right]$$

$$= x_0 e^{\mu_0(t-t_0)} e^{-\lambda_0(t-t_0)} \sum_{k=0}^{\infty} \frac{(\lambda_0(t - t_0))^k}{k!} \mathrm{E}_Q^k\left[e^Q\right]$$

$$= x_0 e^{(\mu_0 + \lambda_0(\mathrm{E}_Q[e^Q]-1))(t-t_0)}, \quad (5.52)$$

where $\lambda_0(t - t_0)$ is the Poisson parameter and $Q = (-\infty, +\infty)$ is taken as the mark space for specificity with

$$\mathrm{E}_Q\left[e^Q\right] = \int_Q e^q \phi_Q(q) dq.$$

Little more useful simplification can be obtained analytically, except for infinite expansions or equivalent special functions, when the mark density $\phi_Q(q)$ is specified. Numerical procedures may be more useful for practical purposes. The state expectation in this distributed mark case (5.52) should be compared with the pure constant linear coefficient case (4.81) of Chapter 4.

Exponential Expectations

Sometimes it is necessary to get the expectation of an exponential of the integral of a jump-diffusion process. The procedure is much more complicated for distributed amplitude

Poisson jump processes than for diffusions since the mark-time process is a product process, i.e., the product of the mark process and the Poisson process. For the time-independent coefficient case, as in a prior example, the exponential processes are easily separable by the law of exponents. However, for the time-dependent case, it is necessary to return to using the space-time process $\mathcal{P}$ and the decomposition approximation used in the mean square limit. The h in the following theorem might be the amplitude coefficient in (5.43), i.e., $h(s, q) = q = \ln(1 + \nu(s, q))$.

Theorem 5.11. *Expectation for the Exponential of Space-Time Counting Integrals.*
Assuming finite second order moments for $h(t, q)$ and convergence in the mean square limit,

$$\mathrm{E}\left[\exp\left(\int_{t_0}^t \int_{\mathcal{Q}} h(s, q) \mathcal{P}(\mathbf{ds}, \mathbf{dq})\right)\right] = \exp\left(\int_{t_0}^t \int_{\mathcal{Q}} \left(e^{h(s,q)} - 1\right) \phi_Q(q, s) dq \lambda(s) ds\right)$$

$$\equiv \exp\left(\int_{t_0}^t \overline{\left(e^h - 1\right)}(s) \lambda(s) ds\right). \tag{5.53}$$

Proof. Let the proper partition of the mark space over disjoint subsets be

$$\mathcal{Q}_m = \{\Delta \mathcal{Q}_j \text{ for } j = 1:m | \cup_{j=1}^m \Delta \mathcal{Q}_j = \mathcal{Q}\}.$$

Since the Poisson measure is Poisson distributed,

$$\Phi_{\mathcal{P}_j}(k) = \mathrm{Prob}[\mathcal{P}(\mathbf{dt}, \Delta \mathcal{Q}_j) = k] = e^{-\overline{\mathcal{P}_j}} \frac{(\overline{\mathcal{P}_j})^k}{k!}$$

with Poisson parameter

$$\overline{\mathcal{P}_j} \equiv \mathrm{E}[\mathcal{P}(\mathbf{dt}, \Delta \mathcal{Q}_j)] = \lambda(t) dt \, \Phi_Q(\Delta \mathcal{Q}_j, t_i)$$

for each subset $\{\Delta \mathcal{Q}_j\}$.

Similarly, let the proper partition over the time-interval be

$$\mathcal{T}_n = \{t_i | t_{i+1} = t_i + \Delta t_i \text{ for } i = 0:n, t_0 = 0, t_{n+1} = t, \max_i[\Delta t_i] \to 0 \text{ as } n \to +\infty\}.$$

The disjoint property over subsets and time-intervals means $\mathcal{P}([t_i, t_i + \Delta t_i), \Delta \mathcal{Q}_j)$ and $\mathcal{P}([t_i, t_i + \Delta t_i), \Delta \mathcal{Q}'_j)$ will be pairwise independent provided $j' \neq j$ for fixed i corresponding to property (5.15) for infinitesimals, while $\mathcal{P}([t_i, t_i + \Delta t_i), \Delta \mathcal{Q}_j)$ and $\mathcal{P}([t_i, t_i + \Delta t'_i), \Delta \mathcal{Q}'_j)$ will be pairwise independent provided $i' \neq i$ and $j' \neq j$, corresponding to property (5.16) for infinitesimals.

For brevity, let $h_{i,j} \equiv h(t_i, q_j^*)$, where $q_j^* \in \Delta \mathcal{Q}_j$, $\mathcal{P}_{i,j} \equiv \mathcal{P}_i([t_i, t_i + \Delta t_i), \Delta \mathcal{Q}_j)$, and $\overline{\mathcal{P}_{i,j}} \equiv \lambda_i \Delta t_i \Phi_Q(\Delta \mathcal{Q}_j)$.

Using mean square limits, with $\mathcal{P}_{i,j}$ playing the dual roles of the two increments $(\Delta t_i, \Delta \mathcal{Q}_j)$, and the law of exponents and independence denoted by $\overset{ind}{\underset{inc}{=}}$, we have

$$
\mathrm{E}\left[\exp\left(\int_{t_0}^t \int_{\mathcal{Q}} h\mathcal{P}\right)\right] \overset{ims}{=} \overset{ms}{\underset{m,n\to\infty}{\lim}} \mathrm{E}\left[\exp\left(\sum_{i=0}^n \sum_{j=1}^m h_{i,j}\mathcal{P}_{i,j}\right)\right]
$$

$$
\overset{ind}{\underset{inc}{=}} \overset{ms}{\underset{m,n\to\infty}{\lim}} \Pi_{i=0}^n \Pi_{j=1}^m \mathrm{E}\left[\exp\left(h_{i,j}\mathcal{P}_{i,j}\right)\right]
$$

$$
= \overset{ms}{\underset{m,n\to\infty}{\lim}} \Pi_{i=0}^n \Pi_{j=1}^m \exp\left(-\overline{\mathcal{P}_{i,j}}\right) \sum_{k_{i,j}=0}^\infty \frac{\overline{\mathcal{P}_{i,j}}^{k_{i,j}}}{k_{i,j}!} \exp\left(h_{i,j}k_{i,j}\right)
$$

$$
= \overset{ms}{\underset{m,n\to\infty}{\lim}} \Pi_{i=0}^n \Pi_{j=1}^m \exp\left(\overline{\mathcal{P}_{i,j}}\left(\exp(h_{i,j}) - 1\right)\right)
$$

$$
= \overset{ms}{\underset{m,n\to\infty}{\lim}} \exp\left(\sum_{i=0}^n \sum_{j=1}^m \left(\exp(h_{i,j}) - 1\right) \lambda_i \Delta t_i \Phi_Q(\Delta Q_i, t_i)\right)
$$

$$
\overset{ims}{=} \exp\left(\int_{t_0}^t \int_{\mathcal{Q}} (\exp(h(s,q)) - 1) \phi_Q(q,s)dq\lambda(s)ds\right)
$$

$$
\equiv \exp\left(\int_{t_0}^t \overline{(\exp(h(s,Q)) - 1)}\lambda(s)ds\right).
$$

Thus, the main technique is to disassemble the mean square limit discrete approximation to get at the independent random part, take its expectation, and then reassemble the mean square limit, justifying the interchange of expectation and exponent integration. □

Remarks 5.12.

- *Note that the mark space subset $\Delta\mathcal{Q}_j$ is never used directly as a discrete element of integration, since the subset would be infinite if the mark space were infinite. The mark space element is used only through the distribution which would be bounded. This is quite unlike the time domain, where we can select t to be finite. If the mark space were finite, say, $\mathcal{Q} = [a, b]$, then a concrete partition of $[a, b]$ similar to the time-partition can be used.*

- *Also note that the dependence on $(X(t), t)$ was not used but could be considered suppressed but absorbed into the existing t dependence of h and $\mathcal{P}$.*

Corollary 5.13. *Expectation of X(t) for Linear SDE.*
Let $X(t)$ be the solution (5.45) with $\overline{v}(t) \equiv \mathrm{E}[v(t, Q)]$ of (5.42). Then

$$
\mathrm{E}[X(t)] = x_0 \exp\left(\int_{t_0}^t (\mu_d(s)\lambda(s)\overline{v}(s)) \, ds\right)
$$

$$
= x_0 \exp\left(\int_{t_0}^t \mathrm{E}[dX(s)/X(s)]ds\right). \tag{5.54}
$$

Proof. The jump part, i.e., the main part, follows from exponential theorem, Theorem 5.11, (5.53), and the lesser part for the diffusion is left as an exercise for the reader.

However, note that the exponent is the time integral of $E[dX(t)/X(t)]$, the relative conditional infinitesimal mean, which is independent of $X(s)$ and is valid only for the linear mark-jump-diffusion SDE. $\square$

Remark 5.14. *The relationship in (5.54) is a **quasi-deterministic equivalence** for linear mark-jump-diffusion SDEs and was shown by Hanson and Ryan [115] in 1989. They also produced a nonlinear jump counterexample that has a formal closed-form solution in terms of the gamma function, for which the result does not hold, and a very similar example is given in Exercise 9 in Chapter 4.*

Moments of Log-Jump-Diffusion Process

For the log-jump-diffusion process $dY(t)$ in (5.43), suppose that the jump-amplitude is time-independent and that the mark variable was conveniently chosen as

$$Q = \ln(1 + \nu(t, Q))$$

so that the SDE has the form

$$dY(t) \overset{\text{dt}}{=} \mu_{ld}(t)dt + \sigma_d(t)dW(t) + QdP(t; Q), \qquad (5.55)$$

or, in the case of applications for which the time-step Δt is an increment that is not infinitesimal like dt, there is some probability of more than one jump,

$$\Delta Y(t) = \mu_{ld}(t)\Delta t + \sigma_d(t)\Delta W(t) + \sum_{k=P(t;Q)+1}^{P(t;Q)+\Delta P(t;Q)} Q_k. \qquad (5.56)$$

The results for the infinitesimal case (5.55) are contained in the incremental case (5.56).

The first few moments can be found in general for (5.56), and if up to the fourth moment, then the skew and kurtosis coefficients can be calculated. These calculations can be expedited by the following lemma, concerning sums of zero-mean IID random variables.

Lemma 5.15. *Zero-Mean IID Random Variable Sums.*
Let $\{X_i | i = 1 : n\}$ be a set of zero-mean IID random variables, i.e., $E[X_i] = 0$. Let $M^{(m)} \equiv E[X_i^m]$ be the mth moment and

$$S_n^{(m)} \equiv \sum_{i=1}^{n} X_i^m$$

with $S_n^{(1)} = S_n$ the usual partial sum over the set and

$$E[S_n^{(m)}] = nM^{(m)}; \qquad (5.57)$$

then the expectation of powers of S_n for $m = 1{:}4$ is

$$E[(S_n)^m] = \begin{cases} 0, & m = 1 \\ nM^{(2)}, & m = 2 \\ nM^{(3)}, & m = 3 \\ nM^{(4)} + 3n(n-1)\left(M^{(2)}\right)^2, & m = 4 \end{cases}. \qquad (5.58)$$

Proof. The proof is done first by the linear property of the expectation and the IID properties of the X_i,

$$\mathrm{E}\left[S_n^{(m)}\right] = \sum_{i=1}^{n} \mathrm{E}[X_i^m] = \sum_{i=1}^{n} M^{(m)} = nM^{(m)}. \tag{5.59}$$

The $m = 1$ case is trivial due to the zero-mean property of the X_i's and the linearity of the expectation operator, $\mathrm{E}[S_n] = \sum_{i=1}^{n} \mathrm{E}[X_i] = 0$.

For $m = 2$, the induction hypothesis from (5.58) is

$$\mathrm{E}\left[S_n^2\right] \equiv \mathrm{E}\left[\left(\sum_{i=1}^{n} X_i^2\right)\right] = nM^{(2)},$$

where the initial condition at $n = 1$ is $\mathrm{E}[S_1^2] = \mathrm{E}[X_1^2] = M^{(2)}$ by definition. The hypothesis can be proved easily by partial sum recursion $S_{n+1} = S_n + X_{n+1}$, application of the binomial theorem, expectation linearity, and the zero-mean IID property:

$$\mathrm{E}\left[S_{n+1}^2\right] = \mathrm{E}\left[(S_n + X_{n+1})^2\right] = \mathrm{E}\left[S_n^2 + 2X_{n+1}S_n + X_{n+1}^2\right]$$
$$= nM^{(2)} + 2 \cdot 0 \cdot 0 + M^{(2)} = (n+1)M^{(2)}. \tag{5.60}$$

QED for $m = 2$.

This is similar for the power $m = 3$, again beginning with the induction hypothesis

$$\mathrm{E}\left[S_n^3\right] \equiv \mathrm{E}\left[\left(\sum_{i=1}^{n} X_i\right)^3\right] = nM^{(3)},$$

where the initial condition at $n = 1$ is $\mathrm{E}[S_1^3] = \mathrm{E}[X_1^3] = M^{(3)}$ by definition. Using the same techniques as in (5.60),

$$\mathrm{E}\left[S_{n+1}^3\right] = \mathrm{E}\left[(S_n + X_{n+1})^3\right] = \mathrm{E}\left[S_n^3 + 3X_{n+1}S_n^2 + 3X_{n+1}^2 S_n^2 + X_{n+1}^3\right]$$
$$= nM^{(3)} + 3 \cdot 0 \cdot nM^{(2)} + 3 \cdot M^{(2)} \cdot 0 + M^{(3)} = (n+1)M^{(3)}. \tag{5.61}$$

QED for $m = 3$.

Finally, the case for the power $m = 4$ is a little different since an additional nontrivial term arises from the product of the squares of two independent variables. The induction hypothesis is

$$\mathrm{E}\left[S_n^4\right] \equiv \mathrm{E}\left[\left(\sum_{i=1}^{n} X_i\right)^4\right] = nM^{(4)} + 3n(n-1)(M^{(2)})^2,$$

where the initial condition at $n = 1$ is $\mathrm{E}[S_1^4] = \mathrm{E}[X_1^4] = M^{(4)}$ by definition. Using the same techniques as in (5.60),

$$\mathrm{E}\left[S_{n+1}^4\right] = \mathrm{E}\left[(S_n + X_{n+1})^4\right] = \mathrm{E}\left[S_n^4 + 4X_{n+1}S_n^3 + 6X_{n+1}^2 S_n^2 + 4X_{n+1}^3 S_n^1 + X_{n+1}^4\right]$$
$$= nM^{(4)} + 3n(n-1)(M^{(2)})^2 + 4 \cdot 0 \cdot nM^{(3)} + 6 \cdot M^{(2)} \cdot nM(2)$$
$$+ 4 \cdot M^{(3)} \cdot 0 + M^{(4)}$$
$$= (n+1)M^{(4)} + 3(n+1)((n+1) - 1)(M^{(2)})^2. \tag{5.62}$$

QED for $m = 4$. $\square$

Remark 5.16. *The results here depend on the IID and zero-mean properties but do not otherwise depend on the particular distribution of the random variables. The results are used in the following theorem.*

Theorem 5.17. *Some Moments of the Log-Jump-Diffusion (LJD) Process $\Delta Y(t)$.*
Let $\Delta Y(t)$ satisfy the stochastic difference equation (5.56), and let the marks Q_k be IID with mean $\mu_j \equiv E_Q[Q_k]$ and variance $\sigma_j^2 \equiv \text{Var}_Q[Q_k]$. Then the first four moments, $m = 1:4$, are

$$\mu_{ljd}(t) \equiv E[\Delta Y(t)] = (\mu_{ld}(t) + \lambda(t)\mu_j)\Delta t; \tag{5.63}$$

$$\sigma_{ljd}(t) \equiv \text{Var}[\Delta Y(t)] = \left(\sigma_d^2(t) + \left(\sigma_j^2 + \mu_j^2\right)\lambda(t)\right)\Delta t; \tag{5.64}$$

$$M_{ljd}^{(3)}(t) \equiv E\left[(\Delta Y(t) - E[\Delta Y(t)])^3\right] = \left(M_j^{(3)} + \mu_j\left(3\sigma_j^2 + \mu_j^2\right)\right)\lambda(t)\Delta t, \tag{5.65}$$

where $M_j^{(3)} \equiv E_Q[(Q_i - \mu_j)^3]$;

$$\begin{aligned}
M_{ljd}^{(4)}(t) &\equiv E\left[(\Delta Y(t) - E[\Delta Y(t)])^4\right] \\
&= \left(M_j^{(4)} + 4\mu_j M_j^{(3)} + 6\mu_j^2\sigma_j^2 + \mu_j^4\right)\lambda(t)\Delta t \\
&\quad + 3\left(\sigma_d^2(t) + \left(\sigma_j^2 + \mu_j^2\right)\lambda(t)\right)^2(\Delta t)^2,
\end{aligned} \tag{5.66}$$

where $M_j^{(4)} \equiv E_Q[(Q_i - \mu_j)^4]$.

Proof. One general technique for calculating moments of the log-jump-diffusion process is **iterated expectations**. Thus

$$\begin{aligned}
\mu_{ljd}(t) = E[\Delta Y(t)] &= \mu_{ld}(t)\Delta t + \sigma_d(t) \cdot 0 + E_{\Delta P(t;Q)}\left[E_Q\left[\sum_{i=1}^{\Delta P(t;Q)} Q_i \middle| \Delta P(t;Q)\right]\right] \\
&= \mu_{ld}(t)\Delta t + E_{\Delta P(t;Q)}\left[\sum_{i=1}^{\Delta P(t;Q)} E_Q[Q_i]\right] \\
&= \mu_{ld}(t)\Delta t + E_{\Delta P(t;Q)}[\Delta P(t;Q)E_Q[Q_i]] = \left(\mu_{ld}(t) + \mu_j\lambda(t)\right)\Delta t,
\end{aligned}$$

proving the first moment formula, using the increment jump-count.

For the higher moments, the main key technique for efficient calculation of the moments is decomposing the log-jump-diffusion process deviation into zero-mean deviation factors, i.e.,

$$\Delta Y(t) - \mu_{ljd}(t) = \sigma_d(t)\Delta W(t) + \sum_{i=1}^{\Delta P(t;Q)}(Q_i - \mu_j) + \mu_j(\Delta P(t;Q) - \lambda(t)\Delta t).$$

In addition, the multiple applications of the binomial theorem and the convenient increment power Table 1.1 for $\Delta W(t)$ and Table 1.2 for $\Delta P(t; Q)$ are used.

The incremental process variance is found by

$$\sigma_{ljd}(t) \equiv \text{Var}[\Delta Y(t)]$$

$$= \text{E}\left[\left(\sigma_d(t)\Delta W(t) + \sum_{i=1}^{\Delta P(t;Q)}(Q_i - \mu_j) + \mu_j(\Delta P(t;\ Q) - \lambda(t)\Delta t)\right)^2\right]$$

$$= \sigma_d^2(t)\text{E}_{\Delta W(t)}[(\Delta W)^2(t)] + 2\sigma_d \cdot 0 + \text{E}\left[\left(\sum_{i=1}^{\Delta P(t;Q)}(Q_i - \mu_j) + \mu_j(\Delta P(t;\ Q) - \lambda(t)\Delta t)\right)^2\right]$$

$$= \sigma_d^2(t)\Delta t + \text{E}_{\Delta P(t;Q)}\left[\sum_{i=1}^{\Delta P(t;Q)}\sum_{k=1}^{\Delta P(t;Q)}\text{E}_Q[(Q_i - \mu_j)(Q_k - \mu_j)]\right.$$

$$\left. + 2\mu_j(\Delta P(t;\ Q) - \lambda(t)\Delta t)\sum_{i=1}^{\Delta P(t;Q)}\text{E}_Q[(Q_i - \mu_j)] + \mu_j^2(\Delta P(t;\ Q) - \lambda(t)\Delta t)^2\right]$$

$$= \sigma_d^2(t)\Delta t + \text{E}_{\Delta P(t;Q)}\left[\Delta P(t;\ Q)\sigma_j^2 + 0 + \mu_j^2(\Delta P(t;\ Q) - \lambda(t)\Delta t)^2\right]$$

$$= \left(\sigma_d^2(t) + \left(\sigma_j^2 + \mu_j^2\right)\lambda(t)\right)\Delta t.$$

The case of the third central moment is similarly calculated,

$$M_{ljd}^{(3)}(t) \equiv \text{E}\left[(\Delta Y(t) - \mu_{ljd}(t))^3\right]$$

$$= \text{E}\left[\left(\sigma_d(t)\Delta W(t) + \sum_{i=1}^{\Delta P(t;Q)}(Q_i - \mu_j) + \mu_j(\Delta P(t;\ Q) - \lambda(t)\Delta t)\right)^3\right]$$

$$= \sigma_d^3(t)\text{E}_{\Delta W(t)}\left[(\Delta W)^3(t)\right]$$

$$+ 3\sigma_d^2 \text{E}_{\Delta W(t)}\left[(\Delta W)^2(t)\right]\text{E}\left[\sum_{i=1}^{\Delta P(t;Q)}(Q_i - \mu_j) + \mu_j(\Delta P(t;\ Q) - \lambda(t)\Delta t)\right]$$

$$+ 3\sigma_d \cdot 0 + \text{E}\left[\left(\sum_{i=1}^{\Delta P(t;Q)}(Q_i - \mu_j) + \mu_j(\Delta P(t;\ Q) - \lambda(t)\Delta t)\right)^3\right]$$

$$= \sigma_d^3(t)\cdot 0 + 3\sigma_d^2(t)\Delta t \cdot 0$$

$$+ \text{E}_{\Delta P(t;Q)}\left[\sum_{i=1}^{\Delta P(t;Q)}\sum_{k=1}^{\Delta P(t;Q)}\sum_{\ell=1}^{\Delta P(t;Q)}\text{E}_Q[(Q_i - \mu_j)(Q_k - \mu_j)(Q_\ell - \mu_j)]\right.$$

$$+ 3\mu_j(\Delta P(t;\ Q) - \lambda(t)\Delta t)\sum_{i=1}^{\Delta P(t;Q)}\sum_{k=1}^{\Delta P(t;Q)}\text{E}_Q[(Q_i - \mu_j)(Q_k - \mu_j)]$$

$$\left. + 3\mu_j^2(\Delta P(t;\ Q) - \lambda(t)\Delta t)^2 \cdot 0 + \mu_j^3(\Delta P(t;\ Q) - \lambda(t)\Delta t)^3\right]$$

$$= \text{E}_{\Delta P(t;Q)}\left[\Delta P(t;\ Q)M_j^{(3)} + 3\mu_j(\Delta P(t;\ Q) - \lambda(t)\Delta t)\Delta P(t;\ Q)\sigma_j^2\right.$$

$$\left. + \mu_j^3(\Delta P(t;\ Q) - \lambda(t)\Delta t)^3\right]$$

$$= \left(M_j^{(3)} + \mu_j\left(3\sigma_j^2 + \mu_j^2\right)\right)\lambda(t)\Delta t,$$

depending only on the jump component of the jump-diffusion.

The case of the fourth central moment is similarly calculated,

$$M_{ljd}^{(4)}(t) \equiv \mathrm{E}\left[(\Delta Y(t) - \mu_{ljd}(t))^4\right]$$

$$= \mathrm{E}\left[\left(\sigma_d(t)\Delta W(t) + \sum_{i=1}^{\Delta P(t;Q)}(Q_i - \mu_j) + \mu_j(\Delta P(t;Q) - \lambda(t)\Delta t)\right)^4\right]$$

$$= \sigma_d^4(t)\mathrm{E}_{\Delta W(t)}\left[(\Delta W)^4(t)\right] + 4\sigma_d^3 \cdot 0 + 6\sigma_d^2 \mathrm{E}_{\Delta W(t)}\left[(\Delta W)^2(t)\right]$$

$$\mathrm{E}\left[\left(\sum_{i=1}^{\Delta P(t;Q)}(Q_i - \mu_j) + \mu_j(\Delta P(t;Q) - \lambda(t)\Delta t)\right)^2\right]$$

$$+ 4\sigma_d \cdot 0 + \mathrm{E}\left[\left(\sum_{i=1}^{\Delta P(t;Q)}(Q_i - \mu_j) + \mu_j(\Delta P(t;Q) - \lambda(t)\Delta t)\right)^4\right]$$

$$= 3\sigma_d^4(t)(\Delta t)^2 + 6\sigma_d^2(t)\Delta t \mathrm{E}_{\Delta P(t;Q)}\left[\sum_{i=1}^{\Delta P(t;Q)}\sum_{k=1}^{\Delta P(t;Q)}\mathrm{E}_Q[(Q_i - \mu_j)(Q_k - \mu_j)]\right.$$

$$\left. + 2\mu_j(\Delta P(t;Q) - \lambda(t)\Delta t) \cdot 0 + \mu_j^2(\Delta P(t;Q) - \lambda(t)\Delta t)^2\right]$$

$$+ \mathrm{E}_{\Delta P(t;Q)}\left[\sum_{i=1}^{\Delta P(t;Q)}\sum_{k=1}^{\Delta P(t;Q)}\sum_{\ell=1}^{\Delta P(t;Q)}\sum_{m=1}^{\Delta P(t;Q)}\mathrm{E}_Q[(Q_i - \mu_j)(Q_k - \mu_j)(Q_\ell - \mu_j)(Q_m - \mu_j)]\right.$$

$$+ 4\mu_j(\Delta P(t;Q) - \lambda(t)\Delta t)\sum_{i=1}^{\Delta P(t;Q)}\sum_{k=1}^{\Delta P(t;Q)}\sum_{\ell=1}^{\Delta P(t;Q)}\mathrm{E}_Q[(Q_i - \mu_j)(Q_k - \mu_j)(Q_\ell - \mu_j)]$$

$$+ 6\mu_j^2(\Delta P(t;Q) - \lambda(t)\Delta t)^2\sum_{i=1}^{\Delta P(t;Q)}\sum_{k=1}^{\Delta P(t;Q)}\mathrm{E}_Q[(Q_i - \mu_j)(Q_k - \mu_j)]$$

$$\left. + 4\mu_j^3(\Delta P(t;Q) - \lambda(t)\Delta t)^3 \cdot 0 + \mu_j^4(\Delta P(t;Q) - \lambda(t)\Delta t)^4\right]$$

$$= 3\sigma_d^4(t)(\Delta t)^2 + 6\sigma_d^2(t)\Delta t \mathrm{E}_{\Delta P(t;Q)}\left[\Delta P(t;Q)\sigma_j^2 + \mu_j^2(\Delta P(t;Q) - \lambda(t)\Delta t)^2\right]$$

$$+ \mathrm{E}_{\Delta P(t;Q)}\left[\Delta P(t;Q)M_j^{(4)} + 3\Delta P(t;Q)(\Delta P(t;Q) - 1)\sigma_j^4\right.$$

$$+ 4\mu_j(\Delta P(t;Q) - \lambda(t)\Delta t)\Delta P(t;Q)M_j^{(3)}$$

$$\left. + 6\mu_j^2(\Delta P(t;Q) - \lambda(t)\Delta t)^2\Delta P(t;Q)\sigma_j^2 + \mu_j^4(\Delta P(t;Q) - \lambda(t)\Delta t)^4\right]$$

$$= \left(M_j^{(4)} + 4\mu_j M_j^{(3)} + 6\mu_j^2\sigma_j^2 + \mu_j^4\right)\lambda(t)\Delta t + 3\left(\sigma_d^2(t) + \left(\sigma_j^2 + \mu_j^2\right)\lambda(t)\right)^2(\Delta t)^2,$$

completing the proofs for moments $m = 1:4$.

Also, as used throughout, the expectations of odd powers of $\Delta W(t)$, the single powers of $(Q_i - \mu_j)$, and the single powers of $(\Delta P(t;Q) - \lambda(t)\Delta t)$ were immediately set to zero. In addition, the evaluation of the mark deviation sums of the form $\mathrm{E}\left[\sum_{i=1}^k(Q_i - \mu_j)^m\right]$ for $m = 1:4$ is based upon general formulas of Lemma 5.15. □

Remarks 5.18.

- *Recall that the third and fourth moments are measures of skewness and peakedness (kurtosis), respectively. The normalized representations in the current notation are the coefficient of skewness,*

$$\eta_3[\Delta Y(t)] \equiv M_{ljd}^{(3)}(t)/\sigma_{ljd}^3(t), \tag{5.67}$$

from (B.11), *and the coefficient of kurtosis,*

$$\eta_4[\Delta Y(t)] \equiv M_{ljd}^{(4)}(t)/\sigma_{ljd}^4(t), \tag{5.68}$$

from (B.12).

- *For example, if the marks are normally or uniformly distributed, then*

$$M_j^{(3)} = 0,$$

since the normal and uniform distributions are both symmetric about the mean, so they lack skew, and thus we have

$$\eta_3[\Delta Y(t)] = \frac{\mu_j \left(3\sigma_j^2 + \mu_j^2\right) \lambda(t)\Delta t}{\sigma_{ljd}^3(t)} = \frac{\mu_j \left(3\sigma_j^2 + \mu_j^2\right) \lambda(t)}{\left(\sigma_d^2(t) + \left(\sigma_j^2 + \mu_j^2\right)\lambda(t)\right)^3 (\Delta t)^2},$$

using $\sigma_{ljd}(t)$ *given by* (5.64). *For the uniform distribution, the mean* μ_j *is given explicitly in terms of the uniform interval* $[a, b]$ *by* (B.15) *and the variance* σ_j^2 *by* (B.16), *while for the normal distribution,* μ_j *and* σ_j^2 *are the normal model parameters. In general, the normal and uniform distribution versions of the log-jump-diffusion process will have skew, although the component incremental diffusion and mark processes are skewless.*

In the normal and uniform mark cases, the fourth moment of the jump marks are

$$M_j^{(4)}/\sigma_j^4 - \begin{cases} 3, & normal \ Q_i \\ 1.8, & uniform \ Q_i \end{cases},$$

which are in fact the coefficients of kurtosis for the normal and uniform distributions, respectively, so

$$\eta_4[\Delta Y(t)] = \left(\begin{cases} 3, & normal \ Q_i \\ 1.8, & uniform \ Q_i \end{cases} \sigma_j^4 + 6\mu_j^2\sigma_j^2 + \mu_j^4\right) \lambda(t)\Delta t/\sigma_{ljd}^4(t)$$

$$+ 3\left(\sigma_d^2(t) + \left(\sigma_j^2 + \mu_j^2\right)\lambda(t)\right)^2 (\Delta t)^2/\sigma_{ljd}^4(t).$$

- *The moment formulas for the differential log-jump-diffusion process* $dY(t)$ *follow immediately from Theorem 5.17 by dropping terms* $O((\Delta t)^2)$ *and replacing* Δt *by* dt.

Distribution of Increment Log-Process

Theorem 5.19. *Distribution of the State Increment Logarithm Process for Linear Mark-Jump-Diffusion SDE.*

Let the logarithm-transform jump-amplitude be $\ln(1 + \nu(t, q)) = q$. *Then the increment of the logarithm process* $Y(t) = \ln(X(t))$, *assuming* $X(t_0) = x_0 > 0$ *and the jump-count increment, approximately satisfies*

$$\Delta Y(t) \simeq \mu_{ld}(t)\Delta t + \sigma_d(t)\Delta W(t) + \sum_{j}^{\Delta P(t;Q)} \widehat{Q}_j \tag{5.69}$$

for sufficiently small Δt, where $\mu_{ld}(t) \equiv \mu_d(t) - \sigma_d^2(t)/2$ is the log-diffusion drift, and $\sigma_d > 0$ and the $\widehat{Q}_j$ are pairwise IID jump-marks for $P(s; Q)$ for $s \in (t, t + \Delta t]$, counting only jumps associated with $\Delta P(t; Q)$ given $P(t; Q)$, with common density $\phi_Q(q)$. The $\widehat{Q}_j$ are independent of both $\Delta P(t; Q)$ and $\Delta W(t)$.

Then the distribution of the log-process $Y(t)$ is the Poisson sum of nested convolutions

$$\Phi_{\Delta Y(t)}(x) \simeq \sum_{k=1}^{\infty} p_k(\lambda(t)\Delta t) \left(\Phi_{\Delta G(t)} \left(*\phi_Q\right)^k \right)(x), \tag{5.70}$$

where $\Delta G(t) \equiv \mu_{ld}(t)\Delta t + \sigma_d(t)\Delta W(t)$ is the infinitesimal Gaussian process and $(\Phi_{\Delta G(t)}(*\phi_Q)^k)(x)$ denotes a convolution of one distribution with k identical densities ϕ_Q. The corresponding log-process density is

$$\phi_{\Delta Y(t)}(x) \simeq \sum_{k=1}^{\infty} p_k(\lambda(t)\Delta t) \left(\phi_{\Delta G(t)} \left(*\phi_Q\right)^k \right)(x). \tag{5.71}$$

Proof. By the law of total probability (B.92), the distribution of the log-jump-diffusion $\Delta Y(t) \simeq \Delta G(t) + \sum_{j}^{\Delta P(t;Q)} \widehat{Q}_j$ is

$$\Phi_{\Delta Y(t)}(x) = \text{Prob}[\Delta Y(t) \leq x] = \text{Prob}\left[\Delta G(t) + \sum_{j=1}^{\Delta P(t;Q)} \widehat{Q}_j \leq x \right]$$

$$= \sum_{k=0}^{\infty} \text{Prob}\left[\Delta G(t) + \sum_{j=1}^{\Delta P(t;Q)} \widehat{Q}_j \leq x | \Delta P(t; Q) = k \right] \text{Prob}[\Delta P(t; Q) = k]$$

$$= \sum_{k=0}^{\infty} p_k(\lambda(t)\Delta t)\Phi^{(k)}(x), \tag{5.72}$$

where $p_k(\lambda(t)\Delta t)$ is the Poisson distribution with parameter $\lambda(t)\Delta t$ and we let

$$\Phi^{(k)}(x) \equiv \text{Prob}\left[\Delta G(t) + \sum_{j=1}^{k} \widehat{Q}_j \leq x \right].$$

For each discrete condition $\Delta P(t; Q) = k$, $\Delta Y(t)$ is the sum of $k + 1$ terms, the normally distributed Gaussian diffusion part $\Delta G(t) = \mu_{ld}(t)\Delta t + \sigma_d(t)\Delta W(t)$, and the Poisson counting sum $\sum_{j=1}^{k} \widehat{Q}_j$, where the marks $\widehat{Q}_j$ are assumed to be IID but otherwise distributed with density $\phi_Q(q)$, while independent of the diffusion and the Poisson counting differential process $\Delta P(t; Q)$. Using the fact that $\Delta W(t)$ is normally distributed with zero-mean and Δt-variance,

$$
\begin{aligned}
\Phi_{\Delta G(t)}(x) &= \mathrm{Prob}[\Delta G(t) \leq x] = \mathrm{Prob}[\mu_{ld}(t)\Delta t + \sigma_d(t)\Delta W(t) \leq x] \\
&= \mathrm{Prob}[\Delta W(t) \leq (x - \mu_{ld}(t)\Delta t)/\sigma_d(t)] = \Phi_{\Delta W(t)}((x - \mu_{ld}(t)\Delta t)/\sigma_d(t)) \\
&= \Phi_n((x - \mu_{ld}(t)\Delta t)/\sigma_d(t); 0, \Delta t) = \Phi_n(x; \mu_{ld}(t)\Delta t, \sigma_d^2(t)\Delta t),
\end{aligned}
$$

provided $\sigma_d(t) > 0$, while also using identities for normal distributions, where $\Phi_n(x; \mu, \sigma^2)$ denotes the normal distribution with mean μ and variance σ^2.

Since $\Phi^{(k)}$ is the distribution for the sum of $k + 1$ independent random variables, with one normally distributed random variable and k IID jump-marks $\widehat{Q}_j$ for each k, $\Phi^{(k)}$ will be the nested convolutions as given in (B.100). Upon expanding in convolutions starting from the distribution for the random variable $\Delta G(t)$ and the kth Poisson counting sum

$$
J_k \equiv \sum_{j=1}^{k} \widehat{Q}_j,
$$

we get

$$
\Phi^{(k)}(x) = \left(\Phi_{\Delta G(t)} * \phi_{J_k}\right)(x) = \left(\Phi_{\Delta G(t)} \prod_{i=1}^{k}\left(*\phi_{Q_i}\right)\right)(x) = \left(\Phi_{\Delta G(t)}\left(*\phi_Q\right)^k\right)(x),
$$

using the identically distributed property of the Q_i's and the compact convolution operator notation

$$
\left(\Phi_{\Delta G(t)} \prod_{i=1}^{k}\left(*\phi_{Q_i}\right)\right)(x) = ((\cdots((\Phi_{\Delta G(t)} * \phi_{Q_1}) * \phi_{Q_2}) \cdots * \phi_{Q_{k-1}}) * \phi_{Q_k})(x),
$$

which collapses to the operator power form for IID marks since $\prod_{i=1}^{k} c = c^k$ for some constant c. Substituting the distribution into the law of total probability form (5.72), the desired result is (5.70), which when differentiated with respect to x yields the kth density $\phi_{\Delta Y(t)}(x)$ in (5.71). $\quad\square$

Remark 5.20. *Several specialized variations of this theorem are found in Hanson and Westman* [124, 126], *but corrections to these papers are made here.*

Corollary 5.21. *Density of Linear Jump-Diffusion with Log-Normally Distributed Jump-Amplitudes.*
Let $X(t)$ be a linear jump-diffusion satisfying SDE (5.69) and let the jump-amplitude mark Q be normally distributed such that

$$
\phi_Q(x; t) = \phi_n(x; \mu_j(t), \sigma_j^2(t)) \tag{5.73}
$$

with jump-mean $\mu_j(t) = \mathrm{E}[Q]$ and jump variance $\sigma_j^2(t) = \mathrm{Var}[Q]$. Then the jump-diffusion density of the log-process $Y(t)$ is

$$
\phi_{\Delta Y(t)}(x) = \sum_{k=1}^{\infty} p_k(\lambda(t)\Delta t)\phi_n(x; \mu_{ld}(t)\Delta t + k\mu_j(t), \sigma_d^2(t)\Delta t + k\sigma_j^2(t)). \tag{5.74}
$$

Proof. By (B.101) the convolution of two normal densities is a normal distribution with a mean that is the sum of the means and a variance that is the sum of the variances. Similarly, by the induction exercise result in (B.196), the pairwise convolution of one normally distributed diffusion process $\Delta G(t) = \mu_{ld}(t)\Delta t + \sigma_d(t)\Delta W(t)$ density and k random mark Q_i densities ϕ_Q for $i = 1:k$ will be a normal density whose mean is the sum of the $k + 1$ means and whose variance is the sum of the $k + 1$ variances. Thus starting with the result (5.72) and then applying (B.196),

$$
\begin{aligned}
\phi_{\Delta Y(t)}(x) &= \sum_{k=1}^{\infty} p_k(\lambda(t)\Delta t)\left(\phi_{\Delta G(t)}\left(*\phi_Q\right)^k\right)(x) \\
&= \sum_{k=1}^{\infty} p_k(\lambda(t)\Delta t)\phi_n\left(x; \mu_{ld}(t)\Delta t + \sum_{i=1}^{k}\mu_j(t), \sigma_d^2(t)\Delta t + \sum_{i=1}^{k}\sigma_j^2(t)\right) \\
&= \sum_{k=1}^{\infty} p_k(\lambda(t)\Delta t)\phi_n(x; \mu_{ld}(t)\Delta t + k\mu_j(t), \sigma_d^2(t)\Delta t + k\sigma_j^2(t)). \quad \square
\end{aligned}
$$

Remark 5.22. *The normal jump-amplitude jump-diffusion distribution has been used in financial applications, initially by Merton [202] and then by others such as Düvelmeyer [76], Andersen et al. [6], and Hanson and Westman [124].*

Corollary 5.23. *Density of Linear Jump-Diffusion with Log-Uniformly Distributed Jump-Amplitudes.*
Let $X(t)$ be a linear jump-diffusion satisfying SDE (5.69), and let the jump-amplitude mark Q be uniformly distributed as in (5.28), i.e.,

$$
\phi_Q(q) = \frac{1}{b-a}U(q; a, b),
$$

where $U(q; a, b)$ is the unit step function on $[a, b]$ with $a < b$. The jump-mean is $\mu_j(t) = (b+a)/2$ and the jump-variance is $\sigma_j^2(t) = (b-a)^2/12$.

Then the jump-diffusion density of the increment log-process $\Delta Y(t)$ satisfies the general convolution form (5.71), i.e.,

$$
\phi_{\Delta Y(t)}(x) = \sum_{k=1}^{\infty} p_k(\lambda(t)\Delta t)\left(\phi_{\Delta G(t)}\left(*\phi_Q\right)^k\right)(x) = \sum_{k=1}^{\infty} p_k(\lambda(t)\Delta t)\phi_{ujd}^{(k)}(x), \quad (5.75)
$$

where $p_k(\lambda(t)\Delta t)$ is the Poisson distribution with parameter $\lambda(t)$. The $\Delta G(t) = \mu_{ld}(t)\Delta t + \sigma_d(t)\Delta W(t)$ is the diffusion term and Q is the uniformly distributed jump-amplitude mark. The first few coefficients of $p_k(\lambda(t)\Delta t)$ for the uniform jump distribution (ujd) are

$$
\phi_{ujd}^{(0)}(x) = \phi_{\Delta G(t)}(x) = \phi_n(x; \mu_{ld}(t)\Delta t, \sigma_d^2(t)\Delta t), \quad (5.76)
$$

where $\phi_n(x; \mu_{ld}(t)\Delta t, \sigma_d^2(t)\Delta t)$ denotes the normal density with mean $\mu_{ld}(t)$ and variance $\sigma_d(t)\Delta t$,

$$
\phi_{ujd}^{(1)}(x) = \left(\phi_{\Delta G(t)} * \phi_Q\right)(x) = \phi_{sn}(x - b, x - a; \mu_{ld}(t)\Delta t, \sigma_d^2(t)\Delta t), \quad (5.77)
$$

*where ϕ_{sn} is the **secant-normal density***

$$\phi_{sn}(x_1, x_2; \mu, \sigma^2) \equiv \frac{1}{(x_2 - x_1)} \Phi_n(x_1, x_2; \mu, \sigma^2) \tag{5.78}$$

$$\equiv \frac{\Phi_n(x_2; \mu, \sigma^2) - \Phi_n(x_1; \mu, \sigma^2)}{x_2 - x_1}$$

with normal distribution $\Phi_n(x_1, x_2; \mu, \sigma^2)$ such that

$$\Phi_n(x_i; \mu, \sigma^2) \equiv \Phi_n(-\infty, x_i; \mu, \sigma^2)$$

for $i = 1 : 2$, and

$$\phi_{ujd}^{(2)}(x) = \left(\phi_{\Delta G(t)}(*\phi_Q)^2\right)(x) \tag{5.79}$$

$$= \frac{2b - x + \mu_{ld}(t)\Delta t}{b - a} \phi_{sn}(x - 2b, x - a - b; \mu_{ld}(t)\Delta t, \sigma^2(t)\Delta t)$$

$$+ \frac{x - 2a - \mu_{ld}(t)\Delta t}{b - a} \phi_{sn}(x - a - b, x - 2a; \mu_{ld}(t)\Delta t, \sigma_d^2(t)\Delta t)$$

$$+ \frac{\sigma_d^2(t)\Delta t}{(b - a)^2} \left(\phi_n(x - 2b; \mu_{ld}(t)\Delta t, \sigma_d^2(t)\Delta t)\right.$$

$$\left. - 2\phi_n(x - a - b; \mu_{ld}(t)\Delta t, \sigma_d^2(t)\Delta t) + \phi_n(x - 2a; \mu_{ld}(t)\Delta t, \sigma_d^2(t)\Delta t)\right).$$

Proof. First the finite range of the jump-amplitude uniform density is used to truncate the convolution integrals for each k using existing results for the mark convolutions, such as $\phi_{uq}^{(2)}(x) = (\phi_Q * \phi_Q)(x) = \phi_{Q_1 + Q_2}(x)$ for IID marks when $k = 2$.

The case for $k = 0$ is trivial since it is given in the theorem equations (5.76).

For a $k = 1$ jump,

$$\phi_{ujd}^{(1)}(x) = (\phi_{\Delta G(t)} * \phi_Q)(x) = \int_{-\infty}^{+\infty} \phi_{\Delta G(t)}(x - y)\phi_Q(y)dy$$

$$= \frac{1}{b - a} \int_a^b \phi_n(x - y; \mu_{ld}(t)\Delta t, \sigma_d^2(t)\Delta t)dy$$

$$= \frac{1}{b - a} \int_{x-b}^{x-a} \phi_n(z; \mu_{ld}(t)\Delta t, \sigma_d^2(t)\Delta t)dz$$

$$= \frac{1}{b - a} \Phi_n(x - b, x - a; \mu_{ld}(t)\Delta t, \sigma_d^2(t)\Delta t)$$

$$= \phi_{sn}(x - b, x - a; \mu_{ld}(t)\Delta t, \sigma_d^2(t)\Delta t),$$

where $-\infty < x < +\infty$, upon change of variables and use of identities.

For $k = 2$ jumps, the triangular distribution exercise result (B.197) is

$$\phi_{uq}^{(2)}(x) = (\phi_Q * \phi_Q)(x) = \frac{1}{(b - a)^2} \begin{cases} x - 2a, & 2a \leq x < a + b \\ 2b - x, & a + b \leq x \leq 2b \\ 0, & otherwise \end{cases}. \tag{5.80}$$

Hence,

$$
\begin{aligned}
\phi_{\text{ujd}}^{(2)}(x) &= (\phi_{\Delta G(t)} * (\phi_Q * \phi_Q))(x) = \int_{-\infty}^{+\infty} \phi_{\Delta G(t)}(x-y)(\phi_Q * \phi_Q)(y)dy \\
&= \frac{1}{(b-a)^2}\left(\int_{2a}^{a+b}(y-2a)\phi_{\Delta G(t)}(x-y)dy + \int_{a+b}^{2b}(2b-y)\phi_{\Delta G(t)}(x-y)dy\right) \\
&= \frac{1}{(b-a)^2}\left(\int_{x-a-b}^{x-2a}(x-z-2a)\phi_{\Delta G(t)}(z)dz \right. \\
&\qquad \left. + \int_{x-2b}^{x-a-b}(2b-x+z)\phi_{\Delta G(t)}(z)dz\right) \\
&= \frac{2b-x+\mu_{ld}(t)\Delta t}{b-a}\phi_{sn}(x-2b,x-a-b;\mu_{ld}(t)\Delta t,\sigma_d^2(t)\Delta t) \\
&\quad + \frac{x-2a-\mu_{ld}(t)\Delta t}{b-a}\phi_{sn}(x-a-b,x-2a;\mu_{ld}(t)\Delta t,\sigma_d^2(t)\Delta t) \\
&\quad + \frac{\sigma_d^2(t)\Delta t}{(b-a)^2}\left(\phi_n(x-2b;\mu_{ld}(t)\Delta t,\sigma_d^2(t)\Delta t)\right. \\
&\quad \left. - 2\phi_n(x-a-b;\mu_{ld}(t)\Delta t,\sigma_d^2(t)\Delta t) + \phi_n(x-2a;\mu_{ld}(t)\Delta t,\sigma_d^2(t)\Delta t)\right),
\end{aligned}
$$

where the exact integral for the normal density has been used.   □

Remarks 5.24.

- *This density form ϕ_{sn} in (5.78) is called a **secant-normal density** since the numerator is an increment of the normal distribution and the denominator is the corresponding increment in its state arguments, i.e., a secant approximation, which here has the form $\Delta\Phi_n/\Delta x$.*

- *The uniform jump-amplitude jump-diffusion distribution has been used in financial applications, initially by the authors in [126] as a simple, but appropriate, representation of a jump component of market distributions, and some errors have been corrected here.*

Example 5.25. *Linear SDE Simulator for Log-Uniformly Distributed Jump-Amplitudes.*
The linear SDE jump-diffusion simulator in MATLAB code C.14 in Online Appendix C can be converted from the simple discrete jump process to the distributed jump process here. The primary change is the generation of another set of random numbers for the mark process Q, e.g.,

$$Q = a + (b-a) * \text{rand}(1, n+1)$$

for a set of $n+1$ uniformly distributed marks on (a, b) so that the jump-amplitudes of $X(t)$ are log-uniformly distributed.

   An example is demonstrated in Figure 5.1 for uniformly distributed marks Q on $(a, b) = (-2, +1)$ and time-dependent coefficients $\{\mu_d(t), \sigma_d(t), \lambda(t)\}$. The MATLAB linear mark-jump-diffusion code C.15, called `linmarkjumpdiff06fig1.m` *in Online*

Appendix C, *is a modification of the linear jump-diffusion SDE simulator code* C.14 *illustrated in Figure* 4.3 *for constant coefficients and discrete mark-independent jumps. The state exponent* $Y(t)$ *is simulated as*

$$YS(i+1) = YS(i) + (\mu_d(i) - \sigma_d^2(i)/2) * \Delta t + \sigma_d(i) * DW(i) + Q(i) * DP(i),$$

with $t(i+1) = t0 + i * \Delta t$ *for* $i = 0 : n$ *with* $n = 1000$, $t0 = 0$, $0 \le t(i) \le 2$. *The incremental Poisson jump term* $\Delta P(i) = P(t_i + \Delta t) - P(t_i)$ *is simulated by a uniform random number generator on* $(0, 1)$ *using the acceptance-rejection technique* [230, 97] *to implement the zero-one jump law to obtain the probability of* $\lambda(i)\Delta t$ *that a jump is accepted there. The same random state is used to obtain the simulations of uniformly distributed* Q *on* (a, b) *conditional on a jump event.*

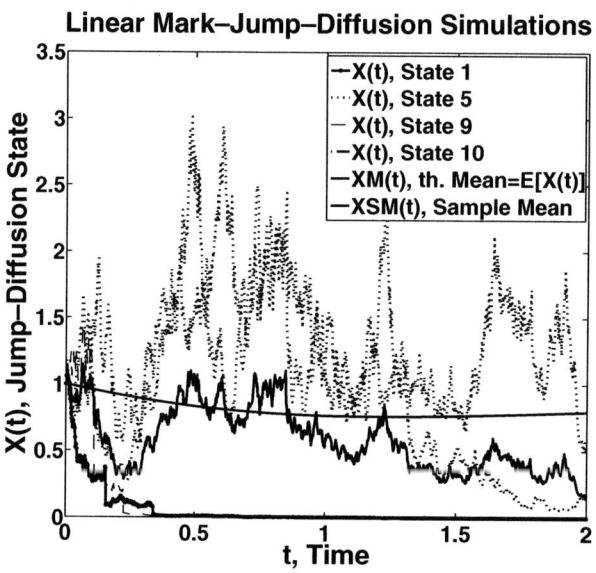

Figure 5.1. *Four linear mark-jump-diffusion sample paths for time-dependent coefficients are simulated using MATLAB* [210] *with* $N = 1000$ *time-steps, maximum time* $T = 2.0$, *and four* `randn` *and four* `rand` *states. Initially,* $x_0 = 1.0$. *Parameter values are given in vectorized functions using vector functions and dot-element operations,* $\mu_d(t) = 0.1 * \sin(t)$, $\sigma_d(t) = 1.5 * \exp(-0.01 * t)$, *and* $\lambda = 3.0 * \exp(-t. * t)$. *The marks are uniformly distributed on* $[-2.0, +1.0]$. *In addition to the four simulated states, the expected state* $E[X(t)]$ *is presented using the quasi-deterministic equivalence* (5.54) *of Hanson and Ryan* [115], *but also presented are the sample mean of the four sample paths.*

5.3   Multidimensional Markov SDE

The general, multidimensional Markov SDE is presented here, along with the corresponding chain rule, establishing proper matrix-vector notation, or extensions where the standard linear algebra is inadequate, for what follows. In the case of the vector[1] state process $\mathbf{X}(t) = [X_i(t)]_{n_x \times 1}$ on some n_x-dimensional state space $\mathcal{D}_x$, the multidimensional SDE can be of the form

$$d\mathbf{X}(t) \overset{\text{sym}}{=} \mathbf{f}(\mathbf{X}(t), t)dt + g(\mathbf{X}(t), t)d\mathbf{W}(t) + h(\mathbf{X}(t), t, \mathbf{Q})d\mathbf{P}(t; Q, \mathbf{X}(t), t), \qquad (5.81)$$

where also

$$\int_Q h(\mathbf{X}(t), t, \mathbf{q})\mathcal{P}(\mathbf{dt}, \mathbf{dq}; \mathbf{X}(t), t) \overset{\text{dt}}{\underset{zol}{=}} h(\mathbf{X}(t), t, \mathbf{Q})d\mathbf{P}(t; Q, \mathbf{X}(t), t) \qquad (5.82)$$

is the notation for the space-time Poisson terms, $\mathbf{W}(t) = [W_i(t)]_{n_w \times 1}$ is an n_w-dimensional vector Wiener process, $\mathbf{P}(t; Q, \mathbf{X}(t), t) = [P_i(t; \mathbf{X}(t), t)]_{n_p \times 1}$ is an n_p-dimensional vector state-dependent Poisson process, the coefficient $\mathbf{f}$ has the same dimension as $\mathbf{X}$, and the coefficients in the set $\{g, h\}$ have dimensions commensurate in multiplication with the set of vectors $\{\mathbf{W}, \mathbf{P}\}$, respectively. Here, $\mathcal{P} = [\mathcal{P}_i]_{n_p \times 1}$ is a vector form of the Poisson random measure with mark random vector $\mathbf{Q} = [Q_i]_{n_p \times 1}$, and $\mathbf{dq} = [(q_i, q_i + dq_i)]_{n_p \times 1}$ is the symbolic vector version of the mark measure notation. The $d\mathbf{P}(t; \mathbf{X}(t), t)$ jump-amplitude coefficient has the component form

$$h(\mathbf{X}(t), t; \mathbf{Q}) = [h_{i,j}(\mathbf{X}(t), t; Q_j)]_{n_x \times n_p}$$

such that the jth Poisson component depends only on the jth mark Q_j since simultaneous jumps are unlikely.

In component and jump-counter form, the SDE is

$$dX_i(t) \overset{\text{dt}}{=} f_i(\mathbf{X}(t), t)dt + \sum_{j=1}^{n_w} g_{i,j}(\mathbf{X}(t), t)dW_j(t)$$

$$+ \sum_{j=1}^{n_p} h_{i,j}(\mathbf{X}(t), t, \mathbf{Q})dP_j(t; Q, \mathbf{X}(t), t) \qquad (5.83)$$

for $i = 1 : n_x$ state components. The jump of the ith state due to the jth Poisson process

$$[X_i](T_{j,k}) = h_{i,j}(\mathbf{X}(T_{j,k}^-), T_{j,k}^-, Q_{j,k}),$$

where $T_{j,k}^-$ is the prejump-time and its k realization with jump-amplitude mark $Q_{j,k}$. The diffusion noise components have zero mean,

$$\text{E}[dW_i(t)] = 0 \qquad (5.84)$$

for $i = 1 : n_w$, while **correlations** are allowed between components,

$$\text{Cov}[dW_i(t), dW_j(t)] = \rho_{i,j}dt = [\delta_{i,j} + \rho_{i,j}(1 - \delta_{i,j})]dt, \qquad (5.85)$$

for $i, j = 1 : n_x$, where $\rho_{i,j}$ is the correlation coefficient between i and j components.

---

[1]Boldface variables or processes denote column vector variables or processes, respectively. The subscript i usually denotes a row index in this notation, while j denotes a column index. For example, $\mathbf{X}(t) = [X_i(t)]_{n_x \times 1}$ denotes that X_i is the ith component for $i = 1 : n_x$ of the single-column vector $\mathbf{X}(t)$.

The jump-noise components, conditioned on $\mathbf{X}(t) = \mathbf{x}$, are Poisson distributed with $\mathcal{P}$ mean assumed to be of the form

$$\mathrm{E}[\mathcal{P}_j(\mathbf{dt}, \mathbf{dq}_j; \mathbf{X}(t), t)|\mathbf{X}(t) = \mathbf{x}] = \phi_{Q_j}^{(j)}(q_j; \mathbf{x}, t)dq_j\lambda_j(t; \mathbf{x}, t)dt \qquad (5.86)$$

for each jump component $j = 1 : n_p$ with jth density $\phi_Q^{(j)}(q_j; \mathbf{x}, t)$ depending only on q_j assuming independence of the marks for different Poisson components but IID for the same component, so that the Poisson mark integral is

$$\begin{aligned}
\mathrm{E}[dP_j(t; Q, \mathbf{X}(t), t)|\mathbf{X}(t) = \mathbf{x}] &= \mathrm{E}\left[\int_{Q_j} \mathcal{P}_j(\mathbf{dt}, \mathbf{dq}_j; \mathbf{x}(t), t)\right] \\
&= \int_{Q_j} \mathrm{E}\left[\mathcal{P}_j(\mathbf{dt}, \mathbf{dq}_j; \mathbf{x}(t), t)\right] \\
&= \int_{Q_j} \phi_Q^{(j)}(q_j; \mathbf{x}, t)dq_i\lambda_j(t; \mathbf{x}, t)dt \\
&= \lambda_j(t; \mathbf{x}, t)dt \qquad (5.87)
\end{aligned}$$

for $i = 1 : n_p$, while the components are assumed to be uncorrelated, with conditioning $\mathbf{X}(t) = \mathbf{x}$ preassumed for brevity,

$$\mathrm{Cov}[\mathcal{P}_j(\mathbf{dt}, \mathbf{dq}_j; \mathbf{x}, t)\mathcal{P}_k(\mathbf{dt}, \mathbf{dq}_k; \mathbf{x}, t)] = \phi_Q^{(j)}(q_j; \mathbf{x}, t)\delta(q_k - q_j)dq_k dq_j\lambda_j(t; \mathbf{x}, t)dt,$$
$$(5.88)$$

generalizing the scalar form (5.15) to vector form, and

$$\begin{aligned}
\mathrm{Cov}[dP_j(t; Q_j, \mathbf{x}, t), dP_k(t; Q_k, \mathbf{x}, t)] &= \int_{Q_j}\int_{Q_k} \mathrm{Cov}[\mathcal{P}_j(\mathbf{dt}, \mathbf{dq}_j; \mathbf{x}, t)\mathcal{P}_k(\mathbf{dt}, \mathbf{dq}_k; \mathbf{x}, t)] \\
&= \lambda_j(t; \mathbf{x}, t)dt\, \delta_{j,k} \qquad (5.89)
\end{aligned}$$

for $j, k = 1 : n_p$, there being enough complexity for most applications. In addition, it is assumed that, as vectors, the diffusion noise $d\mathbf{W}$, Poisson noise $d\mathbf{P}$, and mark random variable $\mathbf{Q}$ are pairwise independent, but the mark random variable depends on the existence of a jump.

This Poisson formulation is somewhat different from others, such as [95, Part 2, Chapter 2]. The linear combination form has been found to be convenient for both jumps and diffusion when there are several sources of noise in the application.

5.3.1 Conditional Infinitesimal Moments in Multidimensions

The conditional infinitesimal moments for the vector state process $\mathbf{X}(t)$ are more easily calculated by component first, using the noise infinitesimal moments (5.84)–(5.89). The conditional infinitesimal mean is

$$\mathrm{E}[dX_i(t)|\mathbf{X}(t) = \mathbf{x}] = f_i(\mathbf{x}, t)dt + \sum_{j=1}^{n_w} g_{i,j}(\mathbf{x}, t)\mathrm{E}[dW_j(t)]$$

$$+ \sum_{j=1}^{n_p} \int_{\mathcal{Q}_j} h_{i,j}(\mathbf{x}, t, q_j) E[\mathcal{P}_j(\mathbf{dt}, \mathbf{dq}_j; \mathbf{x}, t)]$$

$$= f_i(\mathbf{x}, t)dt + \sum_{j=1}^{n_p} \int_{\mathcal{Q}_j} h_{i,j}(\mathbf{x}, t, q_j)\phi_Q^{(j)}(q_j; \mathbf{x}, t)dq_j\lambda_j(t; \mathbf{x}, t)dt$$

$$= \left[f_i(\mathbf{x}, t) + \sum_{j=1}^{n_p} \overline{h}_{i,j}(\mathbf{x}, t)\lambda_j(t; \mathbf{x}, t) \right] dt, \tag{5.90}$$

where $\overline{h}_{i,j}(\mathbf{x}, t) \equiv E_Q[h_{i,j}(\mathbf{x}, t, Q_j)]$. Thus, in vector form,

$$\mathrm{E}[d\mathbf{X}(t)|\mathbf{X}(t) = \mathbf{x}] = \left[\mathbf{f}(\mathbf{x}, t)dt + \overline{h}(\mathbf{x}, t)\boldsymbol{\lambda}(t; \mathbf{x}, t) \right] dt, \tag{5.91}$$

where $\boldsymbol{\lambda}(t; \mathbf{x}, t) = [\lambda_i(t; \mathbf{x}, t)]_{n_p \times 1}$.

For the conditional infinitesimal covariance, again with preassuming conditioning on $\mathbf{X}(t) = \mathbf{x}$ for brevity,

$$\mathrm{Cov}[dX_i(t), dX_j(t)] = \sum_{k=1}^{n_w}\sum_{\ell=1}^{n_w} g_{i,k}(\mathbf{x}, t)g_{j,\ell}(\mathbf{x}, t)\mathrm{Cov}[dW_k(t), dW_\ell(t)]$$

$$+ \sum_{k=1}^{n_p}\sum_{\ell=1}^{n_p} \int_{\mathcal{Q}_k} \int_{\mathcal{Q}_\ell} h_{i,k}(\mathbf{x}, t; q_k)h_{j,\ell}(\mathbf{x}, t; q_\ell)$$

$$\cdot \mathrm{Cov}[\mathcal{P}_k(\mathbf{dt}, \mathbf{dq}_k; \mathbf{x}, t), \mathcal{P}_\ell(\mathbf{dt}, \mathbf{dq}_\ell; \mathbf{x}, t)]$$

$$= \sum_{k=1}^{n_w}\left(g_{i,k}(\mathbf{x}, t)g_{j,k}(\mathbf{x}, t) + \sum_{\ell \neq k} \rho_{k,\ell}g_{i,k}(\mathbf{x}, t)g_{j,\ell}(\mathbf{x}, t) \right) dt$$

$$+ \sum_{k=1}^{n_p} (h_{i,k}h_{j,k})(\mathbf{x}, t)\phi_Q^{(k)}(q_k; \mathbf{x}, t)\lambda_k(t; \mathbf{x}, t)dt$$

$$= \sum_{k=1}^{n_w}\left(g_{i,k}(\mathbf{x}, t)g_{j,k}(\mathbf{x}, t) + \sum_{\ell \neq k} \rho_{k,\ell}g_{i,k}(\mathbf{x}, t)g_{j,\ell}(\mathbf{x}, t) \right) dt$$

$$+ \sum_{k=1}^{n_p} \overline{(h_{i,k}h_{j,k})}(\mathbf{x}, t)\lambda_k(t; \mathbf{x}, t)dt \tag{5.92}$$

for $i = 1 : n_x$ and $j = 1 : n_x$ in precision-dt, where the infinitesimal jump-diffusion covariance formulas (5.85) and (5.88) have been used. Hence, the matrix-vector form of this covariance is

$$\mathrm{Cov}[d\mathbf{X}(t), d\mathbf{X}^\top(t)|\mathbf{X}(t) = \mathbf{x}] \stackrel{\mathrm{dt}}{=} \left[g(\mathbf{x}, t)R'g^\top(\mathbf{x}, t) + \overline{\mathbf{h}\Lambda\mathbf{h}^\top}(\mathbf{x}, t) \right] dt, \tag{5.93}$$

where

$$R' \equiv \left[\rho_{i,j} \right]_{n_w \times n_w} = \left[\delta_{i,j} + \rho_{i,j}(1 - \delta_{i,j}) \right]_{n_w \times n_w}, \tag{5.94}$$

$$\Lambda = \Lambda(t; \mathbf{x}, t) = \left[\lambda_i(t; \mathbf{x}, t) \delta_{i,j} \right]_{n_p \times n_p}. \tag{5.95}$$

The jump in the ith component of the state at jump-time $T_{j,k}$ in the underlying jth component of the vector Poisson process is

$$[X_i](T_{j,k}) \equiv X_i(T_{j,k}^+) - X_i(T_{j,k}^-) = h_{i,j}(X(T_{j,k}^-), T_{j,k}^-; Q_{j,k}) \tag{5.96}$$

for $k = 1 : \infty$ jumps and $i = 1 : n_x$ state components, now depending on the jth mark's kth realization $Q_{j,k}$ at the prejump-time $T_{j,k}^-$ at the kth jump of the jth component Poisson process.

5.3.2   Stochastic Chain Rule in Multidimensions

The stochastic chain rule for a scalar function $\mathbf{Y}(t) = \mathbf{F}(\mathbf{X}(t), t)$, twice continuously differentiable in $\mathbf{x}$ and once in t, comes from the expansion

$$dY(t) = d\mathbf{F}(\mathbf{X}(t), t) = \mathbf{F}(\mathbf{X}(t) + d\mathbf{X}(t), t + dt) - \mathbf{F}(\mathbf{X}(t), t) \tag{5.97}$$

$$= \mathbf{F}_t(\mathbf{X}(t), t) + \sum_{i=1}^{n_x} \frac{\partial \mathbf{F}}{\partial x_i}(\mathbf{X}(t), t) \left(f_i(\mathbf{X}(t), t)dt + \sum_{k=1}^{n_w} g_{i,k}(\mathbf{X}(t), t)dW_k(t) \right)$$

$$+ \frac{1}{2} \sum_{i=1}^{n_x} \sum_{j=1}^{n_x} \sum_{k=1}^{n_w} \sum_{\ell=1}^{n_w} \left(\frac{\partial^2 \mathbf{F}}{\partial x_i \partial x_j} g_{i,k} g_{j,\ell} \right)(\mathbf{X}(t), t)dW_k(t)dW_\ell(t)$$

$$+ \sum_{j=1}^{n_p} \int_{\mathcal{Q}} \left(\mathbf{F} \left(\mathbf{X}(t) + \widehat{\boldsymbol{h}}_j(\mathbf{X}(t), t, q_j), t \right) - \mathbf{F}(\mathbf{X}(t), t) \right)$$

$$\cdot \mathcal{P}_j(\mathbf{dt}, \mathbf{dq}_j; \mathbf{X}(t), t),$$

$$\stackrel{\mathrm{dt}}{=} \left(\mathbf{F}_t(\mathbf{X}(t), t) + \mathbf{f}^\top(\mathbf{X}(t), t) \nabla_x[\mathbf{F}](\mathbf{X}(t), t) \right) dt$$

$$+ \frac{1}{2} \sum_{i=1}^{n_x} \sum_{j=1}^{n_x} \frac{\partial^2 \mathbf{F}}{\partial x_i \partial x_j} \sum_{k=1}^{n_w} \left(g_{i,k} g_{j,k} + \sum_{\ell \neq k}^{n_w} \rho_{k,\ell} g_{i,k} g_{j,\ell} \right)(\mathbf{X}(t), t)dt$$

$$+ \sum_{j=1}^{n_p} \int_{\mathcal{Q}_j} \Delta_j[\mathbf{F}]\mathcal{P}_j$$

$$= \left[\mathbf{F}_t + \mathbf{f}^\top \nabla_x[\mathbf{F}] + \frac{1}{2} \left(g R' g^\top \right) : \nabla_x \left[\nabla_x^\top[\mathbf{F}] \right] \right](\mathbf{X}(t), t)dt$$

$$+ \int_{\mathcal{Q}} \Delta^\top[\mathbf{F}]\mathcal{P}$$

to precision-dt. Here,

$$\nabla_x[\mathbf{F}] \equiv \left[\frac{\partial \mathbf{F}}{\partial x_i}(\mathbf{x}, t) \right]_{n_x \times 1}$$

is the state space gradient (a column n_x-vector),

$$\nabla_x^\top[\mathbf{F}] \equiv \left[\frac{\partial\mathbf{F}}{\partial x_j}(\mathbf{x}, t)\right]_{1\times n_x}$$

is the transpose of the state space gradient (a row n_x-vector),

$$\nabla_x\left[\nabla_x^\top[\mathbf{F}]\right] \equiv \left[\frac{\partial^2\mathbf{F}}{\partial x_i\partial x_j}(\mathbf{x}, t)\right]_{n_x\times n_x}$$

is the Hessian matrix for $\mathbf{F}$, R' is a correlation matrix defined in (5.94),

$$A : B \equiv \sum_{i=1}^{n}\sum_{j=1}^{n} A_{i,j}B_{i,j} = \text{Trace}[AB^\top] \tag{5.98}$$

is the **double-dot product** of two $n \times n$ matrices, related to the trace,

$$\widehat{\boldsymbol{h}}_j(\mathbf{x}, t, q_j) \equiv [h_{i,j}(\mathbf{x}, t, q_j)]_{n_x\times 1} \tag{5.99}$$

is the jth jump-amplitude vector corresponding to the jth Poisson process,

$$\begin{aligned}\Delta^\top[\mathbf{F}] &= \left[\Delta_j[\mathbf{F}](\mathbf{X}(t), t, q_j)\right]_{1\times n_p} \\ &\equiv \left[\mathbf{F}(\mathbf{X}(t) + \widehat{\boldsymbol{h}}_j(\mathbf{X}(t), t, q_j), t) - \mathbf{F}(\mathbf{X}(t), t)\right]_{1\times n_p}\end{aligned} \tag{5.100}$$

is the general jump-amplitude change vector for any t, and

$$\mathcal{P} = \left[\mathcal{P}_i(\mathbf{dt}, \mathbf{dq}_i; \mathbf{X}(t), t)\right]_{n_p\times 1}$$

is the Poisson random measure vector condition. The corresponding jump in $\mathbf{Y}(t)$ due to the jth Poisson component and its kth realization is

$$[\mathbf{Y}]\left(T_{j,k}^-\right) = \mathbf{F}\left(\mathbf{X}\left(T_{j,k}^-\right) + \widehat{\boldsymbol{h}}_j\left(\mathbf{X}\left(T_{j,k}^-\right), T_{j,k}^-, Q_{j,k}\right), T_{j,k}^-\right) - \mathbf{F}\left(\mathbf{X}\left(T_{j,k}^-\right), T_{j,k}^-\right).$$

Example 5.26. *Merton's Analysis of the Black–Scholes Option Pricing Model.*
A good application of multidimensional SDEs in finance is the survey of Merton's analysis [201], [203, Chapter 8] *of the Black–Scholes* [34] *financial options pricing model in Section 10.2 of Chapter 10. This treatment will serve as motivation for the study of SDEs and contains details not in Merton's paper.*

5.4   Distributed Jump SDE Models Exactly Transformable

Here, exactly transformable distributed jump-diffusion SDE models are listed, in the scalar and the vector cases, where conditions are applicable.

5.4.1   Distributed Jump SDE Models Exactly Transformable

- **Distributed scalar jump SDE:**

$$dX(t) = f(X(t), t)dt + g(X(t), t)dW(t) + \int_{Q} h(X(t), t, q)\mathcal{P}(\mathbf{dt}, \mathbf{dq}).$$

- **Transformed scalar process:** $Y(t) = F(X(t), t).$

- **Transformed scalar SDE:**

$$dY(t) = \left(F_t + F_x f + \frac{1}{2} F_{xx} g^2 \right) dt + F_x g \, dW(t)$$
$$+ \int_{Q} (F(X(t) + h(X(t), t, q), t) - F(X(t), t))\mathcal{P}(\mathbf{dt}, \mathbf{dq}).$$

- **Target explicit scalar SDE:**

$$dY(t) = C_1(t)dt + C_2(t)dW(t) + \int_{Q} C_3(t, q)\mathcal{P}(\mathbf{dt}, \mathbf{dq}).$$

5.4.2   Vector Distributed Jump SDE Models Exactly Transformable

- **Vector distributed jump SDE:**

$$d\mathbf{X}(t) = \mathbf{f}(\mathbf{X}(t), t)dt + g(\mathbf{X}(t), t)d\mathbf{W}(t) + \int_{Q} h(\mathbf{X}(t), t, \mathbf{q})\mathcal{P}(\mathbf{dt}, \mathbf{dq}).$$

- **Vector transformed process:** $\mathbf{Y}(t) = \mathbf{F}(\mathbf{X}(t), t).$

- **Transformed component SDE:**

$$dY_i(t) = \left(F_{i,t} + \sum_j F_{i,j} f_j + \frac{1}{2} \sum_j \sum_k \sum_l F_{i,jk} g_{jl} g_{kl} \right) dt$$
$$+ \sum_j F_{i,j} \sum_l g_{jl} dW_l(t)$$
$$+ \sum_\ell \int_{Q} (y_i(\mathbf{X} + \mathbf{h}_\ell, t) - F_i(\mathbf{X}, t))\mathcal{P}_\ell(\mathbf{dt}, \mathbf{dq}_\ell),$$

$$\mathbf{h}_\ell(\mathbf{x}, t, \mathbf{q}_\ell) \equiv [h_{i,\ell}(\mathbf{x}, t, q_\ell)]_{m \times 1}.$$

- **Transformed vector SDE:**

$$d\mathbf{Y}(t) = \left(\mathbf{F}_t + (\mathbf{f}^T \nabla_x)\mathbf{F} + \frac{1}{2}(gg^T : \nabla_x \nabla_x)\mathbf{F} \right) dt + \left((gd\mathbf{W}(t))^T \nabla_x \right)\mathbf{F}$$
$$+ \sum_\ell \int_{Q} (\mathbf{F}(\mathbf{X} + \mathbf{h}_\ell, t) - \mathbf{F}(\mathbf{X}, t))\mathcal{P}_\ell(\mathbf{dt}, \mathbf{dq}_\ell).$$

- **Vector target explicit SDE:**

$$d\mathbf{Y}(t) = \mathbf{C}_1(t)dt + \mathbf{C}_2(t)d\mathbf{W}(t) + \sum_\ell \int_\mathcal{Q} \mathbf{C}_{3,\ell}(t, q_\ell)\mathcal{P}_\ell(\mathbf{dt}, \mathbf{dq}_\ell).$$

- **Original coefficients:**

$$\mathbf{f}(\mathbf{x}, t) = (\nabla_x \mathbf{F}^T)^{-T} \Bigg(\mathbf{C}_1(t) - y_t$$

$$-\frac{1}{2}(\nabla_x \mathbf{F}^T)^{-T} C_2 C_2^T (\nabla_x \mathbf{F}^T)^{-1} : \nabla_x \nabla_x^T \mathbf{F} \Bigg);$$

$$g(\mathbf{x}, t) = (\nabla_x \mathbf{F}^T)^{-T} C_2(t),$$

$$\mathbf{F}(\mathbf{x} + \mathbf{h}_\ell, t) = \mathbf{F}(\mathbf{x}, t) + \mathbf{C}_{3,\ell}(t, q_\ell). \quad \textit{(Note: left in implicit form.)}$$

- **Vector affine transformation example:**

$$\mathbf{F} = A(t)\mathbf{x} + \mathbf{B}(t),$$

$$\mathbf{F}_t = A'\mathbf{x} + \mathbf{B}',$$

$$(\nabla_x \mathbf{F}^T)^T = A,$$

$$\mathbf{f}(\mathbf{x}, t) = A^{-1}(\mathbf{C}_1(t) - A'\mathbf{x} - \mathbf{B}'),$$

$$g(\mathbf{x}, t) = A^{-1} C_2(t),$$

$$\mathbf{h}_\ell(\mathbf{x}, t, q_\ell) = A^{-1} \mathbf{C}_{3,\ell}(t, q_\ell).$$

5.5   Exercises

1. Simulate $X(t)$ for the log-normally distributed jump-amplitude case with mean $\mu_j = \mathrm{E}[Q] = 0.28$ and variance $\sigma_j^2 = \mathrm{Var}[Q] = 0.15$ for the linear jump-diffusion SDE model (5.42) using $\mu_d(t) = 0.82 \sin(2\pi t - 0.75\pi)$, $\sigma_d(t) = 0.88 - 0.44 \sin(2\pi t - 0.75\pi)$, and $\lambda(t) = 8.0 - 1.82 \sin(2\pi t - 0.75\pi)$, $N = 10{,}000$ time steps, $t_0 = 0$, $t_f = 1.0$, $X(0) - x_0$, for $k = 4$ random states, i.e., $\nu(t, Q) = \nu_0(Q) = \exp(Q) - 1$ with Q normally distributed. Plot the k sample states $X_j(t_i)$ for $j = 1 : k$, along with theoretical mean state path, $\mathrm{E}[X(t_i)]$ from (5.49), and the sample mean state path, i.e., $M_x(t_i) = \sum_{j=1}^k X_j(t_i)/k$, all for $i = 1 : N + 1$.

   (Hint: Modify the linear mark-jump-diffusion SDE simulator of Example 5.25 with MATLAB code C.15 from Online Appendix C, and Corollary 5.9 for the discrete exponential expectation.)

2. For the log-double-uniform jump distribution,

$$
\phi_Q(q; t) \equiv
\begin{cases}
0, & -\infty < q < a(t) \\
p_1(t)/|a|(t), & a(t) \le q < 0 \\
p_2(t)/b(t), & 0 \le q \le b(t) \\
0, & b(t) < q < +\infty
\end{cases},
\tag{5.101}
$$

where $p_1(t)$ is the probability of a negative jump and $p_2(t)$ is the probability of a positive jump on $a(t) < 0 \le b(t)$, show that

(a) $\mathrm{E}_Q[Q] = \mu_j(t) = (p_1(t)a(t) + p_2(t)b(t))/2$;

(b) $\mathrm{Var}_Q[Q] = \sigma_j^2(t) = (p_1(t)a^2(t) + p_2(t)b^2(t))/3 - \mu_j^2(t)$;

(c) $\mathrm{E}_Q\!\left[(Q - \mu_j(t))^3\right] = (p_1(t)a^3(t) + p_2(t)b^3(t))/4 - \mu_j(t)(3\sigma_j^2(t) + \mu_j^2(t))$;

(d) $\mathrm{E}[\nu(Q)] = \mathrm{E}[\exp(Q) - 1]$, where the answer needs to be derived.

3. Show that the Itô mean square limit for the integral of the product of two correlated mean-zero, dt-variance, differential diffusion processes, $dW_1(t)$ and $dW_2(t)$, symbolically satisfies the SDE,

$$
dW_1(t)dW_2(t) \overset{\mathrm{dt}}{=} \rho(t)dt,
\tag{5.102}
$$

where

$$
\mathrm{Cov}[\Delta W_1(t_i), \Delta W_2(t_i)] \simeq \rho(t_i)\Delta t_i
$$

for sufficiently small Δt_i. Are any modified considerations required if $\rho = 0$ or $\rho = \pm 1$? You may use the bivariate normal density in (B.144), boundedness theorem, Theorem B.59, Table B.1 of selected moments, and other material in Online Appendix B of preliminaries.

4. Finish the proof of Corollary 5.13 by showing the diffusion part using the techniques of Theorem 5.11, equation (5.53).

5. Prove the corresponding corollary for the variance of $X(t)$ from the solution of the linear SDE:

Corollary 5.27. *Variance of X(t) for Linear SDE.*
Let $X(t)$ be the solution (5.45) with $\overline{v^2}(t) \equiv \mathrm{E}[v^2(t, Q)]$ of (5.42). Then

$$
\mathrm{Var}[dX(t)/X(t)] \overset{\mathrm{dt}}{=} \sigma_d^2(t) + \overline{v^2}(t)
$$

and

$$
\mathrm{Var}[X(t)] = \mathrm{E}^2[X(t)]\left(\exp\left(\int_{t_0}^{t} \mathrm{Var}[dX(s)/X(s)]ds\right) - 1\right).
\tag{5.103}
$$

Be sure to state what extra conditions on processes and precision are needed that were not needed for proving Corollary 5.13 on $\mathrm{E}[X(t)]$.

6. Justify (5.93) for the covariance in multidimensions by giving the reasons for each step in the derivation. See the proof for (5.27).

Suggested References for Further Reading

- Çinlar, 1975 [56]

- Cont and Tankov, 2004 [60]

- Gihman and Skorohod, 1972 [95, Part 2, Chapter 2]

- Hanson, 1996 [109]

- Itô, 1951 [150]

- Kushner and Dupuis, 2001 [179]

- Øksendal and Sulem, 2005 [223]

- Snyder and Miller, 1991 [252, Chapter 4 and 5]

- Westman and Hanson, 1999 [276]

- Westman and Hanson, 2000 [277]

- Zhu and Hanson, 2006 [293]

Chapter 6

Stochastic Optimal Control: Stochastic Dynamic Programming

6.1 Stochastic Optimal Control Problem

This main chapter introduces the optimal stochastic control problem. For many application systems, solving an SDE, or, for that matter an ODE, to obtain its behavior is only part of the problem. The SDE is, in fact, a stochastic ordinary differential equation (SODE). Another, very significant part is finding out how to control the SDE or ODE as a model for controlling the application system.

Thus, the general jump-diffusion SDE (5.81) is reformulated with an additional process, the **vector control process** $\mathbf{U}(t) = [U_i(t)]_{n_u \times 1}$ on some n_u-dimensional control space $\mathcal{D}_u$,

$$d\mathbf{X}(t) \overset{\text{sym}}{=} \mathbf{f}(\mathbf{X}(t), \mathbf{U}(t), t)dt + g(\mathbf{X}(t), \mathbf{U}(t), t)d\mathbf{W}(t)$$
$$+ \int_Q h(\mathbf{X}(t), \mathbf{U}(t), t, \mathbf{q})\mathcal{P}(\mathbf{dt}, \mathbf{dq}; \mathbf{X}(t), \mathbf{U}(t), t), \qquad (6.1)$$

when $t_0 \le t \le t_f$ subject to a given initial state $\mathbf{X}(t_0) = \mathbf{x}_0$, where again $\mathbf{X}(t) = [X_i(t)]_{n_x \times 1}$ is the **vector state process** on some n_x-dimensional state space $\mathcal{D}_x$. The stochastic processes are the n_w-dimensional vector Wiener process or diffusion process $\mathbf{W}(t) = [W_i(t)]_{n_w \times 1}$ and the n_p-dimensional vector Poisson process or jump process $\mathbf{P}(t; \mathbf{Q}, \mathbf{X}(t), \mathbf{U}(t), t) =$

169

$[P_i(t; Q_i, \mathbf{X}(t), \mathbf{U}(t), t)]_{n_p \times 1}$, with IID jump-amplitude mark random vector $\mathbf{Q} = [Q_i]_{n_p \times 1}$ and Poisson random measure

$$\mathcal{P}(\mathbf{dt}, \mathbf{dq}; \mathbf{X}(t), \mathbf{U}(t), t) = [\mathcal{P}_i(\mathbf{dt}, \mathbf{dq}; \mathbf{X}(t), \mathbf{U}(t), t)]_{n_p \times 1}.$$

The n_p-dimensional vector state-dependent compound Poisson process can also be defined as in Chapter 5 in a zero-one law form,

$$\int_{\mathcal{Q}} h(\mathbf{X}(t), \mathbf{U}(t), t, \mathbf{q}) \mathcal{P}(\mathbf{dt}, \mathbf{dq}; \mathbf{X}(t), \mathbf{U}(t), t)$$

$$\underset{zol}{\overset{\mathrm{dt}}{=}} \left[\sum_{j=1}^{n_p} h_{i,j}(\mathbf{X}(t), \mathbf{U}(t), t, Q) dP_j(t; Q_j \mathbf{X}(t), \mathbf{U}(t), t) \right]_{n_x \times 1}$$

with

$$\mathrm{E}[d\mathbf{P}(t; \mathbf{Q}, \mathbf{X}(t), \mathbf{U}(t), t) | \mathbf{X}(t) = \mathbf{x}, \mathbf{U}(t) = \mathbf{u}] = \lambda(t; \mathbf{x}, \mathbf{u}, t) dt,$$

and jump in the ith state component

$$[X_i](T_{j,k}) = h_{i,j}(\mathbf{X}(T_{j,k}^-), \mathbf{U}(T_{j,k}^-), T_{j,k}^-, Q_{j,k}),$$

where $\lambda(t; \mathbf{x}, \mathbf{u}, t)$ is the jump-rate vector, $T_{j,k}^-$ is the kth jump-time of the jth differential Poisson process, and $Q_{j,k}$ is the corresponding mark.

The coefficient functions are the $n_x \times 1$ plant function $\mathbf{f}(\mathbf{x}, \mathbf{u}, t)$, having the same dimension as the state $\mathbf{x}$, the $n_x \times n_w$ volatility function $g(\mathbf{x}, \mathbf{u}, t)$, or square root of the variance of the diffusion term, and the $n_x \times n_p$ jump-amplitude of the jump term $h(\mathbf{x}, \mathbf{u}, t, \mathbf{Q})$, where $\mathbf{Q}$ is the underlying jump-amplitude random mark process, the space part of the space-time Poisson process.

The optimization objective functional for a control formulation may be the combination of a final cost at time t_f and cumulative instantaneous costs, given the initial data $(\mathbf{x}_0, t_0)$. For instance,

$$V[\mathbf{X}, \mathbf{U}, t_f](\mathbf{x}_0, t_0) = \int_{t_0}^{t_f} C(\mathbf{X}(s), \mathbf{U}(s), s) ds + S(\mathbf{X}(t_f), t_f) \qquad (6.2)$$

is a functional of the processes $\mathbf{X}(t)$ and $\mathbf{U}(t)$, where $C(\mathbf{x}, \mathbf{u}, t)$ is the scalar instantaneous or **running cost function** on the **time horizon** $(t_0, t_f]$ given the state at t_0 and $S(\mathbf{x}, t)$ is the **final cost function**; both are assumed continuous. This is the Bolza form of the objective. The objective $V[\mathbf{X}, \mathbf{U}, t_f](\mathbf{x}_0, t_0)$ is a functional of the state $\mathbf{X}$ and control process $\mathbf{U}$, i.e., a function of functions, while it is also dependent on the values of the initial data $(\mathbf{x}_0, t_0)$. The optimal control objective, in this case, is to minimize the expected total costs with respect to the control process on $(t_0, t_f]$. The feedback control of the multibody stochastic dynamical system (6.1) is illustrated in the block diagram displayed in Figure 6.1.

Prior to the optimization step, an averaging step, taking the conditional expectation, conditioned on some initial state, is **essential** to avoid the ill-posed problem of trying to optimize an uncertain, fluctuating objective. It is further assumed here that the running

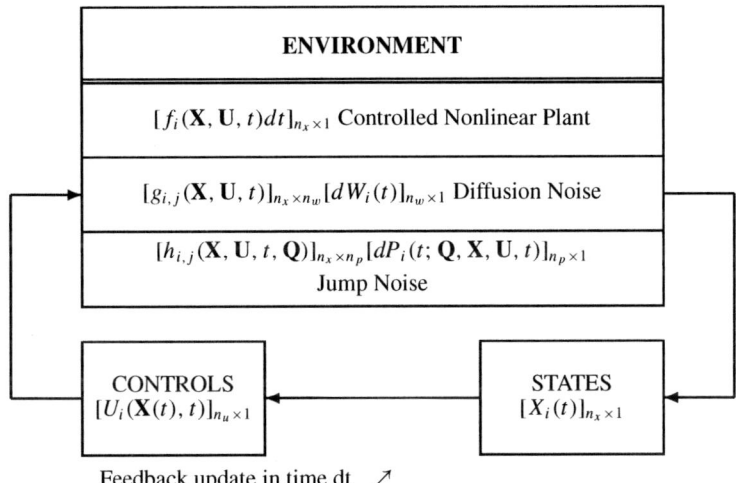

Figure 6.1. *Multibody stochastic dynamical system under feedback control.*

and terminal cost functions permit a unique minimum, subject to stochastic differential dynamics in the multidimensional jump-diffusion case (6.1). Hence, the optimal, expected cost for (6.2) is

$$v^*(\mathbf{x}_0, t_0) \equiv \min_{\mathbf{U}(t_0, t_f]} \left[\underset{(\mathbf{W}, \mathbf{P})(t_0, t_f]}{\mathrm{E}} \left[V[\mathbf{X}, \mathbf{U}, t_f](\mathbf{x}_0, t_0) \bigg| \mathbf{X}(t_0) = \mathbf{x}_0, \mathbf{U}(t_0) = \mathbf{u}_0 \right] \right] \quad (6.3)$$

with the expectation preceding the minimization so that the minimization problem is better posed by smoothing random fluctuations through averaging. In the optimization in (6.4), it is implicit that the stochastic dynamical system (6.1) is a constraint. The minimization over $\mathbf{U}(t_0, t_f]$ denotes the minimization over the control path $\mathbf{U}(t)$ for $t \in (t_0, t_f]$, and similarly the expectation over $\{W, P\}(t_0, t_f]$ denotes expectation over the joint stochastic pair $\{W(t), P(t)\}$ for $t \in (t_0, t_f]$.

Recall that the maximum problem, as in the maximization of profits, portfolio returns, or utility, is an equivalent problem since

$$\max_{\mathbf{U}}[V[\mathbf{X}, \mathbf{U}, t_f](x_0, t_0)] = - \min_{\mathbf{U}}[-V[\mathbf{X}, \mathbf{U}, t_f](\mathbf{x}_0, t_0)]$$

upon reversing the value. However, switching theoretical results from those for a minimum to get those of a maximum basically requires just replacing the minimum function min for the maximum function max, along with replacing positive definiteness conditions for negative definite conditions, in the case of regular optima. For software optimization packages that are designed for minimizations, the user needs to use the negative of the function to be maximized and to take the negative of the final minimum output, for example, the MATLAB built-in function `fminsearch`.

To implement the dynamic part of dynamic programming, the fixed initial condition $\mathbf{X}(t_0) = \mathbf{x}_0$ for the SDE (6.1) needs to be replaced by a more arbitrary start, $\mathbf{X}(t) = \mathbf{x}$, so that the start can be analytically manipulated. This is a small but important step to producing a time-varying objective amenable to analysis. Hence, the optimal expected value is

$$v^*(\mathbf{x}, t) \equiv \min_{\mathbf{U}_{(t,t_f]}} \left[\underset{(\mathbf{W}, \mathbf{P})_{(t,t_f]}}{\mathrm{E}} \left[V[\mathbf{X}, \mathbf{U}, t_f](\mathbf{x}, t) \Big| \mathbf{X}(t) = \mathbf{x}, \mathbf{U}(t) = \mathbf{u} \right] \right]. \quad (6.4)$$

Since the running cost integral vanishes when $t = t_f$, leaving only the terminal cost term conditioned on $\mathbf{X}(t_f) = \mathbf{x}$ and $\mathbf{U}(t_f) = \mathbf{u}$, a simple final condition for the optimal expected cost is

$$v^*(\mathbf{x}, t_f) = S(\mathbf{x}, t_f) \quad (6.5)$$

for any $\mathbf{x}$ in the state domain $\mathcal{D}_x$, assuming that the terminal cost function $S(\mathbf{x}, t_f)$ is a deterministic function. This final condition is the first clue, meaning that dynamic programming will use a backward program in time.

6.2 Bellman's Principle of Optimality

The basic assumption is that the optimization and expectation can be decomposed over increments in time. Bellman's principle of optimality can be systematically derived from optimization in time-steps proceeding backward from the final increment to the initial increment. Also, in the Markov processes case here, the independent increment properties of the Wiener and Poisson processes permit the decomposition of the expectation over time. This decomposition conveniently complements the decomposition of the optimization over time, as in the deterministic case presented in Online Section A.4.

The semi-close-open time-interval $(t, t_f]$ in the optimal expected cost formulation (6.4), given the state at time t, can be decomposed into disjoint increments $(t, t + \delta t]$ and $(t + \delta t, t_f]$ for fixed δt in $t < t + \delta t < t_f$. Symbolically, the **decomposition rules** are written as follows.

Rules 6.1. *Decomposition for Time, Integration, Expectation, and Minimization.*

- *Time domain decompostion into subintervals:*

$$(t, t_f] = (t, t + \delta t] + (t + \delta t, t_f]$$

needs to be further decomposed for discrete approximations into sufficiently small increments Δt_i for $i = 1:n + 1$ such that

$$t_{i+1} = t_i + \sum_{j=1}^{i} \Delta t_j,$$

$t_1 = t, t_\ell = t + \delta t$ *for some integer* $\ell \in [1, n + 1]$, $t_{n+1} = t_f$, $\delta t_n = \max_i[\Delta t_i] \to 0$ *as* $n \to \infty$. *Recall that the approximation to the stochastic dynamics* (6.1) *is*

$$\mathbf{X}_{i+1} \simeq \mathbf{X}_i + \int_{t_i}^{t_i + \Delta t_i} d\mathbf{X}(s) \simeq \mathbf{X}_i + \mathbf{f}_i \Delta t_i + g_i \Delta \mathbf{W}_i + h_i \Delta \mathbf{P}_i$$

for sufficiently small Δt_i, where, for example, $\mathbf{f}_i \equiv \mathbf{f}(\mathbf{X}_i, \mathbf{U}_i, t_i)$, *so that the change from $\mathbf{X}_i$ to $\mathbf{X}_{i+1}$ is due to the control $\mathbf{U}_i$ and random fluctuations $(\Delta \mathbf{W}_i, \Delta \mathbf{P}_i)$ determined from a prior stage.*

- *Integration additive decomposition rule:*

$$\int_t^{t_f} C(\mathbf{X}(s), \mathbf{U}(s), s)ds = \int_t^{t+\delta t} C(\mathbf{X}(s), \mathbf{U}(s), s)ds + \int_{t+\delta t}^{t_f} C(\mathbf{X}(s), \mathbf{U}(s), s)ds \qquad (6.6)$$

for the cumulative running costs by the regular additivity property of regular or Riemann-type integrals, or in terms of small increments in simplified notation. Let

$$V = \int_t^{t_f} Cds + S(\mathbf{X}(t_f), t_f) \simeq \sum_{i=1}^{n+1} \widehat{C}_i$$

be the forward approximation, where $\widehat{C}_i \equiv C_i \Delta t_i = C(\mathbf{X}_i, \mathbf{U}_i, t_i)\Delta t_i$ for $i = 1 : n-1$ and $\widehat{C}_{n+1} \equiv S(\mathbf{X}(t_f), t_f) = S(X_{n+1}, t_{n+1}) = S_{n+1}$.

- *Expectation operator multiplication decomposition rule:*

$$\overline{V} = \underset{(\mathbf{W},\mathbf{P})(t,t_f]}{\mathrm{E}} [V|\mathcal{C}(t)] = \underset{(\mathbf{W},\mathbf{P})(t,t+\delta t]}{\mathrm{E}} \left[\underset{(\mathbf{W},\mathbf{P})(t+\delta t,t_f]}{\mathrm{E}} [V|\mathcal{C}(t+\delta t)] \bigg| \mathcal{C}(t) \right], \qquad (6.7)$$

where V is an objective function and $\mathcal{C}(t) = \{\mathbf{X}(t), \mathbf{U}(t)\}$ is the conditioning at time t. This decomposition relies on the corresponding decomposition of the Markov processes $\mathbf{W}(t)$ and $\mathbf{P}(t; \mathbf{Q}, \mathbf{X}(t), \mathbf{U}(t), t)$ into independent increments, so that the expectation over $\{\mathbf{W}(s), \mathbf{P}(s)\}$ for $s \in (t, t_f]$ is the product of expectation over $\{\mathbf{W}(s), \mathbf{P}(s)\}$ for $s \in (t, t+\delta t]$ and expectation over $\{\mathbf{W}(r), \mathbf{P}(r)\}$ for $r \in (t+\delta t, t_f]$. To compute the expectation over the path of a Markov process, we need to approximate the process by a sum of n independent increments for sufficiently large n to obtain sufficiently small Δt_i and then take the product of the expectations with respect to each of these independent increments, and finally take the limit as $n \rightarrow \infty$ relying on mean square convergence in the result, as in the first two chapters. In simple notation,

$$\overline{V} = \mathrm{E}[V|\mathcal{C}(t)] \simeq \mathrm{E}\left[\sum_{i=1}^{n+1} \widehat{C}_i \bigg| \mathbf{X}_1, \mathbf{U}_1 \right],$$

where $\mathrm{E}[\widehat{C}_1|\mathbf{X}_1, \mathbf{U}_1] \equiv \mathrm{E}_0[\widehat{C}_1] = \widehat{C}_1$ since $\widehat{C}_1 = C(\mathbf{X}_1, \mathbf{U}_1, t_1)\Delta t_1$,

$$\mathrm{E}\left[\widehat{C}_2 \big| \mathbf{X}_1, \mathbf{U}_1\right] = \underset{(\Delta\mathbf{W}_1,\Delta\mathbf{P}_1)}{\mathrm{E}} \left[\widehat{C}_2 \big| \mathbf{X}_1, \mathbf{U}_1\right] \equiv \mathrm{E}_1\left[\widehat{C}_2\right] = \Pi_{j=0}^1 \mathrm{E}_j\left[\widehat{C}\right],$$

$$\mathrm{E}\left[\widehat{C}_3 \big| \mathbf{X}_3, \mathbf{U}_3\right] = \mathrm{E}_1\left[\underset{(\Delta\mathbf{W}_2,\Delta\mathbf{P}_2)}{\mathrm{E}} \left[\widehat{C}_3 \mid \mathbf{X}_2, \mathbf{U}_2\right] \right] \equiv \Pi_{j=0}^2 \mathrm{E}_j\left[\widehat{C}_3\right],$$

so in general,

$$\mathrm{E}\left[\widehat{C}_{i+1} \big| \mathbf{X}_1, \mathbf{U}_1\right] = \Pi_{j=0}^i \mathrm{E}_j\left[\widehat{C}_{i+1}\right]$$

with

$$\mathrm{E}_j\left[\widehat{C}_{i+1}\right] \equiv \underset{(\Delta\mathbf{W}_j,\Delta\mathbf{P}_j)}{\mathrm{E}} \left[\widehat{C}_i \big| \mathbf{X}_j, \mathbf{U}_j\right]$$

for $j = 0 : i$, $\mathrm{E}\big[\widehat{C}_{i+1}\big] = \widehat{C}_{i+1}$ and finally,

$$\overline{V} \simeq \sum_{i=1}^{n+1} \Pi_{j=0}^{i-1} \mathrm{E}_j[\widehat{C}_i] \longrightarrow \mathop{\mathrm{E}}_{(\boldsymbol{W},\boldsymbol{P})(t,t+\delta t]} \left[\int_t^{t+\delta t} C\,ds \right.$$

$$\left. + \mathop{\mathrm{E}}_{(\boldsymbol{W},\boldsymbol{P})(t+\delta t,t_f]} \left[\int_{t+\delta t}^{t_f} C\,ds + S(\mathbf{X}(t_f),t_f) \middle| (\mathbf{X},\mathbf{U})(t+\delta t) \right] \middle| (\mathbf{X},\mathbf{U})(t) \right]$$

as $n \to \infty$, confirming the construction, assuming mean square convergence.

- ***Minimization operator multiplication decomposition rule:***

$$\overline{V}^* = \min_{\boldsymbol{U}(t,t_f]} \big[\, \overline{V} \,\big] = \min_{\boldsymbol{U}(t,t+\delta t]} \left[\min_{\boldsymbol{U}(t+\delta t,t_f]} \big[\, \overline{V} \,\big] \right], \qquad (6.8)$$

where $\overline{V}$ is the expected value of an objective so that the decomposition rule is analogous to the use in deterministic dynamic programming. This decomposition depends on the reasonable heuristic idea that, given a minimum on the later interval $(t + \delta t, t_f]$, taking the minimum of the given minimum over the small earlier interval $(t, t + \delta t]$ yields the minimum over the longer interval $(t, t_f]$. In terms of the small increments (Δt_i) construction,

$$\overline{V}^* \simeq \sum_{i=1}^{n+1} \min_{\boldsymbol{U}(t,t_f]} \big[\Pi_{j=0}^{i-1} \mathrm{E}_j \big[\widehat{C}_i\big] \big] = \sum_{i=1}^{n+1} \left[\Pi_{j=0}^{i-1} \min_{\boldsymbol{U}_j} \mathrm{E}_j \big[\widehat{C}_i\big] \right] = \sum_{i=1}^{n+1} \Pi_{j=0}^{i-1} \mathrm{ME}_j \big[\widehat{C}_i\big],$$

where

$$\mathrm{ME}_0 \equiv \min_{\boldsymbol{U}_1}\big[\mathrm{E}_0\big[\widehat{C}_0 \,\big|\, (\mathbf{X}_0,\mathbf{U}_0)\big]\big]$$

and

$$\mathrm{ME}_j \equiv \min_{\boldsymbol{U}_j}\big[\mathrm{E}_j\big[\widehat{C}_i \,\big|\, \mathbf{X}_j, \mathbf{U}_j\big]\big]$$

for $j = 0 : i - 1$. As $n \to \infty$ and $\delta t_n \to 0$, then

$$\overline{V}^* \to \min_{\boldsymbol{U}(t,t+\delta t]} \left[\mathop{\mathrm{E}}_{(\boldsymbol{W},\boldsymbol{P})(t,t+\delta t]} \left[\int_t^{t+\delta t} C\,ds + \min_{\boldsymbol{U}(t+\delta t,t_f]} \left[\mathop{\mathrm{E}}_{(\boldsymbol{W},\boldsymbol{P})(t+\delta t,t_f]} \right. \right. \right.$$

$$\left. \left. \left. \left[\int_{t+\delta t}^{t_f} C\,ds + S(\mathbf{X}(t_f),t_f) \middle| (\mathbf{X},\mathbf{U})(t+\delta t) \right] \right] \middle| (\mathbf{X},\mathbf{U})(t) \right] \right].$$

The optimal decomposition seems to work for many examples. However, for empirical counterexamples, see Rust [240].

Thus, **optimal expected cost** (6.4) can be decomposed as follows:

$$
v^*(\mathbf{x}, t) = \min_{U(t,t+\delta t]} \left[\mathop{\mathrm{E}}_{(W,P)(t,t+\delta t]} \left[\int_t^{t+\delta t} C(\mathbf{X}(s), \mathbf{U}(s), s)\,ds \right. \right.
$$

$$
+ \min_{5U(t+\delta t, t_f]} \left[\mathop{\mathrm{E}}_{(W,P)(t+\delta t, t_f]} \left[\int_{t+\delta t}^{t_f} C(\mathbf{X}(s), \mathbf{U}(s), s)\,ds + S(\mathbf{X}(t_f), t_f) \right. \right.
$$

$$
\left. \left. \left. \left. \left|\{\mathbf{X}(t + \delta t), \mathbf{U}(t + \delta t)\}\right] \right] \right| \mathbf{X}(t) = \mathbf{x}, \mathbf{U}(t) = \mathbf{u} \right] \right]
$$

$$
= \min_{U(t,t+\delta t]} \left[\mathop{\mathrm{E}}_{(W,P)(t,t+\delta t]} \left[\int_t^{t+\delta t} C(\mathbf{X}(s), \mathbf{U}(s), s)\,ds \right. \right.
$$

$$
\left. \left. + v^*(\mathbf{X}(t + \delta t), t + \delta t) \right| \mathbf{X}(t) = \mathbf{x}, \mathbf{U}(t) = \mathbf{u} \right] \right], \tag{6.9}
$$

where the definition (6.4) for v^* has been reused with the arguments shifted by the time-step δt, since the inner part of the decomposition that is on $(t + \delta t, t_f]$ is precisely the definition of v^* in (6.4) but with arguments shifted from $(\mathbf{x}, t)$ to $(\mathbf{X}(t + dt), t + dt)$. Thus, (6.9) is a backward recursion relation for v^*. The subscript notation $\mathbf{U}(t, t + \delta t]$ under the min operator means that the minimum is with respect to $\mathbf{U}$ in the range $(t, t + \delta t]$, with similar subscript notation $\{\mathbf{W}, \mathbf{P}\}(t, t_f]$ for the expectation operator. Thus, we have formally derived the fundamental recursive formula of *stochastic dynamic programming*.

Lemma 6.2. *Bellman's Principle of Optimality.*
Under the assumptions of the decomposition rules (6.6), (6.7), (6.8) *and the properties of jump-diffusions,*

$$
v^*(\mathbf{x}, t) = \min_{U(t,t+\delta t]} \left[\mathop{\mathrm{E}}_{(W,P)(t,t+\delta t]} \left[\int_t^{t+\delta t} C(\mathbf{X}(s), \mathbf{U}(s), s)\,ds \right. \right.
$$

$$
\left. \left. + v^*(\mathbf{X}(t + \delta t), t + \delta t) \right| \mathbf{X}(t) = \mathbf{x}, \mathbf{U}(t) = \mathbf{u} \right] \right]. \tag{6.10}
$$

The argument of the minimum when it exists, within the control domain $\mathcal{D}_u$, is the optimal control $\mathbf{u}^* = \mathbf{u}^*(\mathbf{x}, t)$. Although the SDE is a forward differential equation integrated forward from the initial condition, the optimal control problem is a backward general or functional equation integrated backward from the final time. The backward equation is quite basic when one has a final objective, here optimal costs. Then the primary question is where to start initially to get that optimum. People do backward calculations all the time, such as when going to a scheduled meeting or a class—the meeting time is fixed and the problem is to estimate what time one should leave to get to the meeting. However, when economic decisions are made, the decision makers' behavior may not follow Bellman's principle of optimality, according to the studies of Rust [240].

In general, capital letters are used for stochastic processes and lowercase letters for conditioned or realized variables.

6.3 Hamilton–Jacobi–Bellman (HJB) Equation of Stochastic Dynamic Programming (SDP)

Using the **principle of optimality** (6.10) and by taking the limit of small δt, replacing δt by dt, we can systematically derive the PDE of stochastic dynamic programming (SDP), also called the stochastic **Hamilton–Jacobi–Bellman** (HJB) equation, for the general, multidimensional Markov dynamics case. From the increment form of the state differential $d\mathbf{X}(t) = \mathbf{X}(t+dt) - \mathbf{X}(t)$, we consider the expansion of the state argument

$$\mathbf{X}(t+dt) = \mathbf{X}(t) + d\mathbf{X}(t)$$

about $\mathbf{X}(t)$ for small $d\mathbf{X}(t)$ and about the explicit time argument $t+dt$ about t in the limit of small time increments dt, using an extension of Taylor approximations extended to include discontinuous (i.e., Poisson) and nonsmooth (i.e., Wiener) processes. Sufficient differentiability of the optimal value function $v^*(\mathbf{x}, t)$, at least to first order in time and second order in state, is assumed except when the function's state argument has Poisson jumps. The spirit of the derivation of the multidimensional chain rule (5.97) is applied to the principle of optimality (6.10), except that the mean square limit substitution for the bilinear Wiener $W_i(t)W_j(t)$ process is not needed here because of the preoptimization expectation operation. Then neglecting $o(dt)$ terms as $dt \to 0^+$ (strictly, we are really working with finite increments δt) and substituting for the conditioning on $\mathbf{X}(t)$ and $\mathbf{U}(t)$, an intermediate reduction of the optimal expected value is

$$
v^*(\mathbf{x}, t) \overset{\mathrm{dt}}{=} \min_{\mathbf{u}} \left[EdPWt \left[C(\mathbf{x}, \mathbf{u}, t)dt + v^*(\mathbf{x}, t) + v_t^*(\mathbf{x}, t)dt \right. \right.
$$
$$
+ \nabla_{\mathbf{x}}^{\top}[v^*](\mathbf{x}, t) \cdot (\mathbf{f}(\mathbf{x}, \mathbf{u}, t)dt + g(\mathbf{x}, \mathbf{u}, t)d\mathbf{W}(t)) \qquad (6.11)
$$
$$
+ \frac{1}{2} d\mathbf{W}^{\top}(t) g^{\top}(\mathbf{x}, \mathbf{u}, t) \nabla_{\mathbf{x}}[\nabla_{\mathbf{x}}^{\top}[v^*]](\mathbf{x}, t)(g(\mathbf{x}, \mathbf{u}, t)d\mathbf{W}(t))
$$
$$
\left. \left. + \sum_{j=1}^{n_p} \int_{\mathcal{Q}} \left(v^*(\mathbf{x} + \widehat{\boldsymbol{h}}_j(\mathbf{x}, \mathbf{u}, t, q_j), t) - v^*(\mathbf{x}, t) \right) \mathcal{P}_j(\mathbf{dt}, \mathbf{dq}_j; \mathbf{x}, \mathbf{u}, t) \right] \right],
$$

where it has been assumed that the random mark variables $Q_j = q_j$ are pairwise independently distributed and the jump-amplitude is separable in the marks. So

$$h(\mathbf{x}, \mathbf{u}, t, \mathbf{q}) = [h_{i,j}(\mathbf{x}, \mathbf{u}, t, q_j)]_{n_x \times n_p} \qquad (6.12)$$

with a corresponding multiplicative factoring of the Poisson random measure. Recall from Chapter 5 (5.99) that the jth vector component of the jump-amplitude is

$$\widehat{\boldsymbol{h}}_j(\mathbf{x}, \mathbf{u}, t, q_j) \equiv \left[h_{i,j}(\mathbf{x}, \mathbf{u}, t, q_j) \right]_{n_x \times 1} \qquad (6.13)$$

for $j = 1 : n_p$, corresponding to the jth Poisson process

$$dP_j(t; \mathbf{x}, \mathbf{u}, t) = \int_{\mathcal{Q}} \mathcal{P}_j(\mathbf{dt}, \mathbf{dq}_j; \mathbf{x}, \mathbf{u}, t)$$

in terms of the jth Poisson mark-time random measure $\mathcal{P}_j$. Note that the first t argument of dP_j is the time implicit to the Poisson process, while the second t argument is an explicit time corresponding to the implicit state and control parametric dependence.

The next step is to take the conditional expectation over the now isolated differential Wiener and Poisson processes but to do so by expanding them in components to facilitate understanding of the step and suppressing some arguments for simplicity:

$$
v^*(\mathbf{x}, t) \overset{dt}{=} v^*(\mathbf{x}, t) + v_t^*(\mathbf{x}, t)dt + \min_{\mathbf{u}} \Bigg[C(\mathbf{x}, \mathbf{u}, t)dt
$$

$$
+ \nabla_{\mathbf{x}}^{\top}[v^*](\mathbf{x}, t) \cdot \left(\mathbf{f}(\mathbf{x}, \mathbf{u}, t)dt + \sum_{i=1}^{n_w} g_i(\mathbf{x}, \mathbf{u}, t) \mathrm{E}_{dW_i}[dW_i(t)] \right)
$$

$$
+ \frac{1}{2} \sum_{i=1}^{n_w} \sum_{j=1}^{n_w} \mathrm{E}_{dW_i, dW_j} \left[dW_i(t) dW_j(t) \right] \left[g^{\top}(\mathbf{x}, \mathbf{u}, t) \nabla_{\mathbf{x}}[\nabla_{\mathbf{x}}^{\top}[v^*]] g(\mathbf{x}, \mathbf{u}, t) \right]_{i,j}
$$

$$
+ \sum_{j=1}^{n_p} \int_{\mathcal{Q}} \left(v^*(\mathbf{x} + \widehat{\mathbf{h}}_j(\mathbf{x}, \mathbf{u}, t, q_j), t) - v^*(\mathbf{x}, t) \right) \mathrm{E}_{\mathcal{P}_j} \left[\mathcal{P}_j(\mathbf{dt}, \mathbf{dq}_j; \mathbf{x}, \mathbf{u}, t) \right] \Bigg]
$$

$$
\overset{ind}{\underset{inc}{=}} v^*(\mathbf{x}, t) + v_t^*(\mathbf{x}, t)dt + \min_{\mathbf{u}} \Bigg[C(\mathbf{x}, \mathbf{u}, t)dt + \nabla_{\mathbf{x}}^{\top}[v^*](\mathbf{x}, t) \left(\mathbf{f}(\mathbf{x}, \mathbf{u}, t)dt + 0 \right)
$$

$$
+ \frac{1}{2} \sum_{i=1}^{n_w} \left[1 + \sum_{j=1}^{n_w} \rho_{i,j}(1 - \delta_{i,j}) \right] \left[g^{\top}(\mathbf{x}, \mathbf{u}, t) \nabla_{\mathbf{x}}[\nabla_{\mathbf{x}}^{\top}[v^*]](\mathbf{x}, t) g(\mathbf{x}, \mathbf{u}, t) \right]_{i,j} dt
$$

$$
+ \sum_{j=1}^{n_p} \lambda_j(t; \mathbf{x}, \mathbf{u}, t) \int_{\mathcal{Q}} \left(v^*(\mathbf{x} + \widehat{\mathbf{h}}_j(\mathbf{x}, \mathbf{u}, t, q_j), t) - v^*(\mathbf{x}, t) \right)
$$

$$
\cdot \Phi_{Q_j}(\mathbf{dq}_j; \mathbf{x}, \mathbf{u}, t)dt \Bigg], \tag{6.14}
$$

where we have used the expectations

- $$\mathrm{E}[dW_i(t)] = 0, \quad \mathrm{E}[dW_i(t)dW_i(t)] = (\delta_{i,j} + \rho_{i,j}(1 - \delta_{i,j}))dt \tag{6.15}$$

with correlation coefficient $\rho_{i,j}$ and

$$
\mathrm{E}[\mathcal{P}_j(\mathbf{dt}, \mathbf{dq}_j; \mathbf{x}, \mathbf{u}, t)] = \lambda_j(t; \mathbf{x}, \mathbf{u}, t) \phi_{Q_j}(q_j; \mathbf{x}, \mathbf{u}, t) dq_j dt. \tag{6.16}
$$

Also, with sufficiently small dt, $\mathbf{U}(t, t + dt]$ has been replaced by the conditioned control vector $\mathbf{u}$ at t.

Note that the $v^*(\mathbf{x}, t)$ value on both sides of the equation cancel and then the remaining common multiplicative factors of dt also cancel, so the **HJB equation** of stochastic dynamic programming (SDP) has been derived for this general case.

Theorem 6.3. *HJB Equation for SDP.*
If $v^*(\mathbf{x}, t)$ is twice differentiable in $\mathbf{x}$ and once differentiable in t, while the operator decomposition rules (6.8)–(6.6) are valid, then

$$
0 = v_t^*(\mathbf{x}, t) + \min_{\mathbf{u}} [\mathcal{H}(\mathbf{x}, \mathbf{u}, t)] \equiv v_t^*(\mathbf{x}, t) + \mathcal{H}^*(\mathbf{x}, t), \tag{6.17}
$$

*where the **Hamiltonian** (technically, a pseudo-Hamiltonian) functional is given by*

$$
\mathcal{H}(\mathbf{x}, \mathbf{u}, t) \equiv C(\mathbf{x}, \mathbf{u}, t) + \nabla_{\mathbf{x}}^{\top}[v^*](\mathbf{x}, t) \cdot \mathbf{f}(\mathbf{x}, \mathbf{u}, t)
$$
$$
+ \frac{1}{2} \left(g R' g^{\top} \right) (\mathbf{x}, \mathbf{u}, t) : \nabla_{\mathbf{x}} \left[\nabla_{\mathbf{x}}^{\top}[v^*] \right] (\mathbf{x}, t)
$$
$$
+ \sum_{j=1}^{n_p} \lambda_j(t; \mathbf{x}, \mathbf{u}, t) \int_{\mathcal{Q}} \left[v^* \left(\mathbf{x} + \widehat{h}_j(\mathbf{x}, \mathbf{u}, t, q_j), t \right) - v^*(\mathbf{x}, t) \right]
$$
$$
\cdot \phi_{Q_j}(q_j; \mathbf{x}, \mathbf{u}, t) dq_j, \tag{6.18}
$$

where the correlation modified indentity R' is defined in (5.94) as

$$
R' = R'(t) \equiv \left[\delta_{i,j} + \rho_{i,j}(t)(1 - \delta_{i,j}) \right]_{n_w \times n_w} \tag{6.19}
$$

and where the correlation coefficient between i and j components is

$$
\rho_{i,j}(t) dt = \text{Cov}[dW_i(t), dW_j(t)] \tag{6.20}
$$

provided $j \neq i$ for $i, j = 1 : n_x$. The double-dot product $A : B$ is defined in (5.98). The optimal control, if it exists, is given by

$$
\mathbf{u}^* = \mathbf{u}^*(\mathbf{x}, t) = \underset{\mathbf{u} \in \mathcal{D}_u}{\text{argmin}} \ [\mathcal{H}(\mathbf{x}, \mathbf{u}, t)], \tag{6.21}
$$

subject to any control constraints, but if there are no control constraints when $\mathcal{D}_u = \mathbb{R}^{n_u}$, then the regular control $\mathbf{u}^{(\text{reg})}(\mathbf{x}, t)$ is obtained, provided appropriate second order conditions are satisfied for the given optimal specification.

This **HJB equation** (6.17) of SDP is no ordinary PDE, but has the following properties.

Properties 6.4.

- *The HJB equation is a **functional PDE or partial integral differential equation (PIDE)** due to the presence of the minimum operator* min *and the Poisson integral term (the last term) with steps in the state argument of the optimal value function v^* due to the jump-amplitude.*

- *The HJB equation is a scalar valued equation but has an $(\boldsymbol{nu} + 1)$-**dimensional solution** consisting of the scalar optimal value function $v^* = v^*(\mathbf{x}, t)$ and the optimal control vector $\mathbf{u}^* = \mathbf{u}^*(\mathbf{x}, t)$ as well. These **dual solutions** are generally tightly coupled in functional dependence. In general, this tight coupling requires a number of iterations between v^* and $\mathbf{u}^*$ to obtain a reasonable approximation to the $(nu + 1)$-dimensional solution. However, it should be noted that the optimal control $\mathbf{u}(\mathbf{x}, t)$ in (6.21) is deterministic, and if the $\mathbf{x}$ dependence is genuine, then it is also feedback optimal control. In fact, the HJB equation is a deterministic equation as well.*

- *A further complication in this functional PDE or PIDE is that the HJB equation of SDP (6.17) has **global state dependence** due to the Poisson jump functional integral term,* *whereas the HJB equation for purely Gaussian or Wiener processes is essentially a diffusion equation that has only **local state dependence** since it depends only on the values $v^*(\mathbf{x}, t)$, $\mathbf{u}^*(\mathbf{x}, t)$, the gradient vector $\nabla_{\mathbf{x}}[v^*](\mathbf{x}, t)$, and the Hessian matrix of second order derivatives $\nabla_{\mathbf{x}}[\nabla_{\mathbf{x}}^{\top}[v^*]](\mathbf{x}, t)$ at $(\mathbf{x}, t)$. Contrast this with the random noise case including the Poisson random measure disturbance, with local dependence at $\mathbf{x}$, but global dependence on a range of points at $\mathbf{x} + \widehat{\boldsymbol{h}}_j(\mathbf{x}, \mathbf{u}, t, q_j)$ depending on the Poisson mark distribution.*

- *Notice that due to the conditional expectation with respect to the current state at time t, only the mean of the stochastic noise is taken, i.e., the mean of the Wiener noise with correlation in (6.15) and Poisson noise in (6.16).*

While letting $C^*(\mathbf{x}, t) \equiv C(\mathbf{x}, \mathbf{u}^*, t)$, $\mathbf{f}^*(\mathbf{x}, t) \equiv \mathbf{f}(\mathbf{x}, \mathbf{u}^*, t)$, $g^*(\mathbf{x}, t) \equiv g(\mathbf{x}, \mathbf{u}^*, t)$, $\widehat{\boldsymbol{h}}_j^*(\mathbf{x}, t, q_j) \equiv \widehat{\boldsymbol{h}}_j(\mathbf{x}, \mathbf{u}^*, t, q_j)$, and so forth for all control-dependent functions, the HJB equation takes the form of a backward parabolic PDE, except with an additional integral term:

$$
\begin{aligned}
0 &= v_t^*(\mathbf{x}, t) + \mathcal{H}(\mathbf{x}, \mathbf{u}^*(\mathbf{x}, t), t) \\
&= v_t^*(\mathbf{x}, t) + C^*(\mathbf{x}, t) + \nabla_{\mathbf{x}}^{\top}[v^*](\mathbf{x}, t) \cdot \mathbf{f}^*(\mathbf{x}, t) \\
&\quad + \frac{1}{2}\left(g^* R' g^{*\top}\right)(\mathbf{x}, t) : \nabla_{\mathbf{x}}\left[\nabla_{\mathbf{x}}^{\top}[v^*]\right](\mathbf{x}, t) \\
&\quad + \sum_{j=1}^{n_p} \lambda_j^*(t; \mathbf{x}, t) \int_{\mathcal{Q}} \Delta_j[v^*](\mathbf{x}, t, q_j) \phi_{Q_j}^*(q_j; \mathbf{x}, t) dq_j,
\end{aligned}
\tag{6.22}
$$

where the jth jump-increment is defined as

$$
\Delta_j[v^*](\mathbf{x}, t, q_j) \equiv v^*\left(\mathbf{x} + \widehat{\boldsymbol{h}}_j^*(\mathbf{x}, t, q_j), t\right) - v^*(\mathbf{x}, t)
\tag{6.23}
$$

and the double-dot product $(A : B)$ is defined in (5.98). The final condition is given by $v^*(\mathbf{x}, t_f) = S(\mathbf{x}, t_f)$.

The name of the Hamilton–Jacobi–Bellman equation comes from the facts that Bellman [25, 26] was the founding developer of dynamic programming and that the general evolution equation, $v_t^*(\mathbf{x}, t) + \mathcal{H}^*(\mathbf{x}, t) = 0$, is called a Hamilton–Jacobi equation, and $\mathcal{H}(\mathbf{x}, \mathbf{u}, t)$ is like a classical Hamiltonian. Sometimes, the HJB equation (6.17) is called simply the Bellman equation, or the stochastic dynamic programming equation, or the PDE of stochastic dynamic programming, or, in particular, the PIDE of stochastic dynamic programming.

6.4 Linear Quadratic Jump-Diffusion (LQJD) Problem

The linear quadratic jump-diffusion (LQJD) problem is also called a linear quadratic Gaussian–Poisson (LQGP) problem or jump-linear quadratic Gaussian (JLQG) problem. The Markov property of the jump-diffusion processes described in this book leads to an analogous dynamic programming formulation to dynamic programming for deterministic processes, as in the deterministic LQ problem of Online Subsection A.4.4. In this chapter,

the LQJD problem is presented in more generality than in Online Appendix A, on deterministic optimal control.

The linear quadratic problem in both state and control leads to a quadratic decomposition of the optimal value function with respect to the state and a linear or feedback
decomposition of the optimal control. However, first the LQJD problem is examined for a
special case that is linear quadratic in control only to show how much of an advantage is
gained by the control dependence alone. For many applications it is not appropriate to have
the problem be linear quadratic in the state.

6.4.1   LQJD in Control Only (LQJD/U) Problem

A general variant of the LQJD problem is the LQJD/U problem—that is, LQJD in control
only. Just having a control problem linear quadratic in control retains an important feature
of the full linear quadratic control problem in that the optimal control can be solved for
exactly in terms of the optimal value, even though the problem may not be LQJD in the
state. The restricted linear quadratic problem in the control only will be treated first to
examine how far the analysis can be taken before treating the full linear quadratic problem
in the state and the control. In many control problems, the state dependence of the plant
function $\mathbf{f}(\mathbf{x}, \mathbf{u}, t)$ is dictated by the application and may be significantly nonlinear, but the
control dependence of the dynamics is up to the control designer, who might choose to
make the control simple, e.g., linear, so that the control process will be manageable for the
control manager. Hence, the LQJD problem in control may only be more appropriate for
some applications. In the past, linear systems were preferred since linear methods were
well known, but now nonlinear methods and problems have become more prevalent as we
try to make more realistic models for applications.

Let the jump-diffusion linear quadratic model, in the control only, be given with the
plant function for the deterministic or nonnoise dynamics term,

$$\mathbf{f}(\mathbf{x}, \mathbf{u}, t) = \mathbf{f}_0(\mathbf{x}, t) + f_1(\mathbf{x}, t)\mathbf{u}, \tag{6.24}$$

with the diffusion term,

$$g(\mathbf{x}, \mathbf{u}, t) = g_0(\mathbf{x}, t), \tag{6.25}$$

assumed control-independent for simplicity, with a jump term decomposition corresponding
to independent sources of n_p-type jumps,

$$h(\mathbf{x}, \mathbf{u}, t, \mathbf{q}) = h_0(\mathbf{x}, t, \mathbf{q}) = [h_{0,i,j}(\mathbf{x}, t, q_j)]_{n_x \times n_p}, \tag{6.26}$$

also assumed control-independent along with the very simplified Poisson noise,

$$d\mathbf{P}(t; \mathbf{Q}, \mathbf{x}, \mathbf{u}, t) = d\mathbf{P}(t; \mathbf{Q}, \mathbf{x}, \mathbf{u}, t), \quad \mathrm{E}[d\mathbf{P}(t; \mathbf{Q}, \mathbf{x}, \mathbf{u}, t)] = \lambda(t; \mathbf{x}, \mathbf{u}, t)dt, \tag{6.27}$$

and finally with the quadratic running cost function,

$$C(\mathbf{x}, \mathbf{u}, t) = C_0(\mathbf{x}, t) + \mathbf{C}_1^\top(\mathbf{x}, t)\mathbf{u} + \frac{1}{2}\mathbf{u}^\top C_2(\mathbf{x}, t)\mathbf{u}. \tag{6.28}$$

It is assumed that all right-hand-side coefficients are commensurate in multiplication and that the product is of the same type at that on the left-hand side. A crucial assumption in the case of a minimum objective is that the quadratic control $C_2(\mathbf{x}, t)$ is positive definite, but $C_2(\mathbf{x}, t)$ can be assumed to be symmetric without loss of generality by the symmetric property of quadratic forms (B.133).

Thus, the pseudo-Hamiltonian is quadratic in the control,

$$\mathcal{H}(\mathbf{x}, \mathbf{u}, t) = \mathcal{H}_0(\mathbf{x}, t) + \mathcal{H}_1^\top(\mathbf{x}, t)\mathbf{u} + \frac{1}{2}\mathbf{u}^\top \mathcal{H}_2(\mathbf{x}, t)\mathbf{u}, \qquad (6.29)$$

where the scalar coefficient is

$$\mathcal{H}_0(\mathbf{x}, t) = \left[C_0 + \mathbf{f}_0^\top \nabla_\mathbf{x}[v^*] + \frac{1}{2}g_0 g_0^\top : \nabla_\mathbf{x}[\nabla_\mathbf{x}[v^*]] \right](\mathbf{x}, t)$$

$$+ \sum_{j=1}^{n_p} \lambda_j(t; \mathbf{x}, t) \int_{\mathcal{Q}_j} \Delta_j[v^*](\mathbf{x}, t, q_j)\phi_{\mathcal{Q}_j}(q_j)dq_j, \qquad (6.30)$$

where the double-dot product (5.98) is $GG^\top : A = \text{Trace}[G^\top AG]$, while the jump-increment is

$$\Delta_j[v^*](\mathbf{x}, t, q_j) \equiv v^*\left(\mathbf{x} + \widehat{\mathbf{h}}_j(\mathbf{x}, t, q_j), t\right) - v^*(\mathbf{x}, t),$$

the linear control coefficient n_u-vector is

$$\mathcal{H}_1(\mathbf{x}, t) = \left[\mathbf{C}_1 + f_1^\top \nabla_\mathbf{x}[v^*]\right](\mathbf{x}, t), \qquad (6.31)$$

and the quadratic control coefficient $n_u \times n_u$-matrix is simply

$$\mathcal{H}_2(\mathbf{x}, t) = C_2(\mathbf{x}, t), \qquad (6.32)$$

where $\mathcal{H}_2(\mathbf{x}, t)$ is assumed to be symmetric along with $C_2(\mathbf{x}, t)$. If the minimum cost is the objective, then $\mathcal{H}_2(\mathbf{x}, t)$ is positive definite since $C_2(\mathbf{x}, t)$ is assumed to be positive definite.

Thus, in search of a regular control minimum, the critical points of the pseudo-Hamiltonian $\mathcal{H}(\mathbf{x}, \mathbf{u}, t)$ are considered by examining the zeros of its gradient,

$$\nabla_\mathbf{u}[\mathcal{H}](\mathbf{x}, \mathbf{u}, t) = \mathcal{H}_1(\mathbf{x}, t) + \mathcal{H}_2(\mathbf{x}, t)\mathbf{u} = \mathbf{0}, \qquad (6.33)$$

yielding the regular control,

$$\mathbf{u}^{(\text{reg})}(\mathbf{x}, t) = -\mathcal{H}_2^{-1}(\mathbf{x}, t)\mathcal{H}_1(\mathbf{x}, t)$$
$$= -C_2^{-1}(\mathbf{x}, t)\left(\mathbf{C}_1 + f_1^\top \nabla_\mathbf{x}[v^*]\right)(\mathbf{x}, t) \qquad (6.34)$$

with the existence of the inverse being guaranteed by positive definiteness. The fact that the regular control can be solved for exactly in terms of the optimal value $v^*(\mathbf{x}, t)$ is a major benefit of having an LQJD problem that is just quadratic in the control. If the usual LQ assumption is made that the control is unconstrained, then the regular control is also the optimal control,

$$\mathbf{u}^*(\mathbf{x}, t) = \mathbf{u}^{(\text{reg})}(\mathbf{x}, t) \qquad (6.35)$$

and the optimal Hamiltonian using (6.34) is

$$
\begin{aligned}
\mathcal{H}^*(\mathbf{x}, t) &\equiv \mathcal{H}(\mathbf{x}, \mathbf{u}^*, t) \\
&= \left[\mathcal{H}_0 - \mathcal{H}_1^\top \mathcal{H}_2^{-1} \mathcal{H}_1 + \frac{1}{2} \mathcal{H}_1^\top \mathcal{H}_2^{-\top} \mathcal{H}_2 \mathcal{H}_2^{-1} \mathcal{H}_1 \right] (\mathbf{x}, t) \\
&= \left[\mathcal{H}_0 - \frac{1}{2} \mathcal{H}_1^\top \mathcal{H}_2^{-1} \mathcal{H}_1 \right] (\mathbf{x}, t),
\end{aligned}
\tag{6.36}
$$

where by symmetry the inverse transpose $\mathcal{H}_2^{-\top} = \mathcal{H}_2^{-1}$. Since the difference of the quadratic $\mathcal{H}$ in control from the designated minimum using the Taylor approximation form and the critical condition (6.33) is

$$
\begin{aligned}
\mathcal{H}(\mathbf{x}, \mathbf{u}, t) - \mathcal{H}^*(\mathbf{x}, t) &= \mathcal{H}_0 - \mathcal{H}^*(\mathbf{x}, t) + (\mathbf{u} - \mathbf{u}^*)^\top \nabla_\mathbf{u}[\mathcal{H}](\mathbf{x}, \mathbf{u}^*, t) \\
&\quad + \frac{1}{2}(\mathbf{u} - \mathbf{u}^*)^\top \nabla_\mathbf{u}[\nabla_\mathbf{u}^\top[\mathcal{H}]](\mathbf{x}, \mathbf{u}^*, t)(\mathbf{u} - \mathbf{u}^*) \\
&= \frac{1}{2} \mathcal{H}_1^\top \mathcal{H}_2^{-1} \mathcal{H}_1 + \frac{1}{2}(\mathbf{u} - \mathbf{u}^*)^\top \mathcal{H}_2 (\mathbf{u} - \mathbf{u}^*) \\
&= \frac{1}{2} \left(\mathcal{H}_1^\top \mathcal{H}_2^{-1} \mathcal{H}_1 \right)(\mathbf{x}, t) + \frac{1}{2}(\mathbf{u} - \mathbf{u}^*)^\top \mathcal{H}_2(\mathbf{x}, t)(\mathbf{u} - \mathbf{u}^*) \\
&\geq \frac{1}{2} \left(\mathcal{H}_1^\top \mathcal{H}_2^{-1} \mathcal{H}_1 \right)(\mathbf{x}, t) \geq 0,
\end{aligned}
\tag{6.37}
$$

it is always possible to solve the optimal control in the minimum problem if $C_2(\mathbf{x}, t)$ and thus $\mathcal{H}_2(\mathbf{x}, t)$ are symmetric positive definite. This corresponds to the minimum principle discussed for deterministic optimal control problems in Online Appendix A.

Within the generality of this linear quadratic problem in control only, the optimal control will generally be nonlinear in the state, so the corresponding HJB equation of SDP,

$$
v_t^*(\mathbf{x}, t) + \mathcal{H}^*(\mathbf{x}, t) = 0,
\tag{6.38}
$$

will be highly nonlinear in the state, with $\mathcal{H}^*(\mathbf{x}, t)$ given by (6.36) and coefficients given by (6.30), (6.31), and (6.32). This requires careful solution by numerical PDE or PIDE methods, or the computational methods of Chapter 8.

These LQJD/U derived results are summarized in the following theorem.

Theorem 6.5. *LQJD/U Equations.*
Let the problem be the LQJD in control-only problem, so that the deterministic plant function $\mathbf{f}(\mathbf{x}, \mathbf{u}, t)$ *is linear in the control as given in* (6.24)*, the coefficient* $g(\mathbf{x}, \mathbf{u}, t)$ *of the Wiener process* $d\mathbf{W}(t)$ *is given in* (6.25)*, the jump-amplitude* $h(\mathbf{x}, \mathbf{u}, t, \mathbf{q})$ *of the Poisson jump process* $d\mathbf{P}(t; \mathbf{Q}, \mathbf{x}, t)$ *is given by* (6.26)*, and the quadratic running cost* $C(\mathbf{x}, \mathbf{u}, t)$ *is given in* (6.28)*.*

Then, the Hamiltonian $\mathcal{H}(\mathbf{x}, \mathbf{u}, t)$ *is quadratic in the control* (6.29) *with coefficients* $\{\mathcal{H}_0(\mathbf{x}, t), \mathcal{H}_1(\mathbf{x}, t), \mathcal{H}_2(\mathbf{x}, t)\}$ *given in* (6.30)*,* (6.31)*,* (6.32)*, respectively. The optimal control vector, in the absence of control constraints, has the linear feedback control form*

$$
\mathbf{u}^*(\mathbf{x}, t) = \mathbf{u}^{(\text{reg})}(\mathbf{x}, t) = -C_2^{-1}(\mathbf{x}, t) \left[\mathbf{C}_1 + f_1^\top \nabla_\mathbf{x}[v^*] \right](\mathbf{x}, t)
\tag{6.39}
$$

as long as the quadratic control coefficient $C_2(\mathbf{x}, t)$ is positive definite in the case of a minimum expected objective and in the absence of constraints on the control. Assuming that an optimal value $v^*(\mathbf{x}, t)$ solution exists, then $v^*(\mathbf{x}, t)$ satisfies the HJB equation,

$$v_t^*(\mathbf{x}, t) + \left(\mathcal{H}_0 - \frac{1}{2}\mathcal{H}_1^\top \mathcal{H}_2^{-1}\mathcal{H}_1\right)(\mathbf{x}, t) = 0. \tag{6.40}$$

The solution $v^*(\mathbf{x}, t)$ is subject to the final condition

$$v^*(\mathbf{x}, t_f) = S(\mathbf{x}, t_f) \tag{6.41}$$

and any necessary boundary conditions.

For solutions of LQJD/U problems, computational methods are essential; see Hanson's 1996 chapter [109] or Chapter 8.

6.4.2   LLJD/U or the Case $C_2 \equiv 0$

If the quadratic cost coefficient $C_2(\mathbf{x}, t) \equiv 0$, then we have

$$\mathcal{H}(\mathbf{x}, \mathbf{u}, t) = \mathcal{H}_0(\mathbf{x}, t) + \mathcal{H}_1^\top(\mathbf{x}, t)\mathbf{u}, \tag{6.42}$$

which is the linear jump-diffusion (LLJD/U) problem in control only. The minimum with respect to the control depends on the linear cost coefficient

$$\mathcal{H}^*(\mathbf{x}, t) = \min_{\mathbf{u}}\left[\mathcal{H}_0(\mathbf{x}, t) + \mathcal{H}_1^\top(\mathbf{x}, t)\mathbf{u}\right] = \mathcal{H}_0(\mathbf{x}, t) + \min_{\mathbf{u}}\left[\mathcal{H}_1^\top(\mathbf{x}, t)\mathbf{u}\right]. \tag{6.43}$$

Since this is a problem of linear or singular control, it makes sense only if the control is constrained, e.g., has componentwise constraints

$$U_i^{(\min)} \le u_i \le U_i^{(\max)}. \tag{6.44}$$

For this type of constraint the minimum is separable by component and the optimal control is an n_u-dimensional bang-bang control

$$\mathcal{H}^*(\mathbf{x}, t) = \mathcal{H}_0(\mathbf{x}, t) + \sum_{i=1}^{n_u} \min\left[\mathcal{H}_{1,i}(\mathbf{x}, t)u_i\right]$$

$$= \mathcal{H}_0(\mathbf{x}, t) + \sum_{i=1}^{n_u} \begin{Bmatrix} \mathcal{H}_{1,i}(\mathbf{x}, t)U_i^{(\max)}, & \mathcal{H}_{1,i}(\mathbf{x}, t) < 0 \\ 0, & \mathcal{H}_{1,i}(\mathbf{x}, t) = 0 \\ \mathcal{H}_{1,i}(\mathbf{x}, t)U_i^{(\min)}, & \mathcal{H}_{1,i}(\mathbf{x}, t) > 0 \end{Bmatrix}$$

$$= \mathcal{H}_0(\mathbf{x}, t) + \frac{1}{2}\mathcal{H}_1(\mathbf{x}, t). * \left[\mathbf{U}^{(\min)}. * (1 + \mathbf{sgn}_1)\right.$$

$$\left. + \mathbf{U}^{(\max)}. * (1 - \mathbf{sgn}_1)\right], \tag{6.45}$$

where $\mathbf{1} \equiv [1]_{n_u \times 1}$, $\mathbf{sgn}_1 \equiv [\text{sgn}(\mathcal{H}_{1,i}(\mathbf{x}, t))]_{n_u \times 1}$,

$$\text{sgn}(x) \equiv \begin{Bmatrix} -1, & x < 0 \\ 0, & x = 0 \\ +1, & x > 0 \end{Bmatrix} \tag{6.46}$$

is the sign or signum function, $\mathbf{U}^{(\min)} \equiv [U_i^{(\min)}]_{n_u \times 1}$, $\mathbf{U}^{(\max)} \equiv [U_i^{(\max)}]_{n_u \times 1}$, and $\mathbf{v}.*\mathbf{u} \equiv [v_i u_i]_{n_u \times 1}$ is the dot-star or element-by-element product. The optimal control is undefined for components for which $\mathcal{H}_{1,i}(\mathbf{x}, t) = 0$ but otherwise is given in composite form:

$$u_i^*(\mathbf{x}, t) = \begin{cases} U_i^{(\max)}, & \mathcal{H}_{1,i}(\mathbf{x}, t) < 0 \\ U_i^{(\min)}, & \mathcal{H}_{1,i}(\mathbf{x}, t) > 0 \end{cases}. \tag{6.47}$$

If the components of $\mathcal{H}_1$ change sign often, that can lead to **chattering control**.

6.4.3   Canonical LQJD Problem

The standard or canonical LQJD problem is linear in the dynamics and quadratic in the costs with respect to both state and control vectors. This LQJD problem is a special case of the LQJD problem in control only and results in substantial simplifications of the solution with a quadratic state decomposition of the optimal value function and a linear or feedback decomposition of the optimal control vector. The decomposition of optimal value and control is similar to that of the deterministic linear quadratic (LQ) problem, but here the more general quadratic state and linear control decompositions are presented.

Let the more general jump-diffusion linear quadratic model be given with the plant function for the deterministic or nonnoise dynamics term and be linear in both state $\mathbf{X}(t)$ and $\mathbf{U}(t)$,

$$\mathbf{f}(\mathbf{x}, \mathbf{u}, t) = \mathbf{f}_0(t) + f_1^\top(t)\mathbf{x} + f_2^\top(t)\mathbf{u}, \tag{6.48}$$

with the first subscript indicating the degree, and the subsequent subscripts, if present, indicating either state (1) or control (2), with the diffusion term

$$g(\mathbf{x}, \mathbf{u}, t) = g_0(t) \tag{6.49}$$

assumed state-independent and control-independent for simplicity and with the jump term,

$$h(\mathbf{x}, \mathbf{u}, t, \mathbf{q}) = h_0(t, \mathbf{q}) \tag{6.50}$$

also assumed state-independent and control-independent for simplicity. The current form of the linear SDE (6.1) is written here as

$$dX(s) \overset{\text{sym}}{=} \mathbf{f}(\mathbf{X}(s), \mathbf{U}(s), s)ds + g_0(s)dW(s) + h_0(s, \mathbf{Q})dP(s; \mathbf{Q}, s) \tag{6.51}$$

on $t \le s \le t_f$ with $\mathrm{E}[dP(t; \mathbf{Q}, t)] = [\lambda_{0,j}(t)dt]_{n_p \times 1}$.

The quadratic running cost function is

$$C(\mathbf{x}, \mathbf{u}, t) = C_0(t) + \mathbf{C}_1^\top(t)\mathbf{x} + \mathbf{C}_2^\top(t)\mathbf{u}$$
$$+ \frac{1}{2}\mathbf{x}^\top C_{1,1}(t)\mathbf{x} + \mathbf{x}^\top C_{1,2}(t)\mathbf{u} + \frac{1}{2}\mathbf{u}^\top C_{2,2}(t)\mathbf{u} \tag{6.52}$$

and the terminal cost also has a general quadratic form

$$S(\mathbf{X}(t_f), t_f) = S_0(t_f) + \mathbf{S}_1^\top(t_f)\mathbf{X}(t_f) + \frac{1}{2}\mathbf{X}^\top(t_f)S_{1,1}(t_f)\mathbf{X}(t_f) \tag{6.53}$$

in the state vector. It is assumed that all right-hand-side coefficients are commensurate in multiplication and the product is of the same type as that on the left-hand side. It is assumed that all coefficients are well defined but in particular that $C_{2,2}(t)$ is positive definite for the minimum problem, a crucial assumption, and symmetric due to the quadratic form, while $C_{1,1}(t)$ and $C_{1,2}(t)$ need to be positive semidefinite. Also, $S_{1,1}(t_f)$ is symmetric, positive semidefinite.

As in the deterministic LQ problem in Online Subsection A.4.4, a quadratic function of the state vector is sought. However, due to the extra linear terms in the quadratic cost beyond the pure quadratic form in (A.126), a more general quadratic decomposition is heuristically assumed for the optimal value,

$$v^*(\mathbf{x}, t) = v_0(t) + \mathbf{v}_1^\top(t)\mathbf{x} + \frac{1}{2}\mathbf{x}^\top v_{1,1}(t)\mathbf{x}, \tag{6.54}$$

where the optimal value coefficients $\{v_0(t), \mathbf{v}_1(t), v_{1,1}(t)\}$ are compatible in multiplication and any product is scalar valued. Without loss of generality, the quadratic coefficient $v_{1,1}(t)$ is taken to be symmetric. Consequently, the partial derivative with respect to time is

$$v_t^*(\mathbf{x}, t) = \dot{v}_0(t) + \dot{\mathbf{v}}_1^\top(t)\mathbf{x} + \frac{1}{2}\mathbf{x}^\top \dot{v}_{1,1}(t)\mathbf{x},$$

where $\{\dot{v}_0(t), \dot{\mathbf{v}}_1(t), \dot{v}_{1,1}(t)\}$ denote the state time derivatives, the state gradient is

$$\nabla_\mathbf{x}[v^*](\mathbf{x}, t) = \mathbf{v}_1(t) + v_{1,1}(t)\mathbf{x},$$

the state Hessian is

$$\nabla_\mathbf{x}\left[\nabla_\mathbf{x}^\top[v^*]\right](\mathbf{x}, t) = v_{1,1}(t),$$

and the jump-increment is

$$\Delta_j[v^*](\mathbf{x}, t, q_j) = \mathbf{v}_1^\top(t)\widehat{\mathbf{h}}_{0,j}(t, q_j) + \frac{1}{2}\widehat{\mathbf{h}}_{0,j}^\top(t, q_j)v_{1,1}(t)\widehat{\mathbf{h}}_{0,j}(t, q_j)$$
$$+\mathbf{x}^\top v_{1,1}(t)\widehat{\mathbf{h}}_{0,j}(t, q_j),$$

where

$$\widehat{\mathbf{h}}_{0,j}(t, q_j) = [h_{0,i,j}(t, q_j)]_{n_x \times 1}$$

for $j = 1 : n_p$.

With the proposed general quadratic decomposition (6.54) of $v^*(\mathbf{x}, t)$, the pseudo-Hamiltonian has a quadratic decomposition in both state and control vectors like the cost coefficient $C(\mathbf{x}, \mathbf{u}, t)$ decomposition (6.52),

$$\mathcal{H}(\mathbf{x}, \mathbf{u}, t) = \mathcal{H}_0(t) + \mathcal{H}_1^\top(t)\mathbf{x} + \mathcal{H}_2^\top(t)\mathbf{u}$$
$$+ \frac{1}{2}\mathbf{x}^\top \mathcal{H}_{1,1}(t)\mathbf{x} + \mathbf{x}^\top \mathcal{H}_{1,2}(t)\mathbf{u} + \frac{1}{2}\mathbf{u}^\top \mathcal{H}_{2,2}(t)\mathbf{u}, \tag{6.55}$$

where the scalar coefficient is

$$\mathcal{H}_0(t) = C_0(t) + \mathbf{f}_0^\top(t)\mathbf{v}_1(t) + \frac{1}{2}\left(g_0 g_0^\top\right)(t) : v_{1,1}(t)$$
$$+ \mathbf{v}_1^\top(t)\overline{h}_0(t). *\boldsymbol{\lambda}_0(t) + \frac{1}{2}v_{1,1}(t). *\overline{(h_0\Lambda h_0)}(t), \tag{6.56}$$

where

$$\overline{h}_0(t) \equiv \left[\int_{Q_j} h_{0,i,j}(t, q_j) \phi_{Q_j}(q_j; t) dq_j \right]_{n_x \times n_p}, \tag{6.57}$$

$$\boldsymbol{\lambda}_0(t) \equiv \left[\lambda_{0,i}(t) \right]_{n_p \times 1}, \tag{6.58}$$

$$\Lambda_0(t) \equiv \left[\lambda_{0,i}(t) \delta_{i,j} \right]_{n_p \times n_p}, \tag{6.59}$$

$$\overline{h_0 \Lambda_0 h_0^\top}(t) \equiv \left[\sum_{k=1}^{n_p} \lambda_{0,k} \int_{Q_j} h_{0,i,k}(t, q_k) h_{0,j,k}(t, q_k) \phi_{Q_k}(q_k; t) dq_k \right]_{n_x \times n_x}, \tag{6.60}$$

the linear state coefficient is

$$\mathcal{H}_1(t) = \mathbf{C}_1(t) + f_1(t)\mathbf{v}_1(t) + v_{1,1}(t)\mathbf{f}_0(t) + v_{1,1}(t)\overline{h}_0(t). *\boldsymbol{\lambda}_0(t), \tag{6.61}$$

the linear control coefficient is

$$\mathcal{H}_2(t) = \mathbf{C}_2(t) + f_2(t)\mathbf{v}_1(t), \tag{6.62}$$

and the quadratic coefficients are

$$\mathcal{H}_{1,1}(t) = C_{1,1}(t) + 2f_1(t)v_{1,1}(t), \tag{6.63}$$

$$\mathcal{H}_{1,2}(t) = C_{1,2}(t) + v_{1,1}^\top(t)f_2^\top(t), \tag{6.64}$$

$$\mathcal{H}_{2,2}(t) = C_{2,2}(t). \tag{6.65}$$

Since quadratic forms operate only on the symmetric part of the quadratic coefficient (B.133), $\mathcal{H}_{2,2}(t)$ will be symmetric positive definite with $C_{2,2}(t)$.

The optimal control is the same as the regular control in the absence of control constraints, so the zero of

$$\nabla_{\mathbf{u}}[\mathcal{H}](\mathbf{x}, \mathbf{u}, t) = \mathcal{H}_2(t) + \mathcal{H}_{1,2}^\top(t)\mathbf{x} + \mathcal{H}_{2,2}(t)\mathbf{u}$$

results in

$$\begin{aligned}
\mathbf{u}^*(\mathbf{x}, t) &= -\mathcal{H}_{2,2}^{-1}(t) \left(\mathcal{H}_2(t) + \mathcal{H}_{1,2}^\top(t)\mathbf{x} \right) \\
&= -C_{2,2}^{-1}(t) \left(\mathbf{C}_2(t) + f_2(t)\mathbf{v}_1(t) + \left(C_{1,2}^\top(t) + f_2(t)v_{1,1}(t) \right) \mathbf{x} \right).
\end{aligned} \tag{6.66}$$

Hence, the optimal control vector is a linear or affine function of the state vector, the general form of linear feedback control. This completes the preliminary work on the LQJD problem for the feedback control state dependence.

Upon substituting the preliminary reduction of the linear optimal control (6.66) into the HJB equation (6.38), the HJB equation of SDP then becomes

$$\begin{aligned}
0 = {}& \dot{v}_0(t) + \dot{\mathbf{v}}_1^\top(t)\mathbf{x} + \frac{1}{2}\mathbf{x}^\top \dot{v}_{1,1}(t)\mathbf{x} + \mathcal{H}_0(t) + \mathcal{H}_1^\top(t)\mathbf{x} \\
& - \mathcal{H}_2^\top(t)\mathcal{H}_{2,2}^{-1}(t) \left(\mathcal{H}_2(t) + \mathcal{H}_{1,2}^\top(t)\mathbf{x} \right) \\
& + \frac{1}{2}\mathbf{x}^\top \mathcal{H}_{1,2}(t)\mathcal{H}_{2,2}^{-1}(t)\mathbf{x} - \mathbf{x}^\top \mathcal{H}_{1,2}(t)\mathcal{H}_{2,2}^{-1}(t) \left(\mathcal{H}_2(t) + \mathcal{H}_{1,2}^\top(t)\mathbf{x} \right) \\
& + \frac{1}{2} \left(\mathcal{H}_2^\top(t) + \mathbf{x}^\top \mathcal{H}_{1,2}(t) \right) \mathcal{H}_{2,2}^{-1}(t) \left(\mathcal{H}_2(t) + \mathcal{H}_{1,2}^\top(t)\mathbf{x} \right).
\end{aligned} \tag{6.67}$$

Next, separating this LQJD form of the HJB equation (6.67) into purely quadratic terms, purely linear terms, and state-independent terms leads to a set of three unidirectionally coupled ordinary matrix differential equations for the optimal control coefficients $v_{1,1}(t)$, $\mathbf{v}_1(t)$, and $v_0(t)$, which are summarized in the following theorem, which we have just derived.

Theorem 6.6. *LQJD Equations.*

Let the $n_x \times 1$ jump-diffusion state process $\mathbf{X}(t)$ satisfy dynamics linear in both the state and the $n_u \times 1$ control $\mathbf{U}(t)$ with an $n_x \times 1$ linear deterministic plant term

$$\mathbf{f}(\mathbf{x}, \mathbf{u}, t) = \mathbf{f}_0(t) + f_1^\top(t)\mathbf{x} + f_2^\top(t)\mathbf{u}$$

from (6.48), with an $n_x \times n_w$ state- and control-independent diffusion coefficient $g_0(t)$ of the $n_w \times 1$ Wiener process $d\mathbf{W}(t)$, and with an $n_x \times n_p$ state- and control-independent jump-amplitude $h_0(t, q)$ (6.49) of the $n_p \times 1$ Poisson process $d\mathbf{P}(t; \mathbf{Q}, t)$. Let the scalar quadratic running cost be

$$C(\mathbf{x}, \mathbf{u}, t) = C_0(t) + \mathbf{C}_1^\top(t)\mathbf{x} + \mathbf{C}_2^\top(t)\mathbf{u}$$
$$+ \frac{1}{2}\mathbf{x}^\top C_{1,1}(t)\mathbf{x} + \mathbf{x}^\top C_{1,2}(t)\mathbf{u} + \frac{1}{2}\mathbf{u}^\top C_{2,2}(t)\mathbf{u},$$

and let the terminal cost be

$$S(\mathbf{X}(t_f), t_f) = S_0(t_f) + \mathbf{S}_1^\top(t_f)\mathbf{X}(t_f) + \frac{1}{2}\mathbf{X}^\top(t_f)S_{1,1}(t_f)\mathbf{X}(t_f).$$

Then the optimal stochastic control problem admits a solution quadratic in the state vector

$$v^*(\mathbf{x}, t) = v_0(t) + \mathbf{v}_1^\top(t)\mathbf{x} + \frac{1}{2}\mathbf{x}^\top v_{1,1}(t)\mathbf{x}$$

with optimal control vector that is linear in the state vector

$$\mathbf{u}^*(\mathbf{x}, t) = -C_{2,2}^{-1}(t)\left(\mathbf{C}_2(t) + f_2(t)\mathbf{v}_1(t) + \left(C_{1,2}^\top(t) + f_2(t)v_{1,1}(t)\right)\mathbf{x}\right).$$

The optimal value $v^*(\mathbf{x}, t)$ coefficients satisfy a unidirectionally coupled set of matrix ODEs, which are solved starting from the $n_x \times n_x$ quadratic coefficient equation

$$0 = \dot{v}_{1,1}(t) + C_{1,1}(t) + 2f_1(t)v_{1,1}(t) \tag{6.68}$$
$$- \left(C_{1,2}(t) + v_{1,1}(t)f_1^\top(t)\right)C_{2,2}^{-1}(t)\left(C_{1,2}^\top(t) + f_1(t)v_{1,1}(t)\right)$$

for $v_{1,1}(t)$, then the $n_x \times 1$ linear coefficient equation

$$0 = \dot{\mathbf{v}}_1(t) + \mathbf{C}_1(t) + f_1(t)\mathbf{v}_1(t) \tag{6.69}$$
$$- \left(C_{1,2}(t) + v_{1,1}(t)f_1^\top(t)\right)C_{2,2}^{-1}(t)\left(\mathbf{C}_2(t) + f_2(t)\mathbf{v}_1(t)\right)$$
$$+ v_{1,1}(t)\overline{h}_0(t)\lambda_0(t).$$

for $\mathbf{v}_1(t)$ *using the existing solution for* $v_{1,1}(t)$, *and finally the scalar state-independent coefficient equation*

$$0 = \dot{v}_0(t) + C_0(t) + \mathbf{f}_0^\top(t)\mathbf{v}_1(t) + \frac{1}{2}g_0(t)g_0^\top(t) : v_{1,1}(t) \tag{6.70}$$

$$-\frac{1}{2}\left(\mathbf{C}_2^\top(t) + \mathbf{v}_1^\top(t)f_2(t)\right)C_{2,2}^{-1}(t)\left(\mathbf{C}_2(t) + f_2(t)\mathbf{v}_1(t)\right)$$

$$+\mathbf{v}_1^\top(t)\overline{h}_0(t). *\boldsymbol{\lambda}_0(t) + \frac{1}{2}\overline{\left(h_0\Lambda_0 h_0^\top\right)}(t) : v_{1,1}(t).$$

Remarks 6.7.

- *The nonlinear differential equation* (6.68) *for the quadratic coefficient* $v_{1,1}(t)$ *is called a* **matrix Riccati equation** *due to the quadratic linearity in* $v_{1,1}(t)$. *Since* $v_{1,1}(t)$ *can be assumed to be symmetric without loss of generality because it is defined as the coefficient of a quadratic form, computational effort can be reduced to finding just the upper or lower triangular part, i.e., just* $n_x(n_x + 1)/2$ *elements.*

- *Once* $v_{1,1}(t)$ *is known or a reasonable approximation is found,* (6.69) *for the linear coefficient* $\mathbf{v}_1(t)$ *will be a linear matrix equation, which is relatively simpler to solve than the matrix Riccati equation.*

- *Similarly, once both* $v_{1,1}(t)$ *and* $\mathbf{v}_1(t)$ *are found to reasonable approximations,* (6.70) *for the state-independent coefficient* $v_0(t)$ *will be a linear scalar equation.*

- *Once the solutions to the time-dependent coefficients* $v_{1,1}(t)$, $\mathbf{v}_1(t)$, *and* $v_0(t)$ *are obtained, then the optimal value* $v^*(\mathbf{x}, t)$ *quadratic decomposition* (6.54) *is justified, at least heuristically.*

6.5   Exercises

1. For the scalar linear jump-diffusion dynamics with arithmetic rather than geometric diffusion,

$$dX(t) = (\mu_0 X(t) + \beta_0 U(t))dt + \sigma_0 dW(t) + \nu_0 X(t)dP(t)$$

for $0 \le t \le t_f$ and initial state $X(0) = x_0 > 0$, and the control process $-\infty < U(t) < +\infty$ is unconstrained. The coefficients $\mu_0 \ne 0$, $\beta_0 \ne 0$, $\sigma_0 > 0$, $\nu_0 \ge 0$, and $\lambda_0 > 0$ are constants, where $\mathrm{E}[dP(t)] = \lambda_0 dt$. (Note that the jump process here is a discrete Poisson process, since there is no mark process.) The costs are quadratic, i.e.,

$$V[X, U](X(t), t) = \frac{1}{2}\int_t^{t_f}\left(q_0 X^2(s) + r_0 U^2(s)\right)ds + \frac{1}{2}S_f X^2(t_f)$$

for $q_0 > 0$, $r_0 > 0$, and $S_f > 0$. Let the optimal, expected value be

$$v^*(x, t) = \min\left[\mathrm{E}\left[V[X, U](X(t), t) \,|\, X(t) = x, U(t) = u\right]\right].$$

(a) Derive the PDE of stochastic dynamic programming for the optimal expected value,

$$v^*(x, t) = \min_u [E[V[X, U](X(t), t) \mid X(t) = x, U(t) = u]],$$

starting from the principle of optimality.

(b) Specify the final condition for $v^*(x, t)$ fully qualified.

(c) Formally find the optimal (unconstrained) control $u^*(x, t)$ in terms of the *shadow* "*cost*" $v_x^*(x, t)$.

(d) Obtain an LQJD solution form for $v^*(x, t)$ and an explicit linear feedback control law for $u^*(x, t)$.

2. Derive the modifications necessary in the set of Riccati-like equations for the scalar LQJD problem when the dynamics are scalar and linear (affine), i.e.,

$$dX(t) = f(X(t), U(t), t)dt + g(X(t), U(t), t)dW(t) + h(X(t), U(t), t)dP(t),$$

where

$$E[dP(t)] = \lambda(t)dt,$$

$$f(x, u, t) = f_0(t) + f_1(t)x + f_2(t)u,$$

$$g(x, u, t) = g_0(t) + g_1(t)x,$$

$$h(x, u, t) = h_0(t) + h_1(t)x,$$

the jump-amplitude being independent of any mark process. The running and terminal costs for a maximum objective are quadratic,

$$C(x, u, t) = C_0(t) + C_1(t)x + C_2(t)u + 0.5C_{1,1}(t)x^2 + C_{1,2}(t)xu + 0.5C_{2,2}(t)u^2,$$

where $C_{2,2}(t) < 0$, and

$$S(x, t) = S_0(t) + S_1(t)x + 0.5 * S_{1,1}(t)x^2,$$

where $S_{1,1}(t) < 0$.

If the objective is to maximize the expected total utility in the unconstrained control case, then find the Riccati ODEs for the coefficient functions $v_0(t)$, $v_1(t)$, $v_{1,1}(t)$ in the solution form

$$v^*(x, t) = v_0(t) + v_1(t)x + 0.5v_{1,1}(t)x^2$$

and $u_0(t)$ and $u_1(t)$ in the form

$$u^*(x, t) = u_0(t) + u_1(t)x$$

explicitly in terms of the $\{v_0(t), v_1(t), v_{1,1}(t)\}$, dynamical, and cost-coefficient functions. Do not try to solve the Riccati equation system of ODEs for $\{v_0(t), v_1(t), v_{1,1}(t)\}$.

3. Let $\beta(t)$ be the discount rate at time t, and let

$$\exp\left(-\widehat{\beta}(t,s)\right) = \exp\left(-\int_t^s \beta(r)dr\right) = \widehat{\beta}(0,s) - \widehat{\beta}(0,t) \qquad (6.71)$$

be the cumulative discount factor for the time-interval $[t,s]$ so the optimal, expected, discounted costs are

$$v^*(x,t) = \min_u \left[E\left[\int_t^{t_f} e^{-\widehat{\beta}(t,s)} C(X(s),U(s),s)ds + e^{-\widehat{\beta}(t,t_f)} S(X(t_f),t_f) \Big| C(t) \right] \right],$$

where $C(t) = \{X(t) = x, U(t) = u\}$ is the conditioning set. Noting that this $v^*(x,t)$ does not have the form to satisfy the principle of optimality given in (6.10) because of the dual-time dependence of the discount factor on (t,s),

(a) show that $w^*(x,t) = \exp\left(-\widehat{\beta}(t)\right)v^*(x,t)$ properly satisfies the usual form of the principle of optimality (6.10), and hence

(b) show that proper modification of the principle of optimality for discounted costs is

$$v^*(x,t) = \min_u \left[E\left[\int_t^{t+\delta t} e^{-\widehat{\beta}(t,s)} C(X(s),U(s),s)ds \right. \right.$$

$$\left. \left. + e^{-\widehat{\beta}(t,t+\delta t)} v^*(X(t+\delta t),t+\delta t) \Big| C(t) \right] \right]. \qquad (6.72)$$

4. Derive the HJB PDE for the scalar optimal stochastic control problem (a simplified jump-diffusion optimal portfolio and consumption problem) with stochastic dynamical system

$$dX(t) = X(t)\left(\mu_0(t)dt + U_1(t)\left(\mu_1(t)dt + \sigma(t)dW(t) + \left(e^Q - 1\right)dP(t)\right)\right)$$
$$- U_2(t)dt,$$

where $t \in [0,t_f]$, $X(0) = x_0 > 0$, $E[dP(t)] = \lambda(t)dt = \text{Var}[dP(t)]$, $E[dW(t)] = 0$, $\text{Var}[dW(t)] = dt$, Q is an IID uniformly distributed mark on $[a,b]$, $a < 0 < b$,

$$\{\mu_0(t), \mu_1(t), \sigma(t), \lambda(t)\}$$

are specified time-dependent coefficients, $X(t) \geq 0$ is the state, $\{U_1(t),U_2(t)\}$ is the control set, $0 \leq U_2(t) \leq K_2 X(t)$, $K_2 > 0$, $-U_N \leq U_1(t) \leq U_P$, $U_N > 0$, $U_P > 0$, and the optimal objective is

$$v^*(x,t) = dps\max_{\{u_1,u_2\}}\left[E_{\{W,P\}}\left[\int_t^{t_f} e^{-\widehat{\beta}(t,s)} \frac{U_2^\gamma(s)}{\gamma}ds \right. \right.$$

$$\left. \left. + e^{-\widehat{\beta}(t,t_f)} \frac{X^\gamma(t_f)}{\gamma} \Big| C(t) \right] \right],$$

where $C(t) \equiv \{X(t) = x, U_1(t) = u_1(t), U_2(t) = u_2(t)\}$ is the conditioning set, $\beta(t) > 0$ is the discount rate with the cumulative discount $\widehat{\beta}(t,s)$ defined in (6.71), $\gamma \in (0,1)$ is a constant utility power, and the zero-state absorbing boundary condition for this problem is $v^*(0^+,t) = 0$.

(a) If Exercise 3 on the form of the principle of optimality of discounting has not been done, then do it now; otherwise proceed to the next item.

(b) Derive the modified HJB equation for time-discounting from the discount form of the principle of optimality in (6.72), with the minimum merely replaced by a maximum. Be sure to point out the difference from the nondiscount form.

(c) Derive the relationship of the optimal controls to the *shadow utility* $v_x^*(x, t)$, accounting for the control constraints.

(d) Test the validity of the constant relative risk aversion (CRRA) canonical separated form of the regular solution,

$$v^*(x, t) = v_0(t)x^\gamma / \gamma,$$

determining what reduced ODE the time-dependent solution factor satisfies, specifying what side (final and boundary) conditions need to be satisfied for the problem.

Suggested References for Further Reading

- Bellman, 1957 [25]

- Cont and Tankov, 2004 [60]

- Fleming and Rishel, 1975 [86]

- Gihman and Skorohod, 1979 [96]

- Hanson, 1996 [109]

- Jazwinski, 1970 [155]

- Kushner, 1967 [174]

- Kushner and Dupuis, 2001 [179]

- Lewis, 1986 [184]

- Øksendal and Sulem, 2005 [223]

- Runggaldier, 2003 [239]

- Stengel, 1994 [258]

- Yong and Zhou, 1999 [290]

Chapter 7

Kolmogorov Forward and Backward Equations and Their Applications

The theory of probability, as a mathematical discipline, can and should be developed from axioms in exactly the same way as Geometry and Algebra.

—Andrey Nikolaevich Kolmogorov (1903–1987),
in Foundations of the Theory of Probability, 1933

Here, the Kolmogorov forward (Fokker–Planck) and backward equations are treated, including their interrelationship and their use in finding transition distributions, densities, moments, and optimal state trajectories. There is a close relationship between the PDE representations in the Kolmogorov equations and the SDE representation. Unlike the SDE, which is a symbolic representation that requires specification of the stochastic integration rule to be well posed, the Kolmogorov equations are deterministic. They can be derived from an SDE using expectations and a chain rule such as Itô's chain rule. Some investigators prefer to solve problems with the Kolmogorov PDEs rather than directly from the underlying SDEs.

7.1 Dynkin's Formula and the Backward Operator

Before deriving the Kolmogorov PDEs, a useful formula due to Dynkin is derived. Dynkin's formula relates the expectation of a function of a jump-diffusion process and a functional of the backward jump-diffusion operator. There are many variants of Dynkin's formula [78], but here a derivation of Schuss [244] for pure-diffusions is modified for jump-diffusions in the time-inhomogeneous case and in one-dimension to start.

Theorem 7.1. *Dynkin's Formula for Jump-Diffusions on $[t_0, t]$ in One Space Dimension.*
Let $X(t)$ be a jump-diffusion process satisfying the SDE

$$dX(t) \stackrel{\text{sym}}{=} f(X(t), t)dt + g(X(t), t)dW(t) + h(X(t), t, Q)dP(t; Q, X(t), t) \quad (7.1)$$

with smooth (continuously differentiable) coefficients $\{f, g, h\}$ with bounded spatial gradients. The diffusion process is the Wiener process $W(t)$ and the jump process is the Poisson

process $P(t; Q, X(t), t)$ such that $\mathrm{E}[dP(t; Q, X(t), t)|X(t) = x] = \lambda(t; x, t)dt$, and Q
is the jump-amplitude mark random variable with density $\phi_Q(q; X(t), t)$. Let $v(x, t)$ be
twice continuously differentiable in x and once in t, while bounded at infinity. Then the
conditional expectation of the composite process $v(X(t), t)$ satisfies Dynkin's formula in
integral form,

$$u(x_0, t_0) = \overline{v}(x_0, t_0; t) \equiv \mathrm{E}[v(X(t), t)|X(t_0) = x_0]$$

$$= v(x_0, t_0) + \mathrm{E}\left[\int_{t_0}^{t}\left(\frac{\partial v}{\partial t}(X(s), s) + \mathcal{B}_x[v](X(s), s)\right)ds\Big|X(t_0) = x_0\right], \quad (7.2)$$

where the dependence on the parameter t is suppressed in $u(x_0, t_0)$. The jump-diffusion
backward operator $\mathcal{B}_{x_0}$, *with respect to the state x_0 for time t dependent coefficients, in*
backward coordinates, is

$$\mathcal{B}_{x_0}[v](x_0, t_0) \equiv f(x_0, t_0)\frac{\partial v}{\partial x_0}(x_0, t_0) + \frac{1}{2}g^2(x_0, t_0)\frac{\partial^2 v}{\partial x_0^2}(x_0, t_0)$$

$$+ \widehat{\lambda}(x_0, t_0)\int_{\mathcal{Q}}\Delta_h[v](x_0, t_0, q)\phi_Q(q; x_0, t_0)dq, \quad (7.3)$$

where $\widehat{\lambda}(x_0, t_0) \equiv \lambda(t; x_0, t_0)$ suppresses the forward time t and the Poisson h-jump is

$$\Delta_h[v](x_0, t_0, q) \equiv v(x_0 + h(x_0, t_0, q), t) - v(x_0, t_0). \quad (7.4)$$

Note that the subscript x_0 on the backward operator $\mathcal{B}_{x_0}$ denotes only that the operator
operates with respect to the backward state variable x_0 for jump-diffusions and denotes
only partial differentiation in the pure-diffusion ($h(x_0, t_0, q) \equiv 0$) case.
   In the time-homogeneous case, $f(x, t) = f(x)$, $g(x, t) = g(x)$, and $h(x, t, q) = h(x, q)$, so $v(x, t) = v(x)$ and

$$u(x_0) \equiv \mathrm{E}[v(X(t))|X(t_0) = x_0]$$

$$= v(x_0) + \mathrm{E}\left[\int_{t_0}^{t}\mathcal{B}_x[v](X(s))ds\Big|X(t_0) = x_0\right], \quad (7.5)$$

dropping the t dependence of the backward operator here.

Proof. Dynkin's formula follows from Itô's chain rule for jump-diffusions here. Thus,

$$dv(X(t), t) \stackrel{\mathrm{dt}}{=} \left(\frac{\partial v}{\partial t} + f\frac{\partial v}{\partial x} + \frac{1}{2}g^2\frac{\partial^2 v}{\partial x^2}\right)(X(t), t)dt + \left(g\frac{\partial v}{\partial x}\right)(X(t), t)dW(t)$$

$$+ \int_{\mathcal{Q}}\Delta_h[v](X(t), t, q)\mathcal{P}(\mathbf{dt}, \mathbf{dq}; X(t), t), \quad (7.6)$$

where common arguments have been condensed. Upon integrating in t,

$$v(X(t), t) = v(x_0, t_0) + \int_{t_0}^{t}\left(\left(\frac{\partial v}{\partial t} + f\frac{\partial v}{\partial x} + \frac{1}{2}g^2\frac{\partial^2 v}{\partial x^2}\right)(X(s), s)ds\right. \quad (7.7)$$

$$\left. + \left(g\frac{\partial v}{\partial x}\right)(X(s), s)dW(s) + \int_{\mathcal{Q}}\Delta_h[v](X(s), s, q)\mathcal{P}(\mathbf{ds}, \mathbf{dq}; X(s), s)\right).$$

Next, taking expectations while using the facts that follow from the independent increment property of Markov processes,

$$E\left[\int_{t_0}^{t} G(X(s), s)dW(s)\right] = 0$$

after (2.43) and with the zero mean jump process

$$E\left[\int_{t_0}^{t} H(X(s), s)\widehat{\mathcal{P}}(\mathbf{ds}, \mathbf{dq}; X(s), s)\right] = 0,$$

generalized from (3.27) with $d\widehat{P}(s)$, where here the mean-zero Poisson random measure is

$$\widehat{\mathcal{P}}(\mathbf{dt}, \mathbf{dq}; X(t), t) \equiv \mathcal{P}(\mathbf{dt}, \mathbf{dq}; X(t), t) - \lambda(t; X(t), t)\phi_Q(q; X(t), t)dqdt; \quad (7.8)$$

then using the definition of the backward operator $\mathcal{B}_x[v]$, we get

$$E[v(X(t), t)|X(t_0) = x_0] = v(x_0, t_0)$$

$$+ E\left[\int_{t_0}^{t} \left(\frac{\partial v}{\partial t} + \mathcal{B}_x[v]\right)(X(s), s)ds \,|X(t_0) = x_0\right]. \tag{7.9}$$

In the time-homogeneous case, without time-dependent coefficients, we need only use the x-dependent test function $v = v(x)$, and the Dynkin formula reduces to (7.5). □

Example 7.2. *Application of Dynkin's Formula to Final Value Problems.*
Consider the final value problem for the backward problem with PDE

$$\frac{\partial v}{\partial t_0}(x_0, t_0) + \mathcal{B}_{x_0}[v](x_0, t_0) = \alpha(x_0, t_0) \quad x_0 \in \Omega, \quad t_0 < t_f,$$
$$v(x_0, t_f) = \gamma(x_0, t_f) \quad x_0 \in \Omega, \tag{7.10}$$

where the general functions $\alpha(x, t)$ and $\gamma(x, t)$ are given, while $\mathcal{B}_{x_0}[v](x_0, t_0)$ is the jump-diffusion backward operator defined in (7.3). From Dynkin's formula (7.2) with $t = t_f$,

$$E[\gamma(X(t_f), t_f)|X(t_0) = x_0] = v(x_0, t_0) + E\left[\int_{t_0}^{t_f} \alpha(X(s), s) \,\middle|\, X(t_0) = x_0\right],$$

where the jump-diffusion process is given by the SDE (7.1). By simple rearrangement, the formal solution to the final value problem is given by

$$v(x_0, t_0) = E\left[\gamma(X(t_f), t_f) - \int_{t_0}^{t_f} \alpha(X(s), s) \,\middle|\, X(t_0) = x_0\right] \tag{7.11}$$

in a more useful form, suitable for stochastic simulations using the given problem functions and the SDE.

*The final problem (7.10) can be called the **Dynkin's equation** corresponding to **Dynkin's formula** (7.2).*

7.2   Backward Kolmogorov Equations

Many exit and stopping time problems rely on the backward Kolmogorov equations, since they represent perturbations of the initial condition when the final condition for exiting or stopping is known. Another very useful application is a PDE governing the behavior of the transition density as a function of the initial state. First the general backward equation in the sense of Kolmogorov is derived using an infinitesimal form of Dynkin's equation.

Theorem 7.3. *General Backward Kolmogorov Equation for Jump-Diffusions on $[t_0, t]$ in One Space Dimension.*
Let the jump-diffusion process $X(t)$ at time t with $X(t_0) = x_0$ at initial or backward time t_0 satisfy (7.1) along with associated conditions, and let the test function $v(X(t))$ also satisfy relevant conditions. Let

$$u(x_0, t_0) = \overline{v}(x_0, t_0; t) \equiv \mathrm{E}[v(X(t)) | X(t_0) = x_0] = \mathrm{E}_{(t_0, t]}[v(X(t)) | X(t_0) = x_0], \quad (7.12)$$

*suppressing the **forward time** t in favor of the **backward time** t_0. Then $u(x_0, t_0)$ satisfies the backward PDE with backward arguments*

$$0 = \frac{\partial u}{\partial t_0}(x_0, t_0) + \mathcal{B}_{x_0}[u](x_0, t_0), \quad (7.13)$$

where the backward operator with respect to x_0 operating on u is

$$\mathcal{B}_{x_0}[u](x_0, t_0) = f(x_0, t_0) \frac{\partial u}{\partial x_0}(x_0, t_0) + \frac{1}{2} g^2(x_0, t_0) \frac{\partial^2 u}{\partial x_0^2}(x_0, t_0) \quad (7.14)$$

$$+ \widehat{\lambda}(x_0, t_0) \int_Q \Delta_h[u](x_0, t_0, q) \phi_Q(q; x_0, t_0) dq,$$

and the h-jump of u is

$$\Delta_h[u](x_0, t_0, q) \equiv u(x_0 + h(x_0, t_0, q), t_0) - u(x_0, t_0) \quad (7.15)$$

with final condition

$$\lim_{t_0 \uparrow t} u(x_0, t_0) = v(x_0). \quad (7.16)$$

Proof. This formal proof is a modified version of the one for pure-diffusions in Schuss [244] modified to include Poisson jump processes. First, the objective is to calculate the backward time partial derivative

$$u(x_0, t_0) - u(x_0, t_0 - dt) \overset{\mathrm{dt}}{=} \frac{\partial u}{\partial t_0} dt \equiv \left. \frac{\partial u}{\partial t_0} \right|_{x_0 \text{ fixed}} dt$$

to consider the infinitesimal backward difference in the spirit of Dynkin's formula, noting that the initial time t_0 is perturbed one step backward in time to $t_0 - dt$ with fixed x_0. On the other hand, using the representation (7.12), splitting the expectation at t_0 using the new random variable $X(t_0)$, and expanding by the stochastic chain rule,

$$
\begin{aligned}
u(x_0, t_0) - u(x_0, t_0 - dt) &= u(x_0, t_0) - E[v(X(t))|X(t_0 - dt) = x_0] \\
&= u(x_0, t_0) - E[E[v(X(t))|X(t_0)]|X(t_0 - dt) = x_0] \\
&= u(x_0, t_0) - E[u(X(t_0), t_0)|X(t_0 - dt) = x_0] \\
&= E[u(x_0, t_0) - u(X(t_0), t_0)|X(t_0 - dt) = x_0] \\
&\overset{dt}{=} E[\mathcal{B}_{x_0}[u](x_0, t_0)dt + g(x_0, t_0)dW(t_0) \\
&\quad + \int_Q \Delta_h[u](X(s), s, q)\widehat{\mathcal{P}}(\mathbf{ds}, \mathbf{dq}; X(s), s)|X(t_0 - dt) = x_0] \\
&= E[\mathcal{B}_{x_0}[u](x_0, t_0)dt|X(t_0 - dt) = x_0] \\
&= \mathcal{B}_{x_0}[u](x_0, t_0)dt \\
&= \left[f(x_0, t_0)\frac{\partial u}{\partial x_0}(x_0, t_0) + \frac{1}{2}g^2(x_0, t_0)\frac{\partial^2 u}{\partial x_0^2}(x_0, t_0) \right. \\
&\quad \left. + \widehat{\lambda}(x_0, t_0)\int_Q \Delta_h[u](x_0, t_0, q)\phi_Q(q; x_0, t_0)dq \right]dt,
\end{aligned}
$$

where the stochastic chain rule (5.41) was used, marked by the dt-precision step, along with expectations over the zero-mean jump-diffusion differentials. Just equating the two above results for $u(x_0, t_0) - u(x_0, t_0 - dt)$ and eliminating the dt factor yields the backward Kolmogorov equation (7.13). The final condition (7.16) simply follows from the definition of $u(x_0, t_0)$ in (7.12), and taking the indicated limit from the backward time t_0 to the forward time t for fixed x_0, we have

$$
\lim_{t_0 \uparrow t} u(x_0, t_0) = \lim_{t_0 \uparrow t} E[v(X(t))|X(t_0) = x_0] = E[v(X(t))|X(t) = x_0] = v(x_0). \quad \square
$$

Transition Probability Distribution $\Phi_{X(t)}(x, t; x_0, t_0)$

One of the most important applications of the backward Kolmogorov equation is for the transition probability, whose distribution is given by

$$
\Phi_{X(t)}(x, t; x_0, t_0) \equiv \text{Prob}[X(t) \le x|X(t_0) = x_0] \tag{7.17}
$$

with density

$$
\phi_{X(t)}(x, t; x_0, t_0) = \frac{\partial \Phi_{X(t)}}{\partial x}(x, t; x_0, t_0), \tag{7.18}
$$

or alternatively by

$$
\begin{aligned}
\phi_{X(t)}(x, t; x_0, t_0)dx &\overset{dx}{=} \text{Prob}[x < X(t) \le x + dx|X(t_0) = x_0] \tag{7.19} \\
&= \text{Prob}[X(t) \le x + dx|X(t_0) = x_0] \\
&\quad - \text{Prob}[X(t) \le x|X(t_0) = x_0]
\end{aligned}
$$

in dx-precision, provided the density exists, including the case of generalized functions (see Online Section B.12) as assumed in this book. In terms of the transition density, the conditional expectation can be rewritten such that

$$u(x_0, t_0) = \overline{v}(x_0, t_0; t) = \mathrm{E}_{(t_0, t]}[v(X(t))|X(t_0) = x_0]$$

$$= \int_{-\infty}^{+\infty} v(x)\phi_{X(t)}(x, t; x_0, t_0)dx. \tag{7.20}$$

Thus, if we let

$$v(x) \stackrel{\mathrm{gen}}{=} \delta(x - \xi),$$

then

$$u(x_0, t_0) = \overline{v}(x_0, t_0; t) = \phi_{X(t)}(\xi, t; x_0, t_0)$$

by definition of the Dirac delta function, and so the transition density satisfies the general backward Kolmogorov equation (7.13) in the backward or initial arguments (x_0, t_0).

Corollary 7.4. *Backward Kolmogorov Equation for Jump-Diffusion Transition Density. Let $\widehat{\phi}(x_0, t_0) \equiv \phi_{X(t)}(x, t; x_0, t_0)$, suppressing the parametric dependence on the forward coordinates (x, t), where the process satisfies the jump-diffusion SDE (7.1) under the specified conditions. Then*

$$0 = \frac{\partial \widehat{\phi}}{\partial t_0}(x_0, t_0) + \mathcal{B}_{x_0}[\widehat{\phi}](x_0, t_0) \tag{7.21}$$

$$= \frac{\partial \widehat{\phi}}{\partial t_0}(x_0, t_0) + f(x_0, t_0)\frac{\partial \widehat{\phi}}{\partial x_0}(x_0, t_0) + \frac{1}{2}g^2(x_0, t_0)\frac{\partial^2 \widehat{\phi}}{\partial x_0^2}(x_0, t_0) \tag{7.22}$$

$$+ \widehat{\lambda}(x_0, t_0)\int_{\mathcal{Q}} \Delta_h\left[\widehat{\phi}\right](x_0, t_0, q)\phi_Q(q; x_0, t_0)dq,$$

subject to the final condition

$$\lim_{t_0 \uparrow t} \widehat{\phi}(x_0, t_0) = \delta(x_0 - x). \tag{7.23}$$

The final condition (7.23) follows from the alternate, differential definition (7.19) of the transition probability density.

Often the transition density backward equation (7.21) is referred to as the **backward Kolmogorov equation**. It is useful for problems in which the final state is known, such as an exit time problem or a stopping time problem, where a state boundary is reached, in the case of finite state domains. For some stochastic researchers, the backward equation is considered more basic than the forward equation, since in the backward equation some final goal may be reached as in stochastic dynamic programming, or some significant event may occur, such as the extinction of a species. The evolution of the moments or expectation of powers of the state is governed by transition probability density.

7.3 Forward Kolmogorov Equations

In contrast to the backward time problems of the previous section, the forward equation will be needed to find the evolution of the transition density forward in time given an initial state.

The basic idea is that the **forward operator** $\mathcal{F}_x$ and the **backward operator** are (formal) **adjoint** operators, i.e., under suitable conditions on the transition density

$$\phi(x, t) = \phi_{X(t)}(x, t; x_0, t_0)$$

with truncated arguments to focus on forward variables, and a well-behaved test function $v(x)$, well-behaved particularly at infinity. Then the operators are related through an inner product equality,

$$(\mathcal{B}_x[v], \phi) = (\mathcal{F}_x[\phi], v), \qquad (7.24)$$

which is derived in Theorem 7.5 below. The conditional expectations in Dynkin's formula can be considered an inner product over a continuous state space with the transition density such that

$$(v, \phi) = \mathrm{E}[v(X(t))|X(t_0) = x_0] = \int_{-\infty}^{+\infty} v(x)\phi(x, t)dx,$$

emphasizing forward variables (x, t).

Theorem 7.5. *Forward Kolmogorov Equation or Fokker–Planck Equation for the Transition Density $\phi(x, t; x_0, t_0)$.*
*Let $\phi(x, t; x_0, t_0)$ be the transition probability density for the jump-diffusion process $X(t)$ that is symbolically represented by the SDE (7.1) along with the coefficient conditions specified in Dynkin's formula, Theorem 7.1. Let $v(x)$ be a bounded and twice differentiable but otherwise arbitrary test function such that the integrated **conjunct** vanishes, i.e.,*

$$\left[\left((f\phi)(x, t) - \frac{1}{2}\frac{\partial(g^2\phi)}{\partial x}(x, t) \right) v(x) + \frac{1}{2}(g^2\phi)(x, t)v'(x) \right]_{-\infty}^{+\infty} = 0, \qquad (7.25)$$

where $(f\phi)(x, t) \equiv f(x, t)\phi(x, t)$, $g^2(x, t) \equiv (g(x, t))^2$, and $v'(x) \equiv (dv/dx)(x)$. Then, in the weak sense, ϕ satisfies the forward Kolmogorov equation in forward space-time variables (x, t),

$$\frac{\partial\phi}{\partial t}(x, t) = \frac{1}{2}\frac{\partial^2(g^2\phi)}{\partial x^2}(x, t) - \frac{\partial(f\phi)}{\partial x}(x; t) - (\widehat{\lambda}\phi)(x, t) \qquad (7.26)$$
$$+ \int_Q (\widehat{\lambda}\phi)(x - \eta, t)|1 - \eta_x|\phi_Q(q; x - \eta, t)dq,$$

where $\eta = \eta(x; t, q)$ is related to the inverse jump-amplitude such that

$$x = \xi + h(\xi, t, q)$$

is the new state value corresponding to the old state value ξ, such that

$$\eta(x; t, q) = h(\xi, t, q),$$

assuming h is monotonic in ξ so that h is invertible with respect to ξ, that the Jacobian

$$(1 - \eta_x) = \left(1 - \frac{\partial\eta}{\partial x}(x; t, q) \right)$$

is nonvanishing, and that the inverse transformation from ξ to x maps $(-\infty, +\infty)$ onto $(-\infty, +\infty)$.

The transition probability density satisfies the delta function intial condition

$$\phi(x, t_0^+) = \phi_{X(t_0^+)}(x, t_0^+; x_0, t_0) = \delta(x - x_0). \tag{7.27}$$

Proof. The main idea of this proof is to perform several integrations by parts to move the partial differentiation from the backward operator on the arbitrary test function $v(x)$ to differentiation of the jump-diffusion transition probability $\phi(x, t) = \phi_{X(t)}(x, t; x_0, t_0)$, deriving the adjoint backward-forward operator relation (7.24) in principle. Differentiating Dynkin's formula (7.2) in forward time t for fixed initial conditions (x_0, t_0) and for some well-behaved test function $v(x)$, we see that

$$\frac{\partial \overline{v}}{\partial t}(x_0, t_0; t) = E\left[\frac{\partial}{\partial t} \int_{t_0}^{t} \mathcal{B}_x[v](X(s))ds \,\middle|\, X(t_0) = x_0 \right]$$
$$= E[\mathcal{B}_x[v](X(t))| X(t_0) = x_0], \tag{7.28}$$

assuming that differentiation and expectation can be interchanged, where the backward operator $\mathcal{B}$ is given in (7.3). However, the conditional expectation of $\mathcal{B}$ on the right-hand side of (7.28) can be written in terms of the transition probability ϕ (7.20),

$$E[\mathcal{B}_x[v](X(t))|X(t_0) = x_0] = \int_{-\infty}^{+\infty} \mathcal{B}_x[v](x)\phi(x, t)dx. \tag{7.29}$$

Combining (7.28) and (7.29), substituting for $\mathcal{B}$ using (7.3), and using two integration by parts on the spatial derivatives to move the spatial derivatives from v to ϕ, we then have

$$\frac{\partial \overline{v}}{\partial t}(x_0, t_0; t) = \int_{-\infty}^{+\infty} v(x)\frac{\partial \phi}{\partial t}(x, t)dx = \int_{-\infty}^{+\infty} \mathcal{B}_x[v](x)\phi(x, t)dx$$

$$= \int_{-\infty}^{+\infty} \left(f(x, t)v'(x) + \frac{1}{2}g^2(x, t)v''(x) \right.$$
$$\left. + \widehat{\lambda}(x, t) \int_{Q} \Delta_h[v](x, t, q)\phi_Q(q; x, t)dq \right) \phi(x, t)dx$$

$$= \int_{-\infty}^{+\infty} \left(-v(x)\frac{\partial(f\phi)}{\partial x}(x, t) - \frac{1}{2}\frac{\partial(g^2\phi)}{\partial x}(x, t)v'(x) \right.$$
$$\left. + (\widehat{\lambda}\phi)(x, t) \int_{Q} \Delta_h[v](x, t, q)\phi_Q(q; x, t)dq \right) dx$$
$$+ \left[(f\phi)(x, t)v(x) + \frac{1}{2}(g^2\phi)(x, t)v'(x) \right]_{-\infty}^{+\infty}$$

$$= \int_{-\infty}^{+\infty} \left(v(x)\left(\frac{1}{2}\frac{\partial^2(g^2\phi)}{\partial x^2}(x, t) - \frac{\partial(f\phi)}{\partial x}(x, t) \right) \right.$$
$$\left. + (\widehat{\lambda}\phi)(x, t) \int_{Q} \Delta_h[v](x, t, q)\phi_Q(q; x, t)dq \right) dx$$
$$+ \left[\left(f\phi - \frac{1}{2}\frac{\partial(g^2\phi)}{\partial x} \right)(x, t)v(x) + \frac{1}{2}(g^2\phi)(x, t)v'(x) \right]_{-\infty}^{+\infty}.$$

The last term is the integrated **conjunct** from two integrations by parts. By the hypothesis in (7.25), this conjunct is required to be zero, so that the forward and backward operators will be genuine adjoint operators. Otherwise, the forward and backward operators would be called *formal adjoints*.

So far only the adjoint diffusion part of the forward operator has been formed with respect to the test function v as an integration weight. There still remains more work to form the corresponding adjoint jump part and this is done by inverting the jump-amplitude function $h(x, t, q)$ with respect to x, assuming that $h(x, t, q)$ is monotonic x. Let the postjump state value be $y = x + h(x, t, q)$ for each fixed (t, q) with the inverse written as $x = y - \eta(y; t, q)$ relating the prejump state to the postjump state. Technically, with fixed (t, q), if $y = (I + h)(x)$, where here I denotes the identity function so $I(x) = x$, then the inverse argument is $x = (I + h)^{-1}(y) = (I - \eta)(y)$ for convenience and $\eta \overset{op}{=} I - (I + h)^{-1}$. Thus, $dx = (1 - \eta_y(y; t, q))dy$, where $(1 - \eta_y(y; t, q))$ is the Jacobian of the inverse transformation. Further, it is assumed that the state domain $(-\infty, +\infty)$ is transformed back onto itself, modulo the sign of the Jacobian. Consequently, we have

$$\int_{-\infty}^{+\infty} v(x) \frac{\partial \phi}{\partial t}(x, t)dx = \int_{-\infty}^{+\infty} v(x) \left(\frac{1}{2} \frac{\partial^2 (g^2 \phi)}{\partial x^2}(x, t) - \frac{\partial (f\phi)}{\partial x}(x, t) - (\widehat{\lambda \phi})(x, t) \right.$$

$$+ \int_Q (\widehat{\lambda \phi})(x - \eta(x; t, q), t)|1 - \eta_x(x; t, q, t)|$$

$$\left. \cdot \phi_Q(q; x - \eta(x; t, q), t)dq \right)dx,$$

upon transforming y into a dummy variable in the state integral back to x so a common factor of the test function $v(x)$ can be collected. Finally, since the test function is assumed to be arbitrary, the coefficients of $v(x)$ must be equivalent on the left and right sides of the equation *in the weak sense*. The argument is that of the *fundamental lemma of the calculus of variations* [36, 15, 164]. This leads to the forward Kolmogorov equation for the transition density $\phi(x, t) = \phi_{X(t)}(x, t; x_0, t_0)$ given in the concluding equation (7.26) of Theorem 7.5,

$$\frac{\partial \phi}{\partial t}(x, t) = \mathcal{F}_x[\phi](x, t)$$

$$\equiv \frac{1}{2} \frac{\partial^2 (g^2 \phi)}{\partial x^2}(x, t) - \frac{\partial (f\phi)}{\partial x}(x; t) - (\widehat{\lambda \phi})(x, t) \tag{7.30}$$

$$+ \int_Q (\widehat{\lambda \phi})(x - \eta(x; t, q), t)|1 - \eta_x(x; t, q)|\phi_Q(q; x - \eta(x; t, q), t)dq.$$

Note that the subscript x on the forward operator $\mathcal{F}_x$ denotes only that the operator operates with respect to the forward variable x for jump-diffusions and denotes only partial differentiation in the pure-diffusion $(h(x, t, q) \equiv 0)$ case.

The initial condition (7.27), $\phi_{X(t_0^+)}(x, t_0^+; x_0, t_0) = \delta(x - x_0)$, is very obvious for the continuous pure-diffusion process, but the jump-diffusion processes undergo jumps triggered by the Poisson process $P(t; Q, X(t), t)$ and so $X(t)$ can be discontinuous. However, a jump is very unlikely in a small time-interval since by (1.42) modified by replacing $\lambda(t)$ with the composite time dependence $\lambda(t; X(t), t)$,

$$\text{Prob}[dP(t; Q, X(t), t) = 0] = p_0(\lambda(t; X(t), t)dt) = e^{-\lambda(t; X(t), t)dt} = 1 + O(dt) \sim 1,$$

as $dt \to 0^+$, so the initial state is certain with probability one by conditioning, i.e.,

$$\phi(x, t) = \phi_{X(t)}(x, t; x_0, t_0) \to \delta(x - x_0) \text{ as } t \to t_0^+. \quad \Box$$

Remarks 7.6.

- *Another applied approach for deriving the forward equation for pure-diffusions is to use the diffusion approximation as given by Feller [85], but this requires strong assumptions about truncating a Taylor expansion just for diffusion processes alone. This approach does not apply to jump diffusions, since the jump-difference term $D_h[\phi]$ would require an infinite expansion.*

- *For the jump-amplitude, a good illustration could be the affine model that is the sum of a state-independent term plus a term purely linear in the state, i.e., $h(x, t, q) = v_0(t, q) + v_1(t, q)x$ for suitable time-mark coefficients, so the inverse of $y = x + h(x, t, q)$ is $x = (y - v_0(t, q))/(1 + v_1(t, q)) = y - \eta(y; t, q)$ and $\eta(y; t, q) = (v_0(t, q) + v_1(t, q)y)/(1 + v_1(t, q))$. For comparison, different cases of this model are tabulated in Table 7.1.*

Table 7.1. *Some simple jump-amplitude models and inverses.*

State Dependence	Direct $h(x, t, q)$	Forward Arg. $x = y - \eta(y; t, q)$	Inverse $\eta(y; t, q)$
constant	$v_0(t, q)$	$y - v_0(t, q)$	$v_0(t, q)$
pure linear	$v_1(t, q)x$	$\dfrac{y}{1 + v_1(t, q)}$	$\dfrac{v_1(t, q)y}{1 + v_1(t, q)}$
affine	$v_0(t, q) + v_1(t, q)x$	$\dfrac{y - v_0(t, q)}{1 + v_1(t, q)}$	$\dfrac{v_0(t, q) + v_1(t, q)y}{1 + v_1(t, q)}$

A mistake is sometimes made by incorrectly generalizing the inverse of the linear jump case $x + v_1(t, q)x = y$, so that $(1 - v_1(t, q))y$ is incorrectly used for the forward argument (x) in the linear case instead of the correct argument, which is $x = y/(1 + v_1(t, q))$.

- *The difference in the jump argument between the backward and forward equations is that in the backward case the postjump or forward value $y = x + h(x, t, q)$ is used, while in the forward case the prejump or backward value $x = y - h(x, t, q) = y - \eta(y; t, q)$ is used.*

7.4   Multidimensional Backward and Forward Equations

For many applications, there can be multiple state variables and multiple sources of random disturbances. In biological problems there can be several interacting species, each suffering from species-specific and common random changes, which can be detrimental or beneficial in effect and range in magnitude from small to large fluctuations. Such effects may be due to the weather, diseases, natural disasters, or interspecies predation. In finance, there are the usual background fluctuations in market values, and then there is the occasional market crash or buying frenzy. In manufacturing systems, there may be a large number of machines which randomly fail with the time to repair being randomly distributed due to the many causes of failure.

Consider again the multidimensional SDE from Chapter 5 for the n_x-dimensional state process $\mathbf{X}(t) = [X_i(t)]_{n_x \times 1}$,

$$d\mathbf{X}(t) \stackrel{\text{sym}}{=} \mathbf{f}(\mathbf{X}(t), t)dt + g(\mathbf{X}(t), t)d\mathbf{W}(t) + h(\mathbf{X}(t), t, \mathbf{Q})d\mathbf{P}(t; Q, \mathbf{X}(t), t), \quad (7.31)$$

where

$$\mathbf{W}(t) = [W_i(t)]_{n_w \times 1}$$

is an n_w-dimensional vector diffusion process and

$$\mathbf{P}(t; Q, \mathbf{X}(t), t) = [P_i(t; Q_i, \mathbf{X}(t), t)]_{n_p \times 1}$$

is an n_p-dimensional vector state-dependent Poisson jump process. The state-dependent coefficient functions are dimensionally specified by

$$\mathbf{f} = [f_i(\mathbf{X}(t), t)]_{n_x \times 1},$$
$$g(\mathbf{X}(t), t) = [g_{i,j}(\mathbf{X}(t), t)]_{n_x \times n_w},$$
$$h(\mathbf{X}(t), t, \mathbf{Q}) = [h_{i,j}(\mathbf{X}(t), t, Q_j)]_{n_x \times n_p}$$

and have dimensions that are commensurate in multiplication. The mark vector, $\mathbf{Q} = [Q_i)]_{n_p \times 1}$, in the last coefficient function is assumed to have components corresponding to all Poisson vector process components. The coefficient $h(\mathbf{X}(t), t, \mathbf{Q})$ of $d\mathbf{P}(t; Q, \mathbf{X}(t), t)$ is merely the mark $\mathbf{Q}$ dependent symbolic form of the jump-amplitude operator-coefficient $h(\mathbf{X}(t), t, \mathbf{q})$, using similar notation, in the corresponding Poisson random mark integral (5.82), i.e.,

$$h(\mathbf{X}(t), t, \mathbf{Q})d\mathbf{P}(t; Q, \mathbf{X}(t), t) \stackrel{\text{sym}}{=} \int_Q h(\mathbf{X}(t), t, \mathbf{q})\mathcal{P}(\mathbf{dt}, \mathbf{dq}; \mathbf{X}(t), t).$$

Dynkin's formula remains unchanged, except for converting the state variable $X(t)$ to a vector $\mathbf{X}(t)$ and making the corresponding change in the backward operator $\mathcal{B}_{\mathbf{x}}[v]$ using the multidimensional stochastic chain rule (5.97),

$$\bar{v}(\mathbf{x}_0, t_0; t) \equiv \mathrm{E}[v(\mathbf{X}(t))|\mathbf{X}(t_0) = \mathbf{x}_0]$$
$$= v(\mathbf{x}_0) + \mathrm{E}\left[\int_{t_0}^t \mathcal{B}_{\mathbf{x}}[v](\mathbf{X}(s); \mathbf{X}(s), s)ds \middle| \mathbf{X}(t_0) = \mathbf{x}_0\right], \quad (7.32)$$

where the backward operator is given as shown below. The multidimensional backward and forward Kolmogorov equations are summarized in the following theorem, with the justification left as an exercise for the reader.

Theorem 7.7. *Kolmogorov Equations for Jump-Diffusions in Multidimensions on $[t_0, t]$.*
Let

$$u(\mathbf{x}_0, t_0) = \bar{v}(\mathbf{x}_0, t_0; t) = \mathrm{E}[v(\mathbf{X}(t))|\mathbf{X}(t_0) = \mathbf{x}_0].$$

Then $u(\mathbf{x}_0, t_0)$ satisfies the following multidimensional backward Kolmogorov PDE with backward arguments,

$$0 = \frac{\partial u}{\partial t_0}(\mathbf{x}_0, t_0) + \mathcal{B}_{\mathbf{x}_0}[u](\mathbf{x}_0, t_0; \mathbf{x}_0, t_0), \tag{7.33}$$

where the backward Kolmogorov operator is defined as

$$\begin{aligned}
\mathcal{B}_{\mathbf{x}_0}[u](\mathbf{x}_0, t_0; \mathbf{x}_0, t_0) &\equiv \mathbf{f}^\top(\mathbf{x}_0, t_0) \nabla_{\mathbf{x}_0}[u](\mathbf{x}_0, t_0) \\
&+ \frac{1}{2}\left(g R' g^\top\right) : \nabla_{\mathbf{x}_0}\left[\nabla_{\mathbf{x}_0}^\top[u]\right](\mathbf{x}_0, t_0) \\
&+ \sum_{j=1}^{n_p} \widehat{\lambda}_j(\mathbf{x}_0, t_0) \int_Q \Delta_j[u](\mathbf{x}_0, t_0, q_j) \phi_{Q_j}(q_j; \mathbf{x}_0, t_0) dq_j,
\end{aligned} \tag{7.34}$$

where R' is a correlation matrix defined in (5.94), $A : B$ is the double dot product (5.98), and

$$\Delta_j[u](\mathbf{x}_0, t_0, q_j) \equiv u(\mathbf{x}_0 + \widehat{\boldsymbol{h}}_j(\mathbf{x}_0, t_0, q_j), t_0) - u(\mathbf{x}_0, t_0)$$

is the jump of u corresponding to the jump-amplitude

$$\widehat{\boldsymbol{h}}_j(\mathbf{x}, t, q_j) \equiv [h_{i,j}(\mathbf{x}, t, q_j)]_{n_x \times 1}$$

of the jth Poisson process P_j at the jth mark for $j = 1 : n_p$ and with final condition

$$u(\mathbf{x}_0, t^-) = \overline{v}(\mathbf{x}_0, t^-; t) = v(\mathbf{x}_0).$$

Similarly, the forward Kolmogorov PDE in the multidimensional transition density $\phi(\mathbf{x}, t; \mathbf{x}_0, t_0)$ as the adjoint of the backward equation is

$$\frac{\partial \phi}{\partial t}(\mathbf{x}, t) = \mathcal{F}_{\mathbf{x}}[\phi](\mathbf{x}, t), \tag{7.35}$$

where the forward Kolmogorov operator is defined as

$$\begin{aligned}
\mathcal{F}_{\mathbf{x}}[\phi](\mathbf{x}, t) &\equiv \frac{1}{2}\nabla_{\mathbf{x}}\left[\nabla_{\mathbf{x}}^\top : \left[g R' g^\top \phi\right]\right](\mathbf{x}, t) \\
&- \nabla_{\mathbf{x}}^\top[\mathbf{f}\phi](\mathbf{x}; t) - \sum_{j=1}^{n_p}(\widehat{\lambda}_j\phi)(\mathbf{x}, t) \\
&+ \sum_{j=1}^{n_p} \int_Q (\widehat{\lambda}_j\phi)(\mathbf{x} - \boldsymbol{\eta}_j(\mathbf{x}; t, q_j), t)\left|1 - \frac{\partial(\boldsymbol{\eta}_j(\mathbf{x}; t, q_j))}{\partial(\mathbf{x})}\right| \\
&\cdot \phi_{Q_j}(q_j; \mathbf{x} - \boldsymbol{\eta}_j(\mathbf{x}; t, q_j), t) dq_j,
\end{aligned} \tag{7.36}$$

where the backward to forward transformation and its Jacobian are

$$\mathbf{x} - \mathbf{x}_0 = \boldsymbol{\eta}_{j'}(\mathbf{x}, t, q_{j'}) = \widehat{\boldsymbol{h}}_{j'}(\mathbf{x}_0, t, q_{j'});$$

$$\frac{\partial(\boldsymbol{\eta}_{j'}(\mathbf{x}; t, q_{j'}))}{\partial(\mathbf{x})} = \mathrm{Det}\left[\left[\frac{\partial \eta_{j',i}(\mathbf{x}; t, q_{j'})}{\partial x_j}\right]_{n_x \times n_x}\right] = \mathrm{Det}\left[\left(\nabla_{\mathbf{x}}\left[\boldsymbol{\eta}_{j'}^\top\right]\right)^\top\right]$$

for $j' = 1 : n_p$.

7.5 Chapman–Kolmogorov Equation for Markov Processes in Continuous Time

Alternate methods for deriving the Kolmogorov equations are based upon a fundamental functional equation of Chapman and Kolmogorov (see Bharucha-Reid [31] or other references at the end of this chapter). Let $\mathbf{X}(t)$ be an $n_x \times 1$ Markov process in continuous time, i.e., a jump-diffusion, on the state space Ω. The transition probability distribution function is given by

$$\Phi(\mathbf{x}, t; \mathbf{x}_0, t_0) = \text{Prob}[\mathbf{X}(t) < \mathbf{x} \mid \mathbf{X}(t_0) = \mathbf{x}_0], \qquad (7.37)$$

provided $t > t_0$ and $\mathbf{X}(t) < \mathbf{x}$ means $X_i(t) < x_i$ for $i = 1 : n_x$. Assuming the probability density exists even if in the generalized sense, we have

$$\phi(\mathbf{x}, t; \mathbf{x}_0, t_0) = \left(\prod_{i=1}^{n_x} \frac{\partial \phi}{\partial x_i} \right) (\mathbf{x}, t; \mathbf{x}_0, t_0). \qquad (7.38)$$

Expressed as a Markov property for distributions, the *Chapman–Kolmogorov equation* for the transition between the start $(\mathbf{x}_0, t_0)$ and the current position $(\mathbf{x}, t)$ through all possible intermediate positions $(\mathbf{y}, s)$ is

$$\Phi(\mathbf{x}, t; \mathbf{x}_0, t_0) = \int_{\Omega} \Phi(\mathbf{y}, s; \mathbf{x}_0, t_0) \Phi(\mathbf{x}, t; d\mathbf{y}, s)$$

$$= \int_{\Omega} \Phi(\mathbf{y}, s; \mathbf{x}_0, t_0) \phi(\mathbf{x}, t; \mathbf{y}, s) d\mathbf{y}, \qquad (7.39)$$

where $t_0 < s < t$. Alternately, the *Chapman–Kolmogorov equation* solely in terms of transition probability densities is

$$\phi(\mathbf{x}, t; \mathbf{x}_0, t_0) = \int_{\Omega} \phi(\mathbf{y}, s; \mathbf{x}_0, t_0) \phi(\mathbf{x}, t; \mathbf{y}, s) d\mathbf{y} \qquad (7.40)$$

upon differentiating (7.39) according to (7.38), again with $t_0 < s < t$. See Bharucha-Reid [31] or other references at the end of this chapter for applications.

7.6 Jump-Diffusion Boundary Conditions

Many boundary value problems for stochastic diffusion processes are similar to their deterministic counterparts, but the stochastic justifications are different. When jump processes are included, then the situation is even more complicated. Since jump processes are discontinuous, jumps may overshoot the boundary, making it more difficult to construct an auxiliary process that will implement the boundary with proper probability law.

7.6.1 Absorbing Boundary Conditions

If the boundary is absorbing, i.e., the process that hits the boundary stays there [85, 99, 244, 163], it is quite easy to specify since the process cannot re-enter the interior, and the transition probability for the process initially at $\mathbf{X}(0) = \mathbf{x}_0$ on the boundary $\Gamma = \partial\Omega$ cannot reach $\mathbf{X}(t) = \mathbf{y}$ in the interior of the domain Ω. Thus, for pure-diffusions

$$\phi_{\mathbf{X}(t)}(\mathbf{x}, t; \mathbf{x}_0, t_0) = \text{Prob}[\mathbf{X}(t) = \mathbf{x} \in \Omega | \mathbf{X}(t_0) = \mathbf{x}_0 \in \Gamma, t > 0] = 0, \qquad (7.41)$$

whereas for jump-diffusions

$$\phi_{\mathbf{X}(t)}(\mathbf{x}, t; \mathbf{x}_0, t_0) = \text{Prob}[\mathbf{X}(t) = \mathbf{x} \in \Omega | \mathbf{X}(0) = \mathbf{x}_0 \notin \text{Interior}[\Omega], t > 0] = 0, \quad (7.42)$$

since it is assumed that a jump overshoot into the boundary or exterior of the region is absorbed.   Kushner and Dupuis [179] have a more elaborate treatment of the absorbing boundary that stops the process once it hits the boundary, which is assumed to be smooth and reachable in finite time (also called attainable or accessible). These are boundary conditions for the transition probability density backward equations, since they are specified on the backward variable x_0.

7.6.2   Reflecting Boundary Conditions

The reflecting boundary is much more complicated, and the smoothness of the boundary (i.e., the boundary is continuously differentiable), is important for defining the reflection. For a simple reflection at a boundary point, $\mathbf{x}_b$, the reflection will be in the plane of the nearby incoming trajectory from $\mathbf{x}_0$ and the normal vector $\mathbf{N}_b$ to the tangent plane of the boundary at $\mathbf{x}_b$. Let $\delta\mathbf{x} = \mathbf{x}_0 - \mathbf{x}_b$ be the distance vector to the point of contact and let $\mathbf{T}_b$ be a tangent vector in the intersection of the tangent plane and the trajectory-normal plane. Using the *stochastic reflection principle*, similar to the reflection principle used in PDEs, a stochastic reflection process can be constructed such that $\delta\mathbf{x}_r = \mathbf{x}_r - \mathbf{x}_b$ is its current increment at the same time as $\delta\mathbf{x}$. The only difference is the opposite sign of its normal component, i.e., $\delta\mathbf{x}_r = -\delta_n\mathbf{N}_b + \delta_t\mathbf{T}_b$ if $\delta\mathbf{x}_0 = +\delta_n\mathbf{N}_b + \delta_t\mathbf{T}_b$, for sufficiently small and positive components d_n and δ_t. Since the reflected process at $\mathbf{x}_r$ by its construction must have the same probability as the original process at $\mathbf{x}_0$, we then have

$$\mathbf{N}_b^\top\nabla_{x_0}[\phi_{\mathbf{X}(t)}](\mathbf{x}, t\mathbf{x}_b, t_0) = \mathbf{N}_b^\top\nabla_{x_0}[\widehat{\phi}](\mathbf{x}_b, t_0) = 0 \qquad (7.43)$$

upon expanding the difference between the two probability densities

$$\widehat{\phi}(\mathbf{x}_0, t_0') - \widehat{\phi}(\mathbf{x}_r, t_0') = \widehat{\phi}(\mathbf{x}_b + \delta_n\mathbf{N}_b + \delta_t\mathbf{T}_b, t_0') - \widehat{\phi}(\mathbf{x}_b - \delta_n\mathbf{N}_b + \delta_t\mathbf{T}_b, t_0') = 0$$

in simplified backward notation at prehit time t_0' here, to order δ_n. The order δ_t cancels out.

See Kushner and Dupuis [179] for more about reflecting boundary conditions and systematically constructing reflecting jump-diffusion processes. Also, see Karlin and Taylor [163] for a thorough discussion of other boundary conditions, such as sticky and elastic, as well as an extensive boundary classification for pure-diffusion problems.

7.7   Stopping Times: Expected Exit and First Passage Times

In many problems, an *exit time*, also called a *stopping time* or a *first passage time*, is of interest. For instance, when a population falls to the zero level and thus ceases to exist, it is said to be extinct, and the time of extinction is of interest. If it is a stochastic population, then the expected extinction time is of interest (see Hanson and Tuckwell [120, 122]). For a neuron, stochastic fluctuations can be important and then the time to reach a threshold to fire a nerve pulse is of interest and, in particular, the expected firing time can be calculated (see Stein [257], Tuckwell [269], Hanson and Tuckwell [121]). In cancer growth studies, the expected doubling time for the size of a tumor is often calculated (see Hanson and Tier [118]). There are many other examples of stopping times. Deterministic exit time problems are introduced as examples and as a basic reference.

Examples 7.8. *Deterministic Exit Time Problems.*

- ***Forward exit time formulation:***
  Let $X(t)$ be the state of the system at time t and be governed by the ODE

$$\frac{dX}{dt}(t) = f(X(t)), \quad X(0) = x_0 \in (a, b), \tag{7.44}$$

  where $f(x)$ is strictly positive or strictly negative, $f(x)$ is continuous, and $1/f(x)$ is integrable on $[a, b]$. Thus inverting (7.44), the forward running time is

$$dt = dT_F(x) = dx/f(x), \quad T_F(x_0) = 0,$$

  so

$$T_F(x) = \int_{x_0}^{x} dy/f(y),$$

  and the forward exit time is

$$T_F(b) \ \text{ if } \ f(x) > 0 \ \text{ or } \ T_F(a) \ \text{ if } \ f(x) < 0.$$

- ***More relevant backward exit time formulation:***
  Since the stochastic exit time problem is more conveniently formulated as a backward time problem, let $x = c$ be the point of exit, so when $x_0 = c$, we then know that the state $X(t)$ is already at the exit and the final condition is $T_B(c) \equiv 0$. Consequently, the backward exit time $T_B(x)$ problem is formulated with $T_B(x) = T_F(c) - T_F(x)$ or $T_B'(x) = -T_F'(x)$ as

$$dT_B(x) = -dx/f(x), \quad T_B(c) = 0,$$

  or in the more conventional backward form,

$$f(x)T_B'(x) = -1, \quad T_B(c) = 0, \tag{7.45}$$

  so

$$T_B(x) = -\int_{c}^{x} dy/f(y),$$

  or the backward exit time ending at $x = c$ is

$$T_B(x_0) = \int_{x_0}^{c} dy/f(y),$$

  where $c = b$ if $f(x) > 0$ or $c = a$ if $f(x) < 0$.

7.7.1   Expected Stochastic Exit Time

First, the exit time is analytically defined, relevant for the piecewise continuous jump-diffusion. For continuous, pure-diffusion processes, it is sufficient to consider when the process hits a boundary. However, when the stochastic process also includes jumps, then it is possible that the process overshoots the boundary and ends up in the exterior of the domain. Here the domain will simply be an open interval in one state dimension.

Again let $X(t)$ be a jump-diffusion process satisfying the SDE

$$dX(t) \stackrel{\text{sym}}{=} f(X(t), t)dt + g(X(t), t)dW(t) + h(X(t), t, Q)dP(t; Q, X(t), t) \quad (7.46)$$

with smooth (continuously differentiable) coefficients $\{f, g, h\}$ with bounded spatial gradients.

Definition 7.9. *In one state dimension, the exit time for the Markov process $X(t)$ in continuous time (7.46) from the open interval (a, b) is*

$$\tau_e(x_0, t_0) \equiv \inf_t [t \mid X(t) \notin (a, b); X(t_0) = x_0 \in (a, b)] \quad (7.47)$$

if it exists.

Before considering a more general formulation using probability theory, some applications of Dynkin's formula will be used to compute the expected extinction time and some higher moments.

Examples 7.10. *Expected Exit Time Applications of Dynkin's Formula.*

- **Small modification of Dynkin's formula for exit times:**
  Consider the following boundary value problem of an inhomogeneous backward Kolmogorov equation:

$$\frac{\partial v}{\partial t_0}(x_0, t_0) + \mathcal{B}_{x_0}[v](x_0, t_0) = \alpha(x_0, t_0), \quad x_0 \in (a, b), \quad (7.48)$$

$$v(x_0, t_0) = \beta(x_0, t_0), \quad x_0 \notin (a, b), \quad (7.49)$$

*where $\mathcal{B}_{x_0}[v](x_0, t_0)$ from (7.14) is the jump-diffusion backward operator, $\alpha(x_0, t_0)$ is a given general state-independent homogeneous term, and $\beta(x_0, t_0)$ is a given general exit boundary value. Both $\alpha(x_0, t_0)$ and $\beta(x_0, t_0)$ depend on the application. Sometimes (7.48) is called **Dynkin's equation** due to its relationship with Dynkin's formula.*

Prior to taking expectations, the integral form (7.9) of the stochastic chain rule was

$$v(X(t), t) = v(x_0, t_0) + \int_{t_0}^t \left(\left(\frac{\partial v}{\partial t} + f \frac{\partial v}{\partial x} + \frac{1}{2} g^2 \frac{\partial^2 v}{\partial x^2} \right) (X(s), s) ds \right.$$

$$+ \left(g \frac{\partial v}{\partial x} \right) (X(s), s) dW(s) \quad (7.50)$$

$$\left. + \int_Q \Delta_h[v](X(s), s, q) \mathcal{P}(\mathbf{ds}, \mathbf{dq}; X(s), s) \right),$$

but now we make the random exit time substitution $t = \tau_e(x_0, t_0)$ for the deterministic time variable, which is simply abbreviated as $t = \tau_e$, and then take expectations, getting an exit time version of Dynkin's formula,

$$E\left[v(X(\tau_e), \tau_e)|X(t_0) = x_0\right] = v(x_0, t_0) \tag{7.51}$$
$$+ E\left[\int_{t_0}^{\tau_e}\left(\frac{\partial v}{\partial t} + \mathcal{B}_x[v]\right)(X(s), s)ds\right].$$

Upon substituting the Kolmogorov PDE or Dynkin's equation (7.48) along with the boundary condition (7.49) into Dynkin's formula

$$E\left[\beta(X(\tau_e), \tau_e)|X(t_0) = x_0\right] = v(x_0, t_0) + E\left[\int_{t_0}^{\tau_e}\alpha(X(s), s)ds\right]. \tag{7.52}$$

- *Ultimate exit time distribution:*
  Let $\alpha(x_0, t_0) = 0$, while $\beta(X(\tau_e), \tau_e) = 1$ since if x_0 starts at an exit, i.e., $x_0 \notin (a, b)$, then exit is certain and the distribution function is 1. Hence, due to the jump-diffusion, $v(x_0, t_0) = 1 = \Phi_{\tau_e(x_0, t_0)}(+\infty)$ on (a, b) under reasonable conditions for the existence of an exit.

- *Expected exit time:*
  Assuming that exit is certain, $\Phi_{\tau_e(x_0, t_0)}(+\infty) = 1$, let $\alpha(x_0, t_0) = -\Phi_{\tau_e(x_0, t_0)}(+\infty) = -1$ and $\beta(X(\tau_e), \tau_e) = 0$, corresponding to $x_0 \notin (a, b)$ implying zero exit time; then

$$E[\tau_e(x_0, t_0)] = t_0 + v^{(1)}(x_0, t_0), \tag{7.53}$$

  where $v^{(1)}(x_0, t_0)$ is the solution to the problem (7.48)–(7.49) with $\alpha(x_0, t_0) = 0$ and $\beta(X(\tau_e), \tau_e) = 0$.

- *Second moment of exit time:*
  Assuming that exit is certain, let $\alpha(x_0, t_0) = -2t_0$ and $\beta(X(\tau_e), \tau_e) = 0$ again; then

$$E[\tau_e^2(x_0, t_0)] = t_0^2 + v^{(2)}(x_0, t_0), \tag{7.54}$$

  where $v^{(2)}(x_0, t_0)$ is the solution to the problem (7.48)–(7.49) with $\alpha(x_0, t_0) = -2t_0$ and $\beta(X(\tau_e), \tau_e) = 0$. Hence, the variance of the exit time on (a, b) is

$$\begin{aligned}\mathrm{Var}[\tau_e(x_0, t_0)] &= E[\tau_e^2(x_0, t_0)] - E^2[\tau_e(x_0, t_0)] \\ &= v^{(2)}(x_0, t_0) - 2t_0 v^{(1)}(x_0, t_0) - (v^{(1)})^2(x_0, t_0)\end{aligned}$$

  and the coefficient of variation (CV) of the exit time is

$$\begin{aligned}\mathrm{CV}[\tau_e(x_0, t_0)] &= \frac{\sqrt{\mathrm{Var}[\tau_e(x_0, t_0)]}}{E[\tau_e(x_0, t_0)]} \\ &= \frac{\sqrt{v^{(2)}(x_0, t_0) - 2t_0 v^{(1)}(x_0, t_0) - (v^{(1)})^2(x_0, t_0)}}{v^{(1)}(x_0, t_0) + t_0}.\end{aligned}$$

- **Higher moments of exit time:**
  Assuming that exit is certain, let $\alpha(x_0, t_0) = -nt_0^{n-1}$ and again $\beta(X(\tau_e), \tau_e) = 0$;
  then

$$E[\tau_e^n(x_0, t_0)] = t_0^n + v^{(n)}(x_0, t_0), \qquad (7.55)$$

where $v^{(n)}(x_0, t_0)$ is the solution to the problem (7.48)–(7.49) with $\alpha(x_0, t_0) = -nt_0^{n-1}$
and $\beta(X(\tau_e), \tau_e) = 0$.

Often conditional exit time moments are of interest, but then the inhomogeneous term $\alpha(x_0, t_0)$ genuinely depends on the state x_0, which makes the (7.51) form of Dynkin's formula not very useful, since then the $\alpha(X(s), s)$ in the integrand genuinely depends on the stochastic process $X(s)$, and the integral is no longer simple. Hence, for more conditional and more general problems, a more general form is needed. This more general form is based upon a generalization of the time-homogeneous derivations in Schuss [244] and in the appendix of Hanson and Tier [118] to the time-dependent coefficient case, obtaining a hybrid backward or Dynkin equation for the exit time density $\phi_{\tau_e(x_0, t_0)}(t)$.

Lemma 7.11. *Exit Time Distribution and Density.*
Given the exit time $\tau_e(x_0, t_0)$ of (7.47), its probability distribution can be related to the distribution for $X(t)$ by

$$\Phi_{\tau_e(x_0, t_0)}(t) = 1 - \int_a^b \phi_{X(t)}(x, t; x_0, t_0)dx, \qquad (7.56)$$

where $\phi_{X(t)}(x, t; x_0, t_0)$ is the transition probability density for the Markov process $X(t) = x$ in continuous time conditionally starting at $X(t_0) = x_0$, as given in (7.18). The density $\phi_{X(t)}(x, t; x_0, t_0)$ is assumed to exist.
Assuming the exit time distribution and the transition density are differentiable even in a generalized sense, the exit time probability density is

$$\phi_{\tau_e(x_0, t_0)}(t) = \frac{\partial \Phi_{\tau_e(x_0, t_0)}}{\partial t}(t).$$

The $\phi_{X(t)}$ transition density is assumed to be twice differentiable in x_0 and once in t, leading to the Kolmogorov equation in the forward time but with the backward operator $\mathcal{B}_{x_0}$,

$$\frac{\partial \phi_{\tau_e(x_0, t_0)}}{\partial t}(t) = \mathcal{B}_{x_0}\left[\phi_{\tau_e(x_0, t_0)}(t)\right] \qquad (7.57)$$

$$= f(x_0, t_0)\frac{\partial \phi_{\tau_e(x_0, t_0)}}{\partial x_0}(t) + \frac{1}{2}g^2(x_0, t_0)\frac{\partial^2 \phi_{\tau_e(x_0, t_0)}}{\partial x_0^2}(t)$$

$$+ \widehat{\lambda}(x_0, t_0)\int_Q \Delta_h[\phi_{\tau_e(x_0, t_0)}(t)](x_0, t_0, q)\phi_Q(q; x_0, t_0)dq,$$

where the jump-function Δ_h is given in (7.4).

Proof. Equation (7.56) for the exit time distribution follows from the probability definitions

$$\Phi_{\tau_e(x_0,t_0)}(t) = \mathrm{Prob}[\tau_e(x_0,t_0) < t] = \mathrm{Prob}[X(t) \notin (a,b)|X(t_0) = x_0]$$
$$= 1 - \mathrm{Prob}[X(t) \in (a,b)|X(t_0) = x_0]$$
$$= 1 - \int_a^b \phi_{X(t)}(x,t;x_0,t_0)dx,$$

i.e., the fact that the exit time probability is the complement of the probability that the process $X(t)$ is in the interval (a,b) and thus yields the right-hand side of (7.56).

Under differentiability assumptions, the exit time density can be related to an integral of the forward operator $\mathcal{F}_x$ using the forward Kolomogorov

$$\phi_{\tau_e(x_0,t_0)}(t) = \frac{\partial \Phi_{\tau_e(x_0,t_0)}}{\partial t}(t) = -\int_a^b \phi_{X(t),t}(x,t;x_0,t_0)dx$$
$$= -\int_a^b \mathcal{F}_x[\phi](x,t;x_0,t_0)dx.$$

Manipulating partial derivatives, first in forward form,

$$\phi_{X(t),t}(x,t;x_0,t_0) = \phi_{X(t),t-t_0}(x,t;x_0,t_0) = -\phi_{X(t),t_0-t}(x,t;x_0,t_0),$$

and then in backward form,

$$\phi_{X(t),t_0}(x,t;x_0,t_0) = \phi_{X(t),t_0-t}(x,t;x_0,t_0),$$

leads to

$$\phi_{\tau_e(x_0,t_0)}(t) = +\int_a^b \phi_{X(t),t_0}(x,t;x_0,t_0)dx = -\int_a^b \mathcal{B}_{x_0}[\phi](x,t;x_0,t_0)dx.$$

Again assuming sufficient differentiability along with the interchange of integral and differential operators, we have

$$\phi_{\tau_e(x_0,t_0),t}(t) = -\int_a^b \mathcal{B}[\phi_{X(t),t}(x,t;x_0,t_0)]dx$$
$$= -\int_a^b \mathcal{B}_{x_0}[\mathcal{F}[\phi_{X(t)}]](x,t;x_0,t_0)dx$$
$$= -\mathcal{B}_{x_0}\left[\int_a^b \mathcal{F}[\phi_{X(t)}](x,t;x_0,t_0)dx\right] = +\mathcal{B}_{x_0}\left[\phi_{\tau_e(x_0,t_0)}(t)\right].$$

This is a hybrid Kolmogorov form of (7.57), since it is in forward time t on the left and the backward operator is on the far right. □

Examples 7.12. *Conditionally Expected Exit Time Applications.*

- *Ultimate probability of exit:*
  The ultimate probability of exit is

$$\Phi_e(x_0, t_0) \equiv \Phi_{\tau_e(x_0,t_0)}(+\infty) = \int_0^\infty \phi_{\tau_e(x_0,t_0)}(t)dt, \qquad (7.58)$$

assuming that the distribution is bounded for all t. Also under the same conditions,

$$\int_0^\infty \phi_{\tau_e(x_0,t_0),t}(t)dt = \phi_{\tau_e(x_0,t_0)}(t)\Big|_0^{+\infty} = 0$$

and then from the exit time density equation (7.57), integration-operator interchange, and (7.58) for $\Phi_e(x_0, t_0)$, we have

$$\int_0^\infty \mathcal{B}[\phi_{\tau_e(x_0,t_0)}(t)]dt = \mathcal{B}[\Phi_e(x_0, t_0)] = 0. \qquad (7.59)$$

For certain exit at both end points a and b, the obvious boundary conditions are $\Phi_e(a, t_0) = 1$ and $\Phi_e(b, t_0) = 1$ for continuous diffusion processes, but $[\Phi_e(x_0, t_0)] = 1$ for $x_0 \notin (a, b)$ for jump-diffusions. Presuming uniqueness, the solution to the boundary value problem is $\Phi_e(x_0, t_0) = 1$.

- *Conditional exit on the right of (a, b):*
  Now suppose we seek the statistics of the ultimate exit on one side of (a, b), say, $x_0 \in [b, +\infty)$, i.e., on the right. The corresponding random exit time variable is

$$\tau_e^{(b)}(x_0, t_0) = \inf_t[\, t \,|X(t) \geq b, \ X(s) \in (a, b), \ t_0 \leq s < t, \ X(t_0) = x_0]$$

and the exit time distribution function is

$$\Phi_{\tau_e^{(b)}(x_0,t_0)}(t) \equiv \text{Prob}[\tau_e^{(b)}(x_0, t_0) < t]$$

and the corresponding density is $\phi_{\tau_e^{(b)}(x_0,t_0)}(t)$. Thus, the ultimate conditional distribution, for counting only exits on the right

$$\Phi_e^{(b)}(x_0, t_0) \equiv \int_0^{+\infty} \phi_{\tau_e^{(b)}(x_0,t_0)}(t)dt$$

has boundary conditions $\Phi_e^{(b)}(x_0, t_0) = 1$ if $x_0 \in [b, +\infty)$ but $\Phi_e^{(b)}(x_0, t_0) = 0$ if $x_0 \in (-\infty, a]$. (For counting only exits on the left, $(-\infty, a]$, the boundary conditions are interchanged for $\Phi_e^{(a)}(x_0, t_0)$.) In general, the conditional distribution $\Phi_e^{(b)}(x_0, t_0)$ will not be one as in the certain ultimate probability in the prior item, so it is necessary to work in exit time moments rather than expected exit times. Let the conditional exit time first moment be

$$M_e^{(b)}(x_0, t_0) \equiv \int_0^{+\infty} t\phi_{\tau_e^{(b)}(x_0,t_0)}(t)dt, \qquad (7.60)$$

and thus the expected conditional exit time is

$$T_e^{(b)}(x_0, t_0) \equiv M_e^{(b)}(x_0, t_0)/\Phi_e^{(b)}(x_0, t_0) \tag{7.61}$$

if $x_0 > a$. Upon integration of both sides of (7.57), making the reasonable assumption

$$t\phi_{\tau_e^{(b)}(x_0, t_0)}(t)\Big|_0^{+\infty} = 0$$

when applying integration by parts on the left, the conditional moment equation, interchanging left and right sides, is

$$\mathcal{B}_{x_0}\left[M_e^{(b)}\right](x_0, t_0) = -\Phi_e^{(b)}(x_0, t_0) \tag{7.62}$$

with boundary condition $M_e^{(b)}(x_0, t_0) = 0$ if $x_0 \notin (a, b)$. The conditions are zero on either side of (a, b) for different reasons, due to instant exit for $x_0 \in [b, +\infty)$ and to excluded exit for $x_0 \in (-\infty, a]$.

7.8 Diffusion Approximation Basis

Until this point, stochastic diffusions have almost been taken as given. There are many derivations for physical diffusions in physics and engineering, such as the diffusion of a fluid concentration in a liquid or gas according to Fick's law for the flux or flow of concentration or the diffusion of heat in a conduction medium according to Fourier's law for the flux of heat. These types of physical diffusions lead to the same or similar diffusion equations, as seen in this chapter when the jump terms are omitted. However, the stochastic diffusions are usually postulated on a different basis.

A fundamental property that distinguishes the pure-diffusion process from the discontinuous jump process among Markov processes in continuous time is that the diffusion process is a continuous process. Let $\mathbf{X}(t) = [X_i(t)]_{n_x \times 1}$ be a continuous process. Then, given some $\delta > 0$, it must satisfy the continuity condition

$$\lim_{\Delta t \to 0} \frac{\text{Prob}[|\Delta\mathbf{X}(t)| > \delta \mid \mathbf{X}(t) = \mathbf{x}]}{\Delta t} = 0 \tag{7.63}$$

so jumps in the process are unlikely.

In addition, two basic moment properties are needed for the continuous process to have a diffusion limit, and these are that the conditional mean increment process satisfies

$$\mathrm{E}[\Delta\mathbf{X}(t)|\mathbf{X}(t) = \mathbf{x}] = \int_\Omega \phi_{\mathbf{X}(t)}(\mathbf{y}, t + \Delta t; \mathbf{x}, t)d\mathbf{y}) \tag{7.64}$$

$$= \boldsymbol{\mu}(\mathbf{x}, t)\Delta t + \mathrm{o}(\Delta t) \text{ as } \Delta t \to 0,$$

where $\boldsymbol{\mu}(\mathbf{x}, t) = [\mu_i(\mathbf{x}, t)]_{n_x \times 1}$, and that the conditional variance increment process satisfies

$$\text{Cov}[\Delta\mathbf{X}(t), \Delta\mathbf{X}^\top(t)|\mathbf{X}(t) = \mathbf{x}] = \sigma(\mathbf{x}, t)\Delta t + \mathrm{o}(\Delta t) \text{ as } \Delta t \to 0, \tag{7.65}$$

where $\sigma(\mathbf{x}, t) = [\sigma_{i,j}(\mathbf{x}, t)]_{n_x \times n_x} > 0$, i.e., is positive definite, and $\phi_{X(t)}(\mathbf{x}, t; \mathbf{x}_0, x_0)d\mathbf{y})$ is the transition probability density for $\mathbf{X}(t)$. Alternatively, these two *infinitesimal moment conditions* can be written

$$\lim_{\Delta t \to 0} \frac{E[\Delta \mathbf{X}(t)|\mathbf{X}(t) = \mathbf{x}]}{\Delta t} = \mu(\mathbf{x}, t)$$

and

$$\lim_{\Delta t \to 0} \frac{\text{Cov}[\Delta \mathbf{X}(t), \Delta \mathbf{X}^\top(t)|\mathbf{X}(t) = \mathbf{x}]}{\Delta t} = \sigma(\mathbf{x}, t).$$

Other technical conditions are needed, and the reader should consult Feller [85, Chapter 10] or Karlin and Taylor [163, Chapter 15], for example, for the history and variations in these conditions. Another technical condition implies that higher order moments are negligible,

$$\lim_{\Delta t \to 0} \frac{E[|\Delta \mathbf{X}(t)|^m \mid \mathbf{X}(t) = \mathbf{x}]}{\Delta t} = 0 \qquad (7.66)$$

for $m \geq 3$.

Remarks 7.13.

- *Note that since our focus is on diffusion, the mth central moment could be used here as in [85, 163], instead of the uncentered mth moment in (7.66), just as the second moment could have been used in (7.65) instead of the covariance. For high moments, the central moment form may be easier to use since means of deviation are trivially zero.*

- *Karlin and Taylor [163] show that from the Chebyshev inequality (Chapter 1, Exercise 4),*

$$\frac{\text{Prob}[|\Delta \mathbf{X}(t)| > \delta \mid \mathbf{X}(t) = \mathbf{x}]}{\Delta t} \leq \frac{E[|\Delta \mathbf{X}(t)|^m \mid \mathbf{X}(t) = \mathbf{x}]}{\delta^m \Delta t}, \qquad (7.67)$$

  *the high moment condition (7.66) for **any** $m \geq 3$ can imply the continuity condition (7.63) for $\delta > 0$. Depending on the problem formulation, the high moment condition may be easier to demonstrate than estimating the tail of the probability distribution in the continuity condition.*

In terms of the general multidimensional jump-diffusion model (7.31), the corresponding infinitesimal parameters, in the absence of the jump term ($h = 0$), are the infinitesimal vector mean

$$\mu(\mathbf{x}, t) = \mathbf{f}(\mathbf{x}, t)$$

and the infinitesimal matrix covariance

$$\sigma(\mathbf{x}, t) = (gg^\top)(\mathbf{x}, t).$$

These infinitesimal properties by themselves do not make a diffusion process, since adding jump processes to a diffusion process invalidates the continuity condition (7.63).

For instance, examining this continuity condition for the simplest case of a simple Poisson process $X(t) = P(t)$ but with a time-dependent jump-rate $\lambda(t) > 0$ yields

$$\frac{\text{Prob}[|\Delta P(t)| > \delta \mid P(t) = j]}{\Delta t} = \sum_{k=1}^{\infty} e^{-\Delta\Lambda(t)} \frac{(\Delta\Lambda)^k(t)}{k!\Delta t} = \frac{1 - e^{-\Delta\Lambda(t)}}{\Delta t},$$

where we assume for continuity's sake that $0 < \delta < 1$ and where

$$\Delta\Lambda(t) = \int_t^{t+\Delta t} \lambda(s)ds \rightarrow \lambda(t)\Delta t \text{ as } \Delta t \rightarrow 0^+.$$

Thus,

$$\lim_{\Delta t \to 0} \frac{\text{Prob}[|\Delta P(t)| > \delta \mid P(t) = j]}{\Delta t} = \lambda(t) > 0,$$

invalidating the continuity condition as expected, although the two basic infinitesimal moments can be calculated. In general, the higher moment criterion (7.66) will not be valid either, since, for example,

$$\lim_{\Delta t \to 0} \frac{E[|\Delta P(t)|^3 \mid \mathbf{X}(t) = \mathbf{x}]}{\Delta t} = \lim_{\Delta t \to 0} \sum_{k=1}^{\infty} e^{-\Delta\Lambda(t)} \frac{(\Delta\Lambda)^k(t)}{k!\Delta t} k^3$$

$$= \lim_{\Delta t \to 0} \frac{\Delta\Lambda(t)(1 + 3\Delta\Lambda(t) + (\Delta\Lambda)^2(t))}{\Delta t}$$

$$= \lambda(t) > 0,$$

where incremental moment table, Table 1.2, has been used. It is easy to guess that the number of infinitesimal moments of the Poisson process will be infinite, extrapolating from Table 1.2, unlike the limit of two infinitesimal moments for diffusion processes. However, the table can be used only to confirm that cases $m = 3:5$ yield the infinitesimal expectation of $\lambda(t)$.

So far these conditions are merely general formulations of diffusion processes for which similar properties have been derived in the earlier chapters of this book. Where their power lies is when they are used to approximate other stochastic processes, such as in the stochastic tumor application using a diffusion approximation that can be solved for tumor doubling times in Subsection 11.2.1.

7.9   Exercises

1. *Derivation of the forward Kolmogorov equation in the generalized sense.* Let the jump-diffusion process $X(t)$ satisfy the SDE

$$
\begin{aligned}
dX(t) = f(X(t), t)dt + g(X(t), t)dW(t) \\
+ h(X(t), t, Q)dP(t; Q, X(t), t),
\end{aligned}
\tag{7.68}
$$

$X(t_0) = x_0$, where the coefficient functions (f, g, h) are sufficiently well behaved, Q is the jump-amplitude random mark with density $\phi_Q(q; X(t), t)$, and $E[dP(t; Q, X(t), t)|X(t) = x] = \lambda(t; Q, x, t)dt$.

(a) (Easy.) Show that, in the generalized sense,

$$\phi(x,t) \stackrel{\text{gen}}{=} \mathrm{E}[\delta(X(t) - x)|X(t_0) = x_0], \quad t_0 < t,$$

where $\phi(x,t) = \phi_{X(t)}(x,t;x_0,t_0)$ is the transition probability density for the process $X(t)$ conditioned on starting at $X(t_0) = x_0$ and $\delta(x)$ is the Dirac delta function.

(b) Show that the Dirac delta function with a composite argument satisfies

$$\int_{-\infty}^{+\infty} F(y)\delta(\gamma(y) - x)dy \stackrel{\text{gen}}{=} F\left(\gamma^{-1}(x)\right)\left|(\gamma^{-1})'(x)\right|,$$

where $\gamma(y)$ is a monotonic function with a nonvanishing derivative and inverse $y = \gamma^{-1}(z)$, such that $(\gamma^{-1})'(z) = 1/\gamma'(y)$ and $|\gamma^{-1}(\pm\infty)| = \infty$.

(c) Apply the previous two results, and other delta function properties from Online Section B.12, to derive the forward Kolmogorov equation (7.26) in the generalized sense.

(Hint: Regarding the proof of (7.26), the diffusion part is much easier given the delta function properties for the derivation, but the jump part is similar and is facilitated by the fact that $\gamma(y) = y + h(y; t, q)$ for fixed (t, q).)

2. *Derivation of the **Feynman–Kac** Dynkin with integrating factor) formula for jump-diffusions.*
Consider the jump-diffusion process

$$dX(t) = f(X(t), t)dt + g(X(t), t)dW(t) + h(X(t), t, Q)dP(t; Q, X(t), t),$$

$X(t_0) = x_0 \in \Omega$, $t_0 < t < t_f$, and related backward Feynman–Kac (pronounced Fineman–Katz) final value problem

$$\frac{\partial v}{\partial t_0}(x_0, t_0) + \mathcal{B}[v](x_0, t_0) + \theta(x_0, t_0)v(x_0, t_0) = \alpha(x_0, t_0), \qquad (7.69)$$

$x_0 \in \Omega, 0 \le t_0 < t_f$, with final condition

$$v(x_0, t_f) = \gamma(x_0, t_f), \quad x_0 \in \Omega, \quad 0 \le t_0 < t_f,$$

where $\mathcal{B}[v](x_0, t_0)$ is the backward operator corresponding to the jump-diffusion process (7.3). The given coefficients, $\theta(x_0, t_0)$, $\alpha(x, t)$, and $\gamma(x, t)$, are bounded and continuous. The solution $v(x_0, t_0)$ is assumed to be twice continuously differentiable in x_0 while once in t.

(a) In preparation, apply the stochastic chain rule to the auxiliary function

$$w(X(t), t) = v(X(t), t)\exp(\Theta(t_0, t))$$

to use an integrating factor technique to remove the non-Dynkin linear source term $\theta(x_0, t_0)v(x_0, t_0)$ from (7.69) with the integrating factor exponent process

$$\Theta(t_0, t) = \int_{t_0}^{t} \theta(X(s), s)ds.$$

Then show (best done using the usual time-increment form of the stochastic chain rule) that

$$
dw(X(t), t) \stackrel{\mathrm{dt}}{=} e^{\Theta(t_0, t)} \left(\left(\frac{\partial v}{\partial t} + \mathcal{B}[v] + \theta v \right) (X(t), t) dt \right. \tag{7.70}
$$

$$
+ \left(g v \frac{\partial v}{\partial x} \right) (X(t), t) dW(t)
$$

$$
\left. + \int_{\mathcal{Q}} \delta_h[v](X(t), t, q) \widehat{\mathcal{P}}(\mathbf{dt}, \mathbf{dq}; X(t), t) \right),
$$

where $\delta_h[v]$ is defined as in (7.4) and $\widehat{\mathcal{P}}$ is defined as in (7.8).

(b) Next integrate the SDE (7.70) on $[t_0, t_f]$, solve for $v(x_0, t_0)$, then take expectations and finally apply the final value problem to obtain the Feynman–Kac formula corresponding to (7.69),

$$
v(x_0, t_0) = \mathrm{E} \left[e^{+\Theta(t_0, t_f)} \gamma(X(t_f), t_f) \right. \tag{7.71}
$$

$$
\left. - \int_{t_0}^{t_f} e^{+\Theta(t_0, s)} \alpha(X(s), s) ds \,\middle|\, X(t_0) = x_0 \right].
$$

(Hint: Follow the procedure in the derivation proof of Theorem 7.3 for this Feynman–Kac formula. See Schuss [244] or Yong and Zhou [290] for pure-diffusion processes.)

3. *Moments of stochastic dynamical systems.*
   Consider first the linear stochastic dynamical system

$$
dX(t) = \mu_0 X(t) dt + \sigma_0 X(t) dW(t) + \nu_0 X(t) h(Q) dP(t; Q), \quad X(t_0) = x_0,
$$

where $\{\mu_0, \sigma_0, \nu_0\}$ is a set of constant coefficients, x_0 is specified, and $h(q)$ has finite moments with respect to a Poisson mark amplitude density $\phi_Z(z)$. Starting with Dynkin's formula (*or the forward Kolmogorov equation if you like deriving results the hard way*), do the following:

(a) Show that the conditional first moment of the process

$$
\overline{X}(t) = \mathrm{E}[X(t)|X(t_0) = x_0]
$$

satisfies a first order ODE in $\overline{X}(t)$ only, with (x_0, t_0) fixed, corresponding to the mean (quasi-deterministic) analogue of the SDE. Solve the ODE in terms of the given initial conditions.

(b) Derive the ODE for second moment

$$
\overline{X^2}(t) = \mathrm{E}[X^2(t)|X(t_0) = x_0]
$$

for the more general SDE

$$
dX(t) = f(X(t))dt + g(X(t))dW(t) + h(X(t), q)dP(t; Q),
$$

$X(t_0) = x_0$, in terms of expected coefficient values over both state and mark spaces.

(c) Use the general second moment ODE of part (b) to derive the corresponding
ODE for the state variance

$$\mathrm{Var}[X(t)] = \overline{X^2}(t) - (\overline{X})^2(t)$$

for the linear dynamical system in part (a). Your result should show that the
ODE is linear in $\mathrm{Var}[X](t)$ with an inhomogeneous term depending on the $\overline{X}(t)$
first moment solution and constants, so the ODE is closed in the sense that it is
independent of any higher moments beyond the second. Solve the ODE.

Suggested References for Further Reading

- Arnold, 1974 [13]

- Bharucha-Reid, 1960 [31]

- Feller, 1971 [85, II]

- Gihman and Skorohod, 1972 [95]

- Goel and Richter-Dyn, 1974 [99]

- Hanson and Tier, 1982 [118]

- Jazwinski, 1970 [155]

- Karlin and Taylor, 1981 [163, II]

- Kushner and Dupuis, 2001 [179]

- Ludwig, 1975 [188]

- Øksendal, 1998 [222]

- Schuss, 1980 [244]

Chapter 8

Computational Stochastic Control Methods

> *God does not care about our mathematical difficulties.*
> *He integrates empirically.*
> —Albert Einstein (1879–1955)

> *An idea which can be used once is a trick.*
> *If it can be used more than once it becomes a method.*
> —George Polya and Gabor Szego

> *"That's when I realized that research was my true calling,*
> *not software," he says. Developing software so other people*
> *could answer the big questions wasn't for him.*
> *He wanted to get back to answering them himself.*
> —Ajay Royyuruat, IBM Genographer, *in* "Dream Jobs," *IEEE*
> *Spectrum*, Volume 43, Number 2, February 2006, pp. 40–41

> *The use of stochastic models, on the other hand, can result*
> *in gigantic increases in the complexity of data volume, storage,*
> *manipulation, and retrieval requirements.*
> —from *Simulation-Based Engineering Science*, Report of the
> National Science Foundation Blue Ribbon Panel on
> Simulation-Based Engineering Science, February 2006

Stochastic dynamic programming is not easy because the **PDE of stochastic dynamic programming**, or the **Hamilton–Jacobi–Bellman equation** of SDP given in (6.17)–(6.20) of Chapter 6, is not a standard partial differential equation (PDE). In fact, it is a functional PDE with just diffusion owing to the presence of a maximum with respect to the control. Also, for the more general jump-diffusion, the additional jump-integrals make the PDE of stochastic dynamic programming a functional partial integral differential equation (PIDE).

The analytic complexity of this functional PIDE indicates that for the usual finite difference or finite element methods, numerical convergence conditions are unknown or not easily ascertainable.

This chapter discusses PDE-oriented finite difference methods developed by the author and coworkers [107, 108, 109, 277, 111] for solving the PDE of **stochastic dynamic programming (SDP)** (6.17)–(6.20), with a special emphasis on techniques and convergence conditions. The numerical foundations and complexity of computational stochastic control are discussed in [111].

An alternative method, called the **Markov chain approximation (MCA)** and developed by Kushner and coworkers [175, 176, 179], relies on using Markov chain probabilities to construct convergent finite difference approximations that are rigorously convergent in the weak sense.

Some methods use a canonical model formulation whose solution algorithm results in significant reduction in the dimensional complexity, e.g., the **linear-quadratic model (LQ)** for the optimal control of jump-diffusions (LQJD or LQGP) [274] and the **constant relative risk aversion (CRRA) utility model** for the optimal portfolios in finance [123, 124, 130, 293]. In addition, special integration methods for jump-integrals and a least squares approximation for forming simpler LQJD problems are discussed [277]. The LQJD canonical model dimensional reduction algorithm is covered in Section 6.4 on p. 179, while the deterministic linear quadratic model and variants are covered in online Section A.3.

Another canonical model dimensional reduction algorithm is treated in Sections 10.4 on p. 317 and 10.5 on p. 327 for two different optimal portfolio and consumption applications.

For a more historical introduction to computational methods in control, see Larson [182], Polak [227], and Dyer and McReynolds [77].

8.1   Finite Difference PDE Methods of SDP

A decade ago, the author contributed an invited chapter "Computational Stochastic Dynamic Programming" in a *Control and Dynamic Systems* volume discussing the use of finite difference methods of solution [109]. This section is based on the author's past experience with large-scale stochastic control applications using many of the largest vector and parallel computers available for academic use from national centers, such as Argonne National Laboratory, Los Alamos National Laboratory, National Center for Supercomputing Applications, San Diego Supercomputing Center, and Pittsburgh Supercomputing Center. An updated version of the techniques involved is given but simplified to one state dimension for convenience.

Consider the jump-diffusion SDE for state $X(t)$ and control $U(t)$,

$$
dX(t) \stackrel{\text{sym}}{=} f(X(t), U(t), t)dt + g(X(t), U(t), t)dW(t)
$$
$$
+ h(X(t), U(t), t, Q)dP(t; Q, X(t), U(t), t), \tag{8.1}
$$

where $dP(t; Q, X(t), U(t), t)$ and $dW(t)$ are the stochastic differentials of the jump-diffusion process including the compound Poisson mark Q with jump-rate $\lambda(t; x, u, t)$. The SDE coefficients, $(f(x, u, t), g(x, u, t), h(x, u, t, q))$, are assumed to be bounded or at least integrable in their arguments, so as not to overrestrict the problem. Let the objective be the minimum of the expected cumulative running costs $C(x, u, t)$ and terminal cost $S(x_f, t_f)$,

$$v^*(x, t) \equiv \min_{U[t, t_f]} \left[EdPdWttf \left[\int_t^{t_f} C(X(s), U(s), s) ds + S(X(t_f), t_f) \right. \right.$$
$$\left. \left. \left| X(t) = x, U(t) = u \right] \right] \tag{8.2}$$

for $t_0 \le t < t_f$.

The application of **Bellman's principle of optimality** and the stochastic chain rule along with the infinitesimal moments $E[dW(t)] = 0$, $\mathrm{Var}[dW(t)] = dt$, and $E[dP(t; Q, X(t), U(t), t) | X(t) = x, U(t) = u] = \lambda(t; x, u, t) dt$ leads to the SDP PIDE using only order dt terms:

$$0 = v_t^*(x, t) + \min_u \left[\mathcal{H}(x, u, t) \right]$$

$$\equiv v_t^*(x, t) + \min_u \left[C(x, u, t) + f(x, u, t) v_x^*(x, t) + \frac{1}{2} g^2(x, u, t) v_{xx}^*(x, t) \right.$$
$$\left. + \lambda(t; x, u, t) \int_Q \left(v^*(x + h(x, u, t, q), t) - v^*(x, t) \right) \phi_Q(q; x, u, t) \right] \tag{8.3}$$

$$= v_t^*(x, t) + \mathcal{H}^*(x, t).$$

If the regular or unconstrained optimal control exists and is unique, then

$$u^{(\mathrm{reg})}(x, t) = \operatorname*{argmin}_u \left[\mathcal{H}(x, u, t) \right], \tag{8.4}$$

but, in general, the optimal control, $u^*(x, t)$, is subject to any control constraints. The final condition from the minimal conditional expected cost objective (8.2) is

$$v^*(x, t) = S(x, t_f). \tag{8.5}$$

However, the boundary conditions in general are model and domain dependent.

8.1.1 Linear Control Dynamics and Quadratic Control Costs

To keep our focus on basic computations, it will be assumed that the drift of the state dynamics is linear in the control and that the running costs are quadratic in the control, i.e., the **LQJD problem in control only (LQJD/U)** discussed in Subsection 6.4.1. These assumptions are more general than the LQJD problem, but are sufficient to determine optimal control clearly in terms of (x, t). Hence, let

$$f(x, u, t) = f_0(x, t) + f_1(x, t) u,$$

$$g(x, u, t) = g_0(x, t), \quad h(x, u, t, q) = h_0(x, t, q),$$

$$\lambda(t; x, u, t) = \lambda_0(t; x, t), \quad \phi_Q(q; x, u, t) = \phi_Q(q), \tag{8.6}$$

$$C(x, u, t) = c_0(x, t) + c_1(t; x, t) u + c_2(x, t) u^2,$$

$$\mathcal{H}(x, u, t) = \mathcal{H}_0(x, t) + \mathcal{H}_1(x, t) u + \tfrac{1}{2} \mathcal{H}_2(x, t) u^2.$$

Thus, the PDE of SDP in Hamilton–Jacobi form using (6.22) with the current assumptions,

$$
\begin{aligned}
0 &= v_t^*(x, t) + \mathcal{H}^*(x, t) \\
&= v_t^*(x, t) + C_0(x, t) + C_1(x, t)u^* + \frac{1}{2}C_2(x, t)(u^*)^2 \\
&\quad + (f_0(x, t) + f_1(x, t)u^*)v_x^*(x, t) + \frac{1}{2}g_0^2(x, t)v_{xx}^*(x, t) \\
&\quad + \lambda_0(t; x, t)\int_{\mathcal{Q}}\left(v^*(x + h_0(x, t, q), t) - v^*(x, t)\right)\phi_{\mathcal{Q}}(q)dq,
\end{aligned}
\tag{8.7}
$$

and the regular control is from (6.34) after simplifications for the current one-state dimension form,

$$
u^{(\mathrm{reg})}(x, t) = -\left(c_1(x, t) + f_1(x, t)v_x^*(x, t)\right)/c_2(x, t),
\tag{8.8}
$$

provided $c_2(x, t) > 0$, i.e., positive definite, for a minimum. Since real problems have constraints, let $U^{(\min)} \leq u(x, t) \leq U^{(\max)}$. Then the optimal control law can be written

$$
\begin{aligned}
u^*(x, t) &= \min(U^{(\max)}, \max(U^{(\min)}, u^{(\mathrm{reg})}(x, t))) \\
&= \begin{cases}
U^{(\min)}, & u^{(\mathrm{reg})}(x, t) \leq U^{(\min)} \\
u^{(\mathrm{reg})}(x, t), & U^{(\min)} \leq u^{(\mathrm{reg})}(x, t) \leq U^{(\max)} \\
U^{(\max)}, & U^{(\max)} \leq u^{(\mathrm{reg})}(x, t)
\end{cases}.
\end{aligned}
\tag{8.9}
$$

For multidimensional state space problems see Chapter 6 or see Hanson's computational SDP chapter in [109].

8.1.2   Crank–Nicolson, Extrapolation-Predictor-Corrector Finite Difference Algorithm for SDP

The numerical algorithm used here is basically a modification of the work of Douglas and Dupont [73, 74] on nonlinear parabolic equations modified for SDP and the PIDE for jump-diffusions.

First, the problem is discretized in backward time since SDP is a backward problem, but the state space is discretized in a regular grid, with N_t nodes in t on $[t_0, t_f]$ and N_x nodes in x on $[x_0, x_{\max}]$,

$$
\begin{aligned}
t &\to T_k = t_f - (k - 1) \cdot \Delta t \text{ for } k = 1{:}N_t, \quad \Delta t = (t_f - t_0)/(N_t - 1), \\
x &\to X_j = x_0 + (j - 1) \cdot \Delta X \text{ for } j = 1{:}N_x, \quad \Delta X = (x_{\max} - x_0)/(N_x - 1).
\end{aligned}
\tag{8.10}
$$

This grid leads to a corresponding discretization of the dependent variables that follow using a **second order central finite difference (CFD)** for the time derivative, evaluating at the midtime point, and second order CFDs for the state derivatives when $j = 1{:}N_x$ for each

$k = 1 : N_t$ corresponding to the backward time-count with $T_1 = t_f$:

$$v^*(X_j, T_k) \to V_{j,k},$$

$$v_t^*(X_j, T_{k+0.5}) \to (V_{j,k+1} - V_{j,k})/(-\Delta t),$$

$$v_x^*(X_j, T_k) \to DV_{j,k} = 0.5(V_{j+1,k} - V_{j-1,k})/\Delta X,$$

$$v_{xx}^*(X_j, T_k) \to DDV_{j,k} = (V_{j+1,k} - 2V_{j,k} + V_{j-1,k})/(\Delta X)^2, \quad (8.11)$$

$$u^{(reg)}(X_j, T_k) \to UR_{j,k} = -\left(C_{1,j,k} + F_{1,j,k}DV_{j,k}\right)/C_{2,j,k},$$

$$u^*(X_j, T_k) \to US_{j,k} = \min(UMAX, \max(UMIN, UR_{j,k})),$$

$$v^*(X_j + h_0(X_j, T_k, q), T_k) \to VH_{j,k}(q),$$

where $F_{i,j,k} = f_i(X_j, T_k)$ for $i = 0 : 1$, $C_{i,j,k} = c_i(X_j, T_k)$ for $i = 0 : 2$, $UMIN = U^{(min)}$, and $UMAX = U^{(max)}$.

The **Crank–Nicolson implicit (CNI)** method provides central differencing in state and time and so is second order accurate in both independent variables, i.e., $O^2(\Delta X) + O^2(\Delta t)$, and the implicitness provides stability over all positive steps in time, Δt. However, for general problems, such as those that are multidimensional or are nonlinear, the implicit and tridiagonal properties are no longer valid, unless the CNI method can be extended by the **alternating directions implicit (ADI)** method through known splittings of the spatial operators. However, for nonlinear problems, recalling from Chapter 6 that the PDE of SDP is nonlinear, the ADI method is not useful and predictor-corrector methods can be used to preserve the second order accuracy in several dimensions and for nonlinear problems. For these more general applications, the basic structure of the CNI method upon dissection consists of a midpoint integral approximation and an averaging to convert the time-midpoint to integral grid point values. Thus, symbolically using the PDE of SDP in Hamilton–Jacobi form, $0 = v_t^*(x, t) + \mathcal{H}^*(x, t)$, using (8.7), the **midpoint rule approximation** is then

$$V_{j,k+1} - V_{j,k} = \int_{T_k}^{T_k - \Delta t} v_t^*(X_j, t)dt = -\int_{T_k}^{T_k - \Delta t} \mathcal{H}(X_j, t)dt$$

$$\simeq +\Delta t \cdot \mathcal{H}(X_j, T_{k+0.5}) = +\Delta t \cdot \mathcal{H}_{j,k+0.5}, \quad (8.12)$$

which is finally followed by a **second order accuracy preserving averaging step**,

$$V_{j,k+1} \simeq V_{j,k} + 0.5 \cdot \Delta t \cdot \left(\mathcal{H}_{j,k} + \mathcal{H}_{j,k+1}\right), \quad (8.13)$$

where the midpoint (midtime) value of the objective has been replaced by targeted values at given time nodes. While this last step may look like a linear assumption, in most cases this can be extended by quasi-linearization, e.g., the average for a power can be approximated by $(V_{j,k+0.5})^{n+1} \simeq 0.5(V_{j,k})^n(V_{j,k} + V_{j,k+1})$ in the zeroth correction with further refinement in subsequent corrections, always keeping the newest update of $V_{j,k+1}$ as a linear term. It is left as an informal exercise for the reader to show that under second order differentiability, the averaging step is second order accurate in time ($O^2(\Delta t)$) at the midpoint, it being well known that the midpoint rule used here is second order accurate in time. It is the midpoint rule evaluation that makes the seemingly first order approximation for $v_t^*(x, t)$ in (8.11) accurate to $O^2(\Delta t)$ rather than to $O(\Delta t)$.

Integration and Interpolation for Jump-Integrals

Another modification is needed for handling the jump-integrals. One procedure is the use of **Gauss-statistics rules** introduced by Westman and Hanson in [277] as a generalization of the Gaussian quadrature rules but customized for the given mark density $\phi_Q(q)$ in the application. These rules use N_q points Q_i and N_q weights w_i and have a polynomial precision of degree $n_q = N_q - 1$. The weights and nodes satisfy the $2 \cdot N_q$ nonlinear equations,

$$\sum_{i=1}^{N_q} w_i \cdot Q_i^j = \mathrm{E}_Q[Q^j] = \int_Q q^j \phi_Q(q)dq, \tag{8.14}$$

for $j = 0:2N_q - 1$. This leads to the Gauss-statistics approximation for the jump-integral:

$$\mathrm{IVH}_{j,k} \equiv \int_Q \mathrm{VH}_{j,k}(q)\phi_Q(q)dq \simeq \sum_{i=1}^{N_q} w_i \mathrm{VH}_{j,k}(Q_i)$$

$$= \sum_{i=1}^{N_q} w_i v^*(X_j + h_0(X_j, T_k, Q_i), T_k). \tag{8.15}$$

In general, the $\mathrm{VH}_{j,k}(Q_i)$ will be implicit values that are not necessarily at specified state nodes $j' = 1:N_t$ in $V_{j',k}$. Just as in Crank–Nicolson averaging, $\mathrm{O}^2(\Delta X)$ interpolation is needed relative to the nearest neighbor state nodes. Let the ith state argument be

$$X_j + h_0(X_j, T_k, Q_i) = X_{j+\ell_i} + \epsilon_i \Delta X,$$

where the floor integer is

$$\ell_i = \ell_{i,j,k} = \lfloor h_0(X_j, T_k, Q_i)/\Delta X \rfloor$$

and fraction

$$\epsilon_i = \epsilon_{i,j,k} = h_0(X_j, T_k, Q_i)/\Delta X - \ell_i.$$

Thus, the $\mathrm{O}^2(\Delta X)$ interpolation is

$$\mathrm{VH}_{j,k}(Q_i) \simeq (1 - \epsilon_i) \cdot V_{j+\ell_i,k} + \epsilon_i \cdot V_{j+\ell_i+1,k}, \tag{8.16}$$

assuming the jumps are not out of range of the state space or are handled by proper boundary conditions. Thus,

$$\mathrm{IVH}_{j,k} \simeq \sum_{i=1}^{N_q} w_i \left((1 - \epsilon_i) \cdot V_{j+\ell_i,k} + \epsilon_i \cdot V_{j+\ell_i+1,k} \right). \tag{8.17}$$

Example 8.1. *Gauss-Statistics Quadrature for Log-Uniform Jump-Amplitudes.*
For example, in the case that $\phi_Q(q)$ is the density of the uniform distribution on $[a, b]$, we have for $N_q = 1$,

$$n_q = 1, \quad w_1 = 1, \quad Q_1 = 0.5(a + b)$$

and for $N_q = 2$,

$$n_q = 3, \quad w_1 = 0.5, \quad w_2 = 0.5,$$

$$Q_1 = 0.5(a + b) - 0.5(b - a)/\sqrt{3}, \quad Q_2 = 0.5(a + b) + 0.5(b - a)/\sqrt{3}.$$

For higher precision on finite mark domains $[a, b]$, piecewise applications of these rules can be made on subdivisions $[q_i, q_{i+1}]$, where $q_i = a + (i - 1)\Delta q$ for $i = 1 : M_q$ nodes with $\Delta q = (b - a)/(M_q - 1)$. See Westman and Hanson [277] for more information.

In the case when there is a special q-dependence of the jump-amplitude coefficient $h_0(x, t, q)$ for which the moments can be easily or conveniently calculated, it may be possible to use just the interpolation of $\text{VH}_{j,k}(q)$ without Gauss-statistics quadrature in q.

Example 8.2. *Geometric Jump-Diffusion with Log-Uniform Jump-Amplitude Jump-Integral Approximation.*
In the financial geometric jump-diffusion with log-uniform jump-amplitude distribution (10.119), the distribution of q is uniform with respect to the log-return $\ln(x)$, but in the original return values the jump in the return is $h(x, t, q) = x \cdot (e^q - 1)$ by Itô's chain rule. For the financial market q is very small, thus so is $e^q - 1$, while a is small and negative with b small and positive. Provided $|\epsilon| \leq 1$, where $\epsilon = X_j(e^q - 1)/\Delta X$, the appropriate piecewise linear interpolation using the explicit node set $\{V_{j-1,k}, V_{j,k}, V_{j+1,k}\}$ is

$$\text{VH}_{j,k}(q) \simeq \begin{cases} (1 - \epsilon)V_{j,k} + \epsilon V_{j+1,k}, & q \geq 0, \epsilon \geq 0 \\ -\epsilon V_{j-1,k} + (1 + \epsilon)V_{j,k}, & q \leq 0, \epsilon \leq 0 \end{cases}. \tag{8.18}$$

Since the factor $(e^q - 1)$ is now explicit, it can be integrated directly without Gaussian quadrature to produce

$$\int_a^b \text{VH}_{j,k}(q)\phi_Q(q)dq \simeq V_{j,k} + \frac{X_j}{\Delta X}(V_{j,k} - V_{j-1,k})\frac{1 + a - e^a}{b - a}$$
$$+ \frac{X_j}{\Delta X}(V_{j+1,k} - V_{j-,k})\frac{e^b - 1 - b}{b - a}. \tag{8.19}$$

Extrapolation, Prediction, and Correction

Summarizing the above CNI discretizations, the PIDE of SDP of (8.7) can be put in the preliminary form

$$V_{j,k+1} = V_{j,k} + \Delta t \cdot \mathcal{H}_{j,k+0.5}$$

$$= V_{j,k} + \Delta t \left(C_{j,k+0.5} + F_{j,k+0.5} \cdot \text{DV}_{j,k+0.5} \right. \tag{8.20}$$
$$\left. + 0.5 \cdot G_{0,j,k+0.5}^2 \cdot \text{DDV}_{j,k+0.5} + \Lambda_k \cdot (\text{IVH}_{j,k+0.5} - V_{j,k+0.5}) \right),$$

where $C_{j,k} = C_{0,j,k} + C_{1,j,k}\mathrm{US}_{j,k} + 0.5 \cdot C_{2,j,k}\mathrm{US}_{j,k}^2$, $F_{j,k} = F_{0,j,k} + F_{1,j,k}\mathrm{US}_{j,k}$, $G_{0,j,k} = g_0(X_j, T_k)$, $\Lambda_k = \lambda_0(T_k)$, $\mathrm{US}_{j,k} = \min(\mathrm{UMAX}, \max(\mathrm{UMIN}, \mathrm{UR}_{j,k}))$, and $\mathrm{UR}_{j,k} = -\left(C_{1,j,k} + F_{1,j,k} \cdot \mathrm{DV}_{j,k}\right)/C_{2,j,k}$, using (8.11).

Once there are two prior values $V_{j,k-1}$ and $V_{j,k}$, which happens when $k \geq 2$, linear extrapolation (ex) can be used to accelerate the SDP corrections. The first step from the final condition at $k = 1$ to $k = 2$ takes the most corrections since no trend is available, except for $V_{j,1}$. Otherwise the extrapolation (ex) step for the time-midpoint is used for $k \geq 2$ rather than the initial prediction at $k = 1$,

$$V_{j,k+0.5}^{(ex)} = \begin{cases} V_{j,k}, & k = 1 \\ 0.5(3V_{j,k} - V_{j,k-1}), & k \geq 2 \end{cases}, \tag{8.21}$$

which is used to update the derivative $\mathrm{DV}_{j,k+0.5}$, second derivative $\mathrm{DDV}_{j,k+0.5}$, regular control $\mathrm{UR}_{j,k+0.5}$, optimal control $\mathrm{UR}_{j,k+0.5}$, and jump functions $\mathrm{VH}_{j,k+0.5}(q)$ in the list (8.11) for the pseudo-Hamiltonian $\Delta t \cdot \mathcal{H}_{j,k+0.5}^{(ex)}$ in (8.12), (8.20) using quasi-linearization for nonlinear terms. The resulting update of the value is called the predictor or first correction step $(c, 1)$,

$$V_{j,k+1}^{(c,1)} = V_{j,k} + \Delta t \cdot \mathcal{H}_{j,k+0.5}^{(ex)} \tag{8.22}$$

for all j, as long as $k \geq 2$. Otherwise the predicted step uses the current value or $V_{j,k+1}^{(c,1)} = V_{j,k} + \Delta t \cdot \mathcal{H}_{j,k}$ using (8.20). The evaluation step uses the updated average

$$V_{j,k+0.5}^{(c,1)} = 0.5(V_{j,k+1}^{(c,1)} + V_{j,k}), \tag{8.23}$$

which is used to update all the needed values in (8.11) and finally in all of the next correction $(c, 2)$,

$$V_{j,k+1}^{(c,2)} = V_{j,k} + \Delta t \cdot \mathcal{H}_{j,k+0.5}^{(c,1)}. \tag{8.24}$$

The γth correction loop given by $V_{j,k+1}^{(c,\gamma)}$ will contain

$$V_{j,k+0.5}^{(c,\gamma)} = 0.5(V_{j,k+1}^{(c,\gamma)} + V_{j,k}) \tag{8.25}$$

plus the corresponding evaluations of $\mathrm{DV}_{j,k+0.5}^{(c,\gamma)}$, $\mathrm{DDV}_{j,k+0.5}^{(c,\gamma)}$, $\mathrm{UR}_{j,k+0.5}^{(c,\gamma)}$, $\mathrm{US}_{j,k+0.5}^{(c,\gamma)}$, $\mathrm{VH}_{j,k+0.5}^{(c,\gamma)}(q)$ including integration, and $\mathcal{H}_{j,k+0.5}^{(c,\gamma)}$. Then

$$V_{j,k+1}^{(c,\gamma+1)} = V_{j,k} + \Delta t \cdot \mathcal{H}_{j,k+0.5}^{(c,\gamma)}. \tag{8.26}$$

The corrections continue until the stopping criterion is reached, for instance, the relative criteria given tolerance tol_v,

$$\left\| V_{j,k+1}^{(c,\gamma+1)} - V_{j,k+1}^{(c,\gamma)} \right\|_1 < \mathrm{tol}_v \left\| V_{j,k+1}^{(c,\gamma)} \right\|_1 \tag{8.27}$$

for each k, continuing corrections if not satisfied, otherwise stopping the corrections, setting $\gamma_{\max} = \gamma + 1$, and setting the final $(k + 1)$st value at

$$V_{j,k+1} = V_{j,k+1}^{(c,\gamma_{\max})}. \tag{8.28}$$

In (8.27), $\|*\|_1$ denotes the one-norm with respect to the state index j for current time index k, but other norms could be used with the one-norm being less computationally costly.

Stability criteria are another matter due to the complexity of the PIDE of SDP in terms of multistate systems, jump-integrals, nonlinear terms, and optimization terms. A rough criterion focuses on the diffusion term $G_{0,j,k+0.5}^2 \mathrm{DDV}_{j,k+0.5}$ in (8.20), which can be expanded by substituting the CFD form (8.11) for $\mathrm{DV}_{j,k+0.5}$ and $\mathrm{DDV}_{j,k+0.5}$ into (8.20) and produces

$$
\begin{aligned}
V_{j,k+1} = {} & \left(1 - \frac{\Delta t}{\Delta X^2} G_{0,j,k+0.5}^2\right) V_{j,k+0.5} \\
& + 0.5 \frac{\Delta t}{\Delta X^2} \left(G_{0,j,k+0.5}^2 + F_{j,k+0.5}\Delta X\right) V_{j+1,k+0.5} \\
& + 0.5 \frac{\Delta t}{\Delta X^2} \left(G_{0,j,k+0.5}^2 - F_{j,k+0.5}\Delta X\right) V_{j-1,k+0.5} \\
& + \Delta t C_{j,k+0.5} + \Lambda_k \Delta t \cdot (\mathrm{IVH}_{j,k+0.5} - V_{j,k+0.5}),
\end{aligned}
\tag{8.29}
$$

where $C_{j,k} = C_{0,j,k} + C_{1,j,k}\mathrm{US}_{j,k} + 0.5 C_{2,j,k}\mathrm{US}_{j,k}^2$ and $F_{j,k} = F_{0,j,k} + F_{1,j,k}\mathrm{US}_{j,k}$.

Following Kushner and Dupuis [179] and ignoring the jump and cost terms, the positivity of the diffusion with drift terms leads to a **parabolic mesh ratio** of approximately

$$
\max_{j,k} \left(G_{0,j,k+0.5}^2\right) \frac{\Delta t}{(\Delta X)^2} < 1,
\tag{8.30}
$$

but it certainly should be less than one. This assumes that the PIDE is **diffusion-dominated** and accounts for the drift as well as other terms in (8.3). The discrete HJB equation is said to be **diffusion-dominated**, modified for current form from a relation in [179] if

$$
\min_{j,k} \left(G_{0,j,k}^2 - |F_{j,k}|\Delta X\right) \geq 0,
\tag{8.31}
$$

where $F_{j,k} = F_{0,j,k} + F_{1,j,k}\mathrm{US}_{j,k}$, so that the coefficients of the nondiagonal terms, $V_{j+1,k+0.5}$ and $V_{j-1,k+0.5}$, are also positive. Otherwise the discrete problem is either mixed domination or **drift-dominated**, ignoring the jump cost terms. The technique is to decrease Δt and/or increase ΔX if spurious oscillations appear. Not that the diffusion-dominated condition (8.31) is satisfied for sufficiently small state stepsize ΔX as long as the diffusion coefficient $G_{0,j,k+0.5}^2$ is not also sufficiently small. For more information on linear and multistate models, see Hanson [109, 216, 112] or Kushner and Dupuis [179].

The CFDs for state derivatives work quite well in the diffusion-dominated regime but are not useful for specified derivative boundary conditions, such as the convection boundary condition and the no-flux or reflecting boundary condition (7.43), e.g., $v_x^*(x_0, t) = 0$ on the left boundary or $v_x^*(x_{\max}, t) = 0$ on the right boundary, respectively, assuming the diffusion coefficient $g_0^2(x, t)/2 > 0$ for a well-defined flux and nonsingular boundary condition. Using second order forward and backward finite differences, respectively, to maintain consistency in numerical accuracy with the central differences in the interior of $[x_0, x_{\max}]$, the derivatives at the boundaries are

$$
\begin{aligned}
v_x^*(x_0, T_k) &\simeq \mathrm{DV}_{1,k} = -0.5(V_{3,k} - 4V_{2,k} + 3V_{1,k})/\Delta x, \\
v_x^*(x_{\max}, T_k) &\simeq \mathrm{DV}_{N_x,k} = +0.5(V_{N_x-2,k} - 4V_{N_x-1,k} + 3V_{N_x,k})/\Delta x.
\end{aligned}
\tag{8.32}
$$

Now, the signs of these terms are not a problem for stability since these conditions are used as eliminants for $V_{1,k}$ for left boundary values and $V_{N_x,k}$ for right boundary values rather than as replacements for the discrete HJB equations (8.29). An alternate derivative boundary condition implementation is to add an artificial boundary to the domain, but the author has found better performance using only the domain with the derivative boundary values such as (8.32).

For **finite element** versions see Chung, Hanson, and Xu [55] or Hanson [109]. Although not on SDP, the work of Chakrabarty and Hanson [50] uses the hybrid of CNI predictor-corrector methods discussed here with finite elements for a large-scale distributed parameter or PDE-driven system. Finite element methods are better for presenting multi-dimensional systems and systems on irregular domains.

8.1.3   Upwinding Finite Differences If Not Diffusion-Dominated

When the diffusion-dominated condition (8.31) is no longer valid, then the drift term becomes important or the system (8.3) becomes drift-dominant and the coefficients of the nondiagonal terms, $V_{j+1,k}$ and $V_{j-1,k}$, are no longer guaranteed to be positive. In this case the system takes on more hyperbolic PDE characteristics since the drift terms are of hyperbolic type, as are first order PDEs. In the case of drift dominance or near drift dominance, following Kushner [179] and others, the finite difference to the first state partial of the optimal value function $v_x^*(X_j, T_k)$ in (8.11) should be changed from second order CFD to first order **upwinded finite differences (UFDs)**, which use forward or backward finite differences (FFDs or BFDs) to coincide with the sign of the drift coefficient, respectively, i.e.,

$$\text{DV}_{j,k} = \begin{cases} (V_{j+1,k} - V_{j,k})/\Delta x, & F_{j,k} \geq 0 \\ (V_{j,k} - V_{j-1,k})/\Delta x, & F_{j,k} < 0 \end{cases}, \tag{8.33}$$

where again $\bar{F}_{j,k} = F_{0,j,k} + F_{1,j,k}\text{US}_{j,k}$. Thus, upwind is in the direction of the drift. However, upwinding requires a sacrifice of numerical accuracy consistency, going from $O(\Delta X^2)$ CFD to $O(\Delta X)$ UFD for the first state partial, in favor of more stable numerical calculations. Substituting the UFD form (8.33) for $\text{DV}_{j,k}$ in (8.20) produces

$$
\begin{aligned}
V_{j,k+1} = {} & \left(1 - \frac{\Delta t}{\Delta X^2}\left(G_{0,j,k+0.5}^2 + 0.5|F_{j,k+0.5}|\Delta X\right)\right)V_{j,k+0.5} \\
& + 0.5\frac{\Delta t}{\Delta X^2}\left(G_{0,j,k+0.5}^2 + [F_{j,k+0.5}]_+\Delta X\right)V_{j+1,k+0.5} \\
& + 0.5\frac{\Delta t}{\Delta X^2}\left(G_{0,j,k+0.5}^2 + [F_{j,k+0.5}]_-\Delta X\right)V_{j-1,k+0.5} \\
& + \Delta t C_{j,k+0.5} + \Lambda_k\Delta t \cdot (\text{IVH}_{j,k+0.5} - V_{j,k+0.5}),
\end{aligned}
\tag{8.34}
$$

where $[f]_\pm \equiv \max[\pm f] \geq 0$, such that $[f]_+ + [f]_- = |f|$ and $[f]_+ - [f]_- = f$. Hence, for the diffusion terms, all coefficients are positive provided the **drift-adjusted parabolic mesh ratio** condition

$$\max_{j,k}\left(G_{0,j,k+0.5}^2 + 0.5|F_{j,k+0.5}|\right)\frac{\Delta t}{(\Delta X)^2} < 1 \tag{8.35}$$

is satisfied without the extra diffusion-dominated condition in (8.31) being needed.

8.1.4 Multistate Systems and Bellman's Curse of Dimensionality

Generalization to multidimensional state spaces can lead to very-large-scale computational problems, since the size of the computational problem grows with the number of dimensions multiplied by the number of nodes per dimension.

Starting with a version of the PDE of SDP in (6.22) modified for the LQJD/U form in (6.24)–(6.28) and no diffusion process correlations ($R' = I_{n_w \times n_w}$), we have

$$
\begin{aligned}
0 = {}& v_t^*(\mathbf{x}, t) + C_0(\mathbf{x}, t) + \mathbf{C}_1^\top(\mathbf{x}, t)\mathbf{u}^* + \frac{1}{2}(\mathbf{u}^*)^\top C_2(\mathbf{x}, t)\mathbf{u}^* \\
& + \nabla_{\mathbf{x}}^\top[v^*](\mathbf{x}, t) \cdot (\mathbf{f}_0(\mathbf{x}, t) + f_1(\mathbf{x}, t)\mathbf{u}^*) \\
& + \frac{1}{2}\left(g_0 g_0^\top\right)(\mathbf{x}, t) : \nabla_{\mathbf{x}}\left[\nabla_{\mathbf{x}}^\top[v^*]\right](\mathbf{x}, t) \\
& + \sum_{\ell=1}^{n_p} \lambda_\ell(t) \int_{\mathcal{Q}_\ell} \left(v^*\left(\mathbf{x} + \widehat{\mathbf{h}}_{0,\ell}(\mathbf{x}, t, q_\ell), t\right) - v^*(\mathbf{x}, t)\right) \phi_{Q_\ell}(q_\ell) dq_\ell,
\end{aligned}
\tag{8.36}
$$

where the double-dot product (:) is defined as a trace in (5.98) and the ℓth jump-amplitude vector is $\widehat{\mathbf{h}}_{0,\ell}(\mathbf{x}, t, q_\ell) \equiv [h_{0,i,\ell}(\mathbf{x}, t, q_\ell)]_{n_x \times 1}$ for $\ell = 1:n_p$.

Let the state dimension be n_x and the realized state vector be given by $\mathbf{x} = [x_i]_{n_x \times 1}$. In discrete form, the state vector with a common N_x nodes per dimension becomes $\mathbf{x} = [x_i]_{n_x \times 1} \to \mathbf{X_j} = [X_{i,j_i}]_{n_x \times 1}$, representing a single point in state space, given one j_i for each state i from the range $j_i = 1:N_x$ for $i = 1:n_x$ with the $X_{i,j_i} = x_{i,0} + (j_i - 1)\Delta X_i$ and $\Delta X_i = (x_{i,\max} - x_{i,0})/(N_x - 1)$. The entire set of points in state space can be represented by $\mathcal{X} = [X_{i,j}]_{n_x \times N_x}$ with the corresponding vector index $J = [J_{i,j}]_{n_x \times N_x}$. This representation leads to a large-scale expansion of the independent variables of SDP from that in (8.37) for each current $k = 1:N_t$, using a CFD for each state component of state partial derivatives:

$$
\begin{aligned}
v^*(\mathbf{X_j}, T_k) &\to V_{J,k} \equiv [V_{j_1, j_2, \ldots, j_{n_x}, k}]_{N_x \times N_x \times \cdots \times N_x}, \\
v_t^*(\mathbf{X_j}, T_k) &\to (V_{J,k+1} - V_{J,k})/(-\Delta t), \\
\nabla_x[v^*](\mathbf{X_j}, T_k) &\to \mathbf{DV}_{J,k} \equiv [\mathrm{DV}_{i, j_1, \ldots, j_{n_x}, k}]_{n_x \times N_x \times \cdots \times N_x} \\
&= \Big[\big(V_{j_1 + \delta_{i,1}, \ldots, j_{n_x} + \delta_{i,n_x}, k} \\
&\quad - V_{j_1 - \delta_{i,1}, \ldots, j_{n_x} - \delta_{i,n_x}, k}\big) \big/ \Delta X_i\Big]_{n_x \times N_x \times \cdots \times N_x}, \\
\nabla_x\left[\nabla_x^\top[v^*]\right](\mathbf{X_j}, T_k) &\to \mathbf{DDV}_{J,k} \equiv [\mathrm{DDV}_{i, j, j_1, \ldots, j_{n_x}, k}]_{n_x \times n_x \times N_x \times \cdots \times N_x}, \\
u^{(\mathrm{reg})}(\mathbf{X_j}, T_k) &\to \mathbf{UR}_{J,k} \equiv [\mathrm{UR}_{i, j_1, \ldots, j_{n_x}, k}]_{n_x \times N_x \times \cdots \times N_x} \\
&= -\left(C_{1,J,k} + F_{1,J,k}\mathrm{DV}_{j,k}\right) . / C_{2,J,k}, \\
u^*(\mathbf{X_j}, T_k) &\to \mathbf{US}_{J,k} \equiv [\mathrm{US}_{i, j_1, \ldots, j_{n_x}, k}]_{n_x \times N_x \times \cdots \times N_x} \\
&= [\min(\mathrm{UMAX}_i, \max(\mathrm{UMIN}_i, \\
&\quad \mathrm{UR}_{i, j_1, \ldots, j_{n_x}, k}))]_{n_x \times N_x \times \cdots \times N_x}, \\
v^*(\mathbf{X_j} + \widehat{\mathbf{h}}_{0,\ell}(\mathbf{X_j}, T_k, q_\ell), T_k) &\to \mathbf{VH}_{J,k}(q_\ell),
\end{aligned}
\tag{8.37}
$$

where $\delta_{i,j}$ is the Kronecker delta, $F_{i,J,k} = f_i(X_J, T_k)$ for $i = 0:1$, $C_{i,J,k} = C_i(X_J, T_k)$ for $i = 0:2$, the symbol "./" denotes element-wise division, $\mathrm{UMIN}_i = U_i^{(\min)}$ for $i = 1:n_x$, and $\text{UMAX}_i = U_i^{(\text{max})}$ for $i = 1:n_x$. The hypercube form of the control constraints is used here only for a concrete example and can be replaced appropriately in the application of interest.

The Hessian matrix is not necessarily diagonal and is so only if the diffusion coefficient $0.5(g_0 g_0^\top)(\mathbf{x}, t)$ is diagonal, so the full, asymmetric Hessian is given as

$$
\begin{aligned}
\text{DDV}_{J,k} &\equiv \left[\text{DDV}_{i,j,j_1,\ldots,j_{n_x},k} \right]_{n_x \times n_x \times N_x \times \cdots \times N_x} \\
&= \Big[\big(V_{j_1 + \delta_{i,1},\ldots,j_{n_x} + \delta_{i,n_x},k} - 2V_{j_1,\ldots,j_{n_x},k} + V_{j_1 - \delta_{i,1},\ldots,j_{n_x} - \delta_{i,n_x},k} \big) \delta_{i,j} \big/ \Delta X_i^2 \\
&\quad + 0.25 \big(V_{j_1 + \delta_{i,1} + \delta_{j,1},\ldots,j_{n_x} + \delta_{i,n_x} + \delta_{j,n_x},k} \\
&\quad - V_{j_1 - \delta_{i,1} + \delta_{j,1},\ldots,j_{n_x} - \delta_{i,n_x} 1 + \delta_{j,n_x},k} - V_{j_1 + \delta_{i,1} - \delta_{j,1},\ldots,j_{n_x} + \delta_{i,n_x} - \delta_{j,n_x},k} \\
&\quad + V_{j_1 - \delta_{i,1} - \delta_{j,1},\ldots,j_{n_x} - \delta_{i,n_x} - \delta_{j,n_x},k} \big) \\
&\quad \cdot (1 - \delta_{i,j}) \big/ (\Delta X_i \Delta X_j) \Big]_{n_x \times n_x \times N_x \times \cdots \times N_x}
\end{aligned}
\tag{8.38}
$$

in the second order accuracy, CFD form. If the Hessian is diagonal, then only the second line of (8.38) is needed. The off-diagonal terms, i.e., when $i \neq j$, are conveniently calculated as the operator product of two CFDs for the two independent partials. In the case where the off-diagonal terms are significant enough that they can affect stability and convergence, Kushner and Dupuis [179] recommend a better form than that given in (8.38) for the cross term in $\text{DDV}_{J,k}$.

These are the basic numerical ingredients for converting the one-state problem of the **Crank–Nicolson extrapolator-predictor-corrector method** in Subsection 8.1.2 to the multistate problem.

Curse of Dimensionality

In the full Hessian case, the Hessian is the largest array that will be needed in the computation and will basically determine the order of both computing and memory demand for the solution of the PDE of SDP. In this full case the demands per time-step k will then be roughly proportional to the order of the $\text{DDV}_{\mathbf{j},k}$ count or

$$
O(N_{\text{DDV}}) = O\left(n_x^2 \cdot \prod_{i=1}^{n_x} N_x \right) = O\left(n_x^2 \cdot N_x^{n_x} \right) = O\left(n_x^2 \cdot e^{n_x \ln(N_x)} \right),
\tag{8.39}
$$

which is n_x times the size of the vector functions such as $\text{DV}_{\mathbf{j},k}$ and will grow exponentially with state dimension times the logarithm of the common number of nodes per dimension. If the number of nodes per dimension varies, i.e., N_i nodes in dimension i, then the geometric mean $N_x = \left(\sum_{i=1}^{n_x} N_i \right)^{1/n_x}$ can be used in place of the common value N_x in the above exponential estimate. This exponential growth in demand quantifies the exponential complexity in solving the PIDE of SDP and is called **Bellman's curse of dimensionality**. However, the very same exponential complexity (8.39) is found in high-dimensional, second order PDEs. If there are $n_x = 6$ states and there are $N_x = 64$ nodes per state using 8-byte (8B) or double words, then the order of the amount of storage required is $N_{\text{DDV}} = 8 \cdot 6^2 \cdot 64^6 \text{B} = 18$ TB, where 1 TB is a terabyte, or a computer trillion bytes, or 1024^4 bytes.

If the discrete Hessian is diagonal, then the amount of storage needed is reduced to some multiple of

$$
N_{\text{DV}} = 8 \cdot n_x \cdot N_x^{n_x} \text{B},
$$

using 8B words, DDV that has the same size as DV, so in the example with $n_x = 6$ and $N_x = 64$, $N_{DV} = 8 \cdot 6 \cdot 64^6 B = 3$ TB, a more reasonable size for a large-scale problem capable computer. If the number of nodes per dimension is reduced to 32 instead of 64, then the amount of storage needed is some multiple of $8 \cdot 6 \cdot 32^6 B = 49,152$ MB $= 48$ GB, approaching PC desktop capability (1 MB being a megabyte or 1024^2 bytes and 1GB being a gigabyte or 1024^3 bytes). The growth of the curse of dimensionality in the logarithm to the base 2 scale is illustrated in Figure 8.1 for the diagonal Hessian size case N_{DV}. Note the top scale in the figure is about 60 log(B) and $2^{60}B = 1024^6B$ is 1 exabyte (1 EB) or 1024 PB or 1024^2 TB (1 PB = 250, B is one petabyte, 1 TB = 240 B, 1 GB = 230 B, and 1 MB $= 2^{20}$ B), and multiples of petabytes are well within the capabilities of our current largest-scale computers.

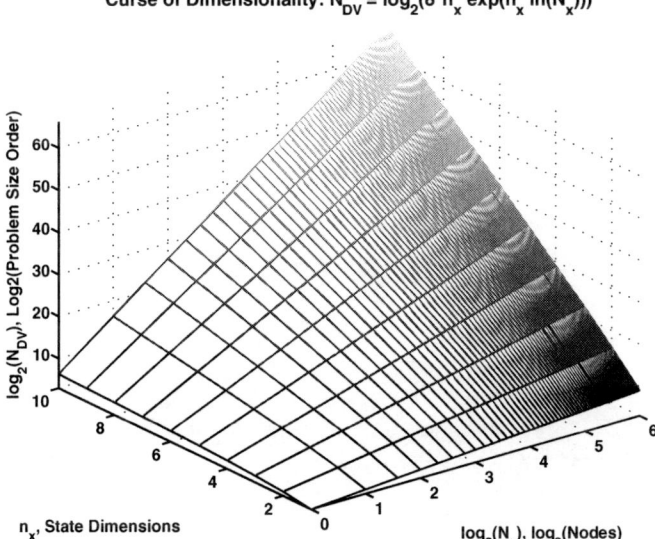

Curse of Dimensionality: $N_{DV} = \log_2(8^*n_x \exp(n_x \ln(N_x)))$

Figure 8.1. *Estimate of the logarithm to base 2 of the order of the growth of memory and computing demands using 8 byte words to illustrate the curse of dimensionality in the diagonal Hessian case for $n_x = 1 : 10$ dimensions and $N_x = 1 : 64 = 1 : 2^6$ nodes per dimension. Note that 1 KB has a base 2 exponent of $10 = \log_2(2^{10})$, while the base 2 exponent is 20 for 1 MB, 30 for 1 GB, 40 for 1 TB, 50 for 1 PB, and 60 for 1 EB.*

For parallel processing techniques in computational stochastic programming, refer to Hanson's chapter in [109]. See also [110] for more general supercomputing techniques that were developed originally to solve computational control application problems.

8.2 Markov Chain Approximation for SDP

Another method for numerically solving SDP problems in continuous time is Kushner's **Markov chain approximation** (MCA) [175, 176], which implicitly provides good convergence properties by normalizing the corresponding finite differences as proper Markov chains. In addition, MCA facilitates the proof of weak convergence using probabilistic arguments. Kushner and Dupuis's [179] method of using an auxiliary stochastic process, so that the composite stochastic process properly satisfies boundary conditions, is also treated. The summary here is in the spirit of this applied text to make the MCA method more accessible, concentrating on the techniques rather than the problems and formal definitions.

8.2.1   MCA Formulation for Stochastic Diffusions

Although MCA is valid for jump-diffusions, only diffusions will be considered here to keep the complexity manageable; the reader can consult [179] for a more complete treatment of MCA. Let the diffusion satisfy the SDE

$$dX(t) \stackrel{\text{sym}}{=} f(X(t), U(t), t)dt + g(X(t), t)dW(t), \tag{8.40}$$

where the notation otherwise is the same as in (8.1) with f and g being bounded, continuous, and Lipschitz continuous in X, while f has the same properties in U, but uniformly. For later reference, we mention the following conditional infinitesimal moments:

$$\begin{aligned}
E[dX(t)|X(t) = x, U(t) = u] &= f(x, u, t)dt, \\
\text{Var}[dX(t)|X(t) = x, U(t) = u] &= g^2(x, t)dt.
\end{aligned} \tag{8.41}$$

Let the minimal, expected costs be defined as

$$v^*(x, t) \equiv \min_{U[t, t_f)} \left[EdPdWttf \left[\int_t^{t_f} C(X(s), U(s), s)ds + S(X(t_f), t_f) \right. \right. \\
\left. \left. \Big| X(t) = x, U(t) = u \right] \right] \tag{8.42}$$

for $t_0 \le t < t_f$. The corresponding PDE of SDP is

$$0 = v_t^*(x, t) + \min_u \left[\mathcal{H}(x, u, t) \right]$$

$$\equiv v_t^*(x, t) + \min_u \left[C(x, u, t) + f(x, u, t)v_x^*(x, t) + \tfrac{1}{2}g^2(x, t)v_{xx}^*(x, t) \right] \tag{8.43}$$

$$= v_t^*(x, t) + \mathcal{H}^*(x, t).$$

The first step of the numerical part of the MCA procedure is to approximate the backward PDE (8.43) by a backward Euler method in time for simplicity. Then using the kth time-step at t_k with optimal value $v_k(x) \simeq v^*(x, t_k)$, the next value is

$$v_{k-1}(x) = v_k(x) + \Delta t_{k-1} \min_u \left[C_k(x, u) + f_k(x, u)v_k'(x) + \frac{1}{2}g_k^2(x)v_k''(x) \right] \tag{8.44}$$

for forward index $k = 1 : N_t$, $t_k \equiv t_{k-1} + \Delta t_{k-1}$, $t_{N_t} = t_f$, $C_k(x, u) = C(x, u, t_k)$, $f_k(x, u, t_k)$, and $g_k(x) = g(x, t_k)$. The final condition is $v_{N_t}(x) = S(x, t_f)$. The time-step Δt_{k-1} is called the MCA **interpolation time increment** and is selected to help form a proper Markov chain for convergence, so the increments are not necessarily constant. Although motivated by an approximation in time, time has been removed from the problem, i.e., the current problem is actually time independent. Finite differences in the state come after specifying diffusion consistency conditions.

8.2.2 MCA Local Diffusion Consistency Conditions

Let ξ_k for $k \geq 0$ denote a Markov chain of discrete stages, intended as a discrete model for the state x, whose spacing is the order of some state mesh measure ΔX, i.e., $|\Delta \xi_k| = O(\Delta X)$, where $\Delta \xi_k \equiv \xi_{k+1} - \xi_k$. Let the Markov chain **transition probability** for diffusions (D) be defined by

$$p^{(D)}(x, y|u) \equiv \text{Prob}[\xi_{k+1} = y | \xi_j, u_j, j < k, \xi_k = x, u_k = u] \qquad (8.45)$$

for transitions from current stage $\xi_k = x$ to the next stage $\xi_{k+1} = y$ using control policy $u_k = u$. (The term *stage* is used to denote a discrete state.) These transitions must satisfy the probability rules of nonnegativity $p^{(D)}(x, y|u) \geq 0$ and probability conservation for transitions, $\sum_\ell p^{(D)}(x, X_\ell|u) = 1$, under current control u and over probable state transitions $y = X_\ell$. The increments $\Delta \xi_k$ must satisfy the MCA **local diffusion consistency** conditions:

$$\text{E}[\Delta \xi_k | x, u] \equiv \sum_\ell (X_\ell - x) \cdot p^{(D)}(x, X_\ell|u) = \Delta t_{k-1} \cdot (f_k(x, u) + \text{o}(1));$$

$$\text{Var}[\Delta \xi_k | x, u] \equiv \sum_\ell (X_\ell - x - \text{E}[\Delta \xi_k | x, u])^2 \cdot p^{(D)}(x, X_\ell|u) \qquad (8.46)$$

$$= \Delta t_{k-1} \cdot (g_k^2(x) + \text{o}(1))$$

with $\Delta \xi_k \to 0^+$ as $\Delta X \to 0^+$ for $k = 0 : N_t - 1$. The conditions are consistent with the first two conditional infinitesimal moments (8.41) of a stochastic diffusion approximation corresponding to the SDE (8.40), so they are neccesary preconditions for convergence of the Markov chain to the diffusion SDE (8.40).

See Section 7.8 p. 213 or Feller [85] for more information. Also, see Kloeden and Platen [166] for stricter definitions of diffusion consistency conditions. The generalization of these diffusion consistency conditions to jump-diffusions is much more complicated but is treated in Subsection 8.2.4.

The discrete process can be used to construct a **piecewise constant** ($pw\cent$) **interpolation** of the state and control processes in continuous time, i.e.,

$$(X^{(\text{pw}\cent)}(t), U^{(\text{pw}\cent)}(t)) = \{(\xi_k, u_k), \quad t_{k-1} \leq t < t_{k-1} + \Delta t_{k-1} = t_k \quad \text{for } k \geq 1\} \qquad (8.47)$$

with the relationship between the interpolation times t_k and interpolation time increments Δt_{k-1} being $t_{k+1} = \sum_{j=0}^k \Delta t_j$. In general, the time increments will depend on ξ_k and u_k, which also depend on the order of state mesh ΔX, so $\Delta t_{k-1} = \Delta t_{k-1}(\xi_k, u_k; \Delta X)$. As the state mesh goes to zero, it is required that the maximal state mesh go to zero, i.e., $\max_{u,x}[\Delta t_{k-1}(x, u; \Delta X)] \to 0^+$.

8.2.3 MCA Numerical Finite Differences for State Derivatives and Construction of Transition Probabilities

Construction of the Markov chain transition probabilities is done by finite differencing the state derivative. The state derivative is upwinded by first order forward or backward differences, i.e., UFDs for greater stability depending on the sign of the drift coefficient $f_k(x, u, t)$ as in (8.33),

$$
v_k'(x) \simeq \begin{cases} \dfrac{v_k(x + \Delta X) - v_k(x)}{\Delta X}, & f_k(x, u) \geq 0 \\[2ex] \dfrac{v_k(x) - v_k(x - \Delta X)}{\Delta X}, & f_k(x, u) < 0 \end{cases} \tag{8.48}
$$

and CFDs of second order accuracy are used for the state second order partial derivative

$$
v_k''(x) \simeq \frac{v_k(x + \Delta X) - 2v_k(x) + v_k(x - \Delta X)}{\Delta X^2}. \tag{8.49}
$$

Alternately, second order upwinding can be used for the state first derivative so that the accuracy is consistent with $O(\Delta X^2)$ accuracy of the second derivative used above, but this leads to a double jump in the state by $2 \pm \Delta X$, so this complication will not be introduced here, although the larger $O(\Delta X)$ error **numerically pollutes** the smaller $O(\Delta X^2)$ error for small ΔX. The $O(\Delta X^2)$ forward and backward finite differences of the form used for derivative boundary conditions in (8.32) would not be useful since the alternating signs would lead to improper, negative transition probabilities for a least one double step transition.

Substituting into (8.44) for $v_{k-1}(x)$ and then collecting the coefficients in terms of transition probabilities, we get

$$
\begin{aligned}
v_{k-1}(x) = \min_{u_{k-1}} \Big[& \Delta t_{k-1} \cdot C_k(x, u_{k-1}) + p_k^{(D)}(x, x | u_{k-1}) \cdot v_k(x) \\
& + p_k^{(D)}(x, x + \Delta X | u_{k-1}) \cdot v_k(x + \Delta X) \\
& + p_k^{(D)}(x, x - \Delta X | u_{k-1}) \cdot v_k(x - \Delta X) \Big],
\end{aligned} \tag{8.50}
$$

and the transition probabilities are found to be

$$
p_k^{(D)}(x, x | u_{k-1}) = 1 - \frac{\Delta t_{k-1}}{\Delta X^2} \cdot \left(g_k^2(x) + \Delta X | f_k(x, u_{k-1})| \right), \tag{8.51}
$$

$$
p_k^{(D)}(x, x + \Delta X | u_{k-1}) = \frac{\Delta t_{k-1}}{\Delta X^2} \cdot \left(0.5 g_k^2(x) + \Delta X [f_k(x, u_{k-1})]_+ \right), \tag{8.52}
$$

$$
p_k^{(D)}(x, x - \Delta X | u_{k-1}) = \frac{\Delta t_{k-1}}{\Delta X^2} \cdot \left(0.5 g_k^2(x) + \Delta X [f_k(x, u_{k-1})]_- \right), \tag{8.53}
$$

where $[f]_\pm \equiv \max[\pm f] \geq 0$. Upwinding ensures that all terms in the coefficients of Δt_{k-1} are nonnegative, so that the up and down transition probabilities, $p_k^{(D)}(x, x + \Delta X | u_{k-1})$ and $p_k^{(D)}(x, x - \Delta X | u_{k-1})$, are nonnegative. Note that on the right-hand side of the conservation law (8.50) for the transition probabilities to get the value function for the past time t_{k-1}, the value function is evaluated at the current time t_k, but the control is for the past time t_{k-1}, which makes it seem like the control is implicit. However, u_{k-1} is thought to be the control to get the state x from time t_{k-1} to time t_k, and the optimization over u_{k-1} will determine u_{k-1} in terms of values at t_k anyway, so it is not really an implicit term. Genuine implicit methods are discussed in Kushner and Dupuis [179].

It is clear that Δt_{k-1} must be sufficiently small so that the state self-transition probability $p_k^{(D)}(x, x|u_{k-1})$ is nonnegative, i.e., is a proper probability. This implies the convergence criteria

$$\frac{\Delta t_{k-1}}{\Delta X^2} \leq \frac{1}{g_k^2(x) + \Delta X|f_k(x, u_{k-1})|} \tag{8.54}$$

or, reformulated in terms of a generalization of the parabolic mesh ratio condition, the criteria are

$$\left(g_k^2(x) + \Delta X|f_k(x, u_{k-1})|\right) \cdot \frac{\Delta t_{k-1}}{(\Delta X)^2} \leq 1, \tag{8.55}$$

including both the diffusion coefficient and the upwinded drift term in the scaling of $\Delta t_{k-1}/(\Delta X)^2$. Since (8.54) should hold for all discrete time-steps k, then we should have

$$\max_{x,u,k}\left[(g_k^2(x) + \Delta X|f_k(x, u)|)\frac{\Delta t_{k-1}}{\Delta X^2}\right] \leq 1. \tag{8.56}$$

The diffusion consistency conditions (8.46) can be directly confirmed in the following three local state case,

$$\begin{aligned}
\mathrm{E}[\Delta\xi_k|x, u_{k-1}] &= p_k^{(D)}(x, x|u_{k-1}) \cdot 0 + p_k^{(D)}(x, x + \Delta X|u_{k-1}) \cdot (+\Delta X) \\
&\quad + p_k^{(D)}(x, x - \Delta X|u_{k-1}) \cdot (-\Delta X) \\
&= \Delta t_{k-1} \cdot ([f_k(x, u_{k-1})]_+ - [f_k(x, u_{k-1})]_-) \\
&\equiv \Delta t_{k-1} \cdot f_k(x, u_{k-1}), \\
\mathrm{Var}[\Delta\xi_k|x, u_{k-1}] &= p_k^{(D)}(x, x|u_{k-1}) \cdot (\Delta t_{k-1}f_k(x, u_{k-1}))^2 \\
&\quad + p_k^{(D)}(x, x + \Delta X|u_{k-1}) \cdot (\Delta X - \Delta t_{k-1}f_k(x, u_{k-1}))^2 \\
&\quad + p_k^{(D)}(x, x - \Delta X|u_{k-1}) \cdot (-\Delta X - \Delta t_{k-1}f_k(x, u_{k-1}))^2 \\
&= \Delta t_{k-1} \cdot \left(g_k^2 + |f_k(x, u_{k-1})|\Delta X - 2\Delta t_{k-1}f_k^2(x, u_{k-1})\right) \\
&= \Delta t_{k-1} \cdot \left(g_k^2 + o(1)\right)
\end{aligned}$$

as $\Delta X \to 0^+$ and consequently $\Delta t_{k-1} \to 0^+$.

Upon proper choice of the time and state grids satisfying (8.56), for example, in the case of regular grids as used in the previous section in (8.10) with N_t nodes in t on $[t_0, t_f]$ and N_x nodes in x on $[x_0, x_{max}]$, $T_k = t_f - (k-1)\Delta t$ for $k = 1 : N_t$, $\Delta t_{k-1} = \Delta t = (t_f - t_0)/(N_t - 1)$ and $X_j = x_0 + (j-1)\Delta X$ for $j = 1 : N_x$, $\Delta X = (x_{max} - x_0)/(N_x - 1)$, then

$$\begin{aligned}
V_{j,k-1} &\equiv v_{k-1}(X_j) \\
&= \Delta t \cdot C_k(X_j, U_{j,k-1}) + p_k^{(D)}(X_j, X_j|U_{j,k-1}) \cdot V_{j,k} \\
&\quad + p_k^{(D)}(X_j, X_{j+1}|U_{j,k-1}) \cdot V_{j+1,k} \\
&\quad + p_k^{(D)}(X_j, X_{j-1}|U_{j,k-1}) \cdot V_{j-1,k}
\end{aligned} \tag{8.57}$$

when the optimal control is

$$
\begin{aligned}
U_{j,k-1} = \operatorname{argmin}_{u_{k-1}} \Big[&\Delta t_{k-1} \cdot C_k(X_j, u_{k-1}) + p_k^{(D)}(X_j, X_j | u_{k-1}) \cdot V_{j,k} \\
&+ p_k^{(D)}(X_j, X_{j+1} | u_{k-1}) \cdot V_{j+1,k} \\
&+ p_k^{(D)}(X_j, X_{j-1} | u_{k-1}) \cdot V_{j-1,k} \Big]
\end{aligned}
\tag{8.58}
$$

for $j = 1 : N_x$ for each stage $k = N_t : -1 : 2$ in backward order. Note that in [179], Kushner and Dupuis suggest a preference for selecting the interpolation time-step Δt_{k-1} so that the self-transition probability $p^{(D)}(x, x | u)$ vanishes, leading to a renormalization of the non-self-transition probabilities such as $p^{(D)}(x, x \pm \Delta X | u)$.

In this section, the **MCA** has only been summarized to convey the main ideas, but those interested in the weak convergence proofs and related theory should consult [177, 179] and additional references therein.

8.2.4 MCA Extensions to Include Jump Processes

In [179, Section 5.6], Kushner and Dupuis briefly present the extensions of the MCA for diffusions to that for jump-diffusions. Earlier, Kushner and DiMasi [178] made contributions to the jump-diffusion optimal control problem, while Kushner [177] more recently gave further results on existence and numerical methods for the problem.

The main idea is based upon the facts that the Poisson process is instantaneous compared to the continuity of the diffusion process and that the Poisson process during short time-intervals Δt can be asymptotically treated as a zero-one Bernoulli process, as mentioned in prior chapters. Starting with the jump-diffusion SDE extension of (8.40), we get

$$
\begin{aligned}
dX(t) \overset{\text{sym}}{=} \ & f(X(t), U(t), t)dt + g(X(t), t)dW(t), \\
&+ h(X(t), U(t), t, Q)dP(t; Q, X(t), U(t), t),
\end{aligned}
\tag{8.59}
$$

where $dP(t; Q, X(t), U(t), t)$ is the differential Poisson process with rate $\lambda(t; x, u, t)$, and $h(x, u, t, q)$ is the state jump-amplitude and generalized probability density $\phi_Q(q)$. The conditional infinitesimal moments are given by

$$
\begin{aligned}
\mathrm{E}[dX(t)|X(t) = x, U(t) = u] &= f(x, u, t)dt + \mathrm{E}_Q[h(x, u, t, Q)]\lambda(t; x, u, t)dt, \\
\mathrm{Var}[dX(t)|X(t) = x, U(t) = u] &= g^2(x, t)dt + \mathrm{E}_Q[h^2(x, u, t, Q)]\lambda(t; x, u, t)dt.
\end{aligned}
\tag{8.60}
$$

By separability of the diffusion and the jumps for sufficiently small time-steps Δt_{k-1}, the diffusion transition probabilities are unchanged, with $p_k^{(D)}(x, y | u)$ for stage k. The probability of zero or one Poisson jump in time-steps of Δt_{k-1} can be written

$$
p_{j,k}^{(J)} = \left\{ \begin{array}{ll} 1 - \lambda \Delta t_{k-1} + \mathrm{o}(\Delta t_{k-1}), & j = 0 \text{ jumps} \\ \lambda \Delta t_{k-1} + \mathrm{o}(\Delta t_{k-1}), & j = 1 \text{ jump} \\ \mathrm{o}(\Delta t_{k-1}), & j \geq 2 \text{ jumps} \end{array} \right\}
\tag{8.61}
$$

as $\Delta t_{k-1} \to 0^+$.

For the discretization jump-amplitude function $h(x, t, q)$ of the corresponding compound Poisson process, a concrete rather than an abstract formulation of Kushner and Dupuis [179] will be given so that the transition of a piecewise constant prejump stage $x = X_j$ for some j to a piecewise constant postjump stage $y = X_\ell$ for some ℓ, where $X_{j+1} = X_j + \Delta X_j$ for $j = 1 : N_x - 1$, $X_1 = x_0$, and $X_{N_x} = x_{\max}$ and the mesh is given by $\Delta X = \max_j (\Delta X_j) \to 0^+$. However, the treatment of jumps is much more complicated than that for diffusion, whose dependence is only local, depending on only nearest neighbor or similarly close nodes, but jump behavior is globally dependent on nodes that may be remote from the current node X_j. Also, the connection of the jump-amplitude function to the jump-amplitude random mark variable q will be clarified. The jump-amplitude may be continuously distributed due to a continuous mark density $\phi_Q(q)$. It is assumed that postjump stage $y = x + h(x, t, q)$ is uniquely invertible with q as a function of y given x, but it is necessary to have a set target $S(X_\ell)$ rather than a point target $y = X_\ell$ so a corresponding set $Q_{j,\ell}(t)$ of positive probability measure can be found. Let $S(X_\ell)$ be a partition of the state domain $[X_1, X_{N_x}]$ such that

$$\sum_{\ell=1}^{N_x} S(X_\ell) = [X_1, X_{N_x}].$$

For each application, the partition $S(X_\ell)$ will depend on the particular boundary conditions, singular points, or related zero points, which could lead to forward or backward shifted intervals or intervals centered about X_ℓ as with rounding. For the discretized mesh, we use piecewise continuous (pwc) step functions, rather than the prior piecewise constant (pw¢) step functions. Given the stage set $S(X_\ell)$, let the jump-amplitude function be

$$H_{j,\ell}^{(\mathrm{pwc})}(t) = h(X_j, t, Q_{j,\ell}(t)) = S(X_\ell) - X_j, \tag{8.62}$$

which implicitly defines the mark set $Q_{j,\ell}(t)$ for $1 \le j < \infty$ and $1 \le \ell < \infty$. This ensures that a jump takes a proper (pwc) stage X_j to a proper (pwc) stage X_ℓ defined by the set $S(X_\ell)$. Given a jump it is also necessary to know the corresponding probability of the transition referenced by (8.62), i.e.,

$$\mathrm{Prob}\left[y = x + h(x, t, q) \in S(X_\ell) \mid x = X_j, \ y \in S(X_\ell)\right]$$
$$= \overline{\Phi}(X_j, X_\ell, t) \equiv \int_{Q_{j,\ell}(t)} \phi_Q(q) dq, \tag{8.63}$$

where $\phi_Q(q)$ is the generalized mark density with corresponding distribution $\Phi_Q(q)$, except that when $h(X_{\hat{j}}, t, q) = 0$ for some $\hat{j}$, i.e., there is a **zero jump** and $y \in S(X_\ell)$ is not achievable for general ℓ, then $\overline{\Phi}_{\hat{j},\ell}(t) \equiv 0$. In the case that $\overline{\Phi}(X_j, X_\ell, t)$ leads to a probabilistically deficient distribution, in general the renormalized form is

$$\widehat{\Phi}(X_j, X_\ell, t) = \overline{\Phi}(X_j, X_\ell, t) \Big/ \overline{\overline{\Phi}}(X_j, t), \tag{8.64}$$

where

$$\overline{\overline{\Phi}}(X_j, t) \equiv \sum_{\ell=1}^{N_x} \overline{\Phi}(X_j, X_\ell, t) = \sum_{\ell=1}^{N_x} \int_{Q_{j,\ell}(t)} \phi_Q(q) dq.$$

Example 8.3. *Geometric Jump-Diffusion Target Mark Set Calculations.*
For the geometric jump-diffusion used in finance, with linear jump-amplitude

$$h(x, t, q) = x J(q, t),$$

it is convenient to choose the log-return jump as the mark, i.e.,

$$q = [\ln(X)](t) = \ln \left(\left(X(t^-) + X(t^-) J(q, t^-) \right) / X(t^-) \right) = \ln(1 + J(q, t^-))$$

*so $h(x, t, q) = x(\exp(q) - 1)$. Hence, $X_1 = x_0 = 0$ is a **zero point** needing special
treatment since there can be no target stage except for $[X_1, X_1] = \{0\}$, so that a proper
partition of $[X_1, X_{N_x}]$ would be $\mathcal{S}(X_1) = \{0\}$ and $\mathcal{S}(X_\ell) = (X_{\ell-1}, X_{\ell-2}]$ for $\ell = 2 : N_x$.
The discrete jump-amplitude $H_{1,\ell}^{(\mathrm{pwc})}(t) \equiv 0$ for definiteness when $X_1 = 0$ and*

$$H_{j,\ell}^{(\mathrm{pwc})}(t) \equiv X_\ell - X_j$$

for $\ell = 2 : N_x$. The target mark set is

$$\mathcal{Q}_{j,\ell}(t) = \left(\ln(X_{\ell-1}/X_j), \ln(X_\ell/X_j) \right]$$

*for $\ell = 2 : N_x$ when $j > 1$. Given a mark density, a renormalized target distribution
$\widehat{\Phi}(X_j, X_\ell, t)$ can be calculated.*

The Markov chain approximation $\xi_k(\Delta X)$ is **locally jump-diffusion consistent** if
there is an **interpolation time-interval** $\Delta t_{k-1} = \Delta t(x, u; \Delta X) \to 0^+$ uniformly in
$(x, u, \Delta X)$ as the mesh gauge $\Delta X \to 0^+$ and so that the following hold:

1. Along with $\Delta t(x, u; \Delta X)$, there is a locally diffusion consistent transition probability
   $p^{(D)}(x, y \mid u; \Delta X)$ satisfying the conditions in (8.46).

2. The **jump-diffusion transition probabilities** $p^{(JD)}(x, y \mid u; \lambda, \Delta X)$ must conserve
   probability over the postjump values $y = X_\ell$ from any given prejump value $x = X_j$,
   i.e.,

   $$\sum_\ell p^{(JD)}(X_j, X_\ell \mid u; \lambda, \Delta X) = 1.$$

3. Markov chain increments $\Delta \xi_k$ satisfy the MCA **jump-diffusion local consistency**
   conditions consistent with the jump-diffusion conditional infinitesimal moments (8.60)
   with replacements $f(x, u, t) \to f_k, (x, u), g(x, t) \to g_k(x), h(x, t, q) \to h_k(x, q)$,
   $H_{j,\ell}^{(\mathrm{pwc})}(t) \to H_{j,\ell,k}^{(\mathrm{pwc})} \widehat{\Phi}(X_j, X_\ell, t) \to \widehat{\Phi}_k(X_j, X_\ell)$, under current control u and over
   probable state transitions

   $$\mathrm{E}[\Delta \xi_k \mid X_j, u_{k-1}] \equiv \sum_\ell (X_\ell - X_j) \cdot p^{(JD)}(X_j, X_\ell \mid u_{k-1}; \lambda, \Delta X)$$
   $$= \Delta t_{k-1} \cdot \left(f_k(X_j, u_{k-1}) + \lambda \mathrm{E}_Q[h_k(X_j, Q)] + \mathrm{o}(1) \right);$$

   $$\mathrm{Var}[\Delta \xi_k \mid X_j, u_{k-1}] \equiv \sum_\ell (X_\ell - X_j - \mathrm{E}[\Delta \xi_k \mid X_j, u_{k-1}])^2 \qquad (8.65)$$
   $$\cdot p^{(JD)}(x, X_\ell \mid u_{k-1}; \lambda, \Delta X)$$
   $$= \Delta t_{k-1} \cdot \left(g_k^2(x) + \lambda \mathrm{E}_Q[h_k^2(X_j, Q)] + \mathrm{o}(1) \right)$$

   with $\Delta \xi_k \to 0^+$ as $\Delta X \to 0^+$ for $k = 0 : N_t - 1$.

4. There is a small error factor $\varepsilon(s, u; \Delta X) = o(\Delta t(x, u; \Delta X))$ that can be used to construct (using the method of Kushner and Dupuis [179], modified for clarification here) the **jump-diffusion transition probability** $p^{(JD)}(x, y \mid u; \lambda, \Delta X)$ and is of the form

$$
p^{(JD)}(X_j, X_\ell \mid u; \lambda, \Delta X)
$$

$$
= (1 - \lambda \Delta t(X_j, u; \Delta X) - \varepsilon(X_j, u; \Delta X)) \cdot p^{(D)}(X_j, X_\ell \mid u; \Delta X) \qquad (8.66)
$$

$$
+ (\lambda \Delta t(X_j, u; \Delta X) + \varepsilon(X_j, u; \Delta X)) \cdot \widehat{\Phi}_k(X_j, X_\ell) \mathbf{1}_{X_\ell \in X_j + H_{j,\ell,k}^{(\mathrm{pwc})}}
$$

for $1 \le j < \infty$ and $1 \le \ell < \infty$, where $\mathbf{1}_{\mathcal{S}}$ is the indicator function for set $\mathcal{S} = \{X_\ell \in X_j + H_{j,\ell,k}^{(\mathrm{pwc})}\}$ and is used so the term it multiplies is used only for a jump.

By using the conservation laws

$$
\sum_{\ell=1}^{N_x} p^{(D)}(X_j, X_\ell \mid u; \Delta X) = 1
$$

and

$$
\sum_{\ell=1}^{N_x} \widehat{\Phi}(X_j, X_\ell, t) = 1,
$$

it is easy to show that the constructed jump-diffusion transition probability in (8.66) is conserved, i.e.,

$$
\sum_{\ell=1}^{N_x} p^{(JD)}(X_j, X_\ell \mid u; \lambda, \Delta X) = 1.
$$

The error factor $\varepsilon(s, u; \Delta X)$ reflects the asymptotically small error terms $o(\Delta t_{k-1})$ in the Poisson counting process definition (8.61) but is selected so that the conservation is exact.

Using the first moment diffusion local consistency condition in (8.46) and a mark density weighted rectangular integration rule, we get

$$
\mathrm{E}_Q[h_k(x, Q)] \simeq \sum_{\ell=1}^{N_x} H_{j,\ell,k}^{(\mathrm{pwc})} \widehat{\Phi}_k(X_j, X_\ell).
$$

Then,

$$
\mathrm{E}[\Delta \xi_k \mid X_j, u] \simeq \Delta t_{k-1}(X_j, u; \Delta X) \cdot \big(f_k(X_j, u) + \mathrm{E}_Q[h_k(x, Q)] + o(1)\big)
$$

$$
= \overline{X}^{(D)} + \overline{X}^{(J)},
$$

splitting the diffusion and jump parts. Similarly, for the second moment jump-diffusion consistency condition, except with more algebra with the above splitting and more small time asymptotics in absorbing all quadratic and smaller time increments into $\Delta t_{k-1} \cdot o(1)$, it can be demonstrated that

$$
\mathrm{Var}[\Delta \xi_k \mid X_j, u] \simeq \Delta t_{k-1}(X_j, u; \Delta X) \cdot \big(g_k^2(X_j) + \mathrm{E}_Q[h_k^2(x, Q)] + o(1)\big).
$$

Further evaluations require knowledge of the mark density, the jump-diffusion coefficients (f, g, h), and the boundary condition on the state domain. Due to the global nature of the compound jump process with jumps beyond the local nodes needed by the diffusion component process, the diffusion mesh ratio criteria (8.56) (or (8.30) in case the CFDs are usable) will have to suffice for practical reasons. See Kushner and Dupius [179] for information on reflected boundary conditions and other techniques for handling boundary conditions when there are jumps.

Suggested References for Further Reading

- Chung, Hanson, and Xu, 1992 [55]

- Douglas and Dupont, 1970 [73]

- Douglas, 1979 [74]

- Dyer and McReynolds, 1979 [77]

- Gunzburger, 2003 [102]

- Hanson, 1989 [107], 1991 [108], 1996 [109], and 2003 [110, 111]

- Hanson and Naimipour, 1993 [112]

- Kushner, 1976 [175], 1990 [176], and 2000 [177]

- Kushner amd DiMasi, 1978 [178]

- Kushner and Dupuis, 2001 [179]

- Kushner and Yin, 1997 [181]

- Larson, 1967 [182]

- Naimipour and Hanson, 1993 [216]

- Polak, 1973 [227]

- Press et al., 2002 [230]

- Westman and Hanson, 1997 [274] and 2000 [277]

- Zhu and Hanson, 2006 [293]

Chapter 9
Stochastic Simulations

Any one who considers arithmetical methods of producing random digits is, of course, in a state of sin.

—John von Neumann (1903–1957), apparently meant as a caution.
http://en.wikiquote.org/wiki/John_von_Neumann

Fast cars, fast women, fast algorithms...
what more could a man want?

—Joe Mattis, http://www.xs4all.nl/~jcdverha/
scijokes/1_5.html#subindex

In this chapter, methods are considered that treat stochastic dynamics, such as direct simulations of SDEs [167, 166], with many numerical techniques offering improvements over the elementary integration methods beyond stochastic versions of Euler's method.

Monte Carlo methods simulate solutions for higher level applications. These applications include many improvements to increase the probable accuracy to reduce the need for large-scale sample sizes. Many of the techniques involve variance reduction and the generation of sample variates for nonuniform distributions [97, 151, 292].

9.1   SDE Simulation Methods

Simulation methods for the dynamics of SDEs are discussed. Basic simulation procedures were introduced in Chapters 2 through 5, but here diffusion and jump-diffusion simulations are discussed and explored much further. Primary references are Kloeden et al. [167], Cyganowski et al. [66, 65, 67], the compact review by D. Higham [140], and D. Higham and Kloeden [144, 145]. Many of these references deal almost entirely with diffusions, and the most comprehensive, theoretically and numerically, on diffusions is the monograph of Kloeden and Platen [166]. Maghsoodi [191], Cyganowski et al. [66, 65], and D. Higham and Kloeden [145] treat jump-diffusions seriously. However, random simulations to solve stochastic optimal control problems are not very useful due to the additional complexity involved in the optimization step, while optimal control problems can be reduced to deterministic ODE or PDE formulations, which can be solved more systematically.

9.1.1   Convergence and Stability for Stochastic Problems and Simulations

Consider the jump-diffusion stochastic differential equation

$$dX(t) = f(X(t), t)dt + g(X(t), t)dW(t) + h(X(t), t)dP(t), \tag{9.1}$$

$X(0) = x_0$ with probability one and $0 \leq t \leq t_f$, where the coefficient functions $f(X(t), t)$, $g(X(t), t)$, and $h(X(t), t)$ are continuously differentiable. (See [166] for tighter conditions; $h(X(t), t)$ could also depend on random marks Q.)

In Subsection 4.3.3, the main concern was formal SDE simulations, but here there will be more attention on convergence of the simulations. Let t_k denote a discrete time such that $t_{k+1} = t_k + \Delta t$ for $k = 0 : N_t - 1$, so $t_{N_t} = t_f$ and $\Delta t = t_f / N_t$. For the state, let X_k denote the discrete approximation at time t_k to the exact value $X(t_k)$, i.e., $X_k \simeq X(t_k)$.

Definition 9.1. *The approximation X_k is said to **converge** to the exact value $X(t_k)$*

- *in the **strong mean absolute error sense** if the conditional expectation*

$$\mathrm{E}\left[|X_k - X(t_k)| \mid X(0) = x_0\right] \to 0^+ \ as \ \Delta t \to 0^+ \tag{9.2}$$

  for fixed time $t_k = k\Delta t$, e.g., $t_f = t_{N_t}$;

  *the strong convergence in the mean absolute error is said to be **order** or with **log-rate** $\gamma_s > 0$ in mean absolute error if*

$$\mathrm{E}\left[|X_k - X(t_k)| \mid X_0 = x_0\right] \leq C_s \cdot (\Delta t)^{\gamma_s} \tag{9.3}$$

  for sufficiently small Δt, for fixed time $t_k = k\Delta t$, e.g., $t_f = t_{N_t}$, and constant $C_s > 0$.

- *in the **weak sense** if the difference in conditional expectations*

$$|\mathrm{E}[X_k \mid X_0 = x_0] - \mathrm{E}[X(t_k) \mid X(0) = x_0]| \to 0^+ \ as \ \Delta t \to 0^+ \tag{9.4}$$

  for fixed time $t_k = k\Delta t$, e.g., $t_f = t_{N_t}$;

  *the weak convergence is said to be **order** or with **log-rate** $\gamma_w > 0$ in mean error if*

$$|\mathrm{E}[X_k \mid X_0 = x_0] - \mathrm{E}[X(t_k) \mid X(0) = x_0]| \leq C_w \cdot (\Delta t)^{\gamma_w} \tag{9.5}$$

  for sufficiently small Δt, for fixed time $t_k = k\Delta t$, e.g., $t_f = t_{N_t}$, and constant $C_w > 0$.

- *Alternately, **strong convergence in mean square error (MSE)**, instead of mean error, can be defined (Maghsoodi [191])*

$$\sup_k \left(\mathrm{E}\left[(X_k - X(t_k))^2 \mid X_0 = x_0\right]\right) \leq C_s^{(\mathrm{mse})} \cdot (\Delta t)^{\gamma_s^{(\mathrm{mse})}} \tag{9.6}$$

  for sufficiently small Δt and constant $C_s^{(\mathrm{mse})} > 0$; thus the maximal RMS error rate is

$$\mathrm{O}\left((\Delta t)^{\gamma_s^{(\mathrm{mse})}/2}\right),$$

  so it is fair to compare the mean absolute error rate γ_s with the RMS error rate $\gamma_s^{(\mathrm{mse})}/2$.

For ODEs, a solution $X(t)$ is **asymptotic stable** as $t \to +\infty$ if

$$\lim_{t \to +\infty} |X(t)| = 0$$

in the continuous time case, and in the discrete time case the approximation X_k is **asymptotic stable** as $k \to +\infty$ if

$$\lim_{k \to +\infty} |X_k| = 0.$$

However, such a definition is not applicable even if the coefficient functions are bounded and otherwise nicely behaved, since for diffusions the range of the random process $W(t)$ is infinite. Thus, the notion of stochastic asymptotic stability has to be modified for stochastic processes.

Definition 9.2.

- *For continuous time, the real stochastic solution $X(t)$ is said to be **asymptotically mean square stable** if*

$$\lim_{t \to +\infty} \mathrm{E}\left[X^2(t) \mid X(0) = x_0 \right] = 0. \tag{9.7}$$

*Alternately, $X(t)$ is **asymptotically stable in probability** if*

$$\mathrm{Prob}\left[\lim_{t \to +\infty} |X(t)| = 0 \;\middle|\; X(0) = x_0 \right]. \tag{9.8}$$

- *For discrete time, the real stochastic approximation X_k is said to **asymptotically mean square stable** if*

$$\lim_{k \to +\infty} \mathrm{E}\left[X_k^2 \mid X_0 = x_0 \right] = 0. \tag{9.9}$$

*Alternately, X_k is **asymptotically stable in probability** if*

$$\mathrm{Prob}\left[\lim_{k \to +\infty} |X_k| = 0 \;\middle|\; X_0 = x_0 \right]. \tag{9.10}$$

As a continuous time example, consider the linear, constant coefficient SDE, letting $(f(x,t), g(x,t), h(x,t)) = (\mu_0, \sigma_0, \nu_0)$ in (9.1),

$$dX(t) = X(t)(\mu_0 dt + \sigma_0 dW(t) + \nu_0 dP(t)),$$

where μ_0, σ_0, ν_0, and λ_0 are constants and where $\mathrm{E}[dP(t)] = \lambda_0 dt$. From (4.80), the exact solution is

$$X(t) = x_0 \exp((\mu_0 - \sigma_0^2/2)t + \sigma_0 W(t))(1 + \nu_0)^{P(t)}. \tag{9.11}$$

Using the independent increment techniques for the expectation in (4.81), the mean square is

$$\mathrm{E}\left[X^2(t_f) \mid X(0) = x_0 \right] = x_0^2 e^{(2(\mu_0 + \lambda_0 \nu_0) + \sigma_0^2 + \lambda_0 \nu_0^2) t_f}.$$

Thus, $X(t_f)$ is asymptotically mean square stable if the exponential is decaying as $t_f \rightarrow +\infty$, so

$$2(\mu_0 + \lambda_0 \nu_0) + \sigma_0^2 + \lambda_0 \nu_0^2 < 0, \tag{9.12}$$

which, in qualitative terms of the relative conditional infinitesimal moments, can be put in the form

$$E[dX(t)/X(t) \mid X(t)] < -0.5 \mathrm{Var}[dX(t)/X(t) \mid X(t)],$$

assuming $x_0 > 0$, so $X(t) > 0$. Hence, the combined jump-diffusion relative infinitesimal mean has to be less than minus one-half the relative infinitesimal variance.

9.1.2   Stochastic Diffusion Euler Simulations

The simplest simulation model using Euler's method for SDEs is more properly called the **Euler–Maruyama (EM) method** to distinguish it from the deterministic Euler method for DEs and the EM method was used in Subsection 4.3.3 in this text and has the stochastic difference form

$$X_{k+1} = X_k + F_k \Delta t + G_k \Delta W_k \tag{9.13}$$

for $k = 0 : N_t - 1$, where $F_k \equiv f(X_k, t_k)$, $G_k \equiv g(X_k, t_k)$, and $\Delta W_k \equiv W(t_{k+1}) - W(t_k)$. For instance, in MATLAB, a fragment of the code for the discrete diffusion approximation for a linear SDE would be like that given in Figure 9.1. Recall that MATLAB is unit based, i.e., array subscripts start at one. In this example, the drift coefficient rate is time-dependent with $f(x, t) = \mu(t)x$, where $\mu(t) = 1/(1 + 0.5t)^2$, but the $dW(t)$-coefficient is time independent with $g(x, t) = \sigma(t)x$, where $\sigma(t) = \sigma_0$ where σ_0 is a constant, i.e.,

$$dX(t) = X(t)(\mu(t)dt + \sigma(t)dW(t)). \tag{9.14}$$

In this case the log-transformation $Y(t) = \ln(X(t))$ by the Itô stochastic chain rule leads to a state-independent SDE solution, $Y(t) = (\mu(t) - \sigma^2(t)/2)dt + \sigma(t)dW(t)$. A simple integration followed by a transformation inversion leads to the general exact stochastic solution

$$X^{(\mathrm{exact})}(t) = x_0 \exp(\overline{\mu}(t) - \overline{\sigma^2}(t)/2 + \overline{(\sigma * W)}(t)), \tag{9.15}$$

where $\overline{\mu}(t) = \int_0^t \mu(s)ds$, $\overline{\sigma^2}(t) = \int_0^t \sigma^2(s)ds$ and $\overline{(\sigma * W)}(t) = \int_0^t \sigma(s)dW(s)$, which in the simpler case here reduces the integral to $\overline{(\sigma * W)}(t) = \sigma_0 W(t)$. Thus, an approximation of this diffusion integral in not necessary. Equation (9.15) is an exact formula, but comparison of the Euler–Maruyama approximation to that of the exact requires an approximate simulation of $W(t)$ in $\overline{(\sigma * W)}(t)$. Following D. Higham's lead [140], a fine grid of N_t sample points is used for the exact formula, and a lumped, coarse grid with $N_t/8$ points is taken from the set for the exact case. This makes for a more accurate comparison. The comparison between the coarse Euler–Maruyama approximation and the fine exact approximation $X^{(\mathrm{exact})}(t)$ in (9.15) is illustrated in Figure 9.2. The error between the Euler–Maruyama approximate path and the exact path at the coarse time points is presented in Figure 9.3. For further computer experiments verifying convergence using path averages, see D. Higham [140]. For the complete sample code used to generate these Euler–Maruyama figures, see code C.17, called `sdeeulersim.m`, in Online Appendix C.

```
function sdeeulersim
% Euler-Maruyama Simulation Test: Linear SDE:
%    dX(t) = X(t)(mu(t)dt+sigma(t)dW(t)),
% Given  Initial data: x0, t0, tf, Nt; functions: f, g, xexact
clc
%
randn('state',8); % Set random state or seed;
x0 = 1; t0 = 0; tf = 5; Nt = 2^14; DT = tf/Nt; sqrtt = sqrt(DT);
X(1) = x0; Xexact(1) = x0;   t = [t0:DT:tf];
DW = randn(1,Nt)*sqrtt; % Simulate DW as sqrt(DT)*randn;
W = cumsum(DW); % Omits initial zero value;
for k = 1:Nt % Exact formula to fine precision}
    Xexact(k+1) = xexact(x0,t(k+1),W(k)); % Calls subfunction;
end
L = 2^3; NL = Nt/L; KL = [0:L:Nt]; DTL = L*DT; tL = [t0:DTL:tf];
for k = 1:NL % Euler formula to lumped, coarse precision:
    DWL = sum(DW(1,KL(k)+1:KL(k+1)));
    Xeul(k+1)=Xeul(k)+f(Xeul(k),tL(k))*DTL+g(Xeul(k),tL(k))*DWL;
    Xdiff(k+1) = Xeul(k+1) - Xexact(KL(k+1));
end
plot(tL,Xeul,'k--','linewidth',3); hold on
plot(t,Xexact,'k-','linewidth',3); hold off
title('SDE Euler-Maruyama and Exact Linear SDE  Simulations');
xlabel('t, Time'); ylabel('X(t), State');
legend('X(t): Euler','Xexact: Exact','Location','Best');
%
function y = f(x,t)
    mu = 1/(1+0.5*t)^2; % Change with application;
    y = mu*x;
%
function y = g(x,t)
    sig = 0.5; % Change with application;
    y = sig*x;
%
function y = xexact(x0,t,w)
% exact solution if available for general linear SDE:
    mubar = 2-2/(1+0.5*t); sig = 0.5; sig2bar = sig^2*t/2;
    y = x0*exp(mubar-sig2bar + sig*w);
%End Code
```

Figure 9.1. *EM SDE simulations code.*

Kloeden and Platen [166, Section 10.2] show for the Euler–Maruyama simulation method, using a level of analysis beyond the scope of this text, that the log-rate of convergence in the strong sense is $\gamma_s = 0.5$, while in the weak sense the rate is $\gamma_w = 1$. Thus, the log-rate for convergence in the weak sense is the same as that for the traditional Euler's method applied to deterministic DEs in the strong or weak sense, i.e., $\gamma = 1$ for the deterministic case, since the expectation operator plays no role.

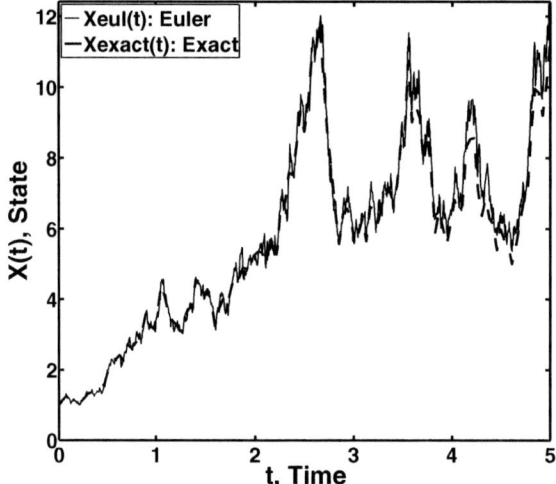

Figure 9.2. *Comparison of coarse Euler–Maruyama and fine exact paths, simulated using MATLAB with $N_t = 1024$ fine sample points for the exact path (9.15) and $N_t/8 = 128$ coarse points for the Euler path (9.13), initial time $t_0 = 0$, final time $t_f = 5$, and initial state $x_0 = 1.0$. Time-dependent parameter values are $\mu(t) = 0.5/(1 + 0.5t)^2$ and $\sigma(t) = 0.5$.*

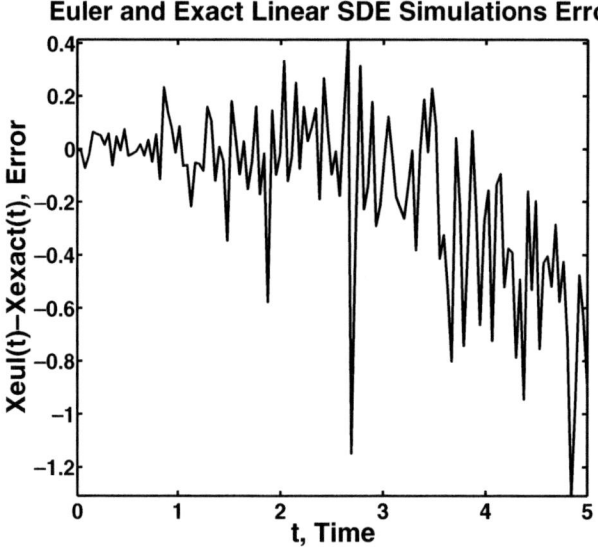

Figure 9.3. *Error in coarse Euler–Maruyama and fine exact paths using the coarse discrete time points. The simulations use MATLAB with the same values and time-dependent coefficients as in Figure 9.2. The Euler maximal-absolute error for this example is $1.3 \simeq 34\Delta t/8$, while for $N_t = 4096$ the maximal error is better at $0.28 \simeq 29\Delta t/8$.*

For convergence in the weak sense, the Euler–Maruyama method and the linear, constant rate SDE,

$$dX(t) = \mu_0 X(t)dt + \sigma_0 X(t)dW(t),$$

where μ_0 and σ_0 are constants, the log-rate result can be shown with a reasonable effort. From (9.15) or (9.11) with $\nu_0 = 0$, the exact solution is

$$X^{(\text{exact})}(t) = x_0 \exp((\mu_0 - \sigma_0^2/2)t + \sigma_0 W(t)).$$

In this case, the EM approximation from (9.13) has the form of a stochastic difference equation (SΔE),

$$X_k = X_{k-1} \cdot (1 + \mu_0 \Delta t + \sigma_0 \Delta W_{k-1}) \tag{9.16}$$

for $k = 1:N_t$, and the expectation of X_k conditioned on the past value X_{k-1} is

$$E[X_k \mid X_{k-1}] = X_{k-1} \cdot (1 + \mu_0 \Delta t),$$

and so by iterated expectations

$$E[X_k \mid X(0) = x_0] = (1 + \mu_0 \Delta t)E[X_{k-1} \mid X_j, j = 0:k - 2] = (1 + \mu_0 \Delta t)^k x_0$$

and finally $E[X_{N_t} \mid X(0) = x_0] = x_0(1 + \mu_0 \Delta t)^{N_t}$ at $t_{N_t} = t_f$. From (4.81), for jump-diffusions but ignoring the jumps, the expectation of the exact solution at the final fixed time is

$$E\left[X^{(\text{exact})}(t_f) \mid X(0) = x_0\right] = x_0 e^{\mu_0 t_f}.$$

The asymptotic evaluation, for sufficiently small Δt, of weak convergence criteria is then

$$\begin{aligned}
\left| E[X_{N_t} \mid X_0 = x_0] - E[X^{(\text{exact})}(t_f) \mid X(0) = x_0] \right| &= |x_0| \cdot \left| (1 + \mu_0 \Delta t)^{N_t} - e^{\mu_0 t_f} \right| \\
&= |x_0| \cdot \left| e^{N_t \ln(1 + \mu_0 \Delta t)} - e^{\mu_0 N_t \Delta t} \right| \\
&\sim |x_0| e^{\mu_0 t_f} \cdot \left| e^{-0.5\mu_0^2 t_f \Delta t} - 1 \right| \\
&\sim |x_0| e^{\mu_0 t_f} \cdot 0.5\mu_0^2 t_f \Delta t = \widetilde{C}_w \Delta t,
\end{aligned}$$

so $\gamma_w = 1$, confirming the weak sense results [166], with $\widetilde{C}_w = 0.5\mu_0^2 |x_0| \exp(\mu_0 t_f)$ for both the linear deterministic and the stochastic Euler methods, although only in the weak sense in the linear stochastic case.

Finally, consider the mean square stability of the EM approximation X_k. Recasting the EM SΔE (9.16) to the recursion form $X_k = A_{k-1} \cdot X_{k-1}$, where $A_k \equiv (1 + \mu_0 \Delta t + \sigma_0 \Delta W_k)$, the solution can be written

$$X_k = x_0 \prod_{\ell=0}^{k-1} A_\ell.$$

Next, considering the mean square,

$$\begin{aligned}
E[X_k^2 \mid X_0 = x_0] &= x_0^2 E\left[\left(\prod_{\ell=0}^{k-1} A_\ell\right)^2\right] = x_0^2 E\left[\prod_{\ell=0}^{k-1} A_\ell^2\right] = x_0^2 \prod_{\ell=0}^{k-1} E\left[A_\ell^2\right] \\
&= x_0^2 \prod_{\ell=0}^{k-1}((1 + \mu_0 \Delta t)^2 + \sigma_0^2 \Delta t) = x_0^2 \left((1 + \mu_0 \Delta t)^2 + \sigma_0^2 \Delta t\right)^k \\
&= x_0^2 \left(1 + 2\mu_0 \Delta t + (\mu_0 \Delta t)^2 + \sigma_0^2 \Delta t\right)^k
\end{aligned} \tag{9.17}$$

by interchanging the power and product operators, interchanging the product and expectation operators due to the independent increments property of the ΔW_k, using $\mathrm{E}[\Delta W_\ell] = 0$ and $\mathrm{E}[\Delta W_\ell^2] = \Delta t$, and the final fact that $\prod_{\ell=0}^{k-1} \theta = \theta^k$. Since as $k \to \infty$, $\theta^k \to 0$ if and only if $\theta < 1$ and in this case obviously $\theta > 0$, so asymptotic mean square stability of the X_k requires that

$$2\mu_0 + \sigma_0^2 + \mu_0^2 \Delta t < 0. \tag{9.18}$$

Note from (9.12) with $\nu_0 = 0$ that the corresponding critical stability condition for the exact solution is $2\mu_0 + \sigma_0^2 < 0$ or that $\mu_0 < -0.5\sigma_0^2$ and that μ_0 must be sufficiently negative, but (9.18) the for EM method is much more restrictive, requiring

$$\mu_0 < -0.5(\sigma_0^2 + \mu_0^2 \Delta t),$$

since the discrete term $\mu_0^2 \Delta t$ has been retained because Δt may not be small enough to be negligible, although $\mu_0^2 dt$ would be negligible compared to one in the dt-precision used in the exact, continuous time case. For numerical consideration, (9.18) could be interpreted as a constraint on the discrete time-step, i.e.,

$$\Delta t < 2 \left| \mu_0 + 0.5\sigma_0^2 \right| / \mu_0^2 \,,$$

valid only if μ_0 is selected to be in the asymptotically mean square stable range, $\mu_0 < -0.5\sigma_0^2$, of the exact solution. For more elaborate discussion of asymptotic stability, see D. Higham [140] for diffusions or D. Higham and Kloeden [146] for jump-diffusions.

9.1.3 Milstein's Higher Order Stochastic Diffusion Simulations

It is difficult to see how to improve on the Euler–Maruyama method (9.13) since it is perfectly consistent with Itô's formulation of forward integration of the diffusion stochastic integral equation

$$X(t) = X(0) + \int_0^t \left(f(X(s), s)ds + g(X(s), s)dW(s) \right), \tag{9.19}$$

corresponding to the diffusion SDE (9.1). Here, only a formal applied mathematical derivation is given, since comprehensive details fill the large volume of Kloeden and Platen [166]. Clues about where to start are the fact that Euler's method has a theoretical log-rate of $\gamma_s = 0.5$ for strong convergence [166] and that the same power obtained for just the expectation of absolute value of the standard diffusion process $\mathrm{E}[|\Delta W_k|] = O(\sqrt{\mathrm{E}[\Delta W_k^2]}) = O(\sqrt{\Delta t})$, as given in Table 1.1 on p. 7. The main idea of expanding the simulation approximation is to expand the coefficient $g(x, t)$ of the term whose expected absolute value would give rise to the $O(\sqrt{\Delta t})$ convergence. A way to do this is to apply iterations with Itô's stochastic chain rule in integrals of $g(X(t), t)$ on $[t_k, t]$, $t \geq t_k$,

$$\begin{aligned} g(X(t), t) = g(X_k, t_k) &+ \int_{t_k}^t \left(\left(g_t + fg_x + 0.5g^2 g_{xx} \right)(X(s), s)ds \right. \\ &\left. + (gg_x)(X(s), s)dW(s) \right), \end{aligned} \tag{9.20}$$

loosely upgrading the $g(x, t)$ requirements needed to twice continuously differential and where **wholesale arguments** have been used, e.g., $(gg_x)(x, t) = g(x, t)g_x(x, t)$.

This stochastic Taylor technique is also called an **Itô–Taylor expansion**. It can be used recursively to obtain very high order approximations, but here (9.20) is substituted into a version of (9.19) rewritten for $[t_k, t_{k+1}]$,

$$
\begin{aligned}
X_{k+1} &= X_k + \int_{t_k}^{t_{k+1}} (f(X(t), t)dt + g(X(t), t)dW(t)) \\
&= X_k + \int_{t_k}^{t_{k+1}} (f(X(t), t)dt + (g(X_k, t_k) \\
&\qquad + \int_{t_k}^{t} ((g_t + fg_x + 0.5gg_{xx})(X(s), s)ds \\
&\qquad + (gg_x)(X(s), s)dW(s)))\, dW(t)) \\
&\simeq X_k + F_k \Delta t + G_k \Delta W_k + G_k G X_k \int_{t_k}^{t_{k+1}} \int_{t_k}^{t} dW(s)dW(t),
\end{aligned}
\tag{9.21}
$$

where $GX_k \equiv g_x(X_k, t_k)$. Next, using the Itô forward integration approximation on coefficient terms and the negligibility of the residual double integral, we obtain

$$
\int_{t_k}^{t_{k+1}} \int_{t_k}^{t} dsdW(t) = \int_{t_k}^{t_{k+1}} (t - t_k)dW(t) = \int_0^{\Delta t} t dW(t) \stackrel{dt}{=} 0
$$

by Itô mean square rules in Δt-precision, which justifies dropping the corresponding terms. The retained double integral is just another form of Itô's fundamental theorem, Theorem 2.10 given in (2.30) on p. 41,

$$
\begin{aligned}
\int_{t_k}^{t_{k+1}} \int_{t_k}^{t} dW(s)dW(t) &= \int_{t_k}^{t_{k+1}} (W(t) - W_k)dW(t) \\
&= \left(\int_0^{\Delta t} W(t)dW(t) \right)_k \stackrel{dt}{=} 0.5 \cdot (\Delta w_k^2 - \Delta t).
\end{aligned}
\tag{9.22}
$$

Thus, **Milstein's approximate method** is the stochastic difference equation (SΔE)

$$
X_{k+1} = X_k + F_k \Delta t + G_k \Delta W_k + 0.5 G_k G X_k \cdot (\Delta W_k^2 - \Delta t)
\tag{9.23}
$$

for the SDE (9.1) and $k = 0 : N_t - 1$, where $F_k \equiv f(X_k, t_k)$, $G_k \equiv g(X_k, t_k)$, $GX_k \equiv g_x(X_k, t_k)$, and $\Delta W_k \equiv W(t_{k+1}) - W(t_k)$. Using the linear, time-dependent SDE model (9.14) as in Figure 9.2 and the same fine-coarse grid numerical procedure, the Milstein and exact simulations are displayed in Figure 9.4, using the code C.18, called `sdemilsteinsim.m` in Online Appendix C. The difference is very slight and hardly noticeable, and the error between the Milstein approximate path and the exact path at the coarse time points is presented in Figure 9.5 using the same code. Finally, Figure 9.6 illustrates the direct difference between the Milstein and Euler–Maruyama approximations, again using the same code to generate all three figures.

The Milstein algorithm converges strongly with log-rate $\gamma_s = 1$, but for the proof and computational justification see Kloeden and Platen [166, Sections 10.3 and 10.6]. Also see

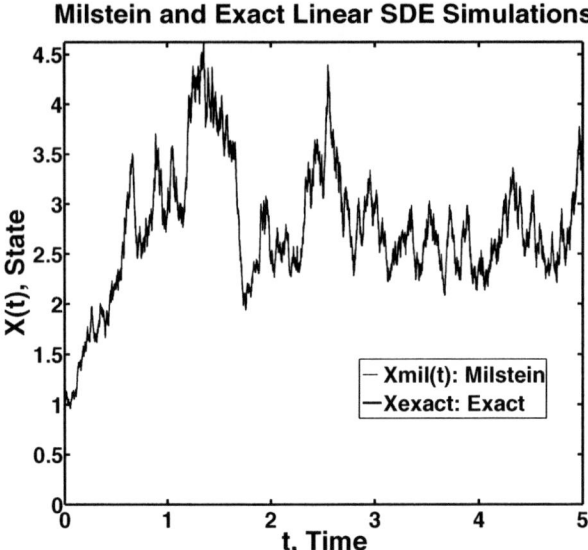

Figure 9.4. *Comparison of coarse Milstein and fine exact paths, simulated using MATLAB with $N_t = 1024$ fine sample points for the exact path (9.15) and $N_t/8 = 128$ coarse points for the Milstein path (9.23), initial time $t_0 = 0$, final time $t_f = 5$, and initial state $x_0 = 1.0$ as in Figure 9.2. Time-dependent parameter values are $\mu(t) = 0.5/(1 + 0.5t)^2$ and $\sigma(t) = 0.5$.*

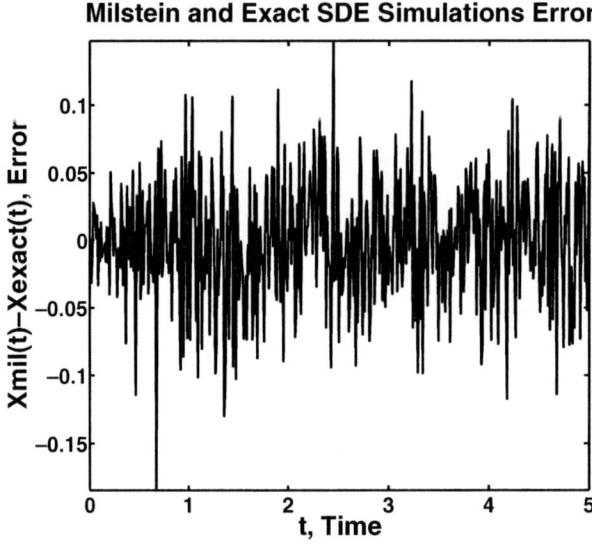

Figure 9.5. *Error in coarse Milstein and fine exact paths using the coarse discrete time points. The simulations use MATLAB with the same values and time-dependent coefficients as in Figure 9.2. The Milstein maximal-absolute error for this example is 1.2, while for $N_t = 4096$ the maximal error is better at 0.95.*

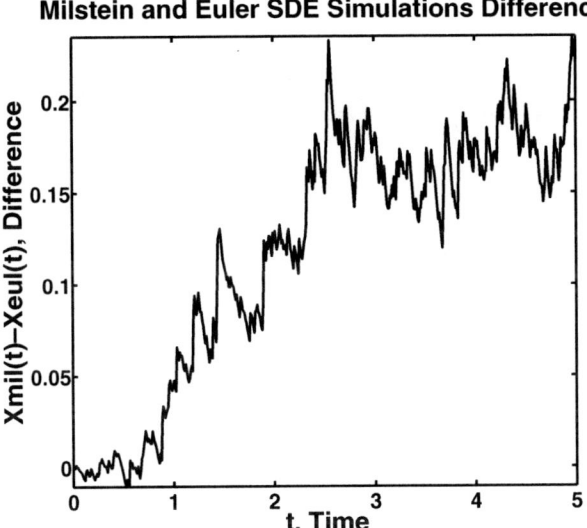

Figure 9.6. *Difference in coarse Milstein and Euler paths using the coarse discrete time points. The simulations use MATLAB with the same values and time-dependent coefficients as in Figure 9.2. The Milstein–Euler maximal-absolute difference for this example is 0.19, while for $N_t = 4096$ the maximal difference is comparable at 0.24.*

D. Higham's very accessible tutorial review [140] for computational justification and a nice Milstein-strong MATLAB code. Maple and MATLAB codes for diffusion SDEs for finance are given in D. Higham and Kloeden [144] along with higher order approximations. Other Maple diffusion codes are found in Cyganowski, Kloeden, and Ombach [67]. *Mathematica* diffusion SDE codes are presented in Stojanovic [259].

However, note that the diffusion factor $0.5(\Delta W_k^2 - \Delta t)$ in the Milstein approximation has the mean $E[0.5(\Delta W_k^2 - \Delta t)] = 0$ and variance

$$\text{Var}[0.5(\Delta W_k^2 - \Delta t)] = 0.25(E[\Delta W_k^4] - (\Delta t)^2) = 0.5(\Delta t)^2,$$

which normally would be negligible in dt-precision. Use of Table 1.1 on p. 7 indicates limited correction possibilities.

9.1.4 Convergence and Stability of Jump-Diffusion Euler Simulations

The stochastic Euler method for jump-diffusions governed by the SDE (9.1) with discrete Poisson jumps at mark-independent amplitudes $h(x, t)$, i.e.,

$$dX(t) = f(X(t), t)dt + g(X(t), t)dW(t) + h(X(t), t)dP(t),$$

in its simplest form using the forward integral approximation of Itô for fixed Δt is

$$X_{k+1} = X_k + F_k \Delta t + G_k \Delta t + H_k \Delta P_k, \tag{9.24}$$

where $(F_k, G_k, H_k) = (f(X_k, t_k), g(X_k, t_k), h(X_k, t_k))$, $\Delta t = t_{k+1} - t_k$, $\Delta W_k = W_{k+1} - W_k$, and $\Delta P_k = P_{k+1} - P_k$ for $k = 0:N_t$. Maghsoodi [191] and Maghsoodi and Harris [192] derived most of the theory behind this method and derived numerous Milstein-like higher order approximations, so sometimes (9.24) is called the **Euler–Maghsoodi method**.

Linear Jump-Diffusion Euler Method Convergence

Following the stochastic diffusion Euler analysis for the linear, constant coefficient case,

$$dX(t) = X(t)(\mu_0 dt + \sigma_0 dW(t) + \nu_0 dP(t)),$$

the discrete Euler is written

$$X_k = B_{k-1} \cdot X_{k-1}; \quad B_k \equiv (1 + (\mu_0 + \lambda_0 \nu_0)\Delta t + \sigma_0 \Delta W_k + \nu_0(\Delta P_k - \lambda_0 \Delta t)),$$

where the discrete Poisson process is written in mean-zero (i.e., martingale) independent increment form for convenience, so that

$$X_k = x_0 \prod_{\ell=0}^{k-1} B_\ell,$$

and by independent increments as well as independent jump-diffusion processes,

$$\mathrm{E}[X_k \mid X_0 = x_0] = x_0 \prod_{\ell=0}^{k-1} \mathrm{E}[B_\ell] = x_0 \prod_{\ell=0}^{k-1}(1 + (\mu_0 + \lambda_0 \nu_0)\Delta t) = x_0(1 + (\mu_0 + \lambda_0 \nu_0)\Delta t)^k.$$

From the exact solution (9.11) using the expectation in (4.81), the final mean square at $t_f = N_t \Delta t$ is

$$\mathrm{E}\left[X(t_f) \mid X(0) = x_0\right] = x_0 e^{(\mu_0 + \lambda_0 \nu_0)t_f}.$$

Next, computing the convergence criteria in the weak sense asymptotically,

$$\left| \mathrm{E}[X_{Nt} \mid X_0 = x_0] - \mathrm{E}[X(t_f) \mid X_0 = x_0] \right|$$

$$= |x_0| \left| (1 + (\mu_0 + \lambda_0 \nu_0)\Delta t)^{N_t} - e^{(\mu_0 + \lambda_0 \nu_0)N_t \Delta t} \right|$$

$$= |x_0| e^{(\mu_0 + \lambda_0 \nu_0)t_f} \left| e^{N_t \ln(1 + (\mu_0 + \lambda_0 \nu_0)\Delta t) - (\mu_0 + \lambda_0 \nu_0)N_t \Delta t} - 1 \right|$$

$$\sim |x_0| e^{(\mu_0 + \lambda_0 \nu_0)t_f} \left| e^{-0.5 N_t(\mu_0 + \lambda_0 \nu_0)^2 \Delta t^2} - 1 \right|$$

$$\sim C_w \Delta t,$$

where $C_w = |x_0|(\mu_0 + \lambda_0 \nu_0)^2 t_f \exp((\mu_0 + \lambda_0 \nu_0)t_f)$ and the convergence in the weak sense is order one in Δt with $\gamma_w = 1$.

The distributed jump case is somewhat similar, except the marks introduce more complications. Let the linear distributed jump-diffusion SDE have constant coefficients except that the relative jump-amplitude depends on the random mark Q and the symbolic product $\nu(Q)dP(t; Q)$ is replaced by the proper jump sum. Thus

$$dX(t) = X(t) \left(\mu_0 dt + \sigma_0 dW(t) + \sum_{\ell=1}^{dP(t;Q)} \nu(Q_\ell) \right),$$

and the discrete Euler processes are written in zero mean form,

$$X_k = \beta_{k-1} \cdot X_{k-1};$$

$$\beta_k \equiv 1 + (\mu_0 + \lambda_0 E[\nu(Q)])\Delta t + \sigma_0 \Delta W_k + E[\nu(Q)](\Delta P_k - \lambda_0 \Delta t)$$

$$+ \sum_{\ell=1}^{\Delta P_k} (\nu(Q_\ell) - E[\nu(Q)]).$$

The exact solution at node t_k upon using the stochastic chain rule and integrating is

$$X(t_k) = x_0 \exp \left((\mu_0 - \sigma_0^2/2)t_k + \sigma_0 W_k + \sum_{\ell=1}^{P_k} Q_\ell \right),$$

where we have again set $Q \equiv \ln(1 + \nu(Q))$ or $\nu(Q) = \exp(Q) - 1$ for the convenience of setting the mark distribution appropriate for the log-process. Using the iterated expectations technique to nest the Poisson and jump-mark expectations, the expectations are

$$E[X_k \mid X_0 = x_0] = x_0 E \left[\prod_{j=0}^{k-1} \beta_j \right] = x_0 \prod_{j=0}^{k-1} E[\beta_j]$$

$$= x_0 \prod_{j=0}^{k-1} (1 + (\mu_0 + \lambda_0(E[\exp(Q)] - 1))\Delta t)$$

$$= x_0 (1 + (\mu_0 + \lambda_0(E[\exp(Q)] - 1))\Delta t)^k$$

for the approximation and

$$E[X(t_k) \mid X(0) = x_0] = x_0 \exp((\mu_0 - \sigma_0^2/2)t_k) E_{W_k}[\exp(\sigma_0 W_k)]$$

$$\cdot E_{P_k} \left[E_Q \left[\exp \left(\sum_{\ell=1}^{P_k} Q_\ell \right) \middle| P_k \right] \right]$$

$$= x_0 \exp((\mu_0 + \lambda_0(E[\exp(Q)] - 1))t_k)$$

for the exact. Again, asymptotic results are derived for weak absolute mean error as $\Delta t \to 0^+$ for fixed t_k,

$$|E[X_k \mid X_0 = x_0] - E[X(t_k) \mid X(0) = x_0]|$$

$$= |x_0| \cdot \left| e^{k \ln(1 + (\mu_0 + \lambda_0(E[\exp(Q)] - 1))\Delta t)} - e^{(\mu_0 + \lambda_0(E[\exp(Q)] - 1))t_k} \right|$$

$$\sim |x_0| e^{(\mu_0 + \lambda_0(E[\exp(Q)] - 1))t_k} \left| e^{-0.5k(\mu_0 + \lambda_0(E[\exp(Q)] - 1))^2 \Delta t^2} - 1 \right| \sim C_w \Delta t,$$

where

$$C_w = 0.5|x_0|t_k(\mu_0 + \lambda_0(E[\exp(Q)] - 1))^2 e^{(\mu_0 + \lambda_0(E[\exp(Q)]-1))t_k}.$$

Again the weak convergence rate is linear with $\gamma_w = 1$.

Maghsoodi [191] shows that the strong mean square error convergence rate (9.6) is $O(\Delta t)$ for the jump-diffusion Euler method for nonlinear coefficients subject to linear Lipschitz bounds, which translates into a strong RMS rate of

$$O\left(\sqrt{\Delta t}\right).$$

A similar result was shown by D. Higham and Kloeden [146] for the implicit jump-diffusion or **stochastic theta method (STM)** with the mean square error based upon piecewise constant interpolation functions rather than the discrete approximate and exact values themselves. (The value $\theta = 0$ yields the explicit, stochastic Euler method, while the theta method is implicit for $0 < \theta \le 1$.) For the jump-diffusion problem, the theta method applies only to the drift term in (9.24),

$$X_{k+1} = X_k + ((1 - \theta)F_k + \theta F_{k+1}) + G_k \Delta W_k + H_k \Delta P_k, \tag{9.25}$$

in order to preserve stochastic consistency with the jump-diffusion conditional infinitesimal moments (8.46), (8.65), by avoiding implicit, backward steps in the diffusion and jump terms. The technical details of STM are beyond the scope of this chapter.

Euler Mean Square Linear Asymptotic Stability for Jump-Diffusions

For the mean square asymptotic stability of the jump-diffusion Euler method, the procedure leading up to the corresponding diffusion critical condition (9.18) is used. Starting with the jump-diffusion linear system recursive form,

$$X_k = B_{k-1} \cdot X_{k-1},$$

the mean square is

$$E[X_k^2|X_0 = x_0] = x_0^2 E\left[\left(\prod_{\ell=0}^{k-1} B_\ell\right)^2\right] = x_0^2 \prod_{\ell=0}^{k-1} E\left[B_\ell^2\right]$$

$$= x_0^2 \prod_{\ell=0}^{k-1} \left((1 + (\mu_0 + \lambda_0 v_0)\Delta t)^2 + (\sigma_0^2 + \lambda_0 v_0^2)\Delta t\right)$$

$$= x_0^2 \left((1 + (\mu_0 + \lambda_0 v_0)\Delta t)^2 + (\sigma_0^2 + \lambda_0 v_0^2)\Delta t\right)^k.$$

Again, as $k \to \infty$, the base of the power k must be less than one since the base is nonnegative, so the mean square asymptotic stability criterion for the linear, constant coefficient, jump-diffusion Euler approximation is

$$2(\mu_0 + \lambda_0 v_0) + \sigma_0^2 + \lambda_0 v_0^2 + (\mu_0 + \lambda_0 v_0)^2 \Delta t < 0, \tag{9.26}$$

which means that $\mu_0 + \lambda_0 \nu_0$ needs to be sufficiently negative (note that $\lambda_0 > 0$ if the jump process is to be genuine),

$$\mu_0 + \lambda_0 \nu_0 < -0.5(\sigma_0^2 + \lambda_0 \nu_0^2 + (\mu_0 + \lambda_0 \nu_0)^2 \Delta t)$$

and when interpreted in terms of the first and second relative conditional infinitesimal moments is

$$E[\Delta X_k / X_k \mid X_k \neq 0] < -0.5E[(\Delta X_k / X_k)^2 \mid X_k \neq 0].$$

If we restrict our attention to when the exact solution is mean square stable, i.e., $2(\mu_0 + \lambda_0 \nu_0) + \sigma_0^2 + \lambda_0 \nu_0^2 < 0$ from (9.12), then (9.26) can be used to construct a constraint on the discrete time-step,

$$\Delta t < 2 \left| \mu_0 + \lambda_0 \nu_0 + 0.5(\sigma_0^2 + \lambda_0 \nu_0^2) \right| / (\mu_0 + \lambda_0 \nu_0)^2 .$$

9.1.5 Jump-Diffusion Euler Simulation Procedures

A simple numerical procedure is given in Subsection 4.3.3 on p. 113ff. for the linear system with discrete jump of size ν_0,

$$dX(t) = X(t)(\mu_0 dt + \sigma_0 dW(t) + \nu_0 dP(t)),$$

using the MATLAB normal random number generator randn and a small time-step zero-one Poisson–Bernoulli jump law using the acceptance-rejection method. Since this zero-one jump law uses the Δt-order asymptotic precision definition of the Poisson process, there is a restriction that $\lambda \Delta t < 1$ so that the one-jump probability is positive. See Program C.14, called linjumpdiff03fig1.m in Online Appendix C, for the MATLAB code used.

However, this $\lambda \Delta t < 1$ condition can be easily rectified by just renormalizing the Poisson distribution, $p_k(\lambda \Delta t) = \exp(-\lambda \Delta t)(\lambda \Delta t)^k / k!$, for a finite number of jumps $k \leq j$ without expanding the $\exp(-\lambda \Delta t)$ factor in the numerator, and so

$$p_k^{(j)}(\lambda \Delta t) \equiv \frac{(\lambda \Delta t)^k / k!}{\sum_{\ell=0}^{j} (\lambda \Delta t)^\ell / \ell!} \qquad (9.27)$$

is valid as long as $\lambda \Delta t > 0$ and conserves probability. This is the same as if the original normalization $\exp(+\lambda \Delta t)$ were expanded by $\lambda \Delta t$ in the denominator to $j + 1$ terms, and the result is called a Padé approximation or rational function. Form (9.27) also exactly preserves the ordering of the Poisson jump probabilities, i.e.,

$$\frac{p_{k+1}^{(j)}(\lambda \Delta t)}{p_k^{(j)}(\lambda \Delta t)} = \frac{\lambda \Delta t}{k + 1}$$

as long as $k = 0 : j - 1$. This form can be used with the acceptance-rejection method as long as the unit interval [0, 1] is partitioned into $j + 1$ subintervals of length $p_k^{(j)}(\lambda \Delta t)$ for $k = 0 : j$ such that a random number generator like the MATLAB rand is used and if the number generated lands in the subinterval corresponding to $p_k^{(j)}(\lambda \Delta t)$, then the realized number of jumps is k. Computer experiment experience indicates that it is best not to put the small subintervals adjacent to the end points of [0, 1] due to the open interval (0, 1) bias of computer random generators.

Distributed Jump Linear Jump-Diffusion Euler Method

In Figure 5.1 on p. 158, the simulations are presented for uniformly distributed marks Q on $(a, b) = (-2, +1)$ and time-dependent linear or geometric jump-diffusion SDE,

$$dX(t) = X(t)(\mu_d(t)dt + \sigma(t)dW(t) + \nu(Q)dP(t; Q)).$$

However, it is more convenient to work with the exponent of the exact solution derived by the stochastic chain rule to obtain the SDE,

$$dY(t) = d\ln(X(t)) = (\mu(t) - \sigma^2(t)/2)dt + \sigma(t)dW(t) + QdP(t; Q),$$

where the mark has been selected as $Q \equiv \ln(1 + \nu(Q))$ for convenience. (This would seem to preclude time-dependence in the jump-amplitude $\nu(Q)$, but time can be included in the mark range $[a, b]$ or the mark density $\phi_Q(q)$.) The MATLAB code C.15 is a modification of the linear jump-diffusion SDE simulator code C.14 illustrated in Figure 4.3 for constant coefficients and discrete mark-independent jumps. The state exponent $Y(t)$ is simulated as

$$YS(k+1) = YS(k) + (\mu(k) - \sigma^2(k)/2) * DT + \sigma(k) * DW(k) + Q(k) * DP(k)$$

with $t(k+1) = t0 + k * DT$ for $k = 0 : NI - 1$ with $NI = 1000$, $t0 = 0$, and $0 \leq t(k) \leq 2$. The incremental Poisson jump term $DP(k) = P(t(k) + DT; Q) - P(t(k); Q)$ is simulated by the MATLAB uniform random number generator rand on $(0, 1)$ using the acceptance-rejection technique [230, 97] (see also Subsection 9.2.3 on p. 270) to implement the zero-one jump law to obtain the probability of $\lambda(i)Dt$ that a jump is accepted; else a jump is rejected. The same random state (seed), but a different set of generated random samples, is used to obtain the simulations of the uniformly distributed Q on (a, b), i.e., $Q = a + (b - a) * \text{rand}(1, NI)$, used only if there is a jump event. Finally, the state itself is computed by a simple exponential inversion of the log-process as

$$X(k+1) = x0 * \exp(Y(k+1))$$

and should be highly accurate for sufficiently small DT since this procedure based upon the exact exponent is the same procedure that is used for producing the exact simulation, say, by Maghsoodi [191]. Clearly, if one has a linear SDE with constant parameter coefficients for an application, then the best strategy is to simulate the exact solution since it is available. However, if the object is just to use the linear SDE for testing a method on more general nonlinear SDEs, related perhaps by similar Lipschitz linear bounds, then simulation of the original linear SDE for $X(t)$ is recommended.

Many deterministic numerical methods are difficult to translate directly into numerical methods of diffusions or jump-diffusions due to the nonsmooth or discontinuous nature of the diffusion process $W(t)$ or the jump process $P(t; Q)$, respectively. Hence, implicit methods or multistep methods (many of these are designed to reduce or eliminate the implicitness of implicit methods) have to be modified to separate the treatment of the deterministic term $(f(x, t)\Delta t)$ from that of the diffusion term $(g(x, t)\Delta W(t))$ or that of the jump term $(h(x, t)\Delta P(t; Q)$ or $h(x, t, q)\Delta P(t; Q))$. It is necessary to preserve stochastic approximation consistency with respect to the jump-diffusion conditional infinitesimal moments (8.46), (8.65).

Stochastic Split-Step Backward Euler Method

One such method is a stochastic modification of the **deterministic backward Euler (DBE)** method ($X_{k+1} = X_k + f(X_{k+1}, t_{k+1})\Delta t$), which for the jump-diffusion problem is split into two stages by Cyganowski and Kloeden [66] and more recently by D. Higham and Kloeden [145]. The first stage is just a backward Euler step, $X_{k+1}^{(dbe)}$, only improved by the deterministic drift and a second stage that adds diffusion and jump term improvement,

$$X_{k+1}^{(dbe)} = X_k + f\left(X_{k+1}^{(dbe)}, t_{k+1}\right)\Delta t,$$

$$X_{k+1}^{(ssbe)} = X_k + g\left(X_{k+1}^{(dbe)}, t_{k+1}\right)\Delta W_k + h\left(X_{k+1}^{(dbe)}, t_{k+1}\right)\Delta P_k, \tag{9.28}$$

which they [66, 145] call a **split-step backward Euler (SSBE)**. The first stage is implicit in $X_{k+1}^{(dbe)}$ and so enhances the stability and convergence, for which some results are given in [66, 145], but no rates of convergence. The coefficients in [145] are autonomous, but time-dependence is added here for generality. An improved refinement is included in [66, 145] and uses the compensated or zero-mean Poisson $\Delta P_k - \lambda_k \Delta t$, a martingale, to obtain the **compensated split-step backward Euler (CSSBE)**,

$$X_{k+1}^{(dbe)} = X_k + \left(f\left(X_{k+1}^{(dbe)}, t_{k+1}\right) + \lambda_k h\left(X_{k+1}^{(dbe)}, t_{k+1}\right)\right)\Delta t,$$

$$X_{k+1}^{(cssbe)} = X_k + g\left(X_{k+1}^{(dbe)}, t_{k+1}\right)\Delta W_k + h\left(X_{k+1}^{(dbe)}, t_{k+1}\right)(\Delta P_k - \lambda_k \Delta t), \tag{9.29}$$

which provides better improvement in the first, deterministic backward Euler, stage. No computational validation is given in [66, 145]. In [147], D. Higham, Mao, and Stuart show $O(\Delta t)$ mean square error convergence rates for SSBE on nonlinear diffusion SDEs with coefficient functions satisfying linear Lipschitz conditions.

Maghsoodi [191] also extended the Milstein algorithm for diffusions to jump-diffusions by expanding the jump coefficient $h(x, t)$ like the diffusion coefficient $g(x, t)$ stochastic Taylor expansion. However, the new and numerous jump terms are much more complicated than the diffusion version, and Cyganowski, Kloeden, and Ombach [67] demonstate by computer experiment that this method works well for discrete jump problems but not for distributed (mark-dependent) jump problems, so the extension will not be discussed here.

Related convergence and stability results for discrete jump-diffusions are given by D. Higham and Kloeden in [145] for the stochastic theta method as previously mentioned in association with the STM algorithm (9.25).

Jump-Adaptive Euler Method

Thus far, methods using constant time-steps $\Delta t = t_{k+1} - t_k$ or a fixed set of variable time-steps $\Delta t_{k-1} t_{k+1} - t_k$ have been discussed, such that the number of jumps of ΔP_k in $[t_k, t_{k+1}]$ have been enumerated and corresponding jump-marks, if present, simulated. An alternate numerical approach, suggested by Maghsoodi [191], is to interlace the set of Poisson random jump-times, T_j for $j = 1 : N_J$ such that $T_{N_J} \leq t_f$, with a fixed set t_ℓ for $\ell = 0 : N_t$ to define a **jump-adaptive (JA) method** grid augmented by initial and final times, such that $\tau_0 \equiv 0 < \tau_k < \tau_{k+1} = \tau_k + \Delta \tau_k < \tau_{N^{(ja)}} = \tau_f$ with subintervals of length

$\Delta \tau_k = \tau_{k+1} - \tau_k$ for $k = 0 : N^{(ja)} - 1$. One restriction is that the mesh measure satisfies $\max_{0 \le k \le N^{(ja)}-1}(\Delta \tau_k) \le \overline{\Delta \tau}$, where $\overline{\Delta \tau} \simeq \Delta t$ plus some leeway.

It is well known that the Poisson subintervals or the time to the next jump $\Delta T_j = T_{j+1} - T_j$ are independent and identically exponentially distributed (1.24) with rate λ. (Unfortunately, the literature on the JA method confuses the IID properties of the interjump-times and the interdependence of the jump-times themselves.) The exponentially distributed Poisson jump-time generation is given on p. 14 using the logarithmic transformation of a uniform random number generator, and a vector version is

```
% log-uniform exponential density:
DT=-log(rand(1,NJ))/lambda;                              (9.30)
T=cumsum(DT);
```

where `rand(1,NJ)` is the MATLAB $1 \times NJ$ vector random generator and `cumsum` is the cumulative sum function, assuming that the total number of jumps is known.

Let the discrete state be denoted as $X_k^{(ja)} \simeq X(\tau_k)$ corresponding to adapted jump-time τ_k, so the jump-diffusion Euler method for discrete jumps is

$$X_{k+1}^{(ja)} = X_k^{(ja)} + F_k^{(ja)} \Delta \tau_k + G_k^{(ja)} \Delta W_k^{(ja)} + H_k^{(ja)} \Delta P_k^{(ja)}, \qquad (9.31)$$

where $\Delta W_k^{(ja)} = W(\tau_{k+1}) - W(\tau_k)$, $\Delta P_k^{(ja)} = P(\tau_{k+1}) - P(\tau_k)$, $F_k^{(ja)} = f(X_k^{(ja)}, \tau_k)$, and similarly for $G_k^{(ja)}$ and $H_k^{(ja)}$. Note that if τ_{k+1} coincides with a jump-time T_j for some j, then $\Delta P_k^{(ja)} = 1$; otherwise $\Delta P_k^{(ja)} = 0$. However, as Maghsoodi [191] warns, when analyzing something like convergence in the mean, it must be recognized that if $\tau_{k+1} = T_j$, then $\Delta W_k^{(ja)} = W(T_j) - W(\tau_k)$ is not statistically independent of $\Delta P_k^{(ja)} = P(T_j) - P(\tau_k)$, if expectations are to be calculated. A sample fragment of the code to compute $\Delta \tau_k$, $\Delta W_k^{(ja)}$, and $\Delta P_k^{(ja)}$ could be as given in Figure 9.7. This code fragment can be patched together with the given application SDE and chosen base numerical algorithm such as the jump-diffusion Euler or SSBE, for instance.

9.2   Monte Carlo Methods

The Monte Carlo method began as a statistical sampling procedure at Los Alamos National Laboratory in 1946 from an idea of Ulam in analogy considering the probability of winning the card game of solitaire, from the idea of von Neumann for programming neutron transport on a newly emerging electronic computer, and from ideas of Metropolis for computer implementation [79, 206, 208]. Without the emergence of electronic computers very few people would attempt to use large-scale statistical sampling to solve large problems. One exception was the famous physicist Fermi, who could calculate very fast using a mechanical calculator and had time to do big calculations because he often could not sleep, so in fact he was using a smaller-scale version of the Monte Carlo method 15 years before it had a name. (For other earlier examples see Hammersley and Handscomb [105], for instance.) The method was named for an uncle of Ulam's who had an obsession about going to gamble at Monaco, the gambling capital of Europe. In a 1949 paper [208], Metropolis and Ulam spelled out the following basic ideas of the method in more or less essay form:

```
function jumpadapt
% Jump adaptive (JA) code fragment:
%    merged regular and jump-times
Nt=10; lambda=9; t0=0; tf=1; Dt = (tf-t0)/Nt;
t = Dt*[0:Nt];   % Regular grid
DT = -log(rand)/lambda; S=DT; j=0;
while S < tf  % Get jump-time grid, T(NJ)<tf
    j=j+1;
    NJ=j;
    T(j)=S; DTJ(j)=DT;
    DT = -log(rand)/lambda;  % Exponential density
    S=S+DT;
end
[tau,ktau]=sort([t T]);  % Concatenate and sort times
Dtau=tau(2:Nja)-tau(1:Nja-1);  % Concatenate and sort times
randn('state',10);
RN=randn(1,Nja-1);  % Std. normal density
DP=zeros(1,Nja-1);
for k=2:Nja
    DW(k-1)=sqrt(Dtau(k-1))*RN(k-1);  % Get DW
    if ktau(k)>Nt+1
        DP(k-1)=1;  % Get final DP
    end
end
```

Figure 9.7. *Jump-adaptive code fragment.*

the potential applications, the statistical approach, the independent random sampling, the frequency distributions, the law of large numbers for convergence, and the asymptotic theorems for probable errors. Although von Neumann is not an author on this paper, it contains his ideas on techniques of random number generation and a hint of his acceptance and rejection method to handle general shaped domains by rejecting those samples which land outside of the domain.

A more major idea of von Neumann was the logical structure of most modern programmable computers, called the **von Neumann computer**. However, the newly emerging electronic computer, called ENIAC, was a very primitive, nonprogrammable, and decidedly "non-von" computer, as non–von Neumann computers are called. Not too long afterward, there was a parallel effort at both Princeton with von Neumann and at Los Alamos with Metropolis to build a von Neumann computer, but Metropolis was able to get the Los Alamos computer, named MANIAC, working first. As it is with most computer advances, faster computers do not save the user time because the user will inevitably present a bigger problem that will take about the same amount of time as the previous problem. The user who thought of the larger Monte Carlo problem for the MANIAC was the physicist Teller, and the problem was calculating the equation of state of an ideal rigid sphere gas. However, the major contribution of the resulting 1953 paper by Metropolis, the Rosenbluths, and the Tellers [207] was the use of **weighted sampling**, now called **importance sampling**, by using the exponential distribution of the energy change as the weight. This version of the Monte Carlo method is called the **Metropolis algorithm** [71] and was selected as one of the **top ten algorithms of the century** [69, 23]. This may be confusing, because the basic Monte Carlo algorithm is sometimes called the **Metropolis algorithm**, too. The 1953 paper [207] contains significantly more detail than the 1949 paper [208]; in both cases Metropolis is the lead author and some would say the lead Monte Carlo computation teacher. The title of the 1953 paper is "Equation of State Calculation by Fast Computing Machines" and the quoted cycle time of the MANIAC translated into 5.6 mHz, i.e., 5.6e-3 cycles per second, which would be extremely slow compared to today's 2 GHz to 4 GHz PCs, or 2.0e+9 to 4.0e+9 cycles per second. The MANIAC was not fast at all.

For general references on the Monte Carlo method, see the classic monograph of Hammersley and Handscomb [105] or the more recent book of Kalos and Whitlock [158]. Many of the more recent advances have come from applications of the Monte Carlo method to finance, so for general references on Monte Carlo with application to finance see Glasserman [97] and Jäckel [151]. For the pioneering and award-winning paper on application of the Monte Carlo method to financial options, see Boyle [39], or for a two-decade update, see Boyle, Broadie, and Glasserman [40].

9.2.1   Basic Monte Carlo Simulations

The benefits of Monte Carlo are realized only in high dimensions and for functionals of stochastic processes with simulation complexities beyond direct simulations of SDEs, as covered in the previous section, or for deterministic problems such as physical diffusions whose solutions can be simulated by Monte Carlo. Many problems can be transformed into an integral form or integral functional such as

$$\mathrm{I}[F] = \int_{\mathcal{V}} F(\mathbf{x})d\mathbf{x}, \tag{9.32}$$

where $\mathbf{x} = [x_i]_{n_x \times 1}$ is an n_x-dimensional vector on volume $\mathcal{V}$ and $F(\mathbf{x})$ is a bounded, integrable, scalar-valued function on $\mathcal{V}$. For instance, if $\mathcal{V}$ is finite, then $\mathrm{I}[F]$ could be interpreted in terms of the expectation

$$\mathrm{I}[F] = V \cdot \mathrm{E}_{\mathbf{X}}[F(\mathbf{X})]$$

of F with respect to uniform variates $\mathbf{X}$ such that $V \equiv \int_{\mathcal{V}} d\mathbf{x} < \infty$ with uniform density $\phi_{\mathbf{X}}(\mathbf{x}) = 1/V$ on domain $\mathcal{V}$.

In general (9.32) can be interpreted to include nonuniform distributions by scaling F by a suitable density $\phi_{\mathbf{X}}(\mathbf{x})$ for variates $\mathbf{X}$ on $\mathcal{V}$ so that

$$F(\mathbf{x}) = f(\mathbf{x})\phi_{\mathbf{X}}(\mathbf{x}),$$

$$\int_{\mathcal{V}} \phi_{\mathbf{X}}(\mathbf{x})d\mathbf{x} = 1,$$

and

$$\mathrm{I}[F] = \mathrm{E}_{\mathbf{X}}[f(\mathbf{X})] = \int_{\mathcal{V}} f(\mathbf{x})\phi_{\mathbf{X}}(\mathbf{x})d\mathbf{x}. \tag{9.33}$$

The general rule for the selection of the density $\phi_X(x)$ is that it capture important character-istics, such as variability, of the integrand $F(x)$ on domain $\mathcal{V}$ such that the function $f(X)$ is bounded and not very variable. The density $\phi_X(x)$ should be known and the generation of its variates should be computable with reasonable effort. In the uniform case, $\phi_X(x) = 1/V$ and $f(X) = V \cdot F(X)$.

Example 9.3. *Risk-Neutral European Call Option Pricing.*
An example of a complex functional is the risk-neutral European call option pricing model of Zhu and Hanson [292] using a jump-diffusion SDE with log-uniformly distributed jump-amplitude marks,

$$\mathcal{C}(S_0, t_f) = \mathrm{E}_{\widetilde{\mathcal{P}}(t_f)}\left[\mathcal{C}^{(BS)}\left(S_0 e^{\widetilde{\mathcal{P}}(t_f)-\lambda\bar{J}t_f}, t_f\right)\right], \tag{9.34}$$

where

$$\widetilde{\mathcal{P}}(t_f) = \sum_{i=1}^{P(t_f;Q)} Q_i \tag{9.35}$$

is the compound Poisson jump process cumulative sum at the strike time t_f with uniformly distributed IID random marks Q_i on $[a, b]$, mean jump-amplitude

$$\bar{J} \equiv \mathrm{E}_Q[J(Q)] \equiv \mathrm{E}[\exp(Q) - 1] = (\exp(b) - \exp(a))/(b - a) - 1, \tag{9.36}$$

and Black–Scholes call option price

$$\mathcal{C}^{(BS)}(s, t_f) \equiv s\Phi(d_1(s)) - Ke^{-rt_f}\Phi(d_2(s)) \tag{9.37}$$

with strike price K, interest rate r, diffusive volatility σ, standardized normal distri-bution function $\Phi(x)$, and Black–Scholes argument functions $d_1(s) \equiv (\ln(s/K) + (r + \sigma^2/2)t_f)/(\sigma\sqrt{t_f})$ and $d_2(s) \equiv d_1(s) - \sigma\sqrt{t_f}$. Refer to [292] for the transformations used to achieve this form, which one would not attempt to evaluate directly, but one would try to estimate the call option price.

Returning to the general integral functional problem (9.33), an estimate $\widehat{\mathrm{I}}_n$ of the value of the integral $\mathrm{I}[F] = \mathrm{E}_X[f(X)]$ is the **sample mean** s_n of n IDD sample points X distributed on $\mathcal{V}$ corresponding to the density $\phi_X(x)$,

$$\widehat{\mathrm{I}}_n = s_n, \tag{9.38}$$

where the sample mean s_n or **Monte Carlo estimator** $\widehat{\mu}_n = s_n$ is

$$\widehat{\mu}_n \equiv s_n = \frac{1}{n}\sum_{i=1}^n f(X_i) \equiv \frac{1}{n}\sum_{i=1}^n f_i, \tag{9.39}$$

the estimate of the mean of f with respect to $\phi_X(x)$. Obviously, the function $f(x)$ must be bounded for the sample mean to exist. The true mean of f is

$$\mu_f = \mathrm{E}_X[f(X)] = \int_{\mathcal{V}} f(x)\phi_X(x)dx.$$

Then, the estimate $\widehat{\mu}_n$ is an **unbiased estimate**, since the **bias** of the estimator from the true mean is zero, i.e.,

$$\beta_{\widehat{\mu}_n} \equiv E_{\mathbf{X}}[\widehat{\mu}_n - \mu_f] = \frac{1}{n}\sum_{i=1}^{n}E_{\mathbf{X}}[f(\mathbf{X}_i)] - \mu_f = E_{\mathbf{X}}[f(\mathbf{X})] - \mu_f = 0, \qquad (9.40)$$

using the IID property of the sample points. Further, by the **strong law of large numbers (SLLN)** (B.115),

$$\widehat{\mu}_n \longrightarrow \mu_f \text{ with probability one as } n \to +\infty.$$

The true variance of f is

$$\sigma_f^2 = \operatorname{Var}_{\mathbf{X}}[f(\mathbf{X})] = \int_{\mathcal{V}} (f(\mathbf{x}) - \mu_f)^2 \phi_{\mathbf{X}}(\mathbf{x})d\mathbf{x}$$

and so the unbiased estimate of the sample variance from (B.109) is

$$\widehat{\sigma}_n^2 = \frac{1}{n-1}\sum_{i=1}^{n}(X_i - \widehat{\mu}_n)^2. \qquad (9.41)$$

Example 9.4. *Choice of Monte Carlo Sampling Distribution.*
A rule of thumb is that while many other distributions may work in generating Monte Carlo estimations, the better density captures more variability of $F(\mathbf{x})$ along with the domain $\mathcal{V}$ and leaves a less variable $f(\mathbf{x})$ to simulate. Thus, the better choice will be the better Monte Carlo results.

It is general numerical practice to choose an integrand weight function that captures most of the variability and can easily be integrated exactly so that the remaining integrand factor can be discretely and well approximated. For example, the truncated normal distribution,

$$I = \frac{1}{\sqrt{2\pi}}\int_a^b e^{-x^2/2}dx, \qquad (9.42)$$

can be Monte Carlo estimated using a uniform (u) density $\phi^{(u)}(x) = 1/(b-a)$ on $[a, b]$ with sampled function

$$f^{(u)}(x) = (b-a)\exp(-x^2/2)/\sqrt{2\pi}$$

or a normal (n) density $\phi^{(n)}(x) = \exp(-x^2/2)/\sqrt{2\pi}$ on $[-\infty, +\infty]$ and

$$f^{(n)}(x) = \mathbf{1}_{x\in[a,b]} = \{1, x \in [a, b]; 0, x \notin [a, b]\}$$

is an indicator function. The exact mean is invariant with respect to the density,

$$\mu_f^{(n)} = I = \Phi_n(a, b; 0, 1) = \mu_f^{(u)},$$

where $\Phi_n(x, y; 0, 1)$ is the usual standard normal distribution in this book on $[x, y]$. However, it is obvious that the exact variance assuming a normal density factor will be much *smaller than the exact variance assuming a uniform density factor and a highly variable* $f(x)$, *if* a *and* b *are not small. In fact,* $\left(\sigma_f^{(n)}\right)^2 = I - I^2$ *for the normal case since* $\mathbf{1}_{x\in[a,b]}^2 = \mathbf{1}_{x\in[a,b]}$ *and* $\left(\sigma_f^{(u)}\right)^2 = (b-a)\Phi_n(\sqrt{2}a, \sqrt{2}b; 0, 1)/(2\sqrt{\pi}) - I^2$ *for the uniform by transformation* $E^{(u)}[(f^{(u)})^2(x)]$ *to the standard normal distribution. As* $a \to -\infty$ *and* $b \to +\infty$, *the standard normal distributions* $\Phi_n \to 1$ *in uniform as well as in normal cases and the difference has the unbounded asymptotic approximation,*

$$\left(\sigma_f^{(u)}\right)^2 - \left(\sigma_f^{(n)}\right)^2 \sim \frac{b-a}{2\sqrt{\pi}} - 1,$$

demonstrating in this extreme case that the choice of the sampling density $\phi_{\mathbf{X}}(\mathbf{x})$ *can make a big difference in the variance* σ_f^2. *A companion computational demonstration code* C.19 *for this problem when* $[a, b] = [-R, R]$ *is given on p.* C27 *of Online Appendix C. Of course, one would not use the uniform distribution on an infinite domain.*

Convergence of Scaled Monte Carlo Esimate Distribution to a Normal Distribution

By the **central limit theorem** (B.116) the **sample mean converges in distribution to a normal distribution**,

$$\text{Prob}\left[\frac{\widehat{\mu}_n - \mu_f}{\sigma_f/\sqrt{n}} \le \xi\right] \longrightarrow \Phi_n(\xi; 0, 1) \text{ as } n \to +\infty, \tag{9.43}$$

or alternately we say $(\widehat{\mu}_n - \mu_f)/(\sigma_f/\sqrt{n}) \xrightarrow{\text{dist}} \xi$, distributed according to $\Phi_n(\xi; 0, 1)$, where $\Phi_n(\xi; 0, 1)$ is the standard normal distribution defined in (B.1.4) and $\sigma_f/\sqrt{n}$ is called the **standard error** or **probable error**. However, this form of the standard error is not very useful since neither σ_f or μ_f are known, else a Monte Carlo approximation would not be neeeded, but $\widehat{\sigma}_n^2$ is an unbiased estimator of σ_f^2 and therefore $\widehat{\sigma}_n^2$ must converge to σ_f^2 in distribution too and thus σ_f will be replaced by $\widehat{\sigma}_n$ relying on continuous extensions of the central limit theorem [151]. However, in general $\widehat{\sigma}_n$ is not necessarily an unbiased estimate of σ_f, since a function of an unbiased estimator of a parameter is not the unbiased estimate of the function of the parameter, as pointed out by Hammersley and Handscomb [105].

Monte Carlo Estimate Confidence Intervals

Following Glasserman's [97] arguments for confidence intervals with variations, the convergence in distribution (9.43) implies as $n \to +\infty$,

$$\text{Prob}\left[\widehat{\mu}_n - \mu_f \le \frac{\widehat{\sigma}_n}{\sqrt{n}}\xi\right] \sim \Phi_n(\xi; 0, 1),$$

so replacing ξ by $-\xi$,

$$\text{Prob}\left[\widehat{\mu}_n - \mu_f \le -\frac{\widehat{\sigma}_n}{\sqrt{n}}\xi\right] \sim \Phi_n(-\xi; 0, 1),$$

and consequently we have an asymptotic formula for confidence intervals about the true mean μ_f,

$$\text{Prob}\left[-\frac{\widehat{\sigma}_n}{\sqrt{n}}\xi \leq \widehat{\mu}_n - \mu_f \leq \frac{\widehat{\sigma}_n}{\sqrt{n}}\xi\right] \sim \Phi_n(\xi; 0, 1) - \Phi_n(-\xi; 0, 1) = 2\Phi_n(\xi; 0, 1) - 1.$$

Putting this in a more useful form, let $\delta > 0$ and $\xi = \xi(\delta)$ such that $2\Phi_n(\xi(\delta); 0, 1) - 1 = 1 - \delta$ or

$$\Phi_n(\xi(\delta); 0, 1) = 1 - \delta/2 \tag{9.44}$$

to simplify the inversion. Thus, a practical, asymptotic **confidence level** $1-\delta$ or $100(1-\delta)\%$ is given by the probability

$$\text{Prob}\left[\widehat{\mu}_n - \frac{\widehat{\sigma}_n}{\sqrt{n}}\xi(\delta) \leq \mu_f \leq \widehat{\mu}_n + \frac{\widehat{\sigma}_n}{\sqrt{n}}\xi(\delta)\right] \sim 1 - \delta \tag{9.45}$$

such that the true mean μ_f is in the **confidence interval**

$$\left(\widehat{\mu}_n - \frac{\widehat{\sigma}_n\xi(\delta)}{\sqrt{n}}, \widehat{\mu}_n + \frac{\widehat{\sigma}_n\xi(\delta)}{\sqrt{n}}\right).$$

If $\xi(\delta) = 1$, the difference between the true value and the estimate is just $\pm$standard error with a confidence level of 68.27% that the simulation will be in the confidence interval, but 32.63% chance that it will be out of it. If the difference is ± 2 standard error, then the level is 95.45% but only a 4.55% "**lack of confidence**" level. Anyway, it will be assumed that the probable error of the Monte Carlo estimator is

$$\widehat{e}_n = |\widehat{\mu}_n - \mu_f| \propto \widehat{\sigma}_n/\sqrt{n}.$$

An important observation is that this probable or standard error is independent of the dimension of the volume n_x, as long as the volume is known. However, if it is necessary to approximate the volume due to its complexity, then this approximation will influence the real error.

Example 9.5. *Convergence and Errors in Monte Carlo Estimators.*
Monte Carlo simulations are illustrated in Figure 9.8 using the uniform density $\phi_X(x) = 1/(b - a)$ on $[a, b]$ for the one-dimensional integral of $F(x) = \sqrt{1 - x^2}$ on $[a, b]$, $-1 \leq a < b \leq +1$, so $f(x) = (b - a) \cdot F(x)$. The computational convergence of the mean $\widehat{\mu}_n$ and standard deviation $\widehat{\sigma}_n$ estimations of $f(x)$ versus the logarithm of sample size $\log_{10}(n)$ are exhibited in Subfigure 9.8(a), while the logarithm of the standard error $\log_{10}(\widehat{\sigma}_n/\sqrt{n})$ is shown versus the logarithm of the actual absolute error $\log_{10}(|\widehat{\mu}_n - \mu_f|)$ in Figure 9.8(a). The computational convergence is somewhat smooth from $n = 10$ to $n = 10,000,000$, but differences in the errors are more dramatic, reflecting the slight variability of $\widehat{\sigma}_n/\sqrt{n}$ and the greater variability of $\widehat{\mu}_n$ compared to the constant exact value μ_f on a log-log plot.

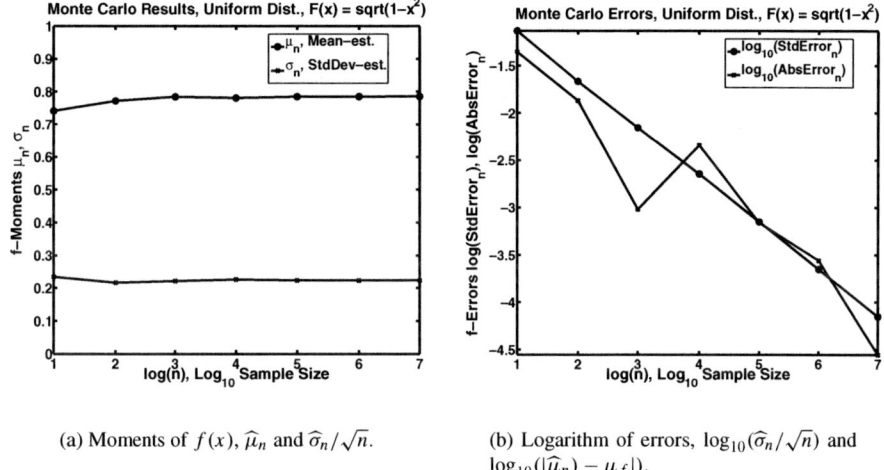

(a) Moments of $f(x)$, $\widehat{\mu}_n$ and $\widehat{\sigma}_n/\sqrt{n}$.   (b) Logarithm of errors, $\log_{10}(\widehat{\sigma}_n/\sqrt{n})$ and $\log_{10}(|\widehat{\mu}_n) - \mu_f|)$.

Figure 9.8. *Monte Carlo simulations for testing use of the uniform distribution to approximate the integral of the integrand $F(x) = \sqrt{1-x^2}$ on $(a, b) = (0, 1)$ using MATLAB code* C.20, *called* mcm1test.m, *on p.* C29 *in Online Appendix* C, *for $n = 10^k$, $k = 1:7$.*

Finite Difference Comparison

Three important characterisitics of Monte Carlo estimators, from Glasserman [97], are bias, variance, and computational effort or time. For computational effort, a primary comparison is with the traditional finite difference methods.

Let the Monte Carlo target integral of (9.32) be over a unit n_x-dimensional hypercube for simplicity, i.e.,

$$\mathcal{V} \equiv [0, 1]^{n_x} = [0, 1] \times [0, 1] \times \cdots \times [0, 1]; \quad V = (1 - 0)^{n_x} = 1,$$

decomposed into a regular grid of m fixed steps $\Delta X = 1/m$ in each dimension so that the grid points in the ith dimension are

$$X_{i, j_i} = j_i/m \text{ for } j_i = 0:m \text{ and } i = 1:n_x.$$

The finite difference approximation will be an expansion of the form,

$$I[F] \simeq I_m^{(fd)} = \sum_{j_1=1}^{m} \cdots \sum_{j_{n_x}=1}^{m} \omega_{j_1} \cdots \omega_{j_{n_x}} \cdot F(j_1/m, \ldots, j_{n_x}/m),$$

where the finite difference method weights are denoted by ω_{j_i} for $j_i = 0:m$ and $i = 1:n_x$ but must at least satisfy the volume conservation consistency condition that

$$\prod_{i=1}^{n_x} \sum_{j_i=1}^{m} \omega_{j_i} \cdot 1 = V = 1,$$

and the higher order method will have even more conditions to be satisfied. There are $m + 1$ grid points per dimension, so the total number of grid points will be $n_{fd} = (m + 1)^{n_x}$ or $m = n_{fd}^{n_x} - 1$. An **rth order finite difference** (FD) **method** will have the following error estimate

$$e_{fd} = I_m^{(fd)} - I[F] = O\left((\Delta X)^r\right) = O\left(m^{-r}\right) = O\left((n_{fd})^{-r/n_x}\right), \tag{9.46}$$

so for n_{fd} and r fixed,

$$e_{fd} \longrightarrow O\left((n_{fd})^{-0}\right) = O(1), \quad \text{as } n_x \to \infty,$$

i.e., in the limit of high problem dimensions, finite difference methods with fixed step sizes become useless, independent of the order r of the method.

A rough theoretical comparison between the computational effort of the Monte Carlo method and fixed spaced finite difference methods (Newton–Cotes rules) can be made by assuming that the gross computational effort will be the order of the total number of points and they will be the same for both types of methods, i.e., $n_{fd} = n$. Also, for a fair comparison, assume that these methods have comparable global errors, i.e., $e_{fd} = O(\hat{e}_n)$ or that the orders of the errors are the same,

$$n^{-r/n_x} = 1/\sqrt{n},$$

which implies that the dimension of $\mathcal{V}$ is related to the order of the finite difference method r,

$$n_x = 2r.$$

Since the Monte Carlo method is a global method, r must be taken to be the global order of the finite difference method. For the simplest integration rule, the left or right rectangular rules (Itô's forward integration is the left rectangular rule), the global order is $r = 1$, so Monte Carlo and finite differences are comparable in computational effort and error when $n_x = 2$. For the trapezoidal or midpoint rule, $r = 2$ and $n_x = 4$ when comparable. For Simpson's (1/3) rule, $r = 4$ and $n_x = 8$ when comparable for even spacing, but for uneven grid spacing $r = 3$ since the cubic bonus due to even spacing symmetry is lost and $n_x = 6$ instead. (Similarly, the midpoint rule order is reduced to that of the other rectangular rules.) See the comments corresponding to Figure 9.10 for comparing results from the trapezoidal and Simpson's rules with the Monte Carlo method using the rejection technique.

Monte Carlo Advantages

- Error is theoretically independent of problem dimension, $n_x = \dim[\mathcal{V}]$.

- Thus, there is no curse of dimensionality, but it is best if $n_x \geq 5$ or so and several random samples are used, i.e.,

$$\left\{ X_{i,j}^{(k)} \middle| i = 1:n_x, \ j = 1:n \text{ sample points}, \ k = 1:K \text{ samples} \right\}.$$

- It works for complex integrands and domains.

- It is not too sensitive to a reasonable sample random number generator.

Monte Carlo Disadvantages

- There are probabilistic error bounds, not strict errors bounds that cannot be exceeded, e.g., 32% of samples can exceed standard error, $\sigma_f / \sqrt{n} \simeq \widehat{\sigma}_n / \sqrt{n}$.

- Irregularity of $F(\mathbf{x})$ or $f(\mathbf{x})$ is not considered, so missed spikes or outliers are possible.

- Generating many large random sample sets for high accuracy can be costly in computer and user time.

- Interplay of functions and volumes can be very complex.

Monte Carlo Ratios and Efficiencies

Any of the advantages and disadvantages listed above are subject to testing and performance evaluation in each case. When comparing two different Monte Carlo methods, say, one with the **basic Monte Carlo method** of Subsection 9.2.1 with variance σ_1^2 and another with variance reduced to variance σ_2^2, both likely to be estimated values, then the user should compare the methods with the **variance reduction ratio**, or simply the **variance ratio**, defined [105] as the improvement ratio from method 1 relative to method 2,

$$\text{VRR}_{1,2} = \frac{\sigma_1^2}{\sigma_2^2}, \tag{9.47}$$

that is, method 2 is the better variance reducer if $\text{VRR}_{1,2} > 1$ and significantly larger.

However, checking for variance reduction alone is not sufficient since the computational costs of the variance reduction should not be excessive, so the **computational cost ratio**

$$\text{CCR}_{1,2} = \frac{\tau_1}{\tau_2} \tag{9.48}$$

should also be checked, where τ_1 is the computational cost (e.g., CPU time) of the first method (usually the basic Monte Carlo method) and τ_2 is the computational cost of the second method.

Hammersley and Handscomb [105] combine both the variance and computational cost ratios into the **efficiency** of method 2 relative to method 1 as

$$\text{Eff}_{1,2} = \text{VRR}_{1,2} \cdot \text{CCR}_{1,2} = \frac{\sigma_1^2 \tau_1}{\sigma_2^2 \tau_2}. \tag{9.49}$$

See also Glasserman [97, pp. 9–12] or Glynn and Whitt [98] for a more thorough description of Monte Carlo efficiency. In addition, Glasserman [97, p. 185] has observed that

The greatest gains in efficiency from variance reduction techniques result from exploiting specific features of a problem, rather than from generic potential variance reduction.

In fact, two primary methods of variance reduction rely on the Monte Carlo user choosing a known factor that represents a significant proportion of the variability of the target function $f(\mathbf{x})$ or the associated density $\phi_{\mathbf{X}}(\mathbf{x})$. **Importance sampling** techniques rely on finding a multiplicative factor that is a better density than the one originally proposed. **Control variate** techniques rely on finding a known additive factor so that when the factor is subtracted from the target function the variance is significantly reduced. In any case, the selection usually depends on good user knowledge of the problem or related model problems.

9.2.2 Inverse Method for Generating Nonuniform Variates

When there is an explicit formula for a distribution of a nonuniform variate in terms of elementary functions, then since the distribution function must lie in [0, 1], an inverse of the distribution function in terms of a uniform variate can transform the nonuniform random variate so that it can be generated by a uniform random variate.

Example 9.6. *Inversion of Exponential to Uniform Distribution.*
This is illustrated in Subsection B.1.7 for the exponential distribution. From (B.40), the exponential distribution for variable $x \geq 0$ and mean $\mu \geq 0$ is

$$\Phi_e(x; \mu) = 1 - \exp(-x/\mu),$$

so equating this to the uniform distribution on [0, 1],

$$\Phi_u(u) = \mathrm{Prob}[0 \leq U \leq u] = u = 1 - \exp(-x/\mu),$$

and inverting yields the inverse relation

$$x = -\mu \cdot \ln(1 - u).$$

However, some computing effort can be saved by eliminating the floating point subtraction in the log-argument by using the complementary property of $\Phi_u(u)$ that $1 - \Phi_u(u) = 1 - u = \mathrm{Prob}[0 \leq U \leq 1 - u]$ is also a uniform distribution for $(1 - u)$ on [0, 1]. (This may seem overly simple, but many students in the sciences without a strong statistics background have difficulty accepting this unless it is spelled out.) Thus, matching the uniform to the exponential distribution can also be formated as

$$\mathrm{Prob}[0 \leq U \leq 1 - u] = 1 - u = 1 - \exp(-x/\mu),$$

leading to a more efficient form for simulations,

$$x = -\mu \cdot \ln(u), \tag{9.50}$$

especially when there is a large number of simulations, $X_i = -\mu \cdot \ln(U_i)$ for $i = 1:n$, e.g., $n = 1.e+6 = 10^6$.

   In general, if it is necessary to generate random variates from a nonuniform random variate X_i with a known distribution function $\Phi_X(x)$ but without an existing random number generator, then an inverse exists:

$$U_i = \Phi_X(X_i) \iff X_i = \Phi_X^{-1}(U_i), \tag{9.51}$$

provided $\Phi_X(x)$ is strictly increasing, $\Phi'_X(x) > 0$. Validation that (9.51) is correct follows from the chain of equations

$$\Phi_X(x) \equiv \text{Prob}[X \le x] = \text{Prob}[\Phi_X^{-1}(U) \le x]$$
$$= \text{Prob}[U \le \Phi_X(x)] = \text{Prob}[U \le u] \equiv \Phi_U(u)$$

using the definition of a probability distribution (9.51) for pairs (X, U) and (u, x), along with the definition of the inverse. For practical purposes, this would mean that $\Phi_X(x)$ is in the form of elementary functions.

Example 9.7. *Use of Built-In Inverses.*
In some special cases, efficient numerical inverses are available, such as the inverse for the error function or complementary error function, erfinv or erfcinv, in MATLAB, which can be used for inverting the normal distribution. (If access to the Statistics Toolbox of MATLAB is available, then norminv built-in function can be used, but the definition norminv(x) = -sqrt(2)*erfcinv(2*x) is trivial, so the toolbox is not necessary.) In Maple, the general procedure using the stats[random] statistics subpackage is based upon its uniform random generator function with the specification of 'inverse' option for nonuniform distributions by the inverse cumulative distribution function ('icdf') method, unless a built-in function is called by name, e.g., normald for normal distribution, or the automatic ('auto') built-in option is specified.

For more general cases, either (1) the distribution $\Phi_X(X_i)$ has a flat subinterval on the interior of its range, say, (c, d), i.e., there is a least one subinterval $c < x_i \le x \le x_{i+1} < d$, where $\Phi'_X(x) = 0$; or (2) the distribution has a jump in the interior of its range, i.e., there is an x_j such that $\Phi_X(x_j^+) > \Phi_X(x_j^-)$. Glasserman [97, Section 2.2.1] is a good reference for these irregular cases and also a good source for many **inverse transform method** examples.

One important example for this book on jump-diffusions is the inversion of the cumulative discrete Poisson distribution with mean Λ to the continuous uniform distribution. The Poisson distribution (B.50) is written as the nth order cumulative distribution with a distribution recursion as

$$P(N) = \sum_{k=0}^{N} p_k(\Lambda); \quad p_0(\Lambda) = 1; \quad p_{k+1}(\Lambda) = \Lambda \cdot p_k(\Lambda)/(k+1).$$

Glasserman's [97] pseudocode is translated to MATLAB code in Figure 9.9. A facsimile of the code in Figure 9.9 has been used successfully by Zhu and Hanson [292] in their Monte Carlo simulation of risk-neutral European call option pricing, cited in Example 9.3. Note that since cumpois takes the jump-count Λ as input, cumpois can be used for temporal Poisson processes, such as in Properties 1.21 on p. 20.

If the components of a vector variate are an independent set of random variates, then it is fairly easy to invert the distribution in favor of a set of independent uniform variates since the joint distribution of independent variates is the product of component marginal distributions (Definition B.35, p. B24), i.e., if

$$\Phi_{\mathbf{X}}(\mathbf{x}) = \prod_{i=1}^{n_x} \Phi_{X_i}(x_i) = \prod_{i=1}^{n_x} u_i, \tag{9.52}$$

```
function N = cumpois(Lambda)
% cumpois function turns uniform point into Poisson jump-count;
U = rand; % generate 1 uniform random point;
% code can be changed to use vector U if needed;
pk = exp(-Lambda); % initialize Poisson distribution;
P = 0; % initialize cumulative distribution;
N = 0; % initialize cumulative jump-counter;
while P < U,  % generate cumulative Poisson count;
     N = N + 1; \% step jump-counter if U too small;
     pk = Lambda*pk/N; \% update Poisson distribution;
     P = P + pk; \% update cumulative distribution;
end
% End function cumpois; returns count N at mean rate Lambda;
```

Figure 9.9. *Inverse Poisson method code to generate jump-counts using the uniform distribution* [97, *Figure* 3.9].

then

$$\mathbf{x} = [x_i]_{n_x \times 1} = \left[\Phi_{X_i}^{-1}(u_i) \right]_{n_x \times 1} \tag{9.53}$$

using the inversion transform method component by component.

For instance, if the $\mathbf{X}_j$ are IID exponentially distributed random vectors with vector mean $\boldsymbol{\mu}$ and the $\mathbf{U}_j$ are generated IID uniformly distributed random vectors for each sample point $j = 1:n$, then the $\mathbf{X}_j$ can be generated by

$$\mathbf{X}_j = \left[X_{i,j} \right]_{n_x \times 1} = \left[-\mu_i \ln(U_{i,j}) \right]_{n_x \times 1} = -\boldsymbol{\mu} .* \log(\mathbf{U}_j)$$

for all j, where $.*$ is the elementswise multiplication symbol and $\log(\mathbf{U})$ is the vectorized natural logarithm function of vector $\mathbf{U}$ as in MATLAB. (The function $\log 10$ is the corresponding MATLAB base 10 logarithm.)

9.2.3   Acceptance and Rejection Method of von Neumann

The method of acceptance and rejection is due to von Neumann [273] and is one of the earliest techniques introduced into the Monte Carlo method. It can be applied to produce samples for unusual probability distributions as well as for unusual domains since the method uses simpler problems for which it is easier to draw variates in simpler domains. In two dimensions, it is just a matter of finding the proportion of points from the simpler, bounding area which lie in the more complicated, interior area.

Note that knowing the formula for a density function $\phi_{\mathbf{X}}^{(1)}(\mathbf{x})$ on domain $\mathcal{V}$ does not mean we know how to generate random variates $\mathbf{X}_i$ for it. Let $\phi_{\mathbf{X}}^{(2)}(\mathbf{x})$ be another density function, such as a uniform, normal, or exponential density function, which is simpler (else not useful), for which there is a known method for generating the corresponding random variates, $\mathbf{X}_i^{(2)}$, and suppose there is a positive constant c for the relative bound

$$\phi_{\mathbf{X}}^{(1)}(\mathbf{x}) \le c \cdot \phi_{\mathbf{X}}^{(2)}(\mathbf{x}) \tag{9.54}$$

for $\mathbf{x}$ in $\mathcal{V}$. For consistency, the target density $\phi_{\mathbf{X}}^{(1)}(\mathbf{x})$ should have a zero when the known comparison generating density $\phi_{\mathbf{X}}^{(2)}(\mathbf{x})$ does, so the relative bound can be written

$$\frac{\phi_{\mathbf{X}}^{(1)}(\mathbf{x})}{\left(c\phi_{\mathbf{X}}^{(2)}(\mathbf{x})\right)} \leq 1,$$

assuming that $0/0 \leq 1$ has been defined. The unit bound indicates that a scalar uniform density will be useful. Since both are densities, the relative bound means that $1 \leq c \cdot 1$ upon integrating both sides of (9.54), so $c \geq 1$ is required. The procedure for the acceptance-rejection method or technique on the ith step is as follows:

1. Generate a random variate $\mathbf{X}_i^{(2)}$ for the comparison density $\phi_{\mathbf{X}}^{(2)}(\mathbf{x})$ (e.g., this comparison density could also be a uniform density for one-dimension, in which case X(i) = rand in MATLAB).

2. Compute the relative ordinate

$$Y_i = \frac{\phi_{\mathbf{X}}^{(1)}(\mathbf{X}_i^{(2)})}{\left(c\phi_{\mathbf{X}}^{(2)}(\mathbf{X}^{(2)})\right)} \tag{9.55}$$

with generated $\mathbf{X}_i^{(2)}$, assuming the relative bound constant c has already been calculated.

3. Generate a scalar uniform random variate U_i and use it to accept or reject the relative ordinate Y_i such that

   - If $U_i \leq Y_i$, then **accept** $\mathbf{X}_i^{(1)} = \mathbf{X}_i^{(2)}$ as a variate for the target density $\phi_{\mathbf{X}}^{(1)}$ and get another point $\mathbf{X}_{i+1}$, stepping i, unless $i + 1 > n$.
   - Else, if $U_i > Y_i$, then **reject** the current $\mathbf{X}_i^{(2)}$ and try another ith generated variate from comparison density $\phi_{\mathbf{X}}^{(2)}$.

Note that

$$\text{Prob}\left[\mathbf{X}^{(2)} \text{ Rejected}\right] = \frac{\text{TotalArea}\left[c\phi_{\mathbf{X}}^{(2)}(\mathbf{x}) - \phi_{\mathbf{X}}^{(1)}(\mathbf{x})\right]}{\text{TotalArea}\left[c\phi_{\mathbf{X}}^{(2)}(\mathbf{x})\right]} = \frac{c-1}{c} \leq 1.$$

In addition, the user wants $(c - 1)$ to be small and positive, i.e., c should be a tight bound constant, to reduce the amount of computation by avoiding too many rejected attempts and thus increasing efficiency. Also, the target distribution $\Phi_{\mathbf{X}}^{(1)}(\mathbf{x})$ for $\mathbf{X} = \mathbf{X}^{(1)}$ (vector inequalities are shorthand notation for a set of component equalities) is

$$\text{Prob}[\mathbf{X} \leq \mathbf{x}] = \frac{\text{TotalArea}\left[\phi_{\mathbf{X}}^{(1)}(\mathbf{y})\,\big|\,\mathbf{y} \leq \mathbf{x}\right]}{\text{TotalArea}\left[c\phi_{\mathbf{X}}^{(2)}(\mathbf{x})\right]} + \text{Prob}\left[\mathbf{X}^{(2)} \text{ Rejected}\right] \cdot \text{Prob}\left[\mathbf{X} \leq \mathbf{x}\right]$$

$$= \frac{1}{c}\Phi_{\mathbf{X}}^{(1)}(\mathbf{x}) + \frac{c-1}{c}\Phi_{\mathbf{X}}^{(1)}(\mathbf{x}) = \Phi_{\mathbf{X}}^{(1)}(\mathbf{x}),$$

consistent with the definition of a distribution.

Example 9.8. *Application of Acceptance-Rejection with Normal Distribution.*
In Figure 9.10, a computational application of the acceptance-rejection technique is illus-
trated for the truncated normal distribution $\Phi_n(a, b; 01)$ defined for a previous uniform-
normal comparison in (9.42) of Example 9.4. The computation converges nicely, with
standard errors of 2.1e-4 when $n = 10^6$ sample points and 6.59e-5 when $n = 10^7$. How-
ever, when these one-dimensional results are compared to standard finite difference meth-
ods the results are not so impressive, e.g., the trapezoidal rule has an absolute error of
2.88e-05 using 101 points and Simpson's (1/3) rule has an absolute error of 3.09e-9 using
the same 101 points, although, as we have said, the finite difference methods are better for
low dimensions.

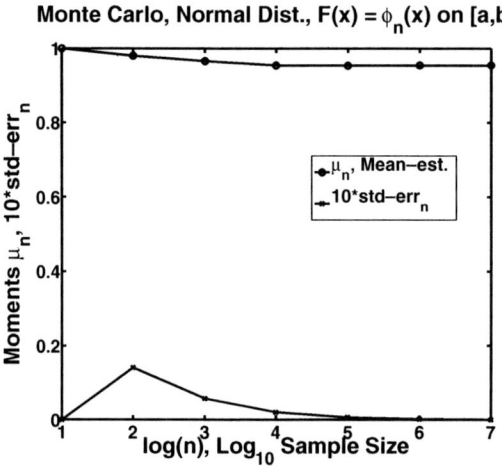

Figure 9.10. *Monte Carlo simulations shown apply the acceptance and rejection*
technique and the normal distribution to compute the estimates for the mean $\widehat{\mu}_n$ and the
magnified standard error $10 \cdot \widehat{\sigma}_n / \sqrt{n}$ for the integral of the truncated normal distribution with
$F(x) = \phi_n(x)$ on $[a, b] = [-2, 2]$ using MATLAB code C.21, `mcm2acceptreject.m`,
on p. C30 in Online Appendix C, for $n = 10^k$, $k = 1 : 7$.

Example 9.9. *Multidimensional Application of Acceptance-Rejection Technique.*
Figure 9.11 illustrates the application of Monte Carlo multidimensional simulations with
the von Neumann acceptance-rejection technique similar to the former $n_x = 1$ truncated
normal distribution problem (9.42) in Example 9.4, but here for dimensions $n_x = 2 : 5$.
Figure 9.11(a) exhibits the Monte Carlo mean estimates, $\widehat{\mu}_n$, which roughly settle down
by sample size $n = 10^4$, but definitely by $n = 10^5$ for this problem and sample sets. In
Figure 9.11(b), the Monte Carlo standard error estimates, $\widehat{\sigma}_n / \sqrt{n}$ are displayed, showing
a remarkable similar decay in sample size beyond sample size $n = 10^3$. Note that since the
integrand $F(\mathbf{x}) = \phi_n(\mathbf{x})$ is the normal density restricted to the vector interval $[\mathbf{a}, \mathbf{b}]$, the
normal density scaled integrand is $f(\mathbf{x}) = \mathbf{1}_{\mathbf{x} \in [\mathbf{a}, \mathbf{b}]}$, an indicator function for the set $[\mathbf{a}, \mathbf{b}]$,
so $f^2(\mathbf{x}) = f(\mathbf{x})$ and the estimate of the standard error,

$$\widehat{\sigma}_n / \sqrt{n} = \sqrt{\widehat{\mu}_n (1 - \widehat{\mu}_n) / (n - 1)},$$

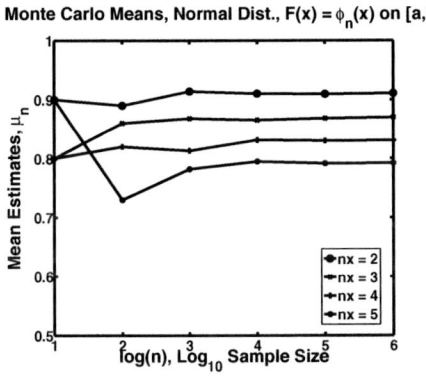

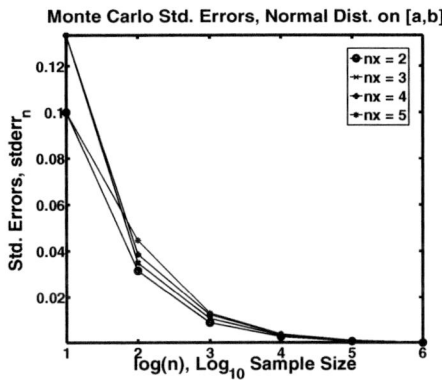

(a) Mean estimates, $\widehat{\mu}_n$, for $f(\mathbf{x})$.   (b) Logarithm of standard errors, $\log_{10}(\widehat{\sigma}_n/\sqrt{n})$.

Figure 9.11. *Monte Carlo simulations for estimating multi-dimensional integrals for the n_x-dimension normal integrand $F(\mathbf{x}) = \phi_n(\mathbf{x})$ on $[\mathbf{a}, \mathbf{b}] = [-2, 2]^{n_x}$ using MATLAB code C.22,* `mcm3multidim.m`, *on p. C32 in Online Appendix C, for $n = 10^k$, $k = 1 : 6$. The acceptance-rejection technique is used to handle the finite domain.*

satisfies the same formula regardless of dimension n_x as suggested by the Monte Carlo theory.

Box–Muller Algorithm for Normal Random Variates

Many of the normal random number generators, if not all, use the algorithm of Box and Muller [38] or updates of it [195] (see also [230, 97]). Since the normal distribution is a special function that cannot be put in terms of elementary functions, it is not exactly invertible by the inverse transform method, except numerically or artificially by defining another special function for the inverse. Box and Muller use pairs of uniform variates and polar coordinates to construct their algorithm to compute a pair of normal variates.

Let U_1 and U_2 be two independent uniform variates on $(0, 1)$, use them to construct a pair of polar coordinates (R, T), and then use those to construct two independent normal variates (X_1, X_2),

$$R = \sqrt{-2\ln(U_1)} \quad \text{and} \quad T = 2\pi U_2,$$
$$X_1 = R\cos(T) \quad \text{and} \quad X_2 = R\sin(T),$$

where $0 < R < \infty$ and $0 < T < 2\pi$ since $0 < U_i < 1$ for $i = 1 : 2$. The inverse transformation is then $\tan(2\pi U_2) = X_2/X_1$ and $-2\ln(U_1) = X_1^2 + X_2^2$ or

$$U_1 = \exp\left(-\left(X_1^2 + X_2^2\right)/2\right) \quad \text{and} \quad U_2 = \tan^{-1}(X_2/X_1)/(2\pi).$$

The Jacobian of the transformation, $(X_1, X_2) \longrightarrow (U_1, U_2)$, is

$$
J = \frac{\partial(U_1, U_2)}{\partial(X_1, X_2)} = \text{Det} \begin{bmatrix} \partial U_1/\partial X_1 & \partial U_1/\partial X_2 \\ \partial U_2/\partial X_1 & \partial U_2/\partial X_2 \end{bmatrix}
$$

$$
= -\exp\left(-\left(X_1^2 + X_2^2\right)/2\right)/(2\pi) = -\Phi_n(X_1, X_2; (0,0), (1,1)),
$$

(9.56)

i.e., the negative of a standard two-dimensional normal distribution for two independent, standard normal variates (X_1, X_2). Thus, only $|J|$ is needed. Conservation of probability consistency is easily verified, since in theory

$$
1 = \int_0^1 \int_0^1 du_1 du_2 = \int_{-\infty}^{+\infty} \int_{-\infty}^{+\infty} |J| dx_1 dx_2 = \frac{1}{2\pi} \int_{-\infty}^{+\infty} \int_{-\infty}^{+\infty} e^{-x_1^2/2} e^{-x_2^2/2} dx_1 dx_2 = 1.
$$

Marsaglia and Bray [195] modified the Box–Muller algorithm to save computing costs by using the acceptance-rejection technique between a square enclosing the unit circle so that the sine and cosine functions would not be needed. They begin by generating two independent uniform variates on the square $(-1, +1) \times (-1, +1)$ rather than on $(0, 1) \times (0, 1)$, i.e., keeping U_1 and U_2 as the initial $(0, 1)$ uniform variates, $U_3 = 2 \cdot U_1 - 1$ and $U_4 = 2 \cdot U_2 - 1$. Next let the squared radius be $R^2 = U_3^2 + U_4^2$ and while $R^2 > 1$, i.e., out of the unit circle, then reject it and try again, but if $R^2 \leq 1$, then compute the normalized Box–Muller radius $R_3 = \sqrt{-2\ln(R_2)/R_2}$ and finally output the independent, standard normal variate pair,

$$
X_3 = R_3 \cdot U_3 \quad \text{and} \quad X_4 = R_3 \cdot U_4.
$$

9.2.4 Importance Sampling

There are two principal ways to reduce the standard error and thus improve the likely accuracy of Monte Carlo simulation relative to the basic Monte Carlo simulation (Subsection 9.2.1; Hammersley and Handscomb call the method in their 1964 compact book [105, Section 5.2] the **crude Monte Carlo method**). One way is to increase the sample size n, but the computational cost is high, e.g., increasing the sample size 100 times is necessary to reduce the standard error by 1/10th of its magnitude due to the weak reciprocal square root order. The other way is to reduce the variance, and a way to do that is to pick a better density to draw samples from that which more closely matches the integrand $F(\mathbf{x})$. **Importance sampling** methods strive to find the better or practical best distribution. As previously mentioned, importance sampling was introduced into the Monte Carlo method in one of the earliest papers [207] on the subject, sometimes called the Metropolis algorithm, in which the desirable sampling distribution was the exponential distribution of energy changes.

Suppose there is an initial density $\phi_{\mathbf{X}}(\mathbf{x})$ such that the mean of $f(x)$ is given by the integral

$$
\mu_f = \text{E}_{\mathbf{X}}[f(\mathbf{X})] = \int_{\mathcal{V}} f(\mathbf{x}) \phi_{\mathbf{X}}(\mathbf{x}) d\mathbf{x} = \int_{\mathcal{V}} F(\mathbf{x}) d\mathbf{x},
$$

(9.57)

but we seek a better density $\widetilde{\phi}_{\mathbf{X}}(\mathbf{x})$ that more closely characterizes the original integrand $F(\mathbf{x})$ and leads to the equivalent formulation

$$
\widetilde{\mu}_{\widetilde{f}} = \text{E}_{\widetilde{\phi}}\left[\left(f\phi/\widetilde{\phi}\right)(\mathbf{X})\right] = \int_{\mathcal{V}} \left(f\phi/\widetilde{\phi}\right)(\mathbf{x}) \widetilde{\phi}_{\mathbf{X}}(\mathbf{x}) d\mathbf{x} = \mu_f,
$$

(9.58)

where $\widetilde{f}(\mathbf{x}) \equiv (f\phi/\widetilde{\phi})(\mathbf{x})$ is a potentially less variable sample target function and $E_{\widetilde{\phi}}$ denotes an expectation with respect to the new density $\widetilde{\phi}_{\mathbf{X}}(\mathbf{x})$, subject to minimal likelihood properties that

$$\widetilde{\phi}_{\mathbf{X}}(\mathbf{x}) \geq 0 \Longleftrightarrow \phi_{\mathbf{X}}(\mathbf{x}) \geq 0,$$

mainly so that any indeterminant $0/0$ form can be defined as 1. The corresponding variance is given by

$$\widetilde{\sigma}_{\widetilde{f}}^2 = \int_{\mathcal{V}} \left(\widetilde{f}(\mathbf{x}) - \widetilde{\mu}_{\widetilde{f}} \right)^2 \widetilde{\phi}_{\mathbf{X}}(\mathbf{x}) d\mathbf{x} = \int_{\mathcal{V}} \widetilde{f}^2(\mathbf{x}) \widetilde{\phi}_{\mathbf{X}}(\mathbf{x}) d\mathbf{x} - \widetilde{\mu}_{\widetilde{f}}^2. \tag{9.59}$$

Since the means are the same, $\widetilde{\mu}_{\widetilde{f}}^2 = \mu_f^2$, reduction of the variance is equivalent to reduction of the new second moment, i.e., making

$$\int_{\mathcal{V}} \widetilde{f}^2(\mathbf{x}) \widetilde{\phi}_{\mathbf{X}}(\mathbf{x}) d\mathbf{x} < \int_{\mathcal{V}} f^2(\mathbf{x}) \phi_{\mathbf{X}}(\mathbf{x}) d\mathbf{x}.$$

The importance sampling goal is to sample at important points of $\widetilde{f}(\mathbf{x})$ such as points of maximum likelihood (see Glasserman [97]).

The Monte Carlo unbiased estimates are the means for n-point samples,

$$\widehat{\mu}_n = \frac{1}{n} \sum_{i=1}^{n} f(\mathbf{X}_i) \text{ and } \widehat{\widetilde{\mu}}_n = \frac{1}{n} \sum_{i=1}^{n} \widetilde{f}(\widetilde{\mathbf{X}}_i), \tag{9.60}$$

where the points $\widetilde{\mathbf{X}}_i$ are sampled from the distribution of the $\widetilde{\phi}_{\mathbf{X}}(\mathbf{x})$ density, while the unbiased sample variances are

$$\widehat{\sigma}_n^2 = \frac{1}{n-1} \sum_{i=1}^{n} (f(\mathbf{X}_i) - \widehat{\mu}_n)^2 \text{ and } \widehat{\widetilde{\sigma}}_n^2 = \frac{1}{n-1} \sum_{i=1}^{n} \left(\widetilde{f}(\widetilde{\mathbf{X}}_i) - \widehat{\widetilde{\mu}}_n \right)^2. \tag{9.61}$$

As with the exact variances or second moments, it is expected that the new second sampled moment is reduced, i.e.,

$$\frac{1}{n} \sum_{i=1}^{n} \widetilde{f}^2(\widetilde{\mathbf{X}}_i) < \frac{1}{n} \sum_{i=1}^{n} f^2(\mathbf{X}_i),$$

since this is equivalent to $\widehat{\widetilde{\sigma}}_n^2 < \widehat{\sigma}_n^2$ for sample variances or $\widehat{\widetilde{\sigma}}_n/\sqrt{n} < \widehat{\sigma}_n/\sqrt{n}$ for the Monte Carlo standard error estimates.

The best choice of a new density obviously would be the normalized absolute value of the full problem,

$$\widetilde{\phi}_{\mathbf{X}}(\mathbf{x}) = |F(\mathbf{x})| \bigg/ \int_{\mathcal{V}} |F(\mathbf{y})| d\mathbf{y},$$

but that would be an absurd circular argument since the normalization factor in the denominator would be the integral we are seeking to estimate if it were the case that $F(\mathbf{x}) > 0$. As Glasserman [97, Figure 4.16] states, importance sampling is the most complex of Monte Carlo techniques for reducing variance, but its effectiveness has the potential to range from the best to the worst. See Glasserman's [97] book for a more advanced treatment.

Analogous concepts arose long ago in the statistically related Gaussian quadrature rules [230], i.e., Gauss-statistics quadrature [275], of numerical analysis. For instance, the Gauss–Legendre rules correspond to integrals weighted in proportion to a uniform density on $[-1, +1]$, Gauss–Laguerre rules to the exponential or gamma densities on $(0, \infty)$, and Gauss–Hermite to the normal distribution $(-\infty, +\infty)$. The criteria for the numerical weights w_i and nodes x_i is that the Gaussian rules give the **best polynomial precision** for the polynomial approximation to the weighted function corresponding to the importance sampled $f(x)$. Practical criteria concern matching the rule with the domain, whether finite, semi-infinite, or full-infinite, but also matching integrand singularities in the case of certain Gaussian rules not mentioned here.

There is a more advanced code, like the adaptive Monte Carlo code called **VEGAS** [183] of Lepage, that primarily uses importance sampling but also uses stratified sampling discussed in the next subsection. The VEGAS algorithm and code is discussed in Press et al. [230].

9.2.5   Stratified Sampling

If the integrands are very variable, then partitioning the domain into disjoint subdomains, computing Monte Carlo estimates on each subdomain, and reassembling the estimates to form a global estimate can usually reduce the global estimated variance, sometimes significantly [105, 97, 151, 230].

Consider a partition of the domain $\mathcal{V}$ into np disjoint parts, called **strata**, such that the union

$$\bigcup_{k=1}^{np} \Delta \mathcal{V}_k = \mathcal{V}$$

and the Monte Carlo integral of interest (9.33)

$$\mu_f = \mathrm{E}_{\mathbf{X}}[f(\mathbf{X})] = \sum_{k=1}^{np} \int_{\Delta \mathcal{V}_k} f(\mathbf{x}) \phi_{\mathbf{X}}(\mathbf{x}) d\mathbf{x} = \sum_{k=1}^{np} p_k \mu_f^{(k)}, \qquad (9.62)$$

where the kth stratum probability is

$$p_k = \int_{\Delta \mathcal{V}_k} \phi_{\mathbf{X}}(\mathbf{x}) d\mathbf{x} > 0 \; \ni \; \sum_{k=1}^{np} p_k = 1,$$

assumed known, and the strata mean

$$\mu_f^{(k)} = \int_{\Delta \mathcal{V}_k} f(\mathbf{x}) \phi_{\mathbf{X}}(\mathbf{x}) d\mathbf{x} / p_k.$$

Let $\mathbf{X}_i^{(k)}$ be a sample point drawn from the density $\phi_{\mathbf{X}}(\mathbf{x})$ on the kth strata $\Delta \mathcal{V}_k$ for $i = 1 : n_k$, where $n_k > 0$ and $\sum_{k=1}^{np} n_k = n$, the sample size. Also let $f_i^{(k)} \equiv f(\mathbf{X}_i^{(k)})$, so that the kth strata Monte Carlo estimate of the mean is

$$\widehat{\mu}_{n_k}^{(k)} = \frac{1}{n_k} \sum_{i=1}^{n_k} f_i^{(k)} \qquad (9.63)$$

and since $\mu_f = \sum_{k=1}^{np} p_k \mu_f^{(k)}$ the total mean estimate is

$$\widehat{\mu}_{n,np} = \sum_{k=1}^{np} p_k \widehat{\mu}_{n_k}^{(k)} = \sum_{k=1}^{np} \frac{p_k}{n_k} \sum_{i=1}^{n_k} f_i^{(k)}. \tag{9.64}$$

This strata sampled estimate is an unbiased estimate since

$$\mu_f^{(k)} = E^{(k)}\left[f\left(\mathbf{X}_i^{(k)}\right)\right] \equiv E\left[f\left(\mathbf{X}_i^{(k)}\right) \middle| \mathbf{X}_i^{(k)} \in \Delta \mathcal{V}\right].$$

Then,

$$E\left[\widehat{\mu}_{n,np}\right] = \sum_{k=1}^{np} \frac{p_k}{n_k} \sum_{i=1}^{n_k} E^{(k)}\left[f_i^{(k)}\right] = \sum_{k=1}^{np} \frac{p_k}{n_k} \sum_{i=1}^{n_k} \mu_f^{(k)} = \sum_{k=1}^{np} p_k \mu_f^{(k)} = \mu_f,$$

independent of the sample distribution n_k. Note that the order of Monte Carlo estimation and stratification are not generally interchangeable if the unbiased property is to be preserved. For instance, if the original simple Monte Carlo estimate $\widehat{\mu}_n$ (9.39) is directly converted to a stratified sum,

$$\widehat{\mu}_n = \frac{1}{n} \sum_{k=1}^{np} \sum_{i=1}^{n_k} f_i^{(k)},$$

and the expectation is calculated as

$$E[\widehat{\mu}_n] = \frac{1}{n} \sum_{k=1}^{np} \sum_{i=1}^{n_k} E^{(k)}\left[f_i^{(k)}\right] = \frac{1}{n} \sum_{k=1}^{np} n_k \mu_f^{(k)},$$

which for general strata means $\mu_f^{(k)}$, this sum will not be μ_f. However, in the special case of **proportional strata sampling** in which $n_k = p_k \cdot n$, then

$$E[\widehat{\mu}_n] = \sum_{k=1}^{np} p_k \mu_f^{(k)} = \mu_f.$$

Recall that the exact variance of f is

$$\sigma_f^2 = \mathrm{Var}_{\mathbf{X}}[f(\mathbf{X})] = \sum_{k=1}^{np} \int_{\Delta \mathcal{V}_k} (f(\mathbf{x}) - \mu_f)^2 \phi_{\mathbf{X}}(\mathbf{x}) d\mathbf{x},$$

but due to the total mean μ_f the total variance does not easily decompose into the strata variances,

$$\begin{aligned}
\left(\sigma_f^{(k)}\right)^2 &= \mathrm{Var}_{\mathbf{X}}^{(k)}[f(\mathbf{X})] = E^{(k)}\left[\left(f(\mathbf{X}) - \mu_f^{(k)}\right)^2\right] \\
&= \int_{\Delta \mathcal{V}_k} \left(f(\mathbf{x}) - \mu_f^{(k)}\right)^2 \phi_{\mathbf{X}}(\mathbf{x}) d\mathbf{x}/p_k.
\end{aligned} \tag{9.65}$$

Following Glasserman [97], the variance is written with the usual second and first moment technique,

$$
\sigma_f^2 = \mathrm{E}\left[f^2(\mathbf{X})\right] - \mu_f^2 = \sum_{k=1}^{np} p_k \mathrm{E}^{(k)}\left[f^2(\mathbf{X})\right] - \mu_f^2
$$

$$
= \sum_{k=1}^{np} p_k \left(\left(\sigma_f^{(k)}\right)^2 + \left(\mu_f^{(k)}\right)^2\right) - \left(\sum_{k=1}^{np} p_k \mu_f^{(k)}\right)^2 .
$$

(9.66)

In contrast, the strata Monte Carlo estimate, the variance, using prior definintions and the IID property of the $X_i^{(k)}$, is

$$
\sigma_{\widehat{\mu}_{n,np}}^2 = \mathrm{Var}\left[\widehat{\mu}_{n,np}\right] = \mathrm{E}\left[\sum_{k=1}^{np} p_k \left(\frac{1}{n_k}\sum_{i=1}^{n_k} f_i^{(k)} - \mu_f^{(k)}\right)\right]
$$

$$
= \sum_{k=1}^{np}\sum_{\ell=1}^{np} \frac{p_k p_\ell}{n_k n_\ell} \sum_{i=1}^{n_k}\sum_{j=1}^{n_k} \mathrm{E}^{(k)}\left[\left(f_i^{(k)} - \mu_f^{(k)}\right)\left(f_j^{(\ell)} - \mu_f^{(\ell)}\right)\right]
$$

(9.67)

$$
= \sum_{k=1}^{np} \frac{p_k^2}{n_k^2} \sum_{i=1}^{n_k}\left(\sigma_f^{(k)}\right)^2 = \sum_{k=1}^{np} \frac{p_k^2}{n_k}\left(\sigma_f^{(k)}\right)^2 .
$$

Thus, the strata reduction of variance will be

$$
\sigma_f^2 - \sigma_{\widehat{\mu}_{n,np}}^2 = \sum_{k=1}^{np} p_k \left(1 - \frac{p_k}{n_k}\right)\left(\sigma_f^{(k)}\right)^2
$$

$$
+ \sum_{k=1}^{np} p_k \left(\mu_f^{(k)}\right)^2 - \left(\sum_{k=1}^{np} p_k \mu_f^{(k)}\right)^2
$$

(9.68)

$$
\geq \sum_{k=1}^{np} p_k \left(1 - \frac{p_k}{n_k}\right)\left(\sigma_f^{(k)}\right)^2 ,
$$

since the second moment majorizes the square of the first moment, here for $\mu_f^{(k)}$ with probability $p_k = 1$. For strata proportional sampling, $n_k = p_k \cdot n$, then

$$
\sigma_f^2 - \sigma_{\widehat{\mu}_{n,np}}^2 \geq \frac{n-1}{n} \sum_{k=1}^{np} p_k \left(\sigma_f^{(k)}\right)^2 ,
$$

(9.69)

so proportional sampling stratification always reduces the variance.

Another form of strata sampling uses $n_k = q_k \cdot n$ so $q_k > 0$ and $\sum_{k=1}^{np} q_k = 1$, but q_k is otherwise arbitrary. This form is called **fractional sampling**. The arbitrariness of the fractions q_k can be used to determine the **optimal sampling allocation** of the stratification with the objective to achieve maximum variance reduction. Since when $n_k = q_k \cdot n$, we have

$$
\sigma_f^2 - \sigma_{\widehat{\mu}_{n,np}}^2 \geq \sum_{k=1}^{np} p_k \left(1 - \frac{p_k}{n q_k}\right)\left(\sigma_f^{(k)}\right)^2 ,
$$

(9.70)

but instead of maximizing the full right-hand side of (9.68) for $\sigma_f^2 - \sigma_{\hat{\mu}_{n,np}}^2$, it is only necessary to minimize the bound in (9.70) containing the variable parameter q_k. This can be done using the **Lagrange multiplier** technique to handle the $\sum_{k=1}^{np} q_k = 1$ constraint with λ as the multiplier by letting

$$S(\mathbf{q}, \lambda) = \sum_{k=1}^{np} p_k \left(1 - \frac{p_k}{nq_k}\right) \left(\sigma_f^{(k)}\right)^2 + \lambda \left(\sum_{k=1}^{np} q_k - 1\right).$$

The reader can easily verify that the optimal allocation solution for the vector of probabilities $\mathbf{q}$ is

$$\mathbf{q}^* = \frac{\mathbf{p}.*\boldsymbol{\sigma}}{(\mathbf{b}^\top \boldsymbol{\sigma})} = \frac{[p_i \sigma_i]_{np \times 1}}{\sum_{k=1}^{np} p_k \sigma_k} \tag{9.71}$$

by taking the gradient of the objective $S(\mathbf{q}, \lambda)$ with respect to $\mathbf{q}$ and eliminating the multiplier λ. Hence, the optimal bound on the variance reduction is

$$\sigma_f^2 - \left(\sigma_{\hat{\mu}_{n,np}}^*\right)^2 \geq \sum_{k=1}^{np} p_k \left(\sigma_f^{(k)} - \frac{1}{n} \sum_{\ell=1}^{np} p_\ell \sigma_f^{(\ell)}\right) \sigma_f^{(k)}. \tag{9.72}$$

See Glasserman [97] for a more advanced treatment of stratified sampling and see Press et al. [229, 230] for a discussion and the advanced **recursive stratified sampling** code called **MISER**.

9.2.6 Antithetic Variates

The **antithetic variate technique** of variance reduction reuses a prior draw, called the **thetic (or thesis) variate**, to construct an opposing random variable, called the antithetic variate, and is usually a mirror image of the thetic variate with the same mean, which has a negative correlation with the thetic variate. The most common antithetic examples are $U_i^{(a)} = 1 - U_i$ for the standard uniform distribution on $[0, 1]$ and $Z_i^{(a)} = -Z_i$ for the standard normal distribution. Hence, $E[U_i^{(a)}] = 0.5 = E[U_i]$, $Var[U_i^{(a)}] = 1/12 = Var[U_i]$, and

$$Cov[U_i, U_i^{(a)}] = -1/12 < 0$$

for the uniform, while $E[Z_i^{(a)}] = 0 = E[Z_i]$, $Var[Z_i^{(a)}] = 1 = Var[Z_i]$, and

$$Cov[Z_i, Z_i^{(a)}] = -1 < 0.$$

The analogous properties hold when the uniform and normal distributions are not standard, i.e, $X_i^{(a)} = b + a - X_i$ on $[a, b]$ for the uniform and $X_i^{(a)} = 2\mu - X_i$ for the normal with mean μ and variance σ. For most other continuous distributions, the samples are drawn from these to standard distributions and are converted by transformation to the target distribution. For example, $X_i = -\mu \ln(U_i)$ and

$$X_i^{(a)} = -\mu \ln(1 - U_i) = -\mu \ln(1 - \exp(-X_i/\mu))$$

for the exponential distribution with mean μ, using Example 9.6.

To keep this section from being too complicated, it will be assumed that the distribution from which the Monte Carlo random variates will be drawn will be from the general uniform in one-dimension ($n_x = 1$) with density $\phi(x) = 1/(b - a)$ on (a, b),

$$\mu_f = \frac{1}{(b - a)} \int_a^b f(x)dx.$$

Note that the antithetic mean will be the same as the thetic mean,

$$\mu_f^{(a)} = \frac{1}{(b - a)} \int_a^b f(b + a - x)dx = \frac{1}{(b - a)} \int_a^b f(y)dy = \mu_f.$$

For the Monte Carlo estimates, we have

$$\widehat{\mu}_n = \frac{1}{n} \sum_{i=1}^n f(X_i); \text{ and } \widehat{\mu}_n^{(a)} = \frac{1}{n} \sum_{i=1}^n f\left(X_i^{(a)}\right),$$

both converging to μ_f by the strong law of large numbers. For the antithetic variate (av) technique, define the thetic-antithetic average mean estimate with limit μ_f as

$$\widehat{\mu}_n^{(av)} = \frac{1}{2}\left(\widehat{\mu}_n + \widehat{\mu}_n^{(a)}\right) \tag{9.73}$$

and note that the Monte Carlo sample size has been doubled to $2n$ using only the original IID sample $\{X_i\}$ sample of n points, but at the computational cost of double the number of function evaluations of $f(x)$. However, if the variance can be reduced substantially, then the original sample size of n could be reduced to compensate for the additional function evaluations.

The new variance is then

$$\text{Var}\left[\widehat{\mu}_n^{(av)}\right] = \tfrac{1}{4}\text{Var}\left[\widehat{\mu}_n + \widehat{\mu}_n^{(a)}\right]$$
$$= \tfrac{1}{4}\text{E}\left[(\widehat{\mu}_n - \mu_f)^2 + \left(\widehat{\mu}_n^{(a)} - \mu_f\right)^2 + 2(\widehat{\mu}_n - \mu_f)\left(\widehat{\mu}_n^{(a)} - \mu_f\right)\right] \tag{9.74}$$
$$= \tfrac{1}{4}\text{Var}[\widehat{\mu}_n] + \tfrac{1}{4}\text{Var}\left[\widehat{\mu}_n^{(a)}\right] + \tfrac{1}{2}\text{Cov}\left[\widehat{\mu}_n, \widehat{\mu}_n^{(a)}\right].$$

If the covariance $\text{Cov}[\widehat{\mu}_n, \widehat{\mu}_n^{(a)}]$ is negative, then a variance reduction ratio of no more than one half would be guaranteed, thus paying for the doubled function evaluations in terms of efficiency (9.49). By a result quoted in Boyle et al. [40], if the target function of f is monotonic, then

$$\text{Cov}[\widehat{\mu}_n, \widehat{\mu}_n^{(a)}] < 0,$$

which is likely true in many applications, e.g., positive payoffs, but perhaps difficult to verify. In multidimensions, negativity conditions will likely have to be replaced by negative semidefinite conditions for practical purposes due to independence across dimensions.

Example 9.10. Antithetic Variates for Compound Poisson Process.
In the jump-diffusion European call option pricing problem of Zhu and Hanson [292], it was necessary to draw a sample from the compound Poisson process with rate λ,

$$S_i = \sum_{j=1}^{N_i} Q_{ij} \text{ for } i = 1:n,$$

estimating the Poisson cumulative sum $\widetilde{P}(t_f)$ in (9.35) of Example 9.3, where the jump-amplitude marks $Q_{i,j}$ are uniformly distributed on $[a, b]$. First the jump-count N_i for $i = 1:n$ sample points is computed by the inverse transform method in Example 9.9, then a set of standard uniform variates $U_{i,j}$ for $j = 1 : N_i$ jumps and $i = 1 : n$ points (i.e., $Q_{i,j} = a + (b - a)U_{i,j}$ and $Q_{i,j}^{(a)} = a + (b - a)(1 - U_{i,j})$). Next the partial sums are computed,

$$S_i = aN_i + (b - a) \sum_{j=1}^{N_i} U_{i,j} \text{ and } S_i^{(a)} = (a + b)N_i - S_i, \qquad (9.75)$$

and these are then used to compute thetic-antithetic averages of jump-shifted Black–Scholes formulas and associated jump-exponentials.

9.2.7   Control Variates

As in importance sampling (Subsection 9.2.4) with its multiplicative factoring of the density by seeking a better density, the **control variate** technique [105] seeks an additive factor, but a known one, that is representative of the variability in the target integrand. This technique was introduced in general by Hammersley and Handscomb [105] in their little book and later introduced to finance along with the antithetic techniques by Boyle [39] in 1977 with a substantial update by Boyle, Broadie, and Glasserman [40] in 1997. See also Glasserman's book [97, Section 4.1] for more recent advances in finance.

Again, consider the target integral, returning back to n_x-dimensional space $\mathcal{V}$ with density $\phi_{\mathbf{X}}(\mathbf{x})$,

$$\mu_f = \int_{\mathcal{V}} f(\mathbf{x})\phi_{\mathbf{X}}(\mathbf{x})d\mathbf{x}$$

and basic Monte Carlo estimate

$$\widehat{\mu}_n = \frac{1}{n} \sum_{i=1}^{n} f(\mathbf{X}_i)$$

converging to μ_f as $n \to \infty$ by the strong law of large numbers, where the set $\{\mathbf{X}_i\}$ of n IID sample points drawn are from the density $\phi_{\mathbf{X}}(\mathbf{x})$.

Next, through knowledge of the problem a simpler function $f^{(c)}(\mathbf{x})$ is found which significantly represents the variability of the target function $f(\mathbf{x})$ and can be used as a control (c) enabler such that

$$\mu_f^{(c)} = \int_{\mathcal{V}} f^{(c)}(\mathbf{x})\phi_{\mathbf{X}}(\mathbf{x})d\mathbf{x}$$

is known or those values can be accurately approximated. Using the same IID set $\{\mathbf{X}_i\}$ of sample points, the basic Monte Carlo estimate is

$$\widehat{\mu}_n^{(c)} = \frac{1}{n} \sum_{i=1}^{n} f^{(c)}(\mathbf{X}_i),$$

which is convergent to and is an unbiased estimate of $\mu_f^{(c)}$. The error $\left(\widehat{\mu}_n^{(c)} - \mu_f^{(c)}\right)$ will be used as a control variable to control the variance reduction of the basic unbiased estimate $\widehat{\mu}_n$ of μ_f by constructing a potentially improved control variate (cv) estimate,

$$\widehat{\mu}_n^{(cv)}(\alpha) \equiv \widehat{\mu}_n - \alpha \left(\widehat{\mu}_n^{(c)} - \mu_f^{(c)}\right), \tag{9.76}$$

where α is a control parameter that will be optimized given the knowledge of the control function $f^{(c)}$. In particular, the partly known error $\left(\widehat{\mu}_n^{(c)} - \mu_f^{(c)}\right)$ will be used to control the control variate estimate error $\left(\widehat{\mu}_n^{(cv)}(\alpha) - \mu_f\right)$, noting from (9.76),

$$\mathrm{E}\left[\widehat{\mu}_n^{(cv)}(\alpha)\right] = \mu_f - \alpha \left(\mu_f^{(c)} - \mu_f^{(c)}\right) = \mu_f,$$

that the unbiased estimation of μ_f is unchanged.

Upon examining the variance of the control variate estimate in terms of α following [97],

$$\mathrm{Var}\left[\widehat{\mu}_n^{(cv)}(\alpha)\right] = \mathrm{Var}\left[\widehat{\mu}_n - \alpha \left(\widehat{\mu}_n^{(c)} - \mu_f^{(c)}\right)\right]$$

$$= \mathrm{E}\left[\left((\widehat{\mu}_n - \mu_f) - \alpha \left(\widehat{\mu}_n^{(c)} - \mu_f^{(c)}\right)\right)^2\right] \tag{9.77}$$

$$= \mathrm{Var}[\widehat{\mu}_n] - 2\alpha\mathrm{Cov}\left[\widehat{\mu}_n, \widehat{\mu}_n^{(c)}\right] + \alpha^2\mathrm{Var}\left[\widehat{\mu}_n^{(c)}\right],$$

a simple quadratic optimization in α produces an optimal control parameter,

$$\alpha^* = \frac{\mathrm{Cov}\left[\widehat{\mu}_n, \widehat{\mu}_n^{(c)}\right]}{\mathrm{Var}\left[\widehat{\mu}_n^{(c)}\right]} = \frac{\rho_{\widehat{\mu}_n, \widehat{\mu}_n^{(c)}}\sqrt{\mathrm{Var}[\widehat{\mu}_n]}}{\sqrt{\mathrm{Var}\left[\widehat{\mu}_n^{(c)}\right]}}, \tag{9.78}$$

where the correlation function is

$$\rho_{X,Y} = \frac{\mathrm{Cov}[X, Y]}{\sqrt{\mathrm{Var}[X]\mathrm{Var}[Y]}}.$$

Thus, the optimal control variate variance is

$$\mathrm{Var}\left[\widehat{\mu}_n^{(cv)}(\alpha^*)\right] = \mathrm{Var}[\widehat{\mu}_n] - \frac{\left(\mathrm{Cov}\left[\widehat{\mu}_n, \widehat{\mu}_n^{(c)}\right]\right)^2}{\mathrm{Var}\left[\widehat{\mu}_n^{(c)}\right]} = \left(1 - \left(\rho_{\widehat{\mu}_n, \widehat{\mu}_n^{(c)}}\right)^2\right)\mathrm{Var}[\widehat{\mu}_n], \quad (9.79)$$

so the absolute value of the correlation $|\rho_{\widehat{\mu}_n, \widehat{\mu}_n^{(c)}}|$ must be less than one for variance reduction. Note that Hammersley and Handscomb in their 1964 book [105, Section 5.5] do not use a control parameter (i.e., $\alpha \equiv 1$) and so require from (9.77) with $\alpha = 1$ that

$$2\mathrm{Cov}\left[\widehat{\mu}_n, \widehat{\mu}_n^{(c)}\right] > \mathrm{Var}\left[\widehat{\mu}_n^{(c)}\right],$$

i.e., the covariance must be sufficiently positive, unlike (9.79). In fact, the optimal variance reduction ratio, from (9.79) and from the definition of VRR (9.47), is

$$\mathrm{VRR}^*_{\widehat{\mu}_n, \widehat{\mu}_n^{(c)}} \equiv \frac{\mathrm{Var}[\widehat{\mu}_n]}{\mathrm{Var}\left[\widehat{\mu}_n^{(cv)}(\alpha^*)\right]} = \frac{1}{\left(1 - \left(\rho_{\widehat{\mu}_n, \widehat{\mu}_n^{(c)}}\right)^2\right)}, \tag{9.80}$$

so the absolute value of the correlation $|\rho_{\widehat{\mu}_n, \widehat{\mu}_n^{(c)}}|$ not only should be less than one but should be sufficiently close to one for significant variance reduction, in theory.

However, the exact statistics represented in the optimal parameter α^* in (9.78) and particularly the related optimal correlation $\rho_{\widehat{\mu}_n, \widehat{\mu}_n^{(c)}}$ are unknown. Hence, in practice, an estimate of α^* is needed, leading to the sample control parameter estimate of α^*,

$$
\widehat{\alpha}_n = \frac{\widehat{c}_n^{(c)}}{\left(\widehat{\sigma}_n^{(c)}\right)^2} \equiv \frac{\dfrac{1}{n-1} \sum_{i=1}^{n} (f_i - \mu_f)\left(f_i^{(c)} - \mu_f^{(c)}\right)}{\dfrac{1}{n-1} \sum_{j=1}^{n} \left(f_j^{(c)} - \mu_f^{(c)}\right)^2}
$$

$$
= \frac{\sum_{i=1}^{n} (f_i - \mu_f)\left(f_i^{(c)} - \mu_f^{(c)}\right)}{\sum_{j=1}^{n} \left(f_j^{(c)} - \mu_f^{(c)}\right)^2},
$$

(9.81)

and the corresponding estimated control variate Monte Carlo estimate

$$
\widehat{\mu}_n^{(cv)}(\widehat{\alpha}_n) = \widehat{\mu}_n - \widehat{\alpha}_n \left(\widehat{\mu}_n^{(c)} - \mu_f^{(c)}\right)
$$

(9.82)

but introducing some bias particularly due to the approximate covariance $\widehat{c}_n^{(c)}$ in (9.81). The bias (9.40) is given by

$$
\beta_{\widehat{\mu}_n^{(cv)}} = E\left[\widehat{\mu}_n^{(cv)}(\widehat{\alpha}_n) - \mu_f\right] = -E\left[\widehat{\alpha}_n \left(\widehat{\mu}_n^{(c)} - \mu_f^{(c)}\right)\right],
$$

(9.83)

which in general will be nonzero due to the nonlinear dependence of $\widehat{\alpha}_n$.

Example 9.11. *Control Variate Adjusted Jump-Diffusion Payoff.*
Zhu and Hanson [292] further reduced the variance of the thetic-antithetic adjusted jump-factor Black–Scholes mentioned in Example 9.10 using the error in the thetic-antithetic adjusted jump-factor,

$$
\Delta Y_i = 0.5 \left(e^{S_i} + e^{S_i^{(a)}}\right) - e^{\lambda t_f \bar{J}},
$$

where the partial sums S_i and $S_i^{(a)}$ are given in (9.75), $\bar{J} \equiv E[J(Q)$ is the asset mean jump-amplitude given in (9.36) of Example 9.3, and t_f is the option exercise time. The complex corrections to the bias $\beta_{\widehat{\mu}_n^{(cv)}}$ in (9.83) are given and proven in [292, 291] along with other results. The combination of antithetic and control variate variance reduction techniques were easy to implement and were efficient despite the theoretical complexity, and the combination was better than either one separately.

For more formation in depth on the control variate technique, see Boyle, Broadie, and Glasserman [40] and Glasserman [97].

Another topic that is important but beyond the scope of this book is the quasi-Monte Carlo method that uses quasi-random or low-discrepancy number sequences which are more genuine deterministic sequences than the pseudo-random number sequences commonly used. Their generation is more complex generally then the pseudo-random sequences, but their big benefit is that convergence is between genuine order $\text{ord}(1/\sqrt{n})$ and $\text{ord}(1/n)$, and so they can outperform the variance reduction techniques just discussed. See Niederreiter [218] for the basic theoretical background to the quasi-Monte Carlo method. For more general information, see Glasserman [97, Chapter 5] and Jäckel [151, Chapter 8]. The Sobol' [253] quasi-random numbers seem to be the best overall performers in various measures as demonstrated in [97, Figures 5.14–5.16] and [151, Figures 8.2–8.9]. Also, see [230, Section 7.7] for the Sobol sequence code `sobseq`.

Suggested References for Further Reading

- Applebaum, 2004 [12]

- Beichl and Sullivan, 2000 [23]

- Boyle, 1977 [39]

- Boyle, Broadie, and Glasserman, 1997 [40]

- Cyganowski and Kloeden, 2000 [66]

- Cyganowski, Grüne, and Kloeden, 2002 [65]

- Cyganowski, Kloeden, and Ombach, 2002 [67]

- Glasserman, 2003 [97]

- Glynn and Whitt, 1992 [98]

- Hammersley and Handscomb, 1964 [105]

- D. Higham, 2001 [140] and 2004 [141]

- D. Higham and Kloeden, 2002 [144] and 2005 [145]

- D. Higham, Mao, and Stuart, 2002 [147]

- Jäckel, 2002 [151]

- Kalos and Whitlock, 1986 [158]

- Kloeden and Platen, 1992 [166]

- Kloeden, Platen, and Schurz, 1994 [167]

- Lepage, 1978 [183]

- Maghsoodi, 1996 [191]

- Maghsoodi and Harris, 1987 [192]

- Metropolis et al., 1953 [207]

- Metropolis and Ulam, 1949 [208]

- Niederreiter, 1992 [218]

- Press and Farrar, 1990 [229]

- Press et al., 2002 [230]

- Zhu, 2005 [291]

- Zhu and Hanson, 2005 [292]

Chapter 10
Applications in Financial Engineering

From the point of view of the risk manager, inappropriate use of the normal distribution can lead to an understatement of risk, which must be balanced against the significant advantage of simplification.
—Alan Greenspan, Joint Central Bank Research Conference, 1995 [80]

Merton (1969, 1971, 1973) uses the formula from Itô's lemma and the continuous-time Bellman equation, but otherwise uses none of the concepts and methods of proof developed by Lebesgue and followers.
—Harry M. Markowitz in the forward to [245]

There is never enough time, unless you're serving it.
—Malcolm Forbes (1919–1990) at
http://www.quotationspage.com/quote/957.html

Stochastic effects play a major role in financial engineering applications, either using a combination of financial assets and other instruments to remove stochasticity altogether through hedging or balancing securities, or just accommodating the financial portfolio analysis to stochastic effects. For general background, the formal derivation of the classical Black–Scholes [34] option pricing model is presented, but students already familiar with the Black–Scholes formulation may prefer to skip to the next, more rigorous section on Merton's [203] justification and generalization of the Black–Scholes model. Applying methods previously developed, this chapter presents the derivation of the Black–Scholes–Merton [34, 201, 203] formula for pricing European call and put options from the stock, bond, and option portfolio diffusion model, including Merton fractions and self-similar solutions [203]. A related option pricing Merton study for underlying stock-bond jump-diffusion models is also discussed. In addition, optimal consumption and portfolio policies for constant relative risk aversion (CRRA) utilities of terminal wealth and instantaneous consumption

287

is discussed for marked jump-diffusions. The notion of scheduled event with distributed response (the so-called **Greenspan processes**) [235, 123, 130] is presented. The role of optimal stochastic control in finance is discussed. The stock jump-diffusion probability density is derived for the linear model treating the composite process as a triad of independent random variables [125, 124].

10.1 Classical Black–Scholes Option Pricing Model

The Black–Scholes option pricing model [34, 35] is perhaps the most used financial model in financial engineering, has been called the most seminal work in finance in the last 25 years, and is probably the most cited work in finance.

The Black–Scholes model is for a portfolio containing a stock option, hedged with the stock itself with price $S(t)$ and a riskless bond with price $B(t)$ at time t providing a constant reference market rate of interest r. The option is assumed to be a **European option**, i.e., there is a contract with a fixed time-to-maturity T to either buy a number of shares of the stock at a given **exercise price** K per share at contract expiration time T (called a **European call option**) or to sell a number of shares of a stock at a given price K per share at a number of shares of a stock at a given exercise price K per share at contract expiration time T (called a **European put option**). The call and put options can be considered together, since they share the same financial market model except for different final boundary conditions at expiration $t = T$. The options contract is between the investor (buyer) and maker (writer) of the contract.

At the end of the term, $t = T$, of the contract, the investor's call option payoff or exercise profit is

$$\max[S(T) - K, 0]. \tag{10.1}$$

Thus the profit from exercising the option is positive only if the final stock price $S(T)$ per share exceeds the contract exercise price K, in which case the investor can buy the stocks at price K, i.e., exercise the option, and then sell the stocks in the market for price $S(T)$. Otherwise, the rational investor does nothing, i.e., does not exercise the option contract. At the start $t = 0$, the investor must bet that $S(T)$ will rise above K and the fixed cost of the bet is the option price, Y_0. Hence, the investor net profit is the payoff (10.1) less the call option price Y_0 for the contract. The net profit position of the contract writer is just the opposite to that of the investor or contract buyer. See Hull [148, pp. 5–10] for a simple, concrete example. A simple version of the Black–Scholes model will be given here, following Hull [148], but with our notation and added explanations.

The situation is reversed for the put option. At the end of the term of the contract, the investor's put option payoff or exercise profit is

$$\max[K - S(T), 0]. \tag{10.2}$$

Thus the profit is positive only if the final stock price $S(T)$ drops below the contract exercise price K, in which case the investor can sell the stocks at price K to the contract maker and then buy stocks more cheaply in the market for $S(T)$, or else the rational investor does nothing, i.e., does not exercise the option. At the start $t = 0$, the investor bets that $S(T)$

will fall below K and again the fixed cost of the bet is the option price. Again, the net profit is the payoff (10.2) less the put option price Y_0 for the contract.

Let the stock or other asset price $S(t)$ dynamics satisfy the linear SDE (often called **geometric Brownian motion**),

$$dS(t) = S(t)(\mu dt + \sigma dW(t)), \quad S(0) = S_0, \tag{10.3}$$

where μ is the constant rate of appreciation of the stock price and σ is the constant volatility (standard deviation) in the stock price. The bond price equation is not really needed, only that a riskless investment grows at a constant rate r so that at time t the principal has grown by an exponential factor $\exp(rt)$ from time zero.

Let the option price be given by the function $Y = F(S(t), t)$ with exercise price K at exercise time $t = T$ when the starting stock price is S_0 at $t = 0$. By the stochastic chain rule, the option price changes according to the SDE,

$$dY(t) = dF(S(t), t) = \left(\frac{\partial F}{\partial t} + \mu S(t) \frac{\partial F}{\partial S} + \frac{1}{2} \sigma^2 S^2(t) \frac{\partial^2 F}{\partial S^2} \right) dt$$
$$+ \sigma S(t) \frac{\partial F}{\partial S} dW(t), \tag{10.4}$$

where all partial derivatives are evaluated at $(S(t), t)$, e.g.,

$$\frac{\partial F}{\partial t} = \frac{\partial F}{\partial t}(S(t), t)$$

denotes the partial derivative of $F(s, t)$ with respect to the second argument t with the first argument s held fixed and evaluated at $(S(t), t)$ after differentiation. The major problem evaluating the initial option price $Y_0 = F(S_0, 0)$ is the volatility or uncertainty term

$$\sigma \frac{\partial F}{\partial S} dW(t)$$

in (10.4) makes any pricing decision difficult unless this term can be controlled or eliminated (i.e., hedged in the language of options pricing).

So to control or **hedge** the volitility term, the value of a **portfolio** of the option and stock is defined as

$$V(t) = N_F F(S(t), t) + N_S S(t), \tag{10.5}$$

where N_F is the number of options and N_S is the number of stocks or other assets. Finding the change in the portfolio value is one of the not so clear assumptions in Black–Scholes option pricing derivations that is addressed in more detail in the next section, but there are also many other explanations, such as in D. Higham's nice introductory options book with emphasis on simulations [141]. For the simple description here, it is assumed that the change in the numbers alone, $F dN_F + S dN_S$ is negligible compared to other changes, i.e.,

$$dV(t) = N_F dF(S(t), t) + N_S dS(t). \tag{10.6}$$

This formula is also called a **self-financing strategy**. Other arguments given are that the N_F and N_S are fixed during changes in F and S or that N_F and N_S change slowly compared to F and S. In fact, the seminal paper of Black and Scholes [34] did take a year or more to get published due to this and other questions [204, 52].

Next, we are interested in eliminating the deviation of the portfolio change for fixed F and S,

$$dV(t) - \mathrm{E}[dV(t)|F, S] = \sigma S \left(N_F \frac{\partial F}{\partial S} + N_S \right) dW(t).$$

So the optimal volatility eliminating hedge is to select the stock number to be

$$N_S^* = -N_F^* \frac{\partial F}{\partial S},$$

where

$$\Delta_F = \partial F / \partial S \tag{10.7}$$

is called the portfolio **delta** in finance and the hedge is called a **delta hedge** [148, p. 310ff]. In terms of fractions with $N_S^* + N_F^* = N$ for fixed N,

$$\frac{N_S^*}{N} = \frac{-\frac{\partial F}{\partial S}}{1 - \frac{\partial F}{\partial S}} \quad \text{and} \quad \frac{N_F^*}{N} = \frac{1}{1 - \frac{\partial F}{\partial S}},$$

providing $\partial F / \partial S \neq 1$. At this point, we will ignore any contradiction with the self-financing assumption, relying in the end on Black–Scholes giving a reasonable and successful formula for option pricing.

Thus,

$$dV^*(t) = N_F^* \left(dF - \frac{\partial F}{\partial S} dS \right) = N_F^* \left(\frac{\partial F}{\partial t} + \frac{1}{2} \sigma^2 \frac{\partial^2 F}{\partial S^2} \right) dt \tag{10.8}$$

using (10.3) and (10.4), while the optimal portfolio value becomes

$$V^*(t) = N_F^* \left(F - \frac{\partial F}{\partial S} S \right). \tag{10.9}$$

In addition, it is necessary to avoid arbitrage, taking advantage of price differentials to make a profit without the trader making his or her own investment. So it is required that the portfolio earn a return at the riskless market rate r or

$$dV^*(t) = r V^*(t) dt, \tag{10.10}$$

the no-arbitrage condition. Finally, the Black–Scholes PDE is formed by combining (10.10) with (10.8) and (10.9), then replacing the stock path function $S(t)$ by the independent stock variable S,

$$\frac{\partial F}{\partial t}(S, t) + \frac{1}{2} \sigma^2 \frac{\partial^2 F}{\partial S^2}(S, t) = r \left(F(S, t) - S \frac{\partial F}{\partial S}(S, t) \right). \tag{10.11}$$

Note that the random volatility term and the mean appreciation (μ) term no longer appear, but the volatility coefficient appears due to the Itô diffusion coefficient correction. This

PDE is a backward PDE for t on $[0, T)$ and S on $[0, \infty)$ with final condition at $t = T$ for any nonnegative S,

$$F(S, T) = \mathcal{C}(S, T) = \max[S - K, 0]$$

for a call option from (10.1) and

$$F(S, T) = \mathcal{P}(S, T) = \max[S - K, 0]$$

for a put option from (10.2). The well-known formula [34, 141] for solution to this PDE can be found in terms of the normal distribution function, but only the results are given here since the details are presented for the more general case in the next section. In the case of the European call option, the Black–Scholes formula is

$$\mathcal{C}_0(S_0) \equiv \mathcal{C}(S_0, 0) = S_0 \Phi_n(d_1(S_0); 0, 1) - K e^{-rT} \Phi_n(d_2(S_0); 0, 1), \qquad (10.12)$$

where the variable arguments of the normal distribution function $\Phi_n(w; \mu, \sigma^2)$ are

$$d_1(s) \equiv \frac{\ln(s/K) + (r + \sigma^2/2)T}{\sigma \sqrt{T}}$$

and

$$d_2(s) \equiv d_1(s) - \sigma \sqrt{T}.$$

In the case of the European put option, the Black–Scholes formula can be found by the well-known and very general **put-call parity** that depends basically on the properties of the maximum function [203, 141],

$$\mathcal{P}_0(S_0) \equiv \mathcal{P}(S_0, 0) = \mathcal{C}_0(S_0) + K e^{-rT} - S_0. \qquad (10.13)$$

     In 1900, Bachelier [16, 62], a student of Poincaré, published a theory of option pricing that was derived from his thesis, but his work was little noticed at the time. Unlike the Black–Scholes diffusion option pricing model based upon the geometric Brownian motion stochastic model, Bachelier's option pricing model was based upon additive Brownian motion, i.e., instead of being linear in the stock price as in the multiplicative noise (4.34) case, the noise was independent of stock price and thus is additive noise (4.31). Bachelier's paper was a very early, very complete, and straight-forward application of stochastic processes in finance. One main drawback is the additive noise, since stock price fluctuations are now assumed to act in a compound or multiplicative fashion. Another drawback is that the purely additive noise case can lead to negative stock prices, since prices are not protected by the multiplicative noise that is linear in the state as used in the Black–Scholes model and guarantees positive prices if prices start out positive. (Thanks to an anonymous reviewer who pointed this out.)

10.2 Merton's Three Asset Option Pricing Model Version of Black–Scholes

Sometimes the Black–Scholes model is called the Black–Scholes–Merton model, since Merton [201], in his theory of rational option pricing paper, gave substantial mathematical justification of the seminal Black–Scholes model using stochastic diffusion processes.

Merton's paper includes generalizations of the Black–Scholes model that provide greater foundations and limitations for their model. Both the Black–Scholes and the Merton papers were published in spring 1973, Merton having held up the publication of his paper out of deference to Black and Scholes. Robert C. Merton and Myron Scholes shared the 1997 Nobel prize in economics for the accomplishments, but unfortunately Fischer Black [204, 52] had passed away in 1995.

The version of the model presented here is mainly based on Merton's more general framework [201] (reprinted in Chapter 8 of [203]). The model is for a portfolio containing a European stock option, hedged with the stock itself with price $S(t)$ and a riskless bond with price $B(t)$ at time t, but with more explicit assumptions than for the classical Black–Scholes.

The market model comprises a number of assumptions which will be enumerated and marked with **BSM** here but formulated in the notation and spirit of this book. Multiple assumptions of Merton have been decomposed into single assumptions to make them easier to modify new problems. One of the objectives of this book to offer sufficient detail to enable the reader to become a practitioner making those new modifications. The more general model of Merton is treated here since many readers will be familiar with the simpler, classic versions of the Black–Scholes option pricing model which can be found in many of the references listed at the end of this chapter, e.g., Hull [148] or Wilmott et al. [282].

- **Assumption BSM1. Frictionless Markets.**
  There are no transaction fees for transactions involving the buying or selling of the three assets in the portfolio, excluding the original price of the option contract.

- **Assumption BSM2. No Dividends.**
  There are no dividends paid on the stock asset.

- **Assumption BSM3. Continuous Trading, without Jumps.**
  Trading among the assets is continuous, so discrete aspects of trading such as jumps are neglected. This assumption is consistent with the no transaction fees and no dividends of the prior two assumptions, since those are discrete events.

- **Assumption BSM4. Borrowing and Short Selling Allowed.**
  Short selling of stock or options is allowed within the term of the contract, with funds placed into the bond asset. Borrowing from the bond asset is allowed to increase the number of shares of the other two assets. Also, it is assumed that the borrowing rate is the same as the lending rate.

- **Assumption BSM5. Linear Stock-Price Stochastic Dynamics.**
  Let $S(t)$ be the **price of stock** per share at time t. Then the $S(t)$ satisfies a linear stochastic diffusion differential equation, written in terms of the rate of return or relative change in time dt:

$$dS(t)/S(t) = \mu_S(t)dt + \sigma_S(t)dW_S(t), \qquad (10.14)$$

  where

  - $dW_S(t) =$ stochastic diffusion differential process for the stock price process $S(t)$, such that $\mathrm{E}[dW_S(t)] = 0$ and

$$(dW_S)^2(t) \overset{\mathrm{ims}}{=} dt,$$

else
$$dW_S(t)dW_S(s) \overset{\text{ims}}{=} 0 \quad \text{if} \ s \neq t$$

by independent increments.

- $\mu_S(t) = \mathrm{E}[dS(t)/S(t)]/dt =$ instantaneous **expected rate of return** on the stock in time dt.

- $\sigma_S^2(t) = \mathrm{Var}[dS(t)/S(t)]/dt =$ instantaneous variance of the rate of return on the stock in time dt, while σ_S is the **volatility** of the stock return. Here volatility denotes a measure of uncertainty [148] but is derived from the French word meaning *to fly*.

Here, a stock is considered a **risky asset**, compared to the bond asset. Since the option profit at exercise depends only on the stock price $S(T)$ at the expiration of the option, for some analyses it is more convenient to view the process in backward time $\tau = T - t$, also called the time-to-maturity, and to consider the stock price in that variable, i.e.,
$$\widehat{S}(\tau) \equiv S(T - \tau).$$

- **Assumption BSM6. Linear Bond-Price Stochastic Dynamics.**
Let $B(t)$ be the **price of bond** asset at time t, in particular a default-free zero-coupon bond or discounted loan with time-to-maturity T. Then the $B(t)$ satisfies a linear stochastic diffusion differential equation, written in terms of the rate of return or relative change in time dt:

$$dB(t)/B(t) = \mu_B(t)dt + \sigma_B(t)dW_B(t), \tag{10.15}$$

where

- $dW_B(t) =$ stochastic diffusion differential process for the bond return process $B(t)$, such that $\mathrm{E}[dW_B(t)] = 0$ and
$$(dW_B)^2(t) \overset{\text{ims}}{=} dt,$$

otherwise
$$dW_B(t)dW_B(s) \overset{\text{ims}}{=} 0 \quad \text{if} \ s \neq t$$

by independent increments.

- $\mu_B(t) = \mathrm{E}[dB(t)/B(t)]/dt =$ instantaneous **expected rate of return** on the bond asset.

- $\sigma_B^2(t) = \mathrm{Var}[dB(t)/B(t)]/dt =$ instantaneous variance of the rate of return on the stock, while σ_B is the **volatility** of the stock return.

Here, a bond is usually considered a **lower risk asset**, compared to the higher risk or risky stock asset. Here, the variance or volatility will be taken as a **measure of riskiness**, so we say that the stock is **riskier** or **more risky** than the bond if $\sigma_S > \sigma_B$. We say that the bond is **riskfree** if $\sigma_B = 0$. However, Merton [203] has more precise measures of riskiness, though his are not as easy to apply (see Exercise 2).

In the more classical Black–Scholes model, the bond price is assumed to be deterministic, so $\sigma_B(t) = 0$, and the mean rate is assumed to be constant, so $\mu_B(t) = r$. In this ideal case the bond is called **riskfree** or **riskless**.

In the case where the bond is treated as a discounted loan, then the pay-back is at the final price $B(T)$, the initial discounted loan amount received is $B(0)$, which should be less than $B(T)$, so $(B(T) - B(0)) > 0$ is the amount discounted. Discounting is a backward time version of interest on principal. In the backward time problem, the time-to-maturity or time-to-go $\tau = T - t$ is the natural time variable.

In the nonstochastic interest rate problem, as in the traditional Black–Scholes formulation, $\sigma_B = 0$, $\mu_B = r$, the mean interest rate for borrowing and selling, and the bond price in backward time is

$$\widehat{B}(\tau) \equiv B(T - \tau).$$

So the bond price decays away from expiration

$$d\widehat{B}(\tau) = -r\widehat{B}(\tau)d\tau$$

with the bond price decaying in τ due to discounting,

$$\widehat{B}(\tau) = \widehat{B}(0)e^{-r\tau} = B(T)e^{-r(T-t)}.$$

This backward time view is consistent with the options contract where the profit depends on the final stock price $S(T)$ and the objective is to find the number of shares at the initial price $S(0)$ in the final value problem for a stochastic differential equation.

- **Assumption BSM7. Bond and Stock Price Fluctuations are Correlated but Not Serially.**
  Thus, the correlation properties between the stock price noise and the bond price noise are

$$dW_B(t)dW_S(t) \overset{\text{ims}}{=} \rho dt, \qquad (10.16)$$

$$dW_B(t)dW_S(s) \overset{\text{ims}}{=} 0 \quad \text{if } s \neq t. \qquad (10.17)$$

The former equation (10.16) expresses correlation on the same time increments at t (see Exercise 1 for the proof), while the latter equation (10.17) expresses the lack of serial correlation on disjoint time-intervals when $s \neq t$, also preserving the independent increment property, where

$$\rho \equiv \frac{\text{Cov}[dS(t), dB(t)]}{\sqrt{\text{Var}[dS(t)]\text{Var}[dB(t)]}} = \frac{\text{Cov}[dW_S(t), dW_B(t)]}{dt}, \qquad (10.18)$$

which equals the instantaneous correlation coefficient between stock and bond returns, provided $\sigma_S(t)$ and $\sigma_B(t)$ are positive.

The joint density for $(dW_S(t), dW_B(t))$ is obviously the bivariate normal density in (B.144) of the preliminaries of Online Appendix B,

$$\phi_{(dS(t),dB(t))}(s, b) = \phi_n \left(\begin{bmatrix} s \\ b \end{bmatrix}; \begin{bmatrix} \mu_S(t) \\ \mu_B(t) \end{bmatrix} dt, \right.$$

$$\left. \begin{bmatrix} \sigma_S^2(t) & \rho(\sigma_S\sigma_B)(t) \\ \rho(\sigma_S\sigma_B)(t) & \sigma_B^2(t) \end{bmatrix} dt \right). \qquad (10.19)$$

Merton [201, 203] claims that the lack of serial correlations is consistent with the **efficient markets hypothesis**. In the simpler expositions of the Black–Scholes model, there are no correlations, so $\rho \equiv 0$ with $\sigma_B = 0$ and $\mu_B = r$, the common interest rate. The mean square limit for nonserial correlation (10.16) is left as an exercise for the reader.

- **Assumption BSM8. No Investor Preferences or Expectations, except for Agreement on Parameters.**
  The investors agree on and have reasonable knowledge of the parameters, such as the means μ_S and μ_B, as well as the volatilities σ_S and σ_B.

- **Assumption BSM9. Option Price is a Function of Stock and Bond Prices.**
  The option price per share at time t,

$$Y(t) = F(S(t), B(t), t; T, K), \qquad (10.20)$$

depends on the stock S and bond B price stochastic variables, as well as on time t and parameters such as the time-to-maturity T and the contracted expiration stock price K per share.

Alternatively, the relationship can be cast in terms of the time-to-maturity, $\tau = T - t$,

$$\widehat{Y}(\tau) = F(S(T - \tau), B(T - \tau), T - \tau; T, K).$$

Although we are interested in the initial option price $Y(0) = \widehat{Y}(T)$, considering the time-dependent option price $Y(t) = \widehat{Y}(\tau)$ allows analysis of the problem and yields more general results that permit conversion of the option contract to another investor at the current option price $Y(t) = \widehat{Y}(\tau)$. In the case of constant coefficients, the results will depend on the general time-to-exercise $\tau = T - t$ without restriction to a fixed exercise time T.

Using a two-state-dimensional version of the stochastic diffusion chain rule, the return on the option asset, initially keeping all quadratic terms in the Taylor expansion, is

$$dY(t) = dF(S(t), B(t), t; T, K)$$
$$\overset{\text{ims}}{=} F_t dt + F_S dS(t) + F_B dB(t)$$
$$+ \frac{1}{2} \left(F_{SS} (dS)^2(t) + 2F_{SB} dB(t)dS(t) + F_{BB} (dB)^2(t) \right), \qquad (10.21)$$

omitting higher order terms that obviously have zero mean limits. Here, $\{F_S, F_B, F_{SS}, F_{SB}, F_{BB}\}$ are the set of first and second partial derivatives of

$F(S, B, t; T, K)$ with respect to the underlying portfolio assets S and B. Next, substitution for the return processes $S(t)$ and $B(t)$ is used, along with the quadratic differential forms in the mean square limit,

$$(dS)^2(t) \overset{ims}{=} \sigma_S^2(t) S^2(t) dt,$$

$$(dB)^2(t) \overset{ims}{=} \sigma_B^2(t) B^2(t) dt, \tag{10.22}$$

$$(dBdS)(t) \overset{ims}{=} \rho \cdot \sigma_B(t) \sigma_S(t) B(t) S(t) dt,$$

which simply follow from the corresponding mean square limit differential forms for $(dW_S)^2(t)$, $(dW_B)^2(t)$, and $(dW_B dW_S)(t)$, respectively, given under previous assumptions. This forces the geometric Brownian motion form on the option price,

$$dY(t) \overset{ims}{=} Y(t) \left(\mu_Y(t) dt + \sigma_{YS}(t) dW_S(t) + \sigma_{YB}(t) dW_B(t) \right), \tag{10.23}$$

where the new option instantaneous return moment coefficients are defined as

$$Y(t)\mu_Y(t) \equiv F_t + \mu_S S F_S + \mu_B B F_B \tag{10.24}$$
$$+ \frac{1}{2} \left(\sigma_S^2 S^2 F_{SS} + 2\rho \sigma_S \sigma_B S B F_{SB} + \sigma_B^2 B^2 F_{BB} \right),$$

$$Y(t)\sigma_{YS}(t) \equiv \sigma_S S F_S, \tag{10.25}$$

$$Y(t)\sigma_{YB}(t) \equiv \sigma_B B F_B. \tag{10.26}$$

- **Assumption BSM10. Self-Financing Portfolio Investments.**
  Let $N_S(t)$, $N_Y(t)$, and $N_B(t)$ be the instantaneous number of shares invested in the stock, option, and bond at time t, respectively, such that the instantaneous values of the assets in dollars are

$$V_S(t) = N_S(t)S(t), \quad V_Y(t) = N_Y(t)Y(t), \quad V_B(t) = N_B(t)B(t), \tag{10.27}$$

respectively. However, it is assumed there is a **zero instantaneous aggregate portfolio value**,

$$V_P(t) = V_S(t) + V_Y(t) + V_B(t) = 0, \tag{10.28}$$

so that the bond value variable can be eliminated,

$$V_B(t) = -(V_S(t) + V_Y(t)). \tag{10.29}$$

Merton [203] defines a **self-financing portfolio** as a trading strategy in which no capital is put in or taken out until maturity. Such a strategy avoids an imbalance between the stock and its option, which would soon disappear as other investors took advantage of the imbalance. This strategy is also related to the avoidance of arbitrage profits and in Black–Scholes is $\mu = r$. Further, this strategy includes a **no consumption** of assets assumption.

It is further assumed that the absolute instantaneous return from the value of the portfolio $V_P(t)$ is a linear combination of the instantaneous returns in each of the three assets, (S, Y, B), giving the **portfolio budget equation**

$$dV_P(t) = N_S(t)dS(t) + N_Y(t)dY(t) + N_B(t)dB(t) \tag{10.30}$$
$$= V_S(t)\frac{dS(t)}{S(t)} + V_Y(t)\frac{dY(t)}{Y(t)} + V_B(t)\frac{dB(t)}{B(t)}$$

using (10.27) to convert from number of shares to asset value assuming that none of the divisors are zero for the latter more convenient form in terms of rates of return. Note that the budget equation cannot be expressed as the instantaneous rate of return since $V_P(t) = 0$.

Substituting for the three asset stochastic dynamics from (10.14), (10.15), (10.23) and eliminating the bond value $V_B(t)$ through (10.29),

$$dV_P(t) = V_S\left(\frac{dS}{S} - \frac{dB}{B}\right) + V_Y\left(\frac{dY}{Y} - \frac{dB}{B}\right)$$
$$= \left((\mu_S - \mu_B)V_S + (\mu_Y - \mu_B)V_Y\right)dt$$
$$+ (\sigma_S V_S + \sigma_{YS}V_Y)dW_S(t)$$
$$+ (-\sigma_B V_S + (\sigma_{YB} - \sigma_B)V_Y)dW_B(t). \tag{10.31}$$

See Merton [203, Chapter 5] for more justification.

Note that (10.30) does not really follow the Itô stochastic calculus but states that the absolute return on the portfolio is the number of shares weighted sum of the absolute returns on the portfolio assets. However, Merton [203, Chapter 5] argues that the missing differential product terms, such as $dN_S(t)S(t)$ and $dN_S(t)dS(t)$, represent consumption or external gains to the portfolio, which would violate the self-financing assumption, making the portfolio open rather than closed to just the three assets.

• **Assumption BSM11. Investor Hedging the Portfolio to Eliminate Volatility.**
Since many investors as individuals or as a group act to avoid stochastic effects, they tune or hedge their trading strategy, as a protection against losses, by removing volatility through removing the coefficients of the stock and bond fluctuations. A main purpose of the stock and bond underlying the option in the portfolio is to give sufficient flexibility to leverage or hedge the stock and bond assets to remove volatilities that would not be possible with the option alone. Hence, setting the coefficients of $dW_S(t)$ and $dW_B(t)$, respectively, to zero in (10.31),

$$\left(\sigma_S V_S^* + \sigma_{YS}V_Y^*\right) = 0, \tag{10.32}$$

$$-\sigma_B V_S^* + (\sigma_{YB} - \sigma_B)V_Y^* = 0. \tag{10.33}$$

The optimal system (10.32), (10.33) has a nontrivial solution for the optimal values (V_S^*, V_B^*) provided the system is singular, i.e., the determinant of the system is zero,

$$0 = \text{Det}\begin{bmatrix} \sigma_S & \sigma_{YS} \\ -\sigma_B & \sigma_{YB} - \sigma_B \end{bmatrix} = \sigma_S(\sigma_{YB} - \sigma_B) + \sigma_{YS}\sigma_B, \tag{10.34}$$

which leads to the **Merton volatility fraction**

$$\frac{\sigma_{YS}}{\sigma_S} = -\frac{\sigma_{YB} - \sigma_B}{\sigma_B}, \tag{10.35}$$

provided $\sigma_S \neq 0$ and $\sigma_B \neq 0$. The single optimal option-stock value relation that makes it work is

$$V_S^* = -\frac{\sigma_{YS} V_Y^*}{\sigma_S}, \tag{10.36}$$

recalling the budget constraint on V_B^*, giving

$$V_B^* = -\left(V_S^* + V_Y^*\right) = -\left(1 - \frac{\sigma_{YS}}{\sigma_S}\right) V_Y^*. \tag{10.37}$$

In the case of the nonstochastic, constant rate bond process, as in the more traditional Black–Scholes model, $\mu_b = r$ and $\sigma_B = 0$, so $\sigma_{YB} = 0$ and the option price is assumed to be independent of the bond price B, i.e., $F = F(S(t), t; T, K)$ and $F_B \equiv 0$. Then only the optimal values (10.36) are obtained, i.e., there is no Merton volatility fraction in the traditional Black–Scholes model.

However, taking the Merton volatility fraction as valid and substituting in for the definitions of the option-stock volatility σ_{YS} and the option-bond volatility σ_{YB} from (10.25)–(10.26), respectively, the option price then turns out to be homogeneous [203] in S and B,

$$Y^* = Y_S^* S + Y_B^* B. \tag{10.38}$$

Since this result is based upon the Merton volatility fraction, it does not appear in the classical Black–Scholes model, and the stock and bond dynamics no longer have common stochastic diffusion forms.

• **Assumption BSM12. Zero Expected Portfolio Return.**
  Further, to avoid arbitrage profits, the expected return must be zero as well. Thus, the coefficient of dt in (10.31) must be zero, aside from the assumption that $V_P(t) = 0$ would imply that $dV_P(t) = 0$, i.e.,

$$0 = (\mu_S - \mu_B) V_S^* + (\mu_Y - \mu_B) V_Y^* \tag{10.39}$$

$$= \left(-(\mu_S - \mu_B)\frac{\sigma_{YS}}{\sigma_S} + (\mu_Y - \mu_B)\right) V_Y^*, \tag{10.40}$$

assuming $V_Y^* \neq 0$. Otherwise, there would be no option and no optimal values (10.36) that would follow from the Merton volatility fraction (10.35). This means that the portfolio returns are hedged to complete equilibrium, deterministically and stochastically. Thus, provided the option value $V_Y^* \neq 0$, by setting the coefficient of V_Y^* in (10.39) to zero, **Merton's Black–Scholes fraction** becomes simply **Merton's fraction** for the expected returns, i.e.,

$$\frac{\mu_Y - \mu_B}{\mu_S - \mu_B} = \frac{\sigma_{YS}}{\sigma_S}. \tag{10.41}$$

Since it does not involve either of the bond related volatilities, σ_B or σ_{YB}, this primary Merton fraction holds for the Black–Scholes model as well. The Black–Scholes fraction (10.41) states that the net drift ratio equals the option-stock volatility ratio, where the net drift is relative to the market interest/discount rate μ_B.

10.2.1   PDE of Option Pricing

To derive the PDE of Black–Scholes–Merton option pricing, the definition of the option expected return μ_Y in (10.24) is viewed as a PDE for the option price function with the option trajectory $Y(t)$ replaced by the composite function equivalent $F(S, B, t; T, K)$ as a function of three independent variables (S, B, t), (S, B) having replaced the underlying state trajectories $(S(t), B(t))$,

$$\mu_Y F \equiv F_t + \mu_S S F_S + \mu_B B F_B \tag{10.42}$$
$$+ \frac{1}{2} \left(\sigma_S^2 S^2 F_{SS} + 2\rho\sigma_S\sigma_B S B F_{SB} + \sigma_B^2 B^2 F_{BB} \right).$$

It is conceptually important to separate the view of S and B as deterministic, independent PDE variables and the view of $S(t)$ and $B(t)$ as the random SDE state trajectories in time and to use each view in the appropriate place.

Next, μ_Y is eliminated using the Black–Scholes fraction (10.41) with $\mu_Y = \mu_B + (\mu_S - \mu_B)\sigma_{YS}/\sigma_S$, and the option-stock induced volatility σ_{YS} is eliminated using its definition in (10.25), i.e., $\sigma_{YS} = \sigma_S S F_S/F$, while the option price F can be eliminated by Merton's homogeneous condition (10.38) with Y replaced by F,

$$F = S F_S + B F_B,$$

incidentally eliminating both first partials F_S and F_B, and so

$$0 = F_t + \frac{1}{2} \left(\sigma_S^2 S^2 F_{SS} + 2\rho\sigma_S\sigma_B S B F_{SB} + \sigma_B^2 B^2 F_{BB} \right). \tag{10.43}$$

This Merton **PDE of option pricing** needs side conditions, such as a final condition at the expiration time, and boundary conditions in the asset variables. The PDE and conditions forming a final value problem (FVP). For the FVP, the natural time variable is the time-to-maturity or time-to-go $\tau = T - t$ and $F_t = -F_\tau$, so the backward formulated PDE (10.43) in forward time t can be written as a forward diffusion, or parabolic PDE in backward time τ,

$$F_\tau = \frac{1}{2} \left(\sigma_S^2 S^2 F_{SS} + 2\rho\sigma_S\sigma_B S B F_{SB} + \sigma_B^2 B^2 F_{BB} \right). \tag{10.44}$$

It is conceptually important to remember that the **PDE problem**, (10.44) plus any final and boundary conditions, is a deterministic problem in realized independent variables $(S, B, t = T - \tau)$, all stochasticity being eliminated, in contrast to the **SDE problem** in the stochastic path variables $(S(t), B(t), Y(t))$, which depends on the independent variable t and underlying stochastic diffusion processes.

In the classical Black–Scholes model, the bond price has no volatility $\sigma_B(t) = 0$, so the Merton homogeneous result (10.38) does not hold since it is based upon the Merton volatility fraction, which is invalid if $\sigma_B(t) = 0$. Thus, starting back at the view of the definition of μ_Y as a PDE (10.42) setting all B partial derivatives to zero, but eliminating μ_Y using the Black–Scholes fraction (10.41) and σ_{YS} using (10.25), letting the option price function in backward time be defined as

$$\widehat{F}(S, \tau; T, K) \equiv F(S, T - \tau; T, K),$$

which leads to Merton's **Black–Scholes option pricing PDE**, including a bond term,

$$\widehat{F}_\tau = \frac{1}{2}\sigma_S^2 S^2 \widehat{F}_{SS} + \mu_B(S\widehat{F}_S + B\widehat{F}_B - \widehat{F}). \tag{10.45}$$

If the assumption that the mean interest/discount rate is the constant market rate, $\mu_B = r$ along with constant stock volatility σ_S, then the standard Black–Scholes option pricing PDE is obtained.

However, many texts do not use Merton's elaborate assumptions, which we have decomposed into a larger number of individual assumptions here; these texts use a different hedging argument to produce the Black–Scholes PDE and the constant rate coefficient r. Dropping the zero aggregate assumption, the portfolio value is then

$$V_P(t) = N_S(t)S(t) + N_Y(t)Y(t) \tag{10.46}$$

in terms of the number of shares times the price per share for the option and the underlying stock. Similarly, the change in the portfolio value is given by the budget equation

$$dV_P(t) = N_S(t)dS(t) + N_Y(t)dY(t), \tag{10.47}$$

ignoring the missing differential forms as in Merton's more general version. Upon eliminating the resultant stochastic terms to form a riskless portfolio, the coefficients of $dW_S(t)$, again yields the stock-option relationship, relating the number of stock shares to that of the options

$$N_S = -N_Y \widehat{F}_S, \tag{10.48}$$

called **delta hedging** since $\Delta \equiv \partial \widehat{F}/\partial S$ is called the **delta** of the option [284], where the definition of σ_{YS} in (10.25) has been used.

Thus,

$$V_P = N_Y \cdot (F - S\widehat{F}_S),$$

where the process $Y(t)$ has been replaced by the composite function definition $Y = F$ in (10.20), and

$$dV_P = N_Y \cdot \left(-\widehat{F}_\tau + \frac{1}{2}S^2\widehat{F}_{SS}\right) dt.$$

Finally, it is assumed that the portfolio will earn at the riskless rate, avoiding arbitrage profits without risk,

$$dV_P(t) = r V_p(t)dt, \tag{10.49}$$

which upon eliminating V_P and dV_P leads to the **Black–Scholes option pricing PDE,**

$$\widehat{F}_\tau = \frac{1}{2}\sigma_S^2 S^2 \widehat{F}_{SS} + r(S\widehat{F}_S - \widehat{F}), \qquad (10.50)$$

independent of N_Y as long as $N_Y \neq 0$ and, as typically written, no longer including the bond term as in Merton's version (10.45).

The Black–Scholes option pricing equation (10.45) is a parabolic or diffusion PDE in two asset values, S and B, but degenerate in B since there is no diffusion term in B and only a drift or mean rate term $rB\widehat{F}_B$.

Two elementary solutions of (10.45) can easily be verified:

- Only a stock asset: $\widehat{F}(S, B, \tau; T, K) = S$.

- Only a deterministic bond asset: $\widehat{F}(S, B, \tau; T, K) = B(T)\exp(-r\tau)$.

10.2.2   Final and Boundary Conditions for Option Pricing PDE

In the case of the European call option, the final option price, for any value S of $S(T)$, satisfies the final option profit conditions given in (10.1) for calls or (10.2) for puts, translated directly as

$$F(S(T), B(T), T; T, K) = \begin{cases} \max[S(T) - K, 0], & \text{call} \\ \max[K - S(T), 0], & \text{put} \end{cases}$$
$$= \max[\theta(S(T) - K), 0], \qquad (10.51)$$

where $\theta = 1$ for calls and $\theta = -1$ for puts. Since $S(T)$ and $B(T)$ are arbitrary but nonnegative, we can replace them by the independent variables S and B, respectively, to form the final condition for the PDE,

$$F(S, B, T; T, K) = \max[\theta(S - K), 0], \qquad (10.52)$$

but we will return to the original form (10.51) when transforming to new variables.

For the other boundary conditions, the discussion will be simplified to the riskfree bond case, i.e., $\sigma_B(t) = 0$, as assumed in the classical Black–Scholes case (10.50), except that the time-dependent interest/discount rate, $\mu_B(t) = r(t)$, will be retained. In the case of risky bonds, the boundary conditions are given by diffusion PDEs instead of explicit functions or values, so solving the PDE (10.44) by computational methods, as in Chapter 8 or in [109, 230, 264, 284], is more practical.

The number of boundary conditions depends on the highest order partial derivative for each independent state variable in the PDE, one condition if it is first order and two conditions if it is second order. Thus, for (10.44) it is two boundary conditions in the stock and one in the bond. Time is not a state variable, but there is one final condition (technically an initial condition for the backward time variable τ) since the time derivative is first order.

At the zero stock price, $S = 0$, Merton's Black–Scholes PDE (10.45) reduces to

$$\widehat{F}_\tau(0, B, \tau; T, K) = r(B\widehat{F}_B - \widehat{F}) \qquad (10.53)$$

upon setting S to zero in the coefficients, assuming the derivatives are bounded, which is a risky assumption before finding the solution. This is a first order PDE, all of which are classified as hyperbolic PDEs, and the usual method of constructing a solution is called the **method of characteristics** [251]. Noting that the PDE problem is a deterministic problem, the PDE (10.53) is compared to the deterministic (non-Itô!) chain rule for $\widetilde{F}(B, \tau) \equiv \widehat{F}(0, B, \tau; T, K)$,

$$d\widetilde{F} = \widetilde{F}_\tau d\tau + \widetilde{F}_B dB, \qquad (10.54)$$

assuming that the differentials $d\tau$ and dB can be varied independently, and the ODEs for the characteristic path are written maintaining relative proportions between the differentials of (10.54) and the corresponding coefficients of (10.53),

$$\frac{d\tau}{1} = -\frac{dB}{rB} = -\frac{d\widetilde{F}}{r\widetilde{F}}.$$

Solving these ODEs successively in pairs,

$$B = \widetilde{B}(\tau) = \kappa e^{-R(\tau)}, \qquad (10.55)$$

where κ is a characteristic path constant of integration and the cumulative rate for time-dependent $r(t)$ is

$$R(\tau) \equiv \int_0^\tau r(T - s)ds \equiv \int_0^\tau \widehat{r}(s)ds, \qquad (10.56)$$

and

$$\widetilde{F} = f(\kappa)e^{-R(\tau)},$$

where $f = f(\kappa)$ is an arbitrary function of integration depending on the constant κ from a prior integration. Using the first integral (10.55) to eliminate κ in favor of $\widetilde{B}$ and τ yields

$$\widetilde{F}(\widetilde{B}(\tau), \tau) = f\left(\widetilde{B}(\tau)e^{R(\tau)}\right)e^{-R(\tau)}. \qquad (10.57)$$

It is not necessary to know much about the method of characteristics, since the reader can verify the solution by the usual substitution procedure. The arbitrary function f can be eliminated by applying the final condition (10.52) at $\tau = 0$ with $R(0) = 0$,

$$\widetilde{F}(\widetilde{B}(0), 0) = f(\widetilde{B}(0)) = F(0, \widetilde{B}(0), T; T, K) = \max[\theta(-K), 0] = 0.5(1 - \theta)K.$$

Since $\widetilde{B}(0) = B(T)$ is considered arbitrary at this point, $f(\widetilde{B}) = 0.5(1 - \theta)K$, a constant (**beware**—Merton [201] assumes $B(T) = 1$), leading to the complete particular solution

$$\widetilde{F}(B, \tau) = \widehat{F}(0, B, \tau; T, K) = 0.5(1 - \theta)Ke^{-R(\tau)}, \qquad (10.58)$$

independent of $B = \widetilde{B}(\tau)$. Note that $\widetilde{B}(\tau)$ is a deterministic path function of a deterministic **ODE problem** since it is derived from the deterministic PDE problem, (10.53) plus conditions, so is different from the stochastic path function $\widehat{B}(\tau)$ for the SDE problem, or more precisely the stochastic ODE problem. The boundary condition (10.58) corresponds to a boundary condition used by Wilmott [284] for finite differences applied to Black–Scholes-type models.

However, since we cannot assume the partial derivatives are bounded for the full Merton model (10.44), we will assume only that the option price will be bounded in the limit of zero stock price:

$$\widehat{F}(S, B, \tau; T, K) \text{ is bounded as } S \to 0^+. \tag{10.59}$$

For large S, it is more difficult to find the proper boundary condition. However, one heuristic choice is to assume that for large S the diffusion term will be exponentially small so the drift terms will dominate:

$$\widehat{F}_\tau \simeq r(S\widehat{F}_S + B\widehat{F}_B - \widehat{F}). \tag{10.60}$$

As with the small stock price limit, the conjecture (10.60) needs to be verified for a solution. Again applying the method of characteristics to $\widetilde{F}(S, B, \tau) \equiv \widehat{F}(S, B, \tau; T, K)$, or checking by substitution, but with four variables, $\{\tau, B, S, \widehat{F}\}$, instead of three,

$$\frac{d\tau}{1} = -\frac{dB}{rB} = -\frac{dS}{rS} = -\frac{d\widehat{F}}{r\widehat{F}}.$$

Integration leads to three constants or functions of integration, two of which can be eliminated in favor of the independent variables S and B,

$$\widehat{F}(S, B, \tau; T, K) = g\left(Se^{R(\tau)}, Be^{R(\tau)}\right)e^{-R(\tau)}, \tag{10.61}$$

where $g = g(S\exp(R(\tau)), B\exp(R(\tau)))$ is an arbitrary function of integration obtained by integrating both the stock and bond characteristic ODEs effectively generating two constants of integration, and $R(\tau)$ is given in (10.56). Applying the final condition (10.52) when $S > K$ yields

$$\widehat{F}(S, B, 0; T, K) = \max[\theta(S - K), 0] = 0.5(1 + \theta)(S - K),$$

so that g is a constant function and the complete particular solution

$$\widehat{F}(S, B, \tau; T, K) \simeq 0.5(1 + \theta)(S - Ke^{-R(\tau)}). \tag{10.62}$$

A similar boundary condition is also specified in Wilmott's [284] finite difference applications. However, it turns out we will not need this condition here, but the condition suggests that the option price will not be bounded as $S \to +\infty$.

The bond boundary condition or conditions are not as straightforward, since the final bond price per share does not appear explicitly in the final option profit formula. At the zero bond price, $B = 0$, the Black–Scholes PDE (10.50) reduces to

$$\widehat{F}_\tau(S, 0, \tau; T, E) = r(S\widehat{F}_S - \widehat{F}) + \frac{1}{2}\sigma_S^2 S^2 \widehat{F}_{SS} \tag{10.63}$$

upon setting B to zero in the coefficients, assuming the derivatives are bounded. However, (10.63) is a diffusion equation rather than a boundary value, so there has been very little simplification of the original Black–Scholes PDE except that the dimension has been reduced to one from two state variables. This may still be useful for computational methods. The reduction in dimension is similar for the Merton version (10.44) of the Black–Scholes option pricing PDE, the only difference being that the drift term is absent. For either PDE, setting $B = B(T)$ in the PDE leads to no simplification since $B(T)$ would be arbitrary. There is still hope, since Merton has a way of transforming away $B(T)$ analytically, but this transformation is modified here.

10.2.3   Transforming PDE to Standard Diffusion PDE

Since the underlying stock and bound price models are linear stochastic diffusion equations, the expectation is that the distribution of the option price should be somehow related to the lognormal distribution studied in Chapter 4. However, here we have two state variables instead of one, so it will be useful to get rid of the bond $\widehat{B}$ dependence since the dependence is so weak that the bond does not appear in the final condition. For this purpose, it is noted that the dimensions of $\widehat{B}$, $\widehat{S}$, $\widehat{F}$, and K are all in the price of dollars per share. Thus, according to **Buckingham's pi theorem** [42] of **dimensional analysis**, the solution can be put into the form of intrinsic **dimensionless groups** collecting all powers (the pi groups) of variables and parameters in the problem to eliminate any extraneous scalings. Two such groups for independent and dependent variables that lead to a self-similar solution without B are

$$x = \frac{\widehat{B}(0)S}{K \cdot B},\tag{10.64}$$

$$G(x, \tau) = \frac{\widehat{B}(0)\widehat{F}(S, B, \tau)}{K \cdot B},\tag{10.65}$$

where the scale factor $K \cdot \widehat{B}/\widehat{B}(0)$ is equivalent to Merton's [201] if we set the final bond price $\widehat{B}(0) = 1$ dollar per share. Let $y(\tau) = K \cdot \widehat{B}(\tau)/\widehat{B}(0)$, then $y(0) = K$ is the final payoff at the exercise price and $dy(\tau) = Kd\widehat{B}(\tau)/\widehat{B}(0)$ is the change in the scale with respect to τ. See Wilmott [284] for more on the use of similarity transformations in the financial context.

   The partial derivatives of the proposed self-similar transformation to eliminate the bond explicitly are

$$\frac{\partial x}{\partial S} = \frac{x}{S}, \quad \frac{\partial x}{\partial B} = -\frac{x}{B},$$

$$S\widehat{F}_S = SG_x, \quad B\widehat{F}_B = -\frac{KB}{\widehat{B}(0)}(xG_x - G), \quad \widehat{F}_\tau = \frac{KB}{\widehat{B}(0)}G_\tau,$$

$$S^2\widehat{F}_{SS} = \frac{KB}{\widehat{B}(0)}x^2G_{xx},$$

$$SB\widehat{F}_{SB} = -\frac{KB}{\widehat{B}(0)}x^2G_{xx},$$

$$B^2\widehat{F}_{BB} = \frac{KB}{\widehat{B}(0)}x^2G_{xx}.$$

Upon substitution of the PDE of option pricing (10.44), a singular diffusion equation is obtained with variable coefficients,

$$G_\tau(x, \tau) = \frac{1}{2}\widehat{\sigma}^2(\tau)x^2 G_{xx}, \tag{10.66}$$

where

$$\widehat{\sigma}^2(\tau) = (\sigma_S^2 - 2\rho\sigma_S\sigma_B + \sigma_B^2)(T - \tau) \tag{10.67}$$

is a combined volatility term where all the volatilities on the right-hand side are evaluated at the common argument of $(T - \tau)$, confirming the validity of the conjectures self-similar solution transformation to transform away the bond variable B, subject to consistent boundary and initial conditions. The boundedness boundary condition (10.59) as $S \to 0^+$ is

$$G(0^+, \tau) \text{ is bounded}. \tag{10.68}$$

As $S \to +\infty$, the option boundary condition should also be bounded for a put, but $O(S)$ for a call is expected. For the final condition it is helpful to consult the original forward form

$$F(S(T), B(T), T; T, K) = \max[\theta(S(T) - K), 0],$$

leading to

$$G(x, 0^+) = \max[\theta(x - 1), 0], \tag{10.69}$$

where the factor $B(T) = \widehat{B}(0)$ washes out by our proper scaling or by Merton's unscaled dollar bond. This completely justifies the assumption of a self-similar transformation heuristically, since it works.

Note that the diffusion PDE (10.66) has a variable diffusion coefficient that is quadratic in x and vanishes as $x \to 0^+$, so that the PDE is called a singular diffusion. However, we still have not transformed the backward time variable τ and we have not used a logarithm transformation like the one we used in Chapter 4. To obtain a standard diffusion PDE, with coefficient $\frac{1}{2}$, let

$$u = u(\tau) = \int_0^\tau \widehat{\sigma}^2(s)ds, \tag{10.70}$$

$$w = w(x, \tau) = \ln(x) + \frac{1}{2}u(\tau), \tag{10.71}$$

$$G(x, \tau) = x\Phi(w(x, \tau), u(\tau)), \tag{10.72}$$

combining several of Merton's [201] transformations. The new time variable u is a diffusion time that helps eliminate the correlation coefficient and other terms. The inverse of the independent variable logarithmic transformation is given by $x = \exp(w - u/2)$ with the diffusion time correction. The new dependent variable

$$\Phi(w, u) = \frac{G(x, \tau)}{x} = \frac{\widehat{F}(S, B, \tau)}{S},$$

provided $S > 0$, is thus the dimensionless ratio of the option price $\widehat{F}$ to the stock price S, comprising another self-similar transformation, and that transformation is common to both $\widehat{F}$ and G. Applying this transformation, being easier than the first, the standard diffusion equation is obtained,

$$\Phi_u(w, u) = \frac{1}{2}\Phi_{ww}(w, u), \tag{10.73}$$

where $-\infty < w < +\infty$ and $0 = u(0) < u \leq u(T)$. The partial derivatives have the following meaning:

$$\Phi_u = \left(\frac{\partial\Phi}{\partial u}\right)_w \quad \text{and} \quad \Phi_{ww} = \left(\frac{\partial^2\Phi}{\partial w^2}\right)_u.$$

On the other hand, the side conditions are not so standard with the final condition (10.69) at $\tau = 0$ for G being transformed to

$$\Phi(w, 0^+) = e^{-w}\max[\theta(e^w - 1), 0] = \max[\theta(1 - e^{-w}), 0], \tag{10.74}$$

where the reader should confirm that this is correct in all cases, since it is generally not correct to bring a variable into a maximum argument. However, for the boundary condition a singular limit is avoided by keeping the x factor multiplying $\Phi(w, u)$, so

$$x(w, u)\Phi(w, u) \tag{10.75}$$

should be bounded as $w \to -\infty$ when $x \to 0^+$.

The solution of (10.73) can be written in terms of the complementary error function erfc or the normal distribution Φ_n, but they are related through several identities, two of which are in (B.20), (B.21). Merton [201], [203, Chapter 5] uses erfc, while Black and Scholes [34] use the standard normal distribution, which in our notation is $\Phi_n(x; 0, 1)$. The simplest fundamental solution

$$\Phi_1(w, u) \equiv \Phi_n(w; 0, u) \tag{10.76}$$

of (10.73) can be derived using Fourier transform methods [103, Chapter 9] or can be derived using the self-similar solution technique used here earlier to remove the bond dependence. See also the introduction to the diffusion equation (B.26) in Online Appendix B. However, it may be much easier to verify

$$\Phi_{1,u}(w, u) = \frac{1}{2}\Phi_{1,ww}(w, u) \tag{10.77}$$

using a symbolic computation system such as Maple or Mathematica. The simple diffusion solution Φ_1 in (10.76) is just too simple and does not satisfy the final condition (10.74) at $u = 0^+$ which can be written in terms of either the standard unit step function $H(x)$ in (B.156) or the averaged unit step function $H_a(x)$ in (B.157)

$$\Phi(w, 0^+) = \theta(1 - e^{-w})H(\theta w) = \theta(1 - e^{-w})H_a(\theta w). \tag{10.78}$$

Since either step function will do, the coefficient vanishes at $w = 0$, but instead the simple solution Φ_1 satisfies the final condition

$$\Phi_1(w, 0^+) = H_a(w), \tag{10.79}$$

as the reader can verify by examining the cases $w > 0$, $w = 0$, and $w < 0$ as $u \to 0^+$.

Thus, another solution is needed to provide the extra variable factor e^{-w}. Specializing to the call option when $\theta = +1$, the second solution is

$$\Phi_2(w, u) \equiv e^{-w+u/2}\Phi_n(w; u, u), \tag{10.80}$$

which can be shown to satisfy the standard diffusion equation (10.73) and a different final condition

$$\Phi_2(w, 0^+) = e^{-w}H_a(w) \tag{10.81}$$

than that of Φ_1 in (10.79). The boundedness condition (10.75) is trivial as $w \to -\infty$ since both Φ_1 and $e^{w-u/2}\Phi_2$ vanish by the definition of Φ_n with $u > 0$. Thus, the transformed solution for the call option price is

$$\begin{aligned}\Phi^{(call)}(w, u) &= \Phi_1(w, u) - \Phi_2(w, u) \\ &\equiv \Phi_n(w; 0, u) - e^{-w+u/2}\Phi_n(w; u, u) \end{aligned} \tag{10.82}$$

$$= \Phi_n\left(\frac{w}{\sqrt{u}}; 0, 1\right) - e^{-w+u/2}\Phi_n\left(\frac{w-u}{\sqrt{u}}; 0, 1\right) \tag{10.83}$$

upon transforming to standard normal distributions. Thus, $\Phi^{(call)}$ satisfies the final condition

$$\Phi^{(call)}(w, 0^+) = (1 - e^{-w})H_a(w). \tag{10.84}$$

The solution form resembles solutions of the diffusion equation on a semi-infinite domain found by the classical method of reflection. Transforming back to the original variables, one can compare this to the original Black–Scholes form with $\mu_B = r$ and $\sigma_B^2 = 0$, so $\widehat{B}(\tau) = \widehat{B}(0)\exp(-r\tau)$ and $u = \sigma_S^2\tau$. The correlation term with ρ (10.18) vanishes with σ_B.

The European put option price solution ($\theta = -1$) is somewhat different, relying on normal distributions complementary to those of the European call option with two component solutions,

$$\begin{aligned}\Phi^{(put)}(w, u) &= \Phi_3(w, u) - \Phi_4(w, u) \\ &\equiv e^{-w+u/2}(1 - \Phi_n(w; u, u)) - (1 - \Phi_n(w; 0, u)) \end{aligned} \tag{10.85}$$

$$= e^{-w+u/2}\left(1 - \Phi_n\left(\frac{w-u}{\sqrt{u}}; 0, 1\right)\right) - \left(1 - \Phi_n\left(\frac{w}{\sqrt{u}}; 0, 1\right)\right), \tag{10.86}$$

where again the final form is in terms of standard normal distributions. The reader can verify that $\Phi^{(put)}(w, u)$ satisfies the standard diffusion equation (10.73) and the put option price final condition,

$$\Phi^{(put)}(w, 0^+) = (e^{-w} - 1)H_a(-w). \tag{10.87}$$

Note that the zero stock limit boundedness condition means that $e^{w-u/2}\Phi^{(put)}(w, u)$ is bounded as $w \to -\infty$, not zero as in the call case. Maple and Mathematica are the preferred tools to use. The put and call option prices are related in a general way according to the principle of **put-call parity**, i.e., in transformed variables,

$$\Phi^{(put)}(w, u) - \Phi^{(call)}(w, u) = \exp(-w + u/2) - 1.$$

See also Exercise 6 on p. 336.

The boundary condition limits of the solutions essentially follow from the corresponding extreme limits of the normal distribution function,

$$\Phi_n(w; \mu, \sigma^2) \to \begin{cases} 0, & w \to -\infty \\ 1, & w \to +\infty \end{cases}, \tag{10.88}$$

except in one case. Thus, for the intermediate transformed call option price multiplied by the transformed stock option $x\Phi^{(call)}$ using (10.82) for $\Phi^{(call)}(w, u)$ satisfies the limiting conditions

$$x\Phi^{(call)}(w, u) \to \begin{cases} 0, & w \to -\infty \ \& \ x \to 0^+ \\ x - 1, & w \to +\infty \ \& \ x \to +\infty \end{cases}, \tag{10.89}$$

consistent with the derived limits (10.58) for Black–Scholes call and put option pricing in the boundary conditions subsection. The put option price is formulated in terms of the complementary normal probability distribution, $1 - \Phi_n(w; \mu, \sigma^2)$, which vanishes exponentially as $w \to +\infty$ and $x \to +\infty$, and so results in an indeterminate form, $\infty \cdot 0$, for $x\Phi^{(put)}$. However, this form can be resolved using **l'Hospital's rule** and the fact that $x = \exp(w - u/2)$,

$$x\left(1 - \Phi_n(w; \mu, \sigma^2)\right) = \frac{1 - \Phi_n(w; \mu, \sigma^2)}{e^{w|u/2}} \to \frac{1}{\sqrt{2\pi\sigma^2}} e^{-(w-m)^2/(2\sigma^2)+w} \to 0,$$

since the larger degree monomial in the exponent dominates the smaller one. Finally, the put option price extreme conditions are

$$x\Phi^{(put)}(w, u) \to \begin{cases} 1 - x, & w \to -\infty \ \& \ x \to 0^+ \\ 0, & w \to +\infty \ \& \ x \to +\infty \end{cases}, \tag{10.90}$$

again consistent with prior derived limits (10.62) for Black–Scholes call and put option pricing. Note that the extreme boundary conditions strongly reflect the final condition.

Reversing the transformations used to convert the answers $\Phi^{(call)}(w, u)$ (10.82)–(10.83) for the call option price and $\Phi^{(put)}(w, u)$ (10.85)–(10.86) for the put option price back to the actual option price $Y(t) = F(S(t), B(t), t; T; K)$ is left as Exercise 5 at the end of this chapter.

While the put option pricing results are not in Merton's continuous returns paper [201], [203, Chapter 8], there are many other results and more exploration with the removal of assumptions, such as the no-dividends exclusion. In his companion discontinuous returns paper [202], [203, Chapter 9], Merton presents one of the first treatments of jump-diffusions in finance.

10.3 Jump-Diffusion Option Pricing

Since the 1973 Black–Scholes–Merton option pricing model is based upon the pure-diffusion stochastic model, there is one obvious missing feature that for large market fluctuations or jumps such as crashes or rallies which characterize extreme market psychology are not represented. There are several papers on the statistical importance of including jumps in financial market models, e.g., see Ball and Torous [18] on stocks and options, Jarrow and Rosenfeld [154] on the capital asset pricing model (CAPM) or Jorion [156] on foreign exchange and stocks.

There are other qualitative features that characterize real market log-return distributions that cannot be reproduced by the pure-diffusion model of Black–Scholes–Merton but can be modeled, in part, by adding jumps to the diffusion process. One feature is that real markets have negatively skewed log-return distributions, provided a sufficient number of years of daily return data is used [128], so that the log-return skewness coefficient (B.11),

$$\eta_3[X] \equiv \frac{\mathrm{E}[(X - \mathrm{E}[X])^3]}{(\mathrm{Var}[X])^{3/2}} < 0,$$

where

$$X = \Delta \ln(S(t_i)) = \ln(S(t_{i+1})) - \ln(S(t_i))$$

is the log-return for trading day t_{i+1} for $i = 1 : n_s - 1$ trading days, while $\eta_3[X] = 0$ for the intrinsically skewless normally distributed log-return model on the pure-diffusion process. Hence, real markets in the long run are found to be pessimistic due to more negative log-returns, including crashes, than positive log-returns.

Another feature is that real market distributions are found to be leptokurtic so that the log-return kurtosis coefficient (B.12),

$$\eta_4[X] \equiv \frac{\mathrm{E}[(X - \mathrm{E}[X])^4]}{(\mathrm{Var}[X])^2} > 3,$$

for $X = \Delta \ln(S(t_i)) = \ln(S(t_{i+1})) - \ln(S(t_i))$, whereas the normally distributed pure-diffusion process is **mesokurtic** (also said to have zero excess kurtosis, $\eta_4[X] - 3$) since $\eta_4[X] = 3$. Leptokurtic means that the distribution is more peaked (kurtic is derived from the word for crown) at the maximum and consequently has fatter tails than the normal distribution.

Still another characteristic is the **volatility smile** which refers to the curvature of the implied volatility, volatility implied by the log-normal Black–Scholes formula, versus the strike price. For more information on volatility smiles and their relation to non-lognormal distributions which they signify, see, e.g., Hull [148].

Merton [202] in 1976 pioneered the analysis of option pricing for stock returns governed by a jump-diffusion model. Merton chose the normal distribution for the jump-amplitude distribution for the log-return. Here, the option pricing with jump-diffusions is described in terms of the jump-diffusion formulations in this book. The stock price is assumed to be subject to extreme changes over a very short period due to significant changes in the firm or in the market. Details can be found in Zhu [291] and Zhu and Hanson [292]. Thus, consider the jump-diffusion model for the stock price $S(t)$ at time t,

$$dS(t) = S(t)(\mu_d dt + \sigma_d dW(t) + J(Q)dP(t)), \quad S(0) = S_0 > 0, \qquad (10.91)$$

where μ_d and σ_d are designated as the diffusion parameters for the standard diffusion $dW(t)$. Here $J(Q) = \exp(Q) - 1 > -1$ is the jump-amplitude for the jumps of Poisson process $dP(t)$, such that the symbolic jump form means

$$J(Q)dP(t) = \sum_{k=P(t)+1}^{P(t)+dP(t)} J(Q_k)$$

for integers $k \geq 1$, and otherwise the sum is zero if $k = P(t)$, i.e., $dP(t) = 0$. The marks Q_k are IID normally distributed. Note, unlike Merton in [202], in (10.91) there are not the same notational shifts in the diffusion drift and jump-amplitude, so that $E[dS(t)/S(t)] = \mu_d dt + \lambda dt E[J(Q)]$ and Merton's $Y - 1$ is the same as $J(Q)$ while $\alpha = \mu_d$.

By the stochastic chain rule the log-return satisfies

$$d\ln(S(t)) = \mu_{ld}dt + \sigma_d dW(t) + \sum_{k=P(t)+1}^{P(t)+dP(t)} Q_k \qquad (10.92)$$

when $\mu_{ld} = \mu_d - 0.5\sigma_d^2$ is the diffusion-corrected mean appreciation coefficient. Under the assumption of constant coefficients, the solution of (10.92) is immediate:

$$S(t) = S_0 \exp\left(\mu_{ld}t + \sigma_d W(t) + \sum_{k=1}^{P(t)} Q_k\right). \qquad (10.93)$$

The solution is positive as long as $S_0 > 0$ and Q_k is assumed real, a consequence of the **geometric jump-diffusion** assumptions.

10.3.1   Jump Diffusions with Normal Jump-Amplitudes

Since the marks Q_k are IID normally, the mark density is defined in our notation as

$$\phi_Q(q) = \phi_n(q; \mu_j, \sigma_j^2), \qquad (10.94)$$

where ϕ_n denotes a normal density with mean $\mu_j = E[Q]$ and variance $\sigma_j^2 = \text{Var}[Q]$.

Let the discrete version

$$\Delta\ln(S(t)) = \mu_{ld}\Delta t + \sigma_d \Delta W(t) + \sum_{k=P(t)+1}^{P(t)+\Delta P(t)} Q_k$$

$$= \mu_{ld}\Delta t + \sigma_d \Delta W(t) + \mu_j(\Delta P(t) - \lambda\Delta t) + \sum_{k=P(t)+1}^{P(t)+\Delta P(t)} (Q_k - \mu_j) \qquad (10.95)$$

of the log-return SDE (10.92) approximate the log-return difference,

$$\Delta\ln(S(t)) \equiv \ln(S(t + \Delta t)) - \ln(S(t)).$$

The last line of (10.95) has the stochastic terms collected into independent and zero mean forms to facilitate moments calculations. The standard moments (mean plus central moments for higher moments) can be calculated (see Theorem 5.17 on p. 149, [132], and [291]) using (10.95) rather than the solution (10.93). Thus,

$$M_1 \equiv \mathrm{E}[\Delta \ln(S(t))] = (\mu_{ld} + \lambda \mu_j)\Delta t,$$
$$M_2 \equiv \mathrm{Var}[\Delta \ln(S(t))] = (\sigma_d^2 + \lambda(\mu_j^2 + \sigma_j^2))\Delta t,$$
$$M_3 \equiv \mathrm{E}[(\Delta \ln(S(t)) - M_1)^3] = \lambda \mu_j(\mu_j^2 + 3\sigma_j^2)\Delta t,$$
$$M_4 \equiv \mathrm{E}[(\Delta \ln(S(t)) - M_1)^4] = \lambda(\mu_j^4 + 6\mu_j^2\sigma^2 + 3\sigma_j^4)\Delta t + 3(\sigma_d^2 + \lambda(\mu_j^2 + \sigma_j^2))^2(\Delta t^2).$$

The variance normalized third moment is the skewness coefficient,

$$\eta_3[\Delta \ln(S(t))] = \frac{\lambda \mu_j(\mu_j^2 + 3\sigma_j^2)}{(\sigma_d^2 + \lambda(\mu_j^2 + \sigma_j^2))^{3/2}(\Delta t)^{1/2}}, \qquad (10.96)$$

so $\eta_3[\Delta \ln(S(t))] < 0$ if the lognormal jump-amplitude mean $\mu_j < 0$, since the jump-rate λ must be positive for there to be jumps. The variance normalized fourth moment is the kurtosis coefficient,

$$\eta_4[\Delta \ln(S(t))] = \frac{\lambda(\mu_j^4 + 6\mu_j^2\sigma_j^2 + 3\sigma_j^4)}{(\sigma_d^2 + \lambda(\mu_j^2 + \sigma_j^2))^2 \Delta t} + 3, \qquad (10.97)$$

so $\eta_4[\Delta \ln(S(t))] > 3$ provided $\mu_j \neq 0$ and $\sigma_j \neq 0$. Thus, the jump-diffusion with lognormally distributed jump-amplitudes provides more realistic log-return distribution properties with skewness whose direction depends on the sign of the mark mean μ_j and leptokurtosis for nontrivial mark distributions.

Another advantage, particularly in analysis, follows from the convolution result that the sum of normals is normally distributed. This is expressed in Corollary 5.21 on p. 154, so for the jump-diffusion with lognormally distributed jump-amplitude, the density with a small modification for the difference and constant coefficients is given as an infinite sum of translated normal densities over all Poisson jumps by

$$\phi_{\Delta \ln(S(t))}(x) = \sum_{k=1}^{\infty} p_k(\lambda \Delta t)\phi_n(x; \mu_{ld}\Delta t + k\mu_j, \sigma_d^2 \Delta t + k\sigma_j^2), \qquad (10.98)$$

where $p_k(\lambda \Delta t)$ is the Poisson distribution (B.50) with parameter $\lambda \Delta t$, and $\phi_n(x; \mu, \sigma^2)$ denotes the normal density with general parameters μ and σ^2.

10.3.2 Risk-Neutral Option Pricing for Jump-Diffusions

Rather than follow Merton's 1976 [202] paper to directly explain his approach using the PDE formulation of the previous section, we will approach option pricing in the presence of both diffusion and jumps by directly applying a **risk-neutral** assumption that the discounted earnings on a European call option is at the existing market rate r, i.e., the risk-neutral call option price has the form

$$C^{\mathrm{rn}}(S_0, T) \equiv e^{-rT}\mathrm{E}^{\mathrm{rn}}[\max[S(T) - K, 0]], \qquad (10.99)$$

where T is the option exercise time, K is the strike price, $\exp(-rT)$ is the discount factor, and E^{rn} denotes the risk-neutral expected value [148, pp. 248–250], depending on the initial asset price S_0 as well.

As Merton points out, the classical Black–Scholes hedge or the **delta hedge** (10.7) is no longer sufficient to eliminate all risk when there are jumps in the underlying asset price that result in nonmarginal changes. This could come from nonsystematic information about the firm to cause extreme changes in value. There are special cases that are of little interest and there is always the possibility of using the Black–Scholes hedge to eliminate the diffusive **volatility-risk** during the quiet period between jumps, but when a jump event arrives there is the possibility of a large loss or other unexpected change in value of the option, i.e., the so-called **jump-risk**, that will not be covered. In short, there are too many random variables in a jump-diffusion to **delta hedge** away with a single stock. For instance, in the compound Poisson process there is the pure counting part of the process and then there is the uncountable IID log-jump-amplitudes or marks Q_k that would need to be hedged.

Letting

$$\mu_J \equiv \mathrm{E}[J(Q)] = \int_{-\infty}^{+\infty} \phi_n(q)\left(e^q - 1\right) dq = e^{\mu_j + 0.5\sigma_j^2} - 1 \qquad (10.100)$$

be the mean jump-amplitude of the asset price, then the mean asset price at the strike time T using iterated expectations on the closed form solution (10.93) is

$$\mathrm{E}[S(T)] = S_0 e^{(\mu_d - 0.5\sigma_d^2)T} \mathrm{E}\left[e^{\sigma_d W(T)} e^{\sum_{k=1}^{N(T)} Q_k}\right]$$

$$= S_0 e^{(\mu_d - 0.5\sigma_d^2)T} \mathrm{E}_{W(T)}\left[e^{\sigma_d W(T)}\right] \mathrm{E}_{N(T)}\left[\prod_{k=0}^{N(T)} \mathrm{E}_{Q_k|N(T)}\left[e^{Q_k} \big| N(T)\right]\right]$$

$$= S_0 e^{(\mu_d - 0.5\sigma_d^2)T} e^{0.5\sigma_d^2 T} \mathrm{E}_{N(T)}\left[\prod_{k=1}^{N(T)} (\mu_J + 1)\right]$$

$$= S_0 e^{\mu_d T} \mathrm{E}_{N(T)}\left[(\mu_J + 1)^{N(t)}\right] = S_0 e^{\mu_d T} \sum_{k=1}^{\infty} p_k(\lambda T)(\mu_J + 1)^k$$

$$= S_0 e^{\mu_d T} e^{-\lambda T} \sum_{k=1}^{\infty} (\lambda T(\mu_J + 1))^k / k! = S_0 e^{\mu_d T + \lambda T \mu_J}$$

$$= S_0 e^{(\mu_d + \lambda \mu_J)T}, \qquad (10.101)$$

where the IID property of the Q_k and the Poisson distribution $p_k(\lambda T)$ (B.50) have also been used.

In the risk-neutral world (see Hull [148, pp. 248–250]),

$$\mathrm{E}[S(T)] = S_0 e^{(\mu_d + \lambda \mu_J)T} = S_0 e^{rT},$$

so the jump-diffusion rate in a risk-neutral world must be

$$\mu_d + \lambda \mu_J = r,$$

the sum of the diffusive and jump mean rates. For consistency with the benchmark Black–Scholes model, this relation will be used to eliminate the diffusive mean rate in a risk-neutral world

$$\mu_d = \mu_d^{\text{rn}} \equiv r - \lambda \mu_J, \tag{10.102}$$

allowing the following formulation of the risk-neutral option.

Definition 10.1. *Jump-Diffusion Risk-Neutral European Call Option.*
Applying the general jump-diffusion solution (10.93) with (10.102) to the risk-neutral European call option payoff (10.99) using the general jump partial sum random variable

$$\widehat{S}_k = \sum_{i=1}^{k} Q_i$$

with density $\phi_{\widehat{S}_k}(s_k)$ yields the form

$$
\begin{aligned}
\mathcal{C}^{\text{rn}}(S_0, T) &\equiv e^{-rT} \mathrm{E}^{\text{rn}}[\max[S(T) - K, 0]] \\
&\equiv e^{-rT} \mathrm{E}\left[\max\left[S_0 e^{(r - \lambda \mu_J - \sigma_d^2/2)T + \sigma_d W(T) + \sum_{k=1}^{P(T)}} - K, 0 \right] \right] \\
&= e^{-rT} \sum_{k=0}^{\infty} p_k(\lambda T) \int_{-\infty}^{+\infty} dw \, \phi_n(w; 0, T) \int_{-\infty}^{+\infty} ds_k \phi_{\widehat{S}_k}(s_k) \\
&\quad \cdot \max\left[S_0 e^{(r - \lambda \mu_J - \sigma_d^2/2)T + \sigma_d w + s_k} - K, 0 \right] \\
&= e^{-rT} \sum_{k=0}^{\infty} p_k(\lambda T) \int_{-\infty}^{+\infty} dw \, \phi_n(w; 0, T) \\
&\quad \cdot \mathrm{E}_{\widehat{S}_k}\left[\max\left[S_0 e^{(r - \lambda \mu_J - \sigma_d^2/2)T + \sigma_d w + \widehat{S}_k} - K, 0 \right] \right].
\end{aligned}
\tag{10.103}
$$

Remark 10.2. *The random sum $\widehat{S}_k$ is used here, rather that the mark Q_i as in (10.101), since the maximum function in (10.103) needs a different splitting of the expectations.*

Theorem 10.3. *Risk-Neutral Call Prices as an Infinite Poisson Sum of Shifted Black–Scholes Call Prices: General Jump-Diffusion Case.*
For the general jump-diffusion,

$$\mathcal{C}^{\text{rn}}(S_0, T) = \sum_{k=0}^{\infty} p_k(\lambda T) \mathrm{E}_{\widehat{S}_k}\left[\mathcal{C}^{(\text{bs})}\left(S_0 e^{\widehat{S}_k - \lambda \mu_J T}, T; K, \sigma_d^2, r \right) \right], \tag{10.104}$$

where the Black–Scholes call price function

$$\mathcal{C}^{(\text{bs})}(s, T; K, \sigma_d^2, r) = s \Phi(d_1(s)) - K e^{-rT} \Phi(d_2(s)), \tag{10.105}$$

$$\Phi(x) \equiv \Phi_n(x; 0, 1) = \frac{1}{\sqrt{2\pi}} \int_{-\infty}^{x} e^{-y^2/2} dy \tag{10.106}$$

is the standardized normal distribution,

$$d_1(s) = \left(\ln(s/K) + (r + \sigma_d^2/2)T\right)/(\sigma_d\sqrt{T}) \text{ and } d_2(s) = d_1(s) - \sigma_d\sqrt{T} \quad (10.107)$$

are Black–Scholes normal argument functions.

Proof. Note that the argument of the last maximum in (10.103) has a root at $w = w_0(s_k)$ when

$$S_0 e^{(r-\lambda\mu_J-\sigma_d^2/2)T+\sigma_d w+s_k} = K$$

or when

$$w_0(s_k) = (\ln(K/S_0) - (r - \lambda\mu_J - \sigma_d^2/2)T + s_k)/\sigma_d, \quad (10.108)$$

allowing the removal of the maximum function. Some further manipulations with the normal integrals permits the transformation to an infinite Poisson sum over Black–Scholes call functions with shifted arguments,

$$C^{rn}(S_0, T) = e^{-rT} \sum_{k=0}^{\infty} p_k(\lambda T)\mathrm{E}_{\widehat{S}_k}\left[\int_{w_0(\widehat{S}_k)}^{+\infty} dw \phi_n(w; 0, T)\right.$$

$$\left. \cdot \left(S_0 e^{(r-\lambda\mu_J-\sigma_d^2/2)T+\sigma_d w+\widehat{S}_k} - K\right)\right],$$

$$= \sum_{k=0}^{\infty} p_k(\lambda T)\mathrm{E}_{\widehat{S}_k}\left[S_0 e^{\widehat{S}_k-\lambda\mu_J T} A(S_0 e^{\widehat{S}_k-\lambda\mu_J T}) - K e^{-rT} B(S_0 e^{\widehat{S}_k-\lambda\mu_J T})\right],$$

where the intermediate functions $A(s)$ and $B(s)$ are derived as

$$A(s) = e^{-\sigma_d^2 T/2} \int_{w_0(s)}^{\infty} dw\, \phi_n(w; 0, T)e^{\sigma_d w}$$

$$= e^{-\sigma_d^2 T/2} \frac{1}{\sqrt{2\pi T}} \int_{w_0(s)}^{\infty} dw\, e^{-w^2/(2T)+\sigma_d w}$$

$$= \frac{1}{\sqrt{2\pi T}} \int_{w_0(s)}^{\infty} dw\, e^{-(w-\sigma_d T)^2/(2T)}$$

$$= \frac{1}{\sqrt{2\pi}} \int_{(w_0(s)-\sigma_d T)/\sqrt{T}}^{\infty} dy\, e^{-y^2/2}$$

$$= \left(1 - \Phi\left((w_0(s) - \sigma_d T)/\sqrt{T}\right)\right)$$

$$= \Phi\left((\sigma_d T - w_0(s))/\sqrt{T}\right)$$

$$= \Phi\left(d_1\left(S_0 e^{s-\lambda\mu_J T}\right)\right),$$

since by (10.108) and (10.107) $(\sigma_d T - w_0(s))/\sqrt{T} = d_1(S_0 e^{s-\lambda\mu_J T})$. The simpler second argument quickly follows from

$$B(s) = \int_{w_0(s)}^{\infty} dw\, \phi_n(w; (0, T) = \Phi\left(-w_0(s)/\sqrt{T}\right)$$

$$= \Phi\left(d_1\left(S_0 e^{s-\lambda\mu_J T}\right) + \sigma_d\sqrt{T}\right) = \Phi\left(d_2\left(S_0 e^{s-\lambda\mu_J T}\right)\right)$$

using (10.108) and (10.107) again. Reassembling $A(s)$ and $B(s)$ from the current equation for C^{rn} yields (10.105) from the relation

$$C^{(\text{bs})}(s, T; K, \sigma_d^2, r) = s A(s) - K e^{-rT} B(s),$$

and thus (10.104) follows.   □

Remark 10.4. *The primary argument s of $C^{(\text{bs})}$ is shifted for each jump number k by a factor $\exp(\widehat{S}_k - \lambda \mu_J T)$ that depends only on the jump process. (The result in this form is valid for general jump-diffusions as treated in this book.)*

If the mark density is normal, $\phi_Q(q) = \phi_n(q; \mu_j, \sigma_j^2)$, then the European call option formula can be simplified.

Theorem 10.5. *Risk-Neutral Call Prices as an Infinite Poisson Sum of Shifted Black–Scholes Call Prices: Lognormal-Jump-Amplitude Jump-Diffusion Case.*
For the lognormal-jump-amplitude jump-diffusion,

$$C_n^{\text{rn}}(S_0, T) = \sum_{k=0}^{\infty} p_k(\lambda T) C_n^{(\text{bs})}\left(S_0 e^{k(\mu_j + \sigma_j^2/2) - \lambda \mu_J T}, T; K, \sigma_k^2(T)/T, r \right), \quad (10.109)$$

where the Black–Sholes call price function

$$C_n^{(\text{bs})}(s, T; K, \sigma_k^2(T)/T, r) = s \Phi\big(\widehat{d}_1\left(s; \sigma_k^2(T)\right)\big) - K e^{-rT} \Phi\big(\widehat{d}_2\left(s, \sigma_k^2(T)\right)\big), \quad (10.110)$$

$$\widehat{d}_1(s; \sigma_k^2(T)) = \left(\ln(s/K) + rT + \sigma_k^2(T)/2\right)/\sigma_k(T), \quad (10.111)$$

$$\widehat{d}_2(s; \sigma_k^2(T)) = d_1(s) - \sigma_k(T)$$

are Black–Scholes normal argument functions, and

$$\sigma_k^2(T) = \sigma_d^2 T + k \sigma_j^2 \quad (10.112)$$

is the log-return variance.

Proof. To simplify the expectation calculations, let

$$X = \sigma_d W(T) + \left(\widehat{S}_k - k \mu_j\right)$$

be the zero mean part of the risk-neutral log-return process obtained by subtracting the mean

$$\mu_k(T) = (r - \lambda_J - \sigma_d^2/2)T + k\mu_j \quad (10.113)$$

and leaving the variance (10.112),

$$\sigma_k^2(T) = \sigma_d^2 T + k \sigma_j^2,$$

so by the normal convolution corollary (10.98) the density is reduced to

$$\phi_X(x) = \phi_n(x; 0, \sigma_k^2(T)). \quad (10.114)$$

The payoff cutoff to remove the maximum function in the normal case then is

$$x_k(T) = \ln(K/S_0) - \mu_k(T).$$

Thus, the normal risk-neutral call price is derived using normal integral identities by

$$
\begin{aligned}
\mathcal{C}_n^{\mathrm{rn}}(S_0, T) &= e^{-rT} \sum_{k=0}^{\infty} p_k(\lambda T) \int_{-\infty}^{+\infty} dx \, \phi_n(x; 0, \sigma_k^2(T)) \max\left[S_0 e^{\mu_k(T)+x} - K, 0 \right] \\
&= e^{-rT} \sum_{k=0}^{\infty} p_k(\lambda T) \int_{x_k(T)}^{+\infty} dx \, \phi_n(x; 0, \sigma_k^2(T)) \left(S_0 e^{\mu_k(T)+x} - K \right) \\
&= e^{-rT} \sum_{k=0}^{\infty} p_k(\lambda T) \frac{1}{\sqrt{2\pi \sigma_k^2(T)}} \\
&\quad \cdot \left(S_0 e^{\mu_k(T)+\sigma_k^2(T)/2} \int_{x_k(T)}^{+\infty} dx \, e^{-(x-\sigma_k^2(T))^2/(2\sigma_k^2(T))} \right. \\
&\qquad \left. - K \int_{x_k(T)}^{+\infty} dx \, e^{-x^2/(2\sigma_k^2(T))} \right) \\
&= \sum_{k=0}^{\infty} p_k(\lambda T) \left(S_0 e^{\mu_k(T)-rT+\sigma_k^2(T)/2} \Phi\left(\frac{\sigma_k^2(T) - x_k(T)}{\sigma_k(T)} \right) \right. \\
&\qquad \left. - Ke^{-rT} \Phi\left(\frac{-x_k(T)}{\sigma_k(T)} \right) \right) \\
&= \sum_{k=0}^{\infty} p_k(\lambda T) \left(S_0 e^{k(\mu_j+\sigma_j^2/2)-\lambda \mu_J T} \Phi\left(\widehat{d}_1 \left(S_0 e^{k(\mu_j+\sigma_j^2/2)-\lambda \mu_J T}; \sigma_k^2(T) \right) \right) \right. \\
&\qquad \left. - Ke^{-rT} \Phi\left(\widehat{d}_2 \left(S_0 e^{k(\mu_j+\sigma_j^2/2)-\lambda \mu_J T}, \upsilon_k^2(T) \right) \right) \right)
\end{aligned}
$$

finally by using (10.111) with (10.113) and (10.112).

Note that by several IID and normal identities,

$$\mathrm{E}[e^{\widehat{S}_k}] = \mathrm{E}[e^{\sum_{i=1}^{k} Q_i}] = \prod_{i=1}^{k} \mathrm{E}[e^{Q_i}] = \prod_{i=1}^{k} e^{\mu_j+\sigma_j^2/2} = e^{k(\mu_j+\sigma_j^2/2)},$$

giving the meaning of this exponential term in (10.109) for the final normal jump-diffusion call option result $\mathcal{C}_n^{\mathrm{rn}}(S_0, T)$. $\square$

Option pricing for other jump-diffusions cannot be written in as simple a form and the Poisson terms increase in complexity exponentially. The use of the double-exponential (Laplace) log-jump-amplitude jump-diffusion was developed by Kou [170] and and Kou and Wang [171]. Zhu and Hanson [292] have developed a Monte Carlo estimation of risk-neutral option pricing for uniform log-jump-amplitude jump-diffusions. Zhu [291] has made a comprehensive study and comparison of various exponential and uniform log-jump-amplitude jump-diffusions using refined Monte Carlo estimations of option prices with several variance reduction techniques. Recently, Yan and Hanson [289] have treated option pricing for the uniform log-jump-amplitude jump-diffusion combined with stochastic volatility (SVJD) using characteristic functions and fast Fourier transforms following the general methodology of Carr and Madan [47]. In a companion paper [131], Hanson and Yan computationally solve the SVJD problem using a systematic finite difference formulation of the free-boundary American put partial integrodifferential complementary problem (PIDCP), implemented using a successive overrelaxation (SOR) method projected on the maximum payoff function. For more details see the thesis of Yan [288].

Some other hedging methods for jump-diffusions, like mean-variance hedging, are treated in a more abstract way by Runggaldier [239], Bingham and Kiesel [33], and Cont and Tankov [60] using a generalization of jump-diffusions allowing infinite jump-rates called Lévy processes (see Chapter 12 in this book).

10.4 Optimal Portfolio and Consumption Models

Prior to Merton's 1973 mathematical justification and generalization of the Black–Scholes model [34] in [201], he did pioneering work on the portfolio and consumption problem in continuous time. Beginning in 1969, Merton's paper [198] (see [203, Chapter 4]) on lifetime portfolio selection with constant relative risk-aversion (CRRA) utilities laid out the background for the widely cited 1971 paper [199, 200] (reprinted in [203, Chapter 5]) on the optimal portfolio and consumption theory with the more general hyperbolic absolute risk-aversion (HARA) utilities that exhibit explicit solutions. While the paper was primarily on geometric Brownian motion (pure-diffusion), generalization to jump-diffusions consisting of Brownian motion and compound Poisson processes with general random finite amplitude is discussed very briefly in [199].

While Merton was often on the leading edge of continuous-time finance and pushing generality of financial models by incorporating the latest financial and stochastic theories, one can get cut on the leading edge. There are a number of errors in the 1971 Merton paper [199, 200] due to the lack of proper boundary conditions and problems with the general HARA utilities. In particular, there are difficulties due to enforcing nonnegative wealth, handing zero wealth (bankruptcy), and maintaining the nonnegativity of consumption. These errors are very thoroughly discussed in Sethi's [245] massive assembly of papers that give corrections and generalizations to the consumption and investment portfolios with an emphasis on bankruptcy and pure-diffusion. The basic problems are clearly discussed in Sethi's introduction [245, Chapter 1], while important basic papers are the paper of Karatzas et al. [160] (reprinted in [245, Chapter 2]) on exact solutions in the infinite horizon case and the paper of Sethi and Taksar [246] (reprinted in [245, Chapter 3]) pinpointing the errors in Merton's 1971 paper [199, 200]. The errors were mainly in certain ranges of the HARA utilities, and these difficulties led to a more thorough exploration of the consumption and portfolio problem.

In this section, the jump-diffusion version for the consumption and portfolio problem is treated with a version of the CRRA utilities that avoids the problematic parameter range of the general HARA utilities. In particular, the text-oriented presentation here is partly based on a portfolio optimization paper with time-dependence and uniformly distributed log-jump-amplitudes of Hanson and Westman [127] with some online corrections.

10.4.1   Log-Uniform Amplitude Jump-Diffusion for Log-Returns

Let $S(t)$ be the price of a single financial asset at time t, such as a stock or mutual fund, governed by a geometric jump-diffusion SDE with time-dependent coefficients,

$$dS(t) = S(t)\left(\mu_d(t)dt + \sigma_d(t)dG(t)\sum_{k=P(t)+1}^{P(t)+dP(t)} J\left(T_k^-, Q_k\right)\right) \qquad (10.115)$$

with $S(0) = S_0$, $S(t) > 0$, where $\mu_d(t)$ is the mean appreciation return rate, $\sigma_d(t)$ is the volatility, $G(t)$ is a continuous Gaussian process with zero-mean and t-variance (G is used for the diffusion component of the noise since W in this section will denote the wealth), $P(t)$ is a discontinuous, Poisson process with jump-rate $\lambda(t)$, and associated jump-amplitude $J(t, Q)$, $-1 < J(t, Q) < \infty$ to avoid bankruptcy at a single jump, with log-return mark Q mean $\mu_j(t)$ and variance $\sigma_j^2(t)$. At the kth Poisson jump, T_k^- is the prejump-time and Q_k is the corresponding IID random pick for the mark. The stochastic processes $G(t)$ and $P(t)$ are assumed to be Markov and pairwise independent. The jump-amplitude mark Q, given that a Poisson jump in time occurs, is also independently distributed. The stock price SDE (10.115) is similar to our prior work [125, 124], except that time-dependent coefficients introduce more realism.

Since the stock price process is a geometric jump-diffusion, the common multiplicative factor of $S(t)$ on the right can be removed by a logarithmic transformation yielding the SDE of the stock price log-return,

$$d\ln(S(t)) = \mu_{ld}(t)dt + \sigma_d(t)dG(t) + \sum_{k=P(t)+1}^{P(t)+dP(t)} \ln\left(1 + J\left(T_k^-, Q_k\right)\right), \qquad (10.116)$$

where $\mu_{ld}(t) \equiv \mu_d(t) - \sigma_d^2(t)/2$ is the log-diffusion drift and $\ln(1 + J(t, q))$, the stock log-return jump-amplitude, is the logarithm of the relative postjump-amplitude.

Since $J(t, q) > -1$, it is convenient to select the mark process to be the log-jump-amplitude random variable,

$$Q = \ln\left(1 + J(t, Q)\right), \qquad (10.117)$$

on the mark space $\mathcal{Q} = (-\infty, +\infty)$. Though this is a convenient mark selection, it implies the independence of the jump-amplitude in time but not of the log-jump-amplitude distribution $\Phi_Q(q; t)$ for Q. For comparison to the Standard and Poor's (S&P500) log-return data, the discrete log-return difference form

$$\Delta\ln(S_i) \equiv \ln(S_{i+1}) - \ln(S_i) = \ln(S_{i+1}/S_i)$$

will be used at time $t_{i+1} = t_i + \Delta t_i$. The corresponding log-return differential $d\ln(S(t))$ in SDE (10.116) is written in the approximate, mean-zero, independent process, discrete form,

$$\Delta\ln(S(t_i)) \simeq \left(\mu_{ld}(t_i) + \lambda(t_i)\mu_j(t_i)\right)\Delta t_i + \sigma_d(t_i)\Delta G(t_i)$$

$$+ \mu_j(t_i)\left(\Delta P(t_i) - \lambda(t_i)\Delta t_i\right) + \sum_{k=P(t)+1}^{P(t)+\Delta P(t)} \left(Q_k - \mu_j(t_i)\right), \qquad (10.118)$$

where $\Delta G(t_i) \equiv G(t_{i+1}) - G(t_i)$ and $\Delta P(t_i) \equiv P(t_{i+1}) - P(t_i)$.

10.4.2   Log-Uniform Jump-Amplitude Model

Extreme jumps in the market are rare events, making it difficult or impossible to separate out the jumps from a background of continuous diffusive changes (see Aït-Sahalia [5]) to determine their distribution. Extreme jumps are limited by **circuit breakers** [11] introduced by the New York Stock Exchange in 1988 as a response to the crash of 1987, so a finite jump-amplitude distribution like the uniform distribution is appropriate. Thus, consider the uniform density on $[a(t), b(t)]$ for the marks Q,

$$\phi_Q(q; t) \equiv \begin{cases} \dfrac{1}{b(t) - a(t)}, & a(t) \leq q \leq b(t) \\ 0, & \text{otherwise} \end{cases}, \tag{10.119}$$

where $a(t) < 0 < b(t)$ to allow for both crashes ($q < 0$) and rallies ($q > 0$).

The basic moments of the uniformly Q (uq) density $\phi_Q(q; t_i)$ yields the mean

$$E_Q[Q] = \mu_j(t_i) = (a(t_i) + b(t_i))/2, \tag{10.120}$$

variance

$$\text{Var}_Q[Q] = \sigma_j^2(t_i) = (b(t_i) - a(t_i))^2/12, \tag{10.121}$$

third central moment

$$M_3^{(uq)}(t_i) \equiv E_Q\left[(Q - \mu_j(t_i))^3\right] = 0,$$

and fourth central moment

$$M_4^{(uq)}(t_i) \equiv E_Q\left[(Q - \mu_j(t_i))^3\right] = 9\sigma_j^4(t_i)/5.$$

In terms of the original jump-amplitude $J(t, Q)$, the mean is

$$\bar{J}(t_i) \equiv E_Q\left[J(Q, t_i)\right] = E_Q\left[e^Q - 1\right] = \frac{e^{b(t_i)} - e^{a(t_i)}}{b(t_i) - a(t_i)} - 1.$$

The first four moments of the uniform jump-diffusion (UJD) log-return difference using (10.118) are

$$M_1^{(ujd)}(t_i) \equiv E[\Delta \ln(S(t))] = (\mu_{ld}(t_i) + \lambda(t_i)\mu_j(t_i))\Delta t_i, \tag{10.122}$$

$$M_2^{(ujd)}(t_i) \equiv \text{Var}[\Delta \ln(S(t_i))] = \left(\sigma_d^2(t_i) + \lambda(t_i)\left(\mu_j^2(t_i) + \sigma_j^2(t_i)\right)\right)\Delta t_i, \tag{10.123}$$

$$M_3^{(ujd)}(t_i) \equiv E\left[\left(\Delta \ln(S(t_i)) - M_1^{(ujd)}(t_i)\right)^3\right]$$
$$= \lambda(t_i)\mu_j(t_i)\left(\mu_j^2(t_i) + 3\sigma_j^2(t_i)\right)\Delta t_i, \tag{10.124}$$

$$M_4^{(ujd)}(t_i) \equiv E\left[\left(\Delta \ln(S(t_i)) - M_1^{(ujd)}(t_i)\right)^4\right]$$
$$= \lambda(t_i)\left(\mu_j^4(t_i) + 6\mu_j^2(t_i)\sigma_j^2(t_i) + 9\sigma_j^4(t_i)/5\right)\Delta t_i \tag{10.125}$$
$$+ 3\left(\sigma_d^2(t_i) + \lambda(t_i)\left(\mu_j^2(t_i) + \sigma_j^2(t_i)\right)\right)^2(\Delta t_i)^2.$$

The $M_m^{(ujd)}(t_i)$ moment calculations, in particular, need Lemma 5.15 from Chapter 5 for the four powers of partial sums of zero-mean IID random variables $\widehat{Q}_k = Q_k - \mu_j$, so

$$\mathrm{E}\left[\left(\sum_{k=1}^{n} \widehat{Q}_k\right)^m\right] = \begin{cases} 0, & m = 1 \\ n M_2^{(uq)}(t_i) = n\sigma_j^2(t_i), & m = 2 \\ n M_3^{(uq)}(t_i) = 0, & m = 3 \\ n M_4^{(uq)}(t_i) + 3n(n-1)\left(M_2^{(uq)}(t_i)\right)^2, & m = 4 \end{cases},$$

where $n = \Delta P(t_i)$.

Let the uniform jump-diffusion be denoted by

$$X_i = \mathcal{G}_i + \sum_{k=P(t)+1}^{P(t)+\Delta P(t)} Q_k,$$

where $\mathcal{G}_i = \mu_{ld}(t_i)\Delta t_i + \sigma_d(t_i)\Delta G(t_i)$ is the nonstandard Gaussian process; then the density for the uniform jump-diffusion X_i is derived from the law of total probability (B.92) summing over all Poisson jumps and the nested convolution property (B.100),

$$\phi_{ujd}(x) = \sum_{k=0}^{\infty} p_k(\lambda(t_i)\Delta t_i)\phi_{ujd}^{(k)}(x),$$

where $p_k(\Lambda)$ is the usual Poisson counting distribution with corresponding kth density coefficient $\phi_{ujd}^{(k)}(x)$ given by

$$\phi_{ujd}^{(k)}(x) = \left(\phi_{\mathcal{G}_i}\left(*\phi_Q\right)^k\right)(x)$$

through the nested convolution property. The complexity of these coefficients grows exponentially with k. However, the first few are, using (5.76) for $k = 0$,

$$\phi_{ujd}^{(0)}(x) = \phi_{\mathcal{G}_i}(x) = \phi_n(x; \mu, \sigma^2),$$

where for brevity $\mu = \mu_{ld}(t_i)\Delta t_i$ and $\sigma^2 = \sigma_d^2(t_i)\Delta t_i$, now dropping the (t_i) argument for brevity and using (5.77) for $k = 1$,

$$\phi_{ujd}^{(1)}(x) = \left(\phi_{\mathcal{G}_i} * \phi_Q\right)(x) = \phi_{sn}(x - b, x - a; \mu, \sigma^2) \equiv \frac{\Phi_n(x - b, x - a; \mu, \sigma^2)}{b - a},$$

where $\phi_{sn}(x - b, x - a; \mu, \sigma^2)$ is called the **secant-normal density** (5.78), and finally from (5.79) with the **triangular density** (5.80) for $k = 2$,

$$\phi_{ujd}^{(2)}(x) = \left(\phi_{\mathcal{G}_i}(*\phi_Q)^2\right)(x) = \frac{2b - x + \mu}{b - a}\phi_{sn}(x - 2b, x - a - b; \mu, \sigma^2)$$

$$+ \frac{x - 2a - \mu}{b - a}\phi_{sn}(x - a - b, x - 2a; \mu, \sigma^2)$$

$$+ \frac{\sigma^2}{(b-a)^2}\left(\phi_n(x - 2b; \mu, \sigma^2) - 2\phi_n(x - a - b; \mu, \sigma^2) + \phi_n(x - 2a; \mu, \sigma^2)\right).$$

There are five stochastic jump-diffusion model parameter processes to be estimated,

$$\{\mu_d(t), \sigma_d^2(t), \mu_j(t), \sigma_j^2(t), \lambda(t)\},$$

assuming that the interest rate process $r(t)$ is given deterministically and the time-steps Δt_i over the given market period. Using the definitions of the jump mean-variance parameters $\{\mu_j(t), \sigma_j^2(t)\}$ in (10.120)–(10.121), the uniform jump range $\{a(t), b(t)\}$ can be estimated instead. The parameter estimations using variants of maximum likelihood methods is beyond the scope of this chapter, but the reader can consult our work in [127, 293] in the time-dependent parameter case and [125, 126, 128, 129, 132] for other background in the time-independent parameter case.

10.4.3 Optimal Portfolio and Consumption Policies Application

Let a portfolio contain a riskless asset, the bond, with price $B(t)$ dollars at time t in years, and a risky asset, the stock, with price $S(t)$ at time t. Let the instantaneous portfolio change fractions be $U_0(t)$ for the bond and $U_1(t)$ for the stock, such that the total satisfies $U_0(t) + U_1(t) = 1$. The bounds for $U_0(t)$ and $U_1(t)$ will be developed later from the jump-amplitude distribution and the nonnegativity of wealth condition.

Let the bond price process be deterministic and exponential,

$$dB(t) = r(t)B(t)dt, \quad B(0) = B_0, \tag{10.126}$$

where $r(t)$ is the bond rate of interest at time t. The stock price $S(t)$ has been given in the jump-diffusion SDE (10.115). The portfolio wealth process changes due to changes in the portfolio fractions less the instantaneous consumption of wealth $C(t)dt$,

$$dW(t) = W(t)\left(r(t)dt + U_1(t)\left((\mu_d(t) - r(t))dt \right. \right.$$
$$\left. \left. + \sigma_d(t)dG(t) + \sum_{k=P(t)+1}^{P(t)+dP(t)} J(Q_k)\right)\right) - C(t)dt \tag{10.127}$$

such that $W(t) \geq 0$ and that the consumption rate is constrained relative to wealth $0 \leq C(t) \leq C_0^{(\max)}W(t)$, consistent with nonnegative constraints that Sethi and Taksar [246] show are needed. In addition, the stock fraction is bounded by fixed constants. $U_0^{(\min)} \leq U_1(t) \leq U_0^{(\max)}$, so borrowing and short selling is permitted. In (10.127), $U_0(t) = 1 - U_1(t)$ has been eliminated [124, 127, 293].

The investor's portfolio objective is to maximize the conditional, expected current value of the discounted utility $\mathcal{U}_f(w)$ of final wealth at the end of the investment final time t_f and the discounted utility of instantaneous consumption preferences $\mathcal{U}(c)$, i.e., the optimal value of the portfolio satisfies

$$v^*(t, w) = \max_{\{u,c\}(t,t_f]}\left[\mathrm{E}\left[e^{-\bar{\beta}(t,t_f)}\mathcal{U}_f(W(t_f)) + \int_t^{t_f} e^{-\bar{\beta}(t,s)}\mathcal{U}(C(s))\, ds \,\middle|\, \mathcal{C}\right]\right], \tag{10.128}$$

conditioned on the state-control set $C = \{W(t) = w, U_1(t) = u, C(t) = c\}$, where the time horizon is assumed to be finite, $0 \le t < t_f$, and $\bar{\beta}(t, s)$ is the cumulative time-discount over time in (t, s) with $\bar{\beta}(t, t) = 0$ and discount rate $\beta(t) = (\partial\bar{\beta}/\partial s)(t, t)$ at time t. To avoid Merton's [199] difficulties with HARA utility functions too general for the portfolio and consumption problem, $\mathcal{U}'(C) \to +\infty$ as $C \to 0^+$ will be assumed for the utility of consumption, while a similar form will be used for the final utility $\mathcal{U}_f(W)$. Thus, the instantaneous consumption $c = C(t)$ and stock portfolio fraction $u = U_1(t)$ serve as two control variables, while the wealth $w = W(t)$ is the single state variable.

Absorbing Boundary Condition at Zero Wealth

Equation (10.128) is subject to a zero wealth absorbing natural boundary condition. This avoids arbitrage, as pointed out by Karatzas et al. [160] (see [245, Chapter 2]). It is necessary to enforce nonnegativity feasibility conditions on both wealth and consumption. They derive formally explicit solutions, from a consumption-investment dynamic programming model with an infinite horizon, that qualitatively correct the results of Merton [199, 200] (see [203, Chapter 6]). See also Sethi and Taksar [246] for specific errors in [199, 200] and Sethi's excellent summary [245, Chapter 1].

Here the Merton boundary condition correction in his 1990 text [203, Chap. 6] is used,

$$v^*(t, 0^+) = \mathcal{U}_f(0)e^{-\bar{\beta}(t, t_f)} + \mathcal{U}(0) \int_t^{t_f} e^{-\bar{\beta}(t, s)} ds, \qquad (10.129)$$

since the consumption must be zero when the wealth is zero. The terminal wealth condition, $v^*(t_f, w) = \mathcal{U}_f(w)$, must also be satisfied.

Portfolio Stochastic Dynamic Programming

Assuming the optimal value $v^*(t, w)$ is continuously differentiable in t and twice continuously differentiable in w, then the stochastic dynamic programming equation (see our papers [124, 127, 293]) follows from an application of the (Itô) stochastic chain rule to the principle of optimality,

$$
\begin{aligned}
0 = {}& v_t^*(t, w) - \beta(t)v^*(t, w) + \mathcal{U}(c^*) \\
& + [(r(t) + (\mu_d(t) - r(t))u^*)w - c^*] v_w^*(t, w) \\
& + \frac{1}{2}\sigma_d^2(t)(u^*)^2 w^2 v_{ww}^*(t, w) \\
& + \frac{\lambda(t)}{b(t) - a(t)} \int_{a(t)}^{b(t)} \left(v^*(t, \alpha(u^*, q)w) - v^*(t, w)\right) dq,
\end{aligned}
\qquad (10.130)
$$

where $u^* = u^*(t, w) \in [U_0^{(min)}, U_0^{(max)}]$ and $c^* = c^*(t, w) \in [0, C_0^{(max)}w]$ are the optimal controls if they exist, while $v_w^*(t, w)$ and $v_{ww}^*(t, w)$ are the partial derivatives with respect to wealth w when $0 \le t < t_f$. Upon a jump, the wealth changes by a factor

$$\alpha(u, q) \equiv 1 + (e^q - 1)u$$

in the postjump wealth argument of (10.130).

Nonnegativity of Wealth and Jump Distribution

Nonnegativity of wealth implies an additional consistency condition for the control since the jump in wealth argument $\alpha(u^*, q)w = (1 + (e^q - 1)u^*)w$ in the stochastic dynamic programming equation (10.130) requires $\alpha(u, q) \geq 0$ on the support interval of the jump-amplitude mark density $\phi_Q(q)$. Hence, it will make a difference in the optimal portfolio stock fraction u^* bounds if the support interval $[a(t), b(t)]$ is finite or if the support interval is $(-\infty, +\infty)$, i.e., had infinite range. Our results will be restricted to the usual case, the $a(t) < 0 < b(t)$, i.e., both crashes and rallies are modeled.

Lemma 10.6. *Bounds on Optimal Stock Fraction due to Nonnegativity of Wealth Jump Argument [293].*
If the support of $\phi_Q(q)$ is the finite interval $q \in [a(t), b(t)]$ with $a(t) < 0 < b(t)$, then $u^*(t, w)$ is restricted by (10.130) to

$$\frac{-1}{\left(e^{b(t)} - 1\right)} \leq u^*(t, w) \leq \frac{1}{\left(1 - e^{a(t)}\right)}, \tag{10.131}$$

but if the support of $\phi_Q(q; t)$ is fully infinite, i.e., $(-\infty, +\infty)$, then $u^*(t, w)$ is restricted by (10.130) to

$$0 \leq u^*(t, w) \leq 1. \tag{10.132}$$

Proof. It is necessary that $\alpha(u, q) \geq 0$ so that $\alpha(u, q)w \geq 0$ when the wealth and its jump-in-wealth in the HJB equation (10.130) argument need to be nonnegative, $w \geq 0$. The borderline case is when the instantaneous stock fraction case is zero, i.e., $u = 0$, so $\alpha(0, q) = 1 > 0$.

Next consider the case when the support, $a(t) \leq q \leq b(t)$, is finite. When $u > 0$, then

$$0 \leq 1 - \left(1 - e^{a(t)}\right) u \leq \alpha(u, q) \leq 1 + \left(e^{b(t)} - 1\right) u.$$

Since $e^{a(t)} < 1 < e^{b(t)}$, the worse case for enforcing $\alpha(u, q) \geq 0$ is on the left, so

$$u \leq \frac{+1}{\left(1 - e^{a(t)}\right)} = \frac{-1}{J(t, a(t))}.$$

When $u < 0$, then

$$0 \leq 1 - \left(e^{b(t)} - 1\right)(-u) \leq \alpha(u, q) \leq 1 + \left(1 - e^{a(t)}\right)(-u).$$

The worse case for enforcing $\alpha(u, q) \geq 0$ is again on the left, so upon reversing signs,

$$u \geq \frac{-1}{\left(e^{b(t)} - 1\right)} = \frac{-1}{J(t, b(t))},$$

completing both sides of the finite case (10.131), which can be written in terms of the original jump-amplitude coefficient $-1/J(t, b(t)) \leq u^*(t, w) \leq -1/J(t, a(t))$.

In the infinite range jump model case when $-\infty < q < +\infty$, then $0 < e^q < \infty$. Thus, when $u > 0$,

$$0 \leq 1 - u < \alpha(u, q) < \infty,$$

so $u \leq 1$. However, when $u < 0$, then

$$-\infty < \alpha(u, q) < 1 - u,$$

so $u < 0$ leads to a contradiction in that $\alpha(u, q)$ is unbounded below and $u \geq 0$, proving (10.132), which is the limiting case of (10.131) when $a(t) \to -\infty$ and $b(t) \to -\infty$. □

Remark 10.7. *This lemma gives the constraints on the instantaneous stock fraction $u^*(t, w)$ that limit the jumps to the jumps that at most just wipe out the investor's wealth. Unlike the case of pure-diffusion where the functional term has local dependence on the wealth mainly through partial derivatives, the case of jump-diffusion has global dependence through jump-integrals over finite differences with jump modified wealth arguments, leading to additional constraints under nonnegative wealth conditions that do not appear for pure-diffusions. The additional constraint comes not from the current wealth or nearby wealth but from the discontinuous wealth created by a jump.*

In the case of the fitted log-uniform jump-amplitude model, the range of the jump-amplitude marks $[a(t), b(t)]$ is covered by the estimated largest range

$$[a^{(\min)}, b^{(\max)}] = \left[\min_t(a(t)), \max_t(b(t))\right] \simeq [-7.113e\text{-}2, +4.990e\text{-}2]$$

over the period 1992–2001 corresponding to $t = 1:10$ using [127] results. The corresponding overall estimated range of the optimal instantaneous stock fraction $u^*(t, w)$ is then

$$[u^{(\min)}, u^{(\max)}] = \left[\frac{-1}{\left(e^{b^{(\max)}} - 1\right)}, \frac{+1}{\left(1 - e^{a^{(\min)}}\right)}\right] \simeq [-19.54, +14.56],$$

in large contrast to the highly restricted infinite range models where

$$[\min(u^*(t, w)), \max(u^*(t, w))] = [0, 1]$$

is fixed for any t.

Regular Optimal Control Policies

In the absence of constraints on the controls, the maximum controls are the regular optimal controls $u^{(\mathrm{reg})}(t, w)$ and $c^{(\mathrm{reg})}(t, w)$, which are given implicitly, provided they are attainable and there is sufficient differentiability in c and u, by the dual critical conditions

$$\mathcal{U}'(c^{(\mathrm{reg})}(t, w)) = v_w^*(t, w), \tag{10.133}$$

$$\sigma_d^2(t)w^2 v_{ww}^*(t, w)u^{(\mathrm{reg})}(t, w) = -(\mu_d(t) - r(t))w v_w^*(t, w)$$

$$-\lambda(t)w\frac{1}{b(t) - a(t)}\int_{a(t)}^{b(t)} (e^q - 1)v_w^*(t, \alpha(u^{(\mathrm{reg})}(t, w), q)w)\, dq \tag{10.134}$$

for the optimal consumption and portfolio policies with respect to the terminal wealth and instantaneous consumption utilities (6.2). Note that (10.133)–(10.134) define the set of regular controls implicitly.

10.4.4 CRRA Utility and Canonical Solution Reduction

For the risk-averse investor, the utilities are assumed to be the constant relative risk aversion (CRRA) power utilities [203, 123] with the same power for both wealth and consumption,

$$\mathcal{U}(x) = \mathcal{U}_f(x) = x^\gamma/\gamma, \quad x \ge 0, \quad 0 < \gamma < 1. \tag{10.135}$$

The CRRA utility designation arises since the relative risk aversion is the negative of the derivative $(\mathcal{U}''(x))$ in the marginal utility $(\mathcal{U}'(x))$ relative to the average change in the marginal utility $(\mathcal{U}'(x)/x)$, or here

$$RRA(x) \equiv -\mathcal{U}''(x)/(\mathcal{U}'(x)/x) = (1 - \gamma) > 0, \tag{10.136}$$

i.e., a positive constant, and is a special case of the more general HARA utilities.

The CRRA power utilities for the optimal consumption and portfolio problem lead to a **canonical reduction** of the stochastic dynamic programming PDE problem to a simpler ODE problem in time, by the separation of wealth and time-dependence,

$$v^*(t, w) = \mathcal{U}(w)v_0(t), \tag{10.137}$$

where only the time function $v_0(t)$ is to be determined. The regular consumption control is a linear function of the wealth,

$$c^{(\mathrm{reg})}(t, w) \equiv w \cdot c_0^{(\mathrm{reg})}(t) = w/v_0^{1/(1-\gamma)}(t), \tag{10.138}$$

using (10.133) and $\mathcal{U}'(x) = x^{\gamma-1}$ from (10.135). The regular stock fraction u from (10.134) is a wealth independent control but is given in uniform case implicit form:

$$u^{(\mathrm{reg})}(t, w) = u_0^{(\mathrm{reg})}(t) \equiv \frac{1}{(1-\gamma)\sigma_d^2(t)} \left[\mu_d(t) - r(t) + \lambda(t)I_1\left(u_0^{(\mathrm{reg})}(t)\right) \right], \tag{10.139}$$

$$I_1(u) = \frac{1}{b(t) - a(t)} \int_{a(t)}^{b(t)} (e^q - 1)\alpha^{\gamma-1}(u, q)dq. \tag{10.140}$$

The wealth independent property of the regular stock fraction is essential for the separability of the optimal value function (10.137). Since (10.139) only defines $u_0^{(\mathrm{reg})}(t)$ implicitly in fixed point form, $u_0^{(\mathrm{reg})}(t)$ must be found by an iteration such as Newton's method, while the general Gauss-statistics quadrature [277] can be used for jump-integrals (see [124]).

The optimal controls, when there are constraints, are given in piecewise form as

$$c^*(t, w)/w = c_0^*(t) \equiv \max\left[\min\left[c_0^{(\mathrm{reg})}(t), C_0^{(\mathrm{max})}\right], 0\right]$$

provided $w > 0$ and

$$u^*(t, w) = u_0^*(t) \equiv \max\left[\min\left[u_0^{(\mathrm{reg})}(t), U_0^{(\mathrm{max})}\right], U_0^{(\mathrm{min})}\right]$$

is independent of w along with $u_0^{(\mathrm{reg})}(t)$.

Substitution of the separable power solution (10.137) and the regular controls (10.138)–(10.139) into the stochastic dynamic programming equation (10.130) leads to an apparent Bernoulli type ODE,

$$0 = v_0'(t) + (1 - \gamma)\left(g_1(t, u_0^*(t))v_0(t) + g_2(t)v_0^{\frac{\gamma}{\gamma-1}}(t)\right), \tag{10.141}$$

$$g_1(t, u) \equiv \frac{1}{1 - \gamma}\Bigg[-\beta(t) + \gamma\left(r(t) + u(\mu_d(t) - r(t))\right) \\ - \frac{\gamma(1 - \gamma)}{2}\sigma_d^2(t)u^2 + \lambda(t)(I_2(t, u) - 1)\Bigg], \tag{10.142}$$

$$g_2(t) \equiv \frac{1}{1 - \gamma}\left[\left(\frac{c_0^*(t)}{c_0^{(\mathrm{reg})}(t)}\right)^\gamma - \gamma\left(\frac{c_0^*(t)}{c_0^{(\mathrm{reg})}(t)}\right)\right], \tag{10.143}$$

$$I_2(t, u) \equiv \frac{1}{b(t) - a(t)}\int_{a(t)}^{b(t)}\alpha^\gamma(u, q)\,dq \tag{10.144}$$

for $0 \le t < t_f$. The coupling of $v_0(t)$ to the time-dependent part of the consumption term $c_0^{(\mathrm{reg})}(t)$ in $g_2(t)$ and the relationship of $c_0^{(\mathrm{reg})}(t)$ to $v_0(t)$ in (10.138) means that the differential equation (10.141) is implicit and highly nonlinear and thus (10.141) is only of Bernoulli type formally. The apparent Bernoulli equation (10.141) can be transformed to an apparent linear differential equation by using the Bernoulli linearizing transformation $\theta(t) = v_0^{1/(1-\gamma)}(t)$ to obtain

$$0 = \theta'(t) + g_1(t, u_0^*)\theta(t) + g_2(t),$$

whose general solution can be inverse transformed to the general solution for the separated time, but implicit function,

$$v_0(t) = \theta^{1-\gamma}(t) = \left[e^{-g_1(t, u_0^*(t))(t_f - t)}\left(1 + \int_t^{t_f} g_2(\tau)e^{g_1(t, u_0^*(t))(t_f - \tau)}d\tau\right)\right]^{1-\gamma}. \tag{10.145}$$

To illustrate this stochastic application, a computational approximation of the solution is presented. The main computational changes from the procedure used in [124] are that the jump-amplitude distribution is now uniform and the portfolio parameters as well as the jump-amplitude distribution are time-dependent. Parameter time-dependence is approximated by quadratic interpolation over the years 1992–2001. The terminal time is taken to be $t_f = 11$, one year beyond this range. For this numerical study, the economic rates are taken to be time independent, so the bond interest rate is $r(t) = 5.75\%$ and the time-discount rate is $\beta(t) = 5.25\%$. The portfolio stock fraction constraints are

$$[U_0^{(\mathrm{min})}, U_0^{(\mathrm{max})}] = [-10, 10] \quad \text{and} \quad C_0^{(\mathrm{max})} = 0.75$$

for consumption relative to wealth.

In Figure 10.1, the optimal portfolio stock fraction $u^*(t)$ is displayed. The portfolio policy is not monotonic in time and the minimum control constraint at $U_0^{(\mathrm{min})}$ is active during the first half-year in $t \in [0, t_f]$, while the maximum constraint is not activated since $u^*(t)$ remains significantly below that constraint. The $u^*(t)$ nonmonotonic behavior is very interesting compared to the constant behavior in the constant parameter model in [124]. Likely the stock fraction grew initially due to the early relatively quiet period, then peaked at the beginning of the fourth year (1996 in the S&P500 data) as the market became noisier and continued to decline due to the final relatively noisier period. In Figure 10.2 the optimal, expected, cumulative consumption, $c^*(t, w)$, is displayed in three dimensions. The optimal consumption policy $c^*(t, w)$ results in this computational example are qualitatively similar to that of the time-independent parameter case in [124].

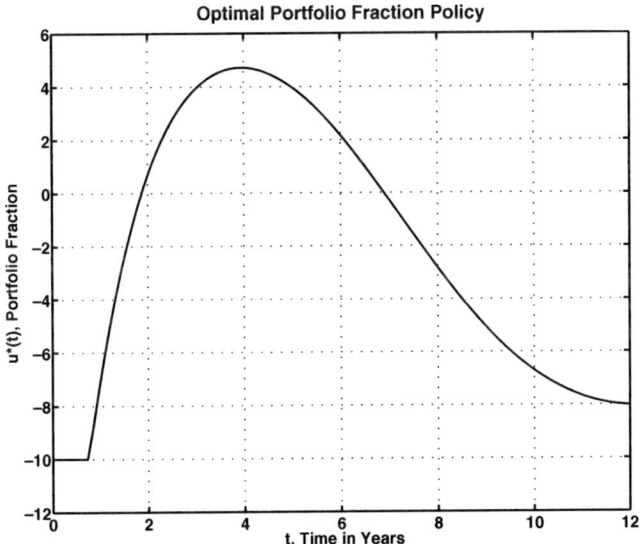

Figure 10.1. *Optimal portfolio stock fraction policy* $u^*(t)$ *on* $t \in [0, 12]$ *subject to the control constraint set* $[U_0^{(min)}, U_0^{(max)}] = [-10, 10]$.

10.5 Important Financial Events Model: The Greenspan Process

Many financially critical announcements can have significant effects in the market, such as those on interest rates, unemployment statistics, budget deficits, trade deficits, prices of supplies such as oil, weather extremes, and many others. Some of these announcements are scheduled like those of the Federal Reserve Board, labor reports, or business earnings. The response to these scheduled announcements is sometimes difficult to predict, because market investors may have already factored in unfavorable or favorable news. On the other hand, unscheduled announcements present both uncertainties in time and response making the compound Poisson processes a reasonable model. The Poisson model would be unsuitable for scheduled announcements. In [130], Hanson and Westman proposed a quasi-deterministic stochastic jump process that resembled the compound Poisson process only in the random jump-amplitude components but otherwise jump at scheduled or deterministic times. The theoretical basis for our paper was the optimal portfolio problem for important external events paper [235] of Rishel. Our contribution was primarily constructing the

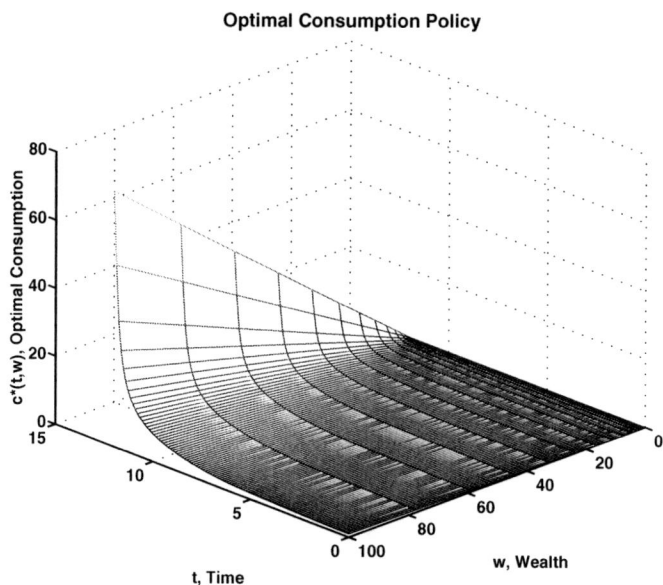

Figure 10.2. *Optimal consumption policy $c^*(t, w)$ for $(t, w) \in [0, 12] \times [0, 100]$.*

intricate computational procedure for the problem and formulating the problem as a full stochastic differential equation model. The formulation appears to be of interest in other financial problems where there are uncertain, scheduled payments such as dividends.

Scheduled jumps affect the market. The response magnitude of the jumps can be random, as described by Rishel [235]. On February 17, 2000, there were large market fluctuations caused by the semiannual economic report of the now former chairman of the Federal Reserve Board, Alan Greenspan, to Congress concerning the raising of interest rates, among other things. The next day was a *double witching day* with the simultaneous expiration of contracts on stock options and indices. Although these events and the market responses to them are quite complicated, these quasi-deterministic processes are strongly motivated by the influential announcement events by Greenspan and thus they might be called *Greenspan processes*.

The optimal portfolio and consumption work [123, 130] will be summarized and reformulated with the constructs of this book. The reformulation uses a more concrete formulation of the quasi-deterministic processes than the more general, abstract Poisson random measure-like formulation in [130]. Also the problem was reduced to a single risky asset model from the multiasset model in [130].

10.5.1   Stochastic Scheduled and Unscheduled Events Model with Stochastic Parameter Processes

Let the usual Poisson process $P(t)$ denote an unscheduled events process whose jumps occur at the random times T_k for $k = 1, 2, \ldots$ with random jump-amplitudes $J(p_k, \mathbf{A}(T_k^-))$, where p is the corresponding random mark and $\mathbf{A}(t)$ is an auxiliary parameter, vector process. Let the quasi-deterministic process or *Greenspan process* $Q(t)$ denote a scheduled events process at fixed times τ_ℓ with random jump-amplitude $K(\widehat{q}; \mathbf{A}(\tau_\ell^-))$, where $\widehat{q}$ is the corresponding random mark. Both processes are right-continuous.

Let the portfolio consist of one almost riskless asset $B(t)$ at time t and one risky asset $S(t)$. The riskless asset $B(t)$ satisfies the familiar form

$$dB(t) = r(\mathbf{A}(t))B(t)dt, \quad B(0) = B_0, \tag{10.146}$$

where the almost riskless asset interest $r(\mathbf{a})$ depends on a mildly random parameter vector $\mathbf{A}(t) = [A_1(t); A_2(t)]$ associated with unscheduled and scheduled event processes. Here, $A_1(t))$ is a parameter for unscheduled events driven by the Poisson process $dP(t)$ with jump-amplitude $J_1(q)$ and random mark p and satisfied by

$$dA_1(t) = A_1(t)J_1(q)dP(t) = \sum_{k=P(t)+1}^{P(t)+dP(t)} A_1(T_k^-)J_1(q_k), \tag{10.147}$$

where $(T_k; q_k)$ are the kth Poisson time-mark parameters. The process $A_1(t)$ can be called a *geometric Poisson process* since the noise is linear in $A_1(t)$, making the noise multiplicative. For reasons cited in the previous section, the range of the unscheduled process mark is finite, so $a \le q \le b$. Also, $A_2(t)$ is a parameter process for scheduled events driven by the quasi-deterministic process $dQ(t)$, with jump-amplitude $K_2(\widehat{q})$ with random mark $\widehat{q}$, and satisfied by

$$dA_2(t) = A_2(t)K_2(\widehat{q})dQ(t) = \sum_{\ell=Q(t)+1}^{Q(t)+dQ(t)} A_2(\tau_\ell^-)K_2(\tau_\ell^-; \widehat{q}_\ell), \tag{10.148}$$

where τ_ℓ is a scheduled event time such that $\tau_{\ell+1} > \tau_\ell$ and $\widehat{q}_\ell$ is the ℓth realized jump-amplitude mark for $\ell = 1{:}M$, where $\tau_M = \max(\tau_\ell) < t_f$ with t_f being the portfolio final time and $\widehat{a} \le \widehat{q} \le \widehat{b}$. The process $A_2(t)$ is also a multiplicative or geometric noise process.

The risky portfolio asset with price $S(t)$ satisfies the SDE

$$dS(t) = S(t)(\mu(\mathbf{A}(t)) + \sigma(\mathbf{A}(t))dG(t) + J(q; \mathbf{A}(t))dP(t) \tag{10.149}$$
$$+ K(\widehat{q}; \mathbf{A}(t))dQ(t)),$$

where $S(0) = S_0, 0 \le S(t) < t_f$, $\mu(\mathbf{a})$ is the mean stock appreciation rate, $G(t)$ is a standard Wiener or Gaussian process, $\sigma(\mathbf{a})$ is the standard deviation coefficient corresponding to $dG(t)$, $\mathbf{A}(t)$ is an auxiliary parameter process, while the compound unscheduled and scheduled jump processes with jump-amplitude shorthand notation can properly be defined as

$$S(t)J(q; \mathbf{A}(t))dP(t) = \sum_{k=P(t)+1}^{P(t)+dP(t)} S(T_k^-)J(q_k; \mathbf{A}(T_k^-)) \tag{10.150}$$

and

$$S(t)K(\widehat{q}; \mathbf{A}(t))dQ(t) = \sum_{\ell=Q(t)+1}^{Q(t)+dQ(t)} S(\tau_\ell^-)K(\widehat{q}_\ell; \mathbf{A}(\tau_\ell^-)). \tag{10.151}$$

The primary difference between forms (10.150) and (10.151) is that in the former $dP(t)$ and T_k^- are stochastic with $\mathrm{E}[dP(t)] = \lambda dt$ and $T_{k+1} - T_k$ exponentially distributed (B.56), while in the latter $dQ(t)$ and τ_ℓ are deterministic so $\mathrm{E}[dQ(t)] = dQ(t)$ and $\mathrm{E}[\tau_\ell] = \tau_\ell$.

10.5.2   Further Properties of Quasi-Deterministic or Scheduled Event Processes: $K(\widehat{q}; A(t))dQ(t)$

The scheduled jump of the $dQ(t)$ of (10.151) is scheduled at prescribed times τ_ℓ and jump-counts $\ell = 1{:}M$, such that $\tau_{\ell+1} > \tau_\ell$ and $\tau_M = \max(\tau_\ell) < t_f$. At these times, the $K(\widehat{q}_\ell; A(\tau_\ell^-))$ are the random jump-amplitudes, where $\widehat{q}_\ell$ is the random mark or background random variable for which the probability distribution can be more conveniently specified. The $A(t)$ is an auxiliary parameter process that is optional for the jump-amplitude function K associated with $dQ(t)$. The $dQ(t)$ is a pure deterministic counting process that triggers the random jump-amplitude.

The expectation of the event response jump-amplitude $K(\widehat{q}; A(t))$ conditioned on the parameter process is

$$\mathrm{E}[K(\widehat{q}; A(t))|A(t) = a] = \mathrm{E}[K(\widehat{q}; a)] \equiv \overline{K}(a).$$

The jump in the ith stock at a jump of the ith scheduled event process is given by

$$[S](\tau_\ell) = S(\tau_\ell^+) - S(\tau_\ell^-) = K(\widehat{q}_\ell; A(\tau_\ell^-))S(\tau_\ell^-)$$

for $\tau_\ell < t_f$, where t_f is the terminal time. The stocks inherit the right-continuity property of the scheduled jump processes.

Similarly, for the scheduled parameter process $A_2(t)$, the jump at τ_ℓ is given by

$$[A_2](\tau_\ell) = A_2(\tau_\ell^+) - A_2(\tau_\ell^-) = K_2(\widehat{q}_\ell)A_2(\tau_\ell^-),$$

which in turn is similar to the jump of the unscheduled parameter process $A_1(t)$,

$$[A_1](T_k) = A_1(T_k^+) - A_2(T_k^-) = J_1(q_k)A_2(T_k^-).$$

10.5.3   Optimal Portfolio Utility, Stock Fraction, and Consumption

The setup of this optimal portfolio problem is similar to that of the prior section, so the focus will be mainly on differences arising from the quasi-deterministic scheduled event processes, while skipping similar intermediate steps. Let $W(t)$ be the portfolio wealth at time t, $U_1(t)$ is the vector of the **instantaneous** fraction of wealth invested in the risky assets at vector price $S(t)$, such that the riskless asset fraction at price $B(t)$ satisfies

$$U_0(t) = 1 - U_1(t),$$

and $C(t)$ is the consumption of wealth. As in the prior sections, the jump-amplitude distributions will be assumed to be of finite range, so that the risky asset fractions will not be restricted to $[0, 1]$ but will be restricted to some larger and reasonable range $[U^{(\min)}, U^{(\max)}]$.

Following (10.127), the portfolio wealth process, relative changes due to relative changes in the portfolio fractions less the instantaneous consumption of wealth $C(t)dt$, is governed by the SDE

$$
\begin{aligned}
dW(t) = {} & (W(t)\,(r(\mathbf{A}(t)) + U_1(t)(\mu(\mathbf{A}(t)) - r(\mathbf{A}(t))) - C(t)))\,dt \\
& + W(t)U_1(t)\sigma(\mathbf{A}(t))dG(t) + W(t)U_1(t)J(q;\mathbf{A}(t))dP(t) \\
& + W(t)U_1(t)K(\widehat{q};\mathbf{A}(t))dQ(t)
\end{aligned}
\tag{10.152}
$$

with the necessary conditions that $W(t) \geq 0$ and that the consumption rate is constrained relative to wealth $0 \leq C(t) \leq C_0^{(\max)}W(t)$. For the stochastic dynamic programming formulation, it is necessary to know the jumps in the wealth for both unscheduled and scheduled jump-times, which are

$$
[W](T_k) = W(T_k^+) - W(T_k^-) = W(T_k^-)U_1(T_k^-)J_1(q_k)
\tag{10.153}
$$

and

$$
[W](\tau_\ell) = W(\tau_\ell^+) - W(\tau_\ell^-) = W(\tau_\ell^-)U_1(\tau_\ell^-)K_2(\widehat{q}_\ell).
\tag{10.154}
$$

The investor's objective is to maximize the conditional, expected current value of the discounted utility $\mathcal{U}_f(w;\mathbf{a})$ of final wealth at the end of the investment final time t_f and the discounted utility of instantaneous consumption preferences $\mathcal{U}(c)$ so that the optimal value of the portfolio satisfies

$$
\begin{aligned}
v^*(t,w;\mathbf{a}) = \max_{\{u,c\}(t,t_f)} \Bigg[\mathrm{E}\Bigg[& e^{-\overline{\beta}(t,t_f)}\mathcal{U}_f(W(t_f);\mathbf{a}) \\
& + \int_t^{t_f} e^{-\overline{\beta}(t,s)}\mathcal{U}(C(s))\,ds \Bigg| \mathcal{C} \Bigg] \Bigg]
\end{aligned}
\tag{10.155}
$$

conditioned on the state-control set $\mathcal{C} = \{W(t) = w, U_1(t) = u, C(t) = c, \mathbf{A}(t) = \mathbf{a}\}$, where the time horizon is assumed to be finite, $0 \leq t < t_f$, and $\overline{\beta}(t,s) \equiv \int_t^s \beta(\mathbf{A}(z))dz$ is the integral over the instant nominal discount rate $\beta(\mathbf{A}(t))$ on $[t,s]$. The instantaneous consumption $c = C(t)$ and stock portfolio fraction vector $u = U_1(t)$ serve as two control variables, while the wealth $w = W(t)$ is the single state variable.

Again, Merton's zero-wealth boundary condition correction given in his 1990 text [203, Chapter 6] is used, but here it has the extra parameter argument

$$
v^*(t,0^+;\mathbf{a}) = \mathcal{U}_f(0;\mathbf{a})e^{-\overline{\beta}(t,t_f)} + \mathcal{U}(0)\int_t^{t_f} e^{-\overline{\beta}(t,s)}ds
\tag{10.156}
$$

since the consumption must be zero when the wealth is zero. The terminal wealth condition

$$
v^*(t_f,w;\mathbf{a}) = \mathcal{U}_f(w;\mathbf{a})
\tag{10.157}
$$

must also be satisfied and provides the start of the stochastic dynamic programming problem, a backward time problem.

The CRRA power utilities (10.135)–(10.136) are also used here, as in the last section, for the risk-averse investor, with the same power for consumption and wealth but now with parameter values,

$$
\mathcal{U}(c) = c^\gamma / \gamma, \quad c \geq 0, \quad 0 < \gamma < 1,
$$

$$
\mathcal{U}_i(a_i) = |a_i|^{\gamma_i} / \gamma_i, \quad a_i \neq 0, \quad \gamma_i \neq 0, \quad i = 1, 2, \tag{10.158}
$$

$$
\mathcal{U}_f(w; \mathbf{a}) = \mathcal{U}(w)\mathcal{U}_1(a_1)\mathcal{U}(a_2), \quad w \geq 0, \quad \mathbf{a} = [a_1; a_2].
$$

The utilities satisfy general properties, where in the case of consumption, for example, (1) it is nonnegative, $\mathcal{U}(c) \geq 0$, (2) the marginal utility is favorable toward consumption, $\mathcal{U}'(c) > 0$, (3) but at a decreasing rate, $\mathcal{U}''(c) < 0$.

The application of stochastic dynamic programming to the standard jump-diffusion with only Gaussian and Poisson noise leads to a single PDE in time t and wealth, as in the previous section, because the Gaussian and Poisson noise, in particular the Poisson jump-times, average out with the expectation used in the objective. However, in the present problem with scheduled quasi-deterministic jumps, the scheduled jump-times are not averaged out by the expectation operator. Thus, between scheduled jump-times, τ_ℓ for parameters $i = 1{:}2$ and jump-counters ℓ, the optimal value function $v^*(t, w; \mathbf{a})$, using the principle of optimality and expanding using the SDEs and the stochastic chain rule to dt-precision, satisfies

$$
\begin{aligned}
0 = {}& v_t^*(t, w; \mathbf{a}) - \beta(\mathbf{a})v^*(t, w; \mathbf{a}) \\
&+ \max_{u,c}\bigg[\mathcal{U}(c) + ((r(\mathbf{a}) + u(\mu(\mathbf{a}) - r(\mathbf{a})))w - c)\, v_w^*(t, w; \mathbf{a}) \\
&+ \frac{1}{2}(u\sigma(\mathbf{a}))^2 w^2 v_{ww}^*(t, w; \mathbf{a}) \\
&+ \lambda \int_a^b \big(v^*(t, w(1 + uJ(q; \mathbf{a})); a_1(1 + J_1(q)), a_2) - v^*(t, w; \mathbf{a}) \big) \phi_q(q)dq \bigg] \\
= {}& v_t^*(t, w; \mathbf{a}) - \beta(\mathbf{a})v^*(t, w; \mathbf{a}) + \mathcal{U}(c^*) \\
&+ \big((r(\mathbf{a}) + u^*(\mu(\mathbf{a}) - r(\mathbf{a})))w - c^* \big) v_w^*(t, w; \mathbf{a}) \\
&+ \frac{1}{2}(u^*\sigma(\mathbf{a}))^2 w^2 v_{ww}^*(t, w; \mathbf{a}) \\
&+ \lambda \int_a^b \big(v^*(t, w(1 + u^*J(q; \mathbf{a})); a_1(1 + J_1(q)), a_2) - v^*(t, w; \mathbf{a}) \big) \phi_q(q)dq.
\end{aligned} \tag{10.159}
$$

This equation is valid starting from the terminal wealth condition (10.157) and otherwise holding on open time-intervals in backward order determined by the scheduled jump-times from (τ_M, t_f) to $(\tau_{\ell-1}, \tau_\ell)$ for $\ell = M{:}{-}1{:}2$ (the triple construct has the form *start : step : stop* as in MATLAB) and $(0, \tau_1)$. Here, $u^* = u^*(t, w; \mathbf{a})$ and $c^* = c^*(t, w; \mathbf{a})$ are the optimal arguments of the maximum in the first part of (10.159) and are subject to previously stated constraints.

While the unscheduled Poisson jumps are instantaneous and random, the expectation from the objective averages the jumps with $\mathrm{E}[dP(t)] = \lambda dt$ which is the same order as the contributions of the continuous terms in (10.159). The scheduled jumps are instantaneous and deterministic so they do not average. The continuous terms contribute zero in that instant and only the scheduled jump-integral survives. Hence, at the scheduled jump-time

τ_ℓ for $\ell = M{:}{-}1{:}1$ there is a new *stochastic dynamic programming jump-condition*,

$$v^*(\tau_\ell^-, w; \mathbf{a}) = \int_{\widehat{a}}^{\widehat{b}} v^* \left(\tau_\ell^+, w(1 + u_\ell^- K(\widehat{q}; \mathbf{a})); a_1, a_2(1 + K_2(\widehat{q})) \right) \phi_q(q) dq, \quad (10.160)$$

where $u_\ell^- = u^*(\tau_\ell^-, w; \mathbf{a})$. This condition does not arise in the usual jump-diffusion problem with only unscheduled jumps. Note that the value of $v^*(\tau_\ell^+, w; \mathbf{a})$ is to be found from integrating (10.159) from $\tau_{\ell+1}$ to τ_ℓ, so that the jump-condition (10.160) provides the new backward value $v^*(\tau_\ell^-, w; \mathbf{a})$ which is the start for the integration of (10.159) on $(\tau_{\ell-1}, \tau_\ell)$.

Since there is a nonnegativity condition on wealth, that condition also applies to arguments in (10.159) and (10.160), so

$$(1 + u^* J(q; \mathbf{a})) \geq 0$$

and

$$(1 + u^* K(\widehat{q}; \mathbf{a})) \geq 0$$

are additionally required, respectively.

If the consumption and stock fraction are unconstrained, then the regular controls, $c^{(\text{reg})}(t, w; \mathbf{a})$ and $u^{(\text{reg})}(t, w; \mathbf{a})$, are implicitly obtained, assuming sufficient differentiability,

$$\mathcal{U}' \left(c^{(\text{reg})}(t, w; \mathbf{a}) \right) = v_w^*(t, w; \mathbf{a}) \qquad (10.161)$$

and

$$(\sigma(\mathbf{a})w)^2 v_{ww}^*(t, w; \mathbf{a}) u^*(t, w; \mathbf{a}) = -(\mu(\mathbf{a}) - r(\mathbf{a}))w v_w^*(t, w; \mathbf{a})$$
$$- \lambda w \int_a^b J(q) v_w^*(t, w(1 + u^* J(q)); a_1(1 + J_1(q)), a_2) \phi_q(q) dq. \qquad (10.162)$$

Since these regular control policies introduce both implicitness and nonlinearities into the PDE of stochastic dynamic programming (10.159), the solution will require computational iterations. There is also a jump in the regular stock fraction from (10.160) and it is given implicitly by

$$0 = w \int_{\widehat{a}}^{\widehat{b}} K(\widehat{q}; \mathbf{a}) v^*(\tau_\ell^+, w(1 + u_\ell^{(\text{reg})-} K(\widehat{q}; \mathbf{a})); a_1, a_2(1 + K_2(\widehat{q}))), \qquad (10.163)$$

where $u_\ell^{(\text{reg})-} = u^{(\text{reg})}(\tau_\ell^-, w; \mathbf{a})$. The optimal policies (c^*, u^*) are found by applying the constraints to the regular control policies $(c^{(\text{reg})}, u^{(\text{reg})})$.

10.5.4 Canonical CRRA Model Solution

The great advantage of the CRRA power utilities (10.158) for the portfolio and consumption optimization problem is that the solution is separable in the form

$$v^*(t, w; \mathbf{a}) = \mathcal{U}_f(w; \mathbf{a}) v_0(t; \mathbf{a}) \qquad (10.164)$$

so the wealth state can be completely stripped away in terms of a given utility function $\mathcal{U}_f(w; \mathbf{a})$, avoiding the exponential computational complexity of the *curse of dimensionality*. Also, the terminal condition (10.158) is easily satisfied as long as the remaining time-dependent part of the solution satisfies

$$v_0(t_f; \mathbf{a}) = 1.$$

Since $\mathcal{U}(0^+; \mathbf{a}) = 0 = \mathcal{U}(0^+)$, the zero-wealth absorbing boundary condition (10.156) reduced to $v^*(t, 0^+; \mathbf{a}) = 0$.

Substituting the canonical solution into the implicit equation (10.161) for $c^{(\text{reg})}(t, w; \mathbf{a})$ yields a preliminary solution linear in w and in terms of $v_0(t; \mathbf{a})$,

$$c^{(\text{reg})}(t, w; \mathbf{a}) = w \cdot c_0^{(\text{reg})}(t; \mathbf{a}) \equiv \frac{w\psi_2(\mathbf{a})}{v_0^{1/(1-\gamma)}(t; \mathbf{a})}, \qquad (10.165)$$

where $\psi_2(\mathbf{a}) \equiv 1/(\mathcal{U}_1(a_1)\mathcal{U}_2(a_2))^{1/(1-\gamma)}$, using some algebra. The corresponding optimal consumption is given by

$$c^*(t, w; \mathbf{a}) = wc_0^*(t; \mathbf{a}) = w \max \left(c_0^{(\text{reg})}(t; \mathbf{a}), C_0^{(\text{max})} \right). \qquad (10.166)$$

However, the reduction of the $u^{(\text{reg})}(t, w; \mathbf{a})$ does not eliminate the implicitness but yields a solution independent of w, i.e., $u^{(\text{reg})}(t, w; \mathbf{a}) = u_0^{(\text{reg})}(t; \mathbf{a})$, a prime criterion for separability, where

$$u_0^{(\text{reg})}(t; \mathbf{a}) = \frac{1}{(1-\gamma)\sigma^2(\mathbf{a})} \left(\mu(\mathbf{a}) - r(\mathbf{a}) + \frac{\lambda}{\gamma} I_1' \left(t, u_0^{(\text{reg})}(t; \mathbf{a}); \mathbf{a} \right) \right) \qquad (10.167)$$

and where

$$I_1'(t, u; \mathbf{a}) \equiv \gamma^2 \int_a^b J(q; \mathbf{a}) \frac{\mathcal{U}(1 + uJ(q; \mathbf{a}))}{(1 + uJ(q; \mathbf{a}))} \mathcal{U}(1 + J_1(q))\psi_1(t, q; \mathbf{a})\phi_q(q)dq,$$

$$\psi_1(t, q; \mathbf{a}) \equiv \frac{v_0(t; (1 + J_1(q))a_1, a_2)}{v_0(t; a_1, a_2)},$$

noting that $\psi_1(t, q; \mathbf{a})$ is the primary source of implicitness. The corresponding optimal portfolio fraction is given by

$$u^*(t, w; \mathbf{a}) = \max \left(U^{(\text{min})}, \min \left(U^{(\text{max})}, u_0^{(\text{reg})}(t; \mathbf{a}) \right) \right). \qquad (10.168)$$

Substituting the PDE (10.159) and CRRA separated solution (10.164) along with the optimal controls (10.166)–(10.168) leads to an implicit Bernoulli-type ODE,

$$0 = v_0'(t; \mathbf{a}) + (1 - \gamma) \left(\psi_3'(t, u^*(t; \mathbf{a}); \mathbf{a})v_0(t; \mathbf{a}) + \widehat{\psi}_2(t; \mathbf{a})v_0^{\frac{\gamma}{\gamma-1}}(t; \mathbf{a}) \right), \qquad (10.169)$$

where

$$\psi_3'(t, u; \mathbf{a}) = \partial \psi_3(t, u; \mathbf{a})/\partial t \equiv \frac{1}{1-\gamma} \left(-\beta(\mathbf{a}) + \gamma \left(r(\mathbf{a}) - (\mu(\mathbf{a}) - r(\mathbf{a}))u \right) \right.$$

$$\left. - \frac{\gamma(1-\gamma)}{2\sigma^2(\mathbf{a})} u^2 + \lambda(I_1(t, u; \mathbf{a}) - 1) \right),$$

$$\widehat{\psi}_2(t; \mathbf{a}) \equiv \frac{1}{1-\gamma} \left(\left(\frac{c_0^*(t; \mathbf{a})}{c_0^{(\text{reg})}(t; \mathbf{a})} \right)^\gamma - \gamma \left(\frac{c_0^*(t; \mathbf{a})}{c_0^{(\text{reg})}(t; \mathbf{a})} \right) \right) \psi_2(\mathbf{a}),$$

$$I_1(t, u; \mathbf{a}) - 1) \equiv \gamma \int_a^b \mathcal{U}(1 + u J(q; \mathbf{a})) \psi_1(t, q; \mathbf{a}) \mathcal{U}_1(1 + J_1(q)) \psi_1(t, q; \mathbf{a}) \phi_q(q) dq,$$

when t is on $(\tau_{\ell-1}, \tau_\ell)$ for $\ell = (M+1):-1:1$, conveniently defining $\tau_{M+1} \equiv t_f$ and $\tau_0 \equiv 0$. The implicit, nonlinear Bernoulli equation can be linearized by the transformation

$$\theta(t) = v_0^{1/(1-\gamma)}(t; \mathbf{a}),$$

so (10.169) becomes

$$0 = \theta'(t) + \psi_3'(t, u^*(t; \mathbf{a}); \mathbf{a})\theta(t) + \widehat{\psi}_2(t; \mathbf{a}), \tag{10.170}$$

which can easily be solved for $\theta(t)$ but only formerly in terms of the implicit dependence on the controls which requires iteration to obtain a fully explicit solution.

Besides iterations, the computation of the solution has many complications in terms of integrating the jump-integrals embedded in the coefficients, merging a regular time grid with the scheduled jumps, and assembling solutions on the unscheduled subintervals with the jump-conditions at the scheduled jumps in time. A summary for the computational algorithm is given in [130] along with the solutions for a test case of discrete jumps and various parameter values. The merger of the regular-time grid and the jump-time grid is illustrated in the simple jump-adapted code fragment in Figure 9.7 on p. 257. This complication is the most-asked question about this problem, mostly because it has many other applications in finance where jumps are added onto a continuous process, such as discrete transaction costs, dividends, and death benefits.

10.6 Exercises

Depending on the instructor, many of these exercises, whether numerical or theoretical, can be done with MATLAB, Maple, or Mathematica, but if theoretical, the Symbolic Toolbox in MATLAB will be needed.

1. Show that the Itô mean square limit for correlated bond-stock price noise at time t (10.16)

$$dW_B(t)dW_S(t) \stackrel{\text{ims}}{=} \rho dt \tag{10.171}$$

is valid. Are there any special treatments required if $\rho = 0$ or $\rho = \pm 1$? You may use the bivariate normal density in (B.144) or Table B.1 of selected moments in the preliminaries of Online Appendix B.

2. Merton [201] (see [203], p. 266) gives a stricter definition of more risky or riskier,

   Security $X_1(t)$ is **more risky** than security $X_2(t)$ if

   $$X_1(t) = q X_2(t) + \epsilon,$$

   where $(q, X_2(t), \epsilon)$ are mutually independent, $E[q] = 1$, $E[X_2(t)] = \mu_2(t)$, $E[\epsilon] = 0$, $\text{Var}[q] = \sigma_q^2$, $\text{Var}[X_1(t)] = \sigma_1^2(t)$, $\text{Var}[X_2(t)] = \sigma_2^2(t) > 0$, and $\text{Var}[\epsilon] = \sigma_\epsilon^2$.

   (a) Show that

   $$\sigma_1^2(t) = (1 + \sigma_q^2)\sigma_2^2(t) + \mu_2^2(t)\sigma_q^2 + \sigma_\epsilon^2 > \sigma_2^2(t).$$

   (b) Can you demonstrate this for a financial application or critically evaluate the applicability of the definition?

3. Verify that the call option pricing solution $\Phi^{(call)}(w, u)$, (10.82) or (10.83), satisfies the

   (a) Standard diffusion PDE (10.73), and

   (b) Call final condition (10.84).

   Either Maple or Mathematica is recommended.

4. Verify that the put option pricing solution $\Phi^{(put)}(w, u)$, (10.85) or (10.86), satisfies the

   (a) Standard diffusion PDE (10.73), and

   (b) Put final condition (10.87).

   Either Maple or Mathematica is recommended.

5. (a) Reverse the transformations to obtain option pricing solutions for

   $$F^{(call)}(S, B, t; T, K) \quad \text{and} \quad F^{(put)}(S, B, t; T, K)$$

   from the transformed solutions $\Phi^{(call)}(w, u)$ and $\Phi^{(put)}(w, u)$, respectively, through restoring the original variables B, S, F, and $\tau = T - t$.

   (b) Reduce the final restored form to the Black–Scholes assumptions on volatilities and mean rates.

6. Show that the transformed call and put option solutions satisfy a more usual call-put parity principle,

   $$\left(\widehat{F}^{((put))} - \widehat{F}^{((call))}\right)(\widehat{S}, \widehat{B}, \tau; T, K) = K \exp(-R(\tau)) - \widehat{S}(\tau),$$

   if certain conditions are satisfied and specify those conditions.

Suggested References for Further Reading

- Aït-Sahalia, 2004 [5]

- Aourir et al., 2002 [11]

- Bachelier, 1900 [16]

- Ball and Torous, 1985 [18]

- Black, 1989 [35]

- Black and Scholes, 1973 [34]

- Bingham and Kiesel, 2004 [33]

- Bossaerts, 2002, [41]

- Bridgeman, 1963 [42]

- Carr and Madan, 1999 [47]

- Chichilnisky, 1996 [52]

- Cont and Tankov, 2004 [60]

- Courtault et al., 2000 [62]

- Cox and Rubinstein, 1985 [64]

- Duffie, 1992 [75]

- Haberman, 1983 [103]

- Hanson, 1996 [109]

- Hanson and Westman, 2001 [123], 2002 [125, 126, 124, 127], 2003 [128, 129], and 2004 [130, 132]

- Heath and Schweizer, 2000 [136]

- D. Higham, 2004 [141]

- Hull, 2000 [148]

- Jarrow and Rosenfeld, 1984 [154]

- Jorion, 1989 [156]

- Kamien and Schwartz, 1981 [159]

- Karatzas et al., 1986 [160]

- Karatzas and Shreve, 1998 [161]

- Klebaner, 1998 [165]

- Kou, 2002 [170]

- Kou and Wang, 2004 [171]

- Lipton, 2001 [186]

- Merton, 1969 [198], 1971 [199], 1973a [200], 1973b [201], 1976 [202], and 1990 [203]

- Merton and Scholes, 1996 [204]

- Mikosch, 1998 [209]

- Neftci, 2000 [217]

- Pliska, 1997 [225]

- Press et al., 2002 [230]

- Rishel, 1999 [235]

- Rogers and Williams, 2000 [236]

- Runggaldier, 2003 [239]

- Sethi, 1997 [245]

- Sethi and Taksar, 1988 [246]

- Shreve, 2004 [248]

- Sneddon, 1957 [251]

- Tavella and Randall, 2000 [264]

- Westman and Hanson, 2000, [277]

- Wilmott, 2000 [284]

- Wilmott, Howison, and Dewynne, 1996 [282]

- Yan and Hanson, 2006 [289]

- Zhu, 2005 [291]

- Zhu and Hanson, 2005 [292] and 2006 [293]

Chapter 11

Applications in Mathematical Biology and Medicine

Despite assertions in both the lay and the professional literature, it is now known that normal physiology is anything but "regular."...Loss of event-to-event variability occurs during normal aging and also occurs pathologically in critical illness.
—Dr. Timothy G. Buchman, 2004 [45]

Mathematics Is Biology's Next Microscope, Only Better; Biology Is Mathematics' Next Physics, Only Better.
—Joel E. Cohen, 2004 [59]

The application to optimal harvesting in uncertain environments is made in the presence of both background Gaussian noise and catastrophic jump events. Many problems in nature exhibit random effects and undergo catastrophic changes for which the stochastic calculus of continuous Wiener processes alone is inadequate.

11.1 Stochastic Bioeconomics: Optimal Harvesting Applications

For deterministic problems of optimal harvesting of renewable resources, the seminal reference by C. W. Clark is *Mathematical Bioeconomics: The Optimal Management of Renewable Resources* [57]. The book is nicely self-contained with introductions to the necessary economics, calculus of variations, and optimal control theory. An excellent survey of stochastic bioeconomics is given by Anderson and Sutinen in [9].

In this chapter, examples of optimal harvesting problems in random environments are illustrated. The first application is optimal harvesting with random population fluctuations [242]. A second application is optimal harvesting with random population fluctuations but also with price fluctuations [116] and so is a two-dimensional state generalization of the first application.

339

11.1.1 Optimal Harvesting of Logistically Growing Population Undergoing Random Jumps

This problem of natural logistic growth of a renewable resource subject to random disasters and bonanzas was treated by Ryan and Hanson [242]. The parameter data were motivated by the **boom and bust** characteristics of Antarctic pelagic whaling at the time as studied by Clark and Lamberson [58]. The problem is summarized in the notation of this book, so for more information the reader should refer to [242].

Let $X(t)$ be the amount of biomass (mass of the biological species) of the harvested species at time t with stochastic dynamics given by

$$dX(t) = X(t)\left(r(1 - X(t)/K) - qU(t)\right)dt + X(t)\sum_{i=1}^{n_p} v_i dP_i(t), \qquad (11.1)$$

where $X(0) = x_0 > 0$ is the initial biomass, $r > 0$ is the constant intrinsic growth rate, and $K > 0$ is the constant biomass carrying capacity that reflects the size of the population that the environment can support in absence of harvesting and other factors. Hence, the natural growth function $f(x) = rx(1 - x/K)$ is called the logistic function since as $x \to K$ a saturation effect due to crowding limits growth. Under the assumption of linear harvesting, the rate of harvesting is $H(t) = h(X(t), U(t)) = qU(t)X(t)$, where $U(t) \geq 0$ is the harvesting effort or rate and also the control variable, while $q > 0$ is called the **catchability coefficient** and is a measure of the efficiency of the harvest. The population suffers from rare random jumps from various sources for $i = 1 : n_p$ linear in the biomass $X(t)$ with proportions $-1 < v_i$. The negative values $-1 < v_i < 0$ denote disastrous effects but are limited by a lower bound so that the population will not be wiped out by a single disaster, while the positive values $v_i > 0$ denote bonanzas or beneficial effects. The randomness of the jumps is modeled by a set of n_p Poisson processes $P_i(t)$ with common infinitesimal means and variances

$$E[dP_i(t)] = \lambda_i dt = \text{Var}[dP_i(t)]$$

for $i = 1 : n_p$, where $\lambda_i > 0$ is the ith jump-rate. The actual jump at the jth jump-time $t_{i,j}$ of the ith Poisson process is given in jump notation by

$$[X](t_{i,j}) \equiv X(t_{i,j}^+) - X(t_{i,j}^-) = v_i X(t_{i,j}^-).$$

The motivation for the multitude of jump terms in (11.1) is that large random fluctuations can come from many causes, like climatic changes, overfishing, and epizootics (see [212, 139, 250, 242], for instance).

In [241], Ryan and Hanson treated the optimal harvesting case where the natural growth of the biomass is exponential with jumps, i.e., $1/K = 0$ and the natural growth function is linear, $f(x) = rx$, so the overall growth of $X(t)$ is exponential with harvesting and jumps. The model (11.1) is a pure jump model with logistic drift because the focus is on the effects of jumps on the harvesting bioeconomics, although diffusion terms could have been easily added to the model. For $r > 0$ with no harvesting and jumps, the exponential model $dX(t) = rX(t)dt$ leads to unbounded exponential growth, while the logistic model $dX(t) = rX(t)(1 - X(t)/K)dt$ leads to saturated growth as $X(t) \to K^-$ if $x_0 < K$ or limiting decay as $X(t) \to K^+$ if $x_0 > K$. The density dependent (nonlinear) jump case is treated by Hanson and Ryan in [114].

The economic value of the harvest, starting at time t with biomass x and ending at the final time t_f, is given by the expected, discounted present value,

$$\overline{V}[X, U](x, t) = \mathrm{E}\left[\int_t^{t_f} e^{-\delta s} h(X(s), U(s)) R(X(s), U(s)) ds \right| \\ X(t) = x, U(t) = u \right],$$

(11.2)

where δ is the continuous, inflation-corrected discount rate with discounting starting at $t = 0$ and $\exp(-\delta t)$ is the discount factor which accounts for the opportunity costs of investing money elsewhere in a secure investment. The instantaneous net harvest revenue per unit harvest is

$$R(x, u) = (ph(x, u) - C(u)) / h(x, u).$$

It can be assumed that $x > 0$ and $u > 0$ to avoid dividing by zero, but the net revenue always appears in the product form $h(x, u) R(x, u)$, so the divide check is not needed. The price of a unit of a harvested biomass ($h = qux$) is p, and

$$C(u) = c_1 u + c_2 u^2$$

is the total cost of the harvesting effort when the biomass or stock size is x given that $c_2 > 0$ so that $C(u)$ is a genuine quadratic. Note that $C(u)$ is assumed to be quadratic in the effort, which suggests that the effort is more costly the larger it becomes. In the case of fisheries, this might mean that more inefficient fishing boats or less experienced fisherman are used as the fishing effort $U(t)$ increases. The effort is assumed to be bounded, i.e., constrained, so that

$$0 \le U^{(\mathrm{min})} \le U(t) \le U^{(\mathrm{max})} < \infty$$

(11.3)

and the objective is to seek the maximum, expected current value

$$v^*(x, t) = \max_U \left[\overline{V}[X, U](x, t) \right].$$

Thus, the goal is to calculate the optimal value $\overline{V}^*(x, t)$ and the optimal feedback control or effort

$$u^*(x, t) = \operatorname*{argmax}_U \left[\overline{V}[X, U](x, t) \right]$$

for $0 \le t < t_f$. However, the initial optimal expected, current value $\overline{V}^*(x, 0)$ is the optimal expected, discounted present value of future revenues.

To facilitate the application of the Hamilton–Jacobi–Bellman (HJB) equation Theorem 6.3 to the discounted current value form in (11.2) with the so-called **cost function** $C(\mathbf{x}, \mathbf{u}, t) = \exp(-\delta t) h(\mathbf{x}, u) R(\mathbf{x}, u)$ here, the discount factor $\exp(-\delta t)$ can be removed in the pseudo-Hamiltonian by converting from the present value $v^*(x, t)$ of Chapter 6 to the current value $\mathcal{V}^*(x, t)$ by the transformation

$$v^*(x, t) = \exp(-\delta t) \mathcal{V}^*(x, t).$$

Thus, $v_t^*(x, t) = \exp(-\delta t)(V_t^*(x, t) - \delta t \, V^*(x, t))$, where v_t^* and V_t^* are the partial derivatives of the value functions with respect to time. Note that initially both value functions coincide, $v^*(x, 0) = V^*(x, 0)$.

Kamien and Schwartz [159] define the difference between the present and current value in terms of the present and current value Hamiltonians. The current value Hamiltonian $\widehat{\mathcal{H}}(x, u, t)$ is related to the present value Hamiltonian $\mathcal{H}(x, u, t)$ by

$$\widehat{\mathcal{H}}(x, u, t) \equiv e^{+\delta t} \mathcal{H}(x, u, t)$$
$$= (pqux - c_1 u - c_2 u^2) + (rx(1 - x/K) - qux)V_x^*(x, t)$$
$$+ \sum_{i=1}^{n_p} \lambda_i(t) \left(V^*(x + v_i x, t) - V^*(x, t) \right),$$

canceling out the discount factor $\exp(-\delta t)$. Further, separating out the control terms, the HJB equation is

$$0 = V_t^*(x, t) - \delta V^*(x, t) + rx(1 - x/K)V_x^*(x, t) + \widehat{\mathcal{S}}^*(x, t)$$
$$+ \sum_{i=1}^{n_p} \lambda_i(t) \left(V^*(x + v_i x, t) - V^*(x, t) \right), \tag{11.4}$$

where the control switching term contains all control terms in the quadratic form,

$$\widehat{\mathcal{S}}(x, u, t) \equiv ((p - V_x^*(x, t))qx - c_1 - c_2 u)u,$$

including only the control-dependent terms. The interior critical point of $\widehat{\mathcal{S}}(x, u, t)$ with respect to the control u is the regular optimal control,

$$u^{(\text{reg})}(x, t) = \frac{0.5}{c_2} \left((p - V_x^*(x, t)) qx - c_1 \right), \tag{11.5}$$

since $c_2 > 0$, with the regular control being easily computed in terms of the gradient $V_x^*(x, t)$ due to the quadratic cost assumption. As in the case of many applications, the control is constrained as in (11.3), so the constrained optimal control is the composite **bang-regular-bang** control function,

$$u^*(x, t) = \begin{cases} U^{(\min)}, & u^{(\text{reg})}(x, t) \le U^{(\min)} \\ u^{(\text{reg})}(x, t), & U^{(\min)} \le u^{(\text{reg})}(x, t) \le U^{(\max)} \\ U^{(\max)}, & U^{(\max)} \le u^{(\text{reg})}(x, t) \end{cases}. \tag{11.6}$$

Consequently, the optimal control switch term is

$$\widehat{\mathcal{S}}^*(x, t) \equiv \widehat{\mathcal{S}}(x, u^*(x, t), t) = c_2 u^*(x, t) \left(2u^{(\text{reg})}(x, t) - u^*(x, t) \right)$$

after some algebraic manipulations. When $u^{(\text{reg})}(x, t)$ is within the constraints (11.3), the switch term will be quadratic in $u^{(\text{reg})}(x, t)$, i.e., $\mathcal{S}^*(x, t) = c_2(u^{(\text{reg})})^2(x, t)$, and consequently quadratic in the value gradient $V_x^*(x, t)$, so the PDE of stochastic dynamic programming will be PDE with a quadratic nonlinearity. The gradient $V_x^*(x, t)$ is the so-called **shadow price** [57] for the way it directly modifies the price p in (11.5) and represents the expected value of future harvests [57]. The PDE is also a partial differential-difference equation, since the discrete Poisson jumps lead to difference terms in (11.4) rather than the mark integral over difference terms as more generally presented in Chapter 6.

The final condition for the backward HJB equation is $V^*(x, t_f) = 0$ for $x > 0$ in the absence of salvage or terminal costs. Thus, the final regular control or effort at $t = t_f$ is given by

$$u^{(\text{reg})}(x, t_f) = (pqx - c_1)/c_2 = c_1(x - x_f)/(2c_2 x_f),$$

where $x_f \equiv c_1/(pq)$ is also the deterministic equilibrium stock value x_∞ [57]. However, in this stochastic case, if $c_1 \neq 0$, the final minimum control switch point is

$$x_{f,\min} = x_f \left(1 + 2c_2 U^{(\min)}/c_1\right)$$

and the final maximum control switch point is

$$x_{f,\max} = x_f \left(1 + 2c_2 U^{(\max)}/c_1\right).$$

As the biomass approaches extinction levels, $X(t) \to 0^+$, the rate of change $dX(t)$ (11.1) vanishes along with it, but the net revenue $R(x, u)$ should have become negative since costs dominate at low biomass. Hence, it will be assumed in addition that $R(x, u) \geq 0$, i.e., replacing $R(x, u)$ by $\max[R(x, u), 0]$, so that the extinction boundary condition is

$$V^*(0^+, t) = 0$$

for $0 < t < t_f$.

A very reasonable approximation to the solution can be obtained from the **quasi-deterministic approximation**,

$$dX^{(\text{qdet})}(t) \equiv \mathrm{E}\left[dX(t)|\, X(t) = X^{(\text{qdet})}(t), U(t) = U^{(\text{qdet})}(t)\right]$$
$$= r^{(\text{qdet})} X^{(\text{qdet})}(t) \left(1 - X^{(\text{qdet})}(t)/K^{(\text{qdet})}\right) dt,$$

where $r^{(\text{qdet})} \equiv r + \sum_{i=1}^{n_p} \lambda_i \nu_i$ and $K^{(\text{qdet})} \equiv K r^{(\text{qdet})}/r$, comprising an approximate logistic model. For this simplified model, the HJB equation will no longer have difference terms since the jumps have been averaged out, but the optimal control will still be of the form (11.6).

Due to the complexity of the PDE, numerical methods are needed to approximate the solution. The HJB equation can be solved with PDE-oriented finite difference methods [109] or the probability-oriented Markov chain approximation [179]. The finite difference method requires a sufficiently small mesh ratio for a comparison regular parabolic PDE [109] in the jump-diffusion case, while the Markov chain approximation, if the Markov chain probabilities are properly constructed, automatically comes with a weak convergence property [179]. For the current application in [242] and also in [241], Hanson and Ryan used the PDE-oriented finite difference method of [109] with predictor-corrector procedures to iterate on the nonlinear terms and precision-preserving interpolation to approximate the jump terms by values at neighboring finite difference nodes. Both methods are variations of the finite difference method and are summarized in Chapter 8 in Section 8.1 for the PDE-oriented method and Section 8.2 for the Markov chain approximation, respectively.

The primary bioeconomic parameters used in [242] come from [58], i.e., r, K, q, p, and c_1, while other parameters like δ, t_f, λ_i, and ν_i are reasonable estimates. Many of these estimated parameters were subjected to sensitivity tests in [242] in the many numerical results presented there. Some of the parameters are now obsolete, since whaling is no longer permitted in many countries or else is highly restricted. Interest and discount rates are much lower now than then. However, significant sensitivity in u^* and V^* was found to the parameters δ, c_2, and $\lambda_i \nu_i / r$ for both a bonanza-dominated case with $\lambda_i \nu_i / r = 2\delta_{i,1}$ and a disaster-dominated case with $\lambda_i \nu_i / r = -0.5\delta_{i,2}$, where here $\delta_{i,j}$ is the Kronecker delta. In particular, in the **cheap control** limit as $c_2 \to 0^+$, the **bang-regular-bang** control law approaches a **bang-bang** control law in the absence of a regular control component.

11.1.2 Optimal Harvesting with Both Price and Population Random Dynamics

The optimal harvesting problem, under joint population and price fluctuations in a random jump environment of Hanson and Ryan [116], is also an example of a two-dimensional state problem. Here, the problem is briefly summarized in the notation of this book. For a general introduction to stochastic resource modeling, consult Anderson and Sutinen [9] or Mangel [193].

Again, let $X_1(t)$ be the amount of biomass (mass of the biological species) of the harvested species population at time t with stochastic dynamics consisting of logistic deterministic dynamics, discrete Poisson jumps, and now with background stochastic diffusion,

$$dX_1(t) = X_1(t)\left((r_1(1 - X_1(t)/K_1) - q_1U_1(t))\,dt + \sigma_1 dW_1(t) + \sum_{i=1}^{n_1} \nu_{i,1} dP_{i,1}(t)\right), \quad (11.7)$$

$X_1(0) = x_{1,0} > 0$, where the extra subscript 1 designates population parameters or processes, i.e., the essential biological component of the bioeconomic process. The diffusion process $\sigma_1 dW_1(t)$ satisfies zero mean and $\sigma_1^2 dt$ variance properties, with $\sigma_1 > 0$ assumed. For the Poisson processes, $\nu_{i,1} > -1$ and $\lambda_{i,1} > 0$ for $i = 1:n_1$.

The economic process or price process $p(t)$ depends on the time-dependent biomass harvest rate $H(t) = h(X_1(t), U_1(t)) = q_1 U_1(t) X_1(t)$ and other stochastic processes. Since on the average $p(t)$ decreases as $h(t)$ increases [116] following **supply-demand principles**, the price is assumed to satisfy

$$p(t) = (p_0/H(t) + p_1) X_2(t), \quad (11.8)$$

where p_0 is a constant supply-demand coefficient such that $p(t)H(t)$ is the gross harvest return, p_1 is the constant price per unit harvested biomass coefficient, and $X_2(t)$ is a fluctuating inflationary factor [116] satisfying the SDE

$$dX_2(t) = X_2(t)\left(r_2 dt + \sigma_2 dW_2(t) + \sum_{i=1}^{n_2} \nu_{i,2} dP_{i,2}(t)\right), \quad (11.9)$$

$X_2(0) = x_{2,0} > 0$, where the extra subscript 2 designates parameters and processes in the price process SDE, $\sigma_2 > 0$, $\nu_{i,2} > -1$, and $\lambda_{i,2} > 0$ for $i = 1:n_2$. It is assumed that all primary stochastic processes, $P_{i,1}(t)$, $P_{i,2}(t)$, $W_1(t)$, and $W_2(t)$, are pairwise independent.

The economic value of the harvest, starting at time t with biomass x_1 and ending at the final time t_f, is given by the expected, discounted current value

$$\overline{V}[\mathbf{X}, U_1](\mathbf{x}, t) = \mathrm{E}\left[\int_t^{t_f} e^{-\delta s} H(s) R(\mathbf{X}(s), U_1(s)) ds \middle| \mathbf{X}(t) = \mathbf{x}, U_1(t) = u_1\right], \quad (11.10)$$

where δ is the continuous, nominal discount rate, uncorrected by inflation since inflation is included in $X_2(t)$, with discounting starting at t. The random vector state is $\mathbf{X}(t) = [X_1(t)\ X_2(t)]^\top$ and $\mathbf{x} = [x_1\ x_2]^\top$ is a sampled vector state such that

$$R(\mathbf{x}, u_1) = ((p_0 + p_1 h(x_1, u_1)) x_2 - C(u_1)) / h(x_1, u_1)$$

is the instantaneous net harvest revenue rate per unit biomass. Note that $x_1 > 0$ and $u_1 > 0$. The net revenue always appears in the product form $h(x_1, u_1) R(\mathbf{x}, u_1)$ so a zero check for $R(\mathbf{x}, u_1)$ is unneeded. The price of a harvested biomass unit is p and

$$C(u_1) = c_1 u_1 + c_2 u_1^2$$

is the total cost of the harvesting effort given that $c_2 > 0$ so that $C(u_1)$ is a genuine quadratic. The effort control constraints are again assumed to be

$$0 \leq U_1^{(\min)} \leq U_1(t) \leq U_1^{(\max)} < \infty, \quad (11.11)$$

while the objective is to seek the maximum, expected current value

$$v^*(\mathbf{x}, t) = \max_{U_1}\left[\overline{V}[\mathbf{X}, U_1](\mathbf{x}, t)\right]$$

and the optimal feedback effort control

$$u_1^*(\mathbf{x}, t) = \underset{U_1}{\operatorname{argmax}}\left[\overline{V}[\mathbf{X}, U_1](\mathbf{x}, t)\right]$$

for $0 \leq t < t_f$. Again, the present values $v^*(\mathbf{x}, t)$ are transformed current values $V^*(\mathbf{x}, t)$,

$$v^*(\mathbf{x}, t) = \exp(-\delta t) V^*(\mathbf{x}, t).$$

The current value Hamiltonian $\widehat{\mathcal{H}}(\mathbf{x}, u_1, t)$ related to the present value Hamiltonian $\mathcal{H}(\mathbf{x}, u_1, t)$ is

$$\begin{aligned}
\widehat{\mathcal{H}}(\mathbf{x}, u_1, t) &\equiv e^{+\delta t} \mathcal{H}(\mathbf{x}, u_1, t) \\
&= (p_0 + p_1 q_1 u_1 x_1) x_2 - c_1 u_1 - c_2 u_1^2 \\
&\quad + (r_1 x_1 (1 - x_1/K_1) - q_1 u_1 x_1) V_{x_1}^*(\mathbf{x}, t) + r_2 x_2 V_{x_2}^*(\mathbf{x}, t) \\
&\quad + \frac{1}{2} \sigma_1^2 x_1^2 V_{x_1,x_1}^*(\mathbf{x}, t) + \frac{1}{2} \sigma_2^2 x_2^2 V_{x_2,x_2}^*(\mathbf{x}, t) \\
&\quad + \sum_{i=1}^{n_1} \lambda_{i,1}(t) \left(V^*((1 + \nu_{i,1}) x_1, x_2, t) - V^*(\mathbf{x}, t)\right) \\
&\quad + \sum_{i=1}^{n_2} \lambda_{i,2}(t) \left(V^*(x_1, (1 + \nu_{i,2}) x_2, t) - V^*(\mathbf{x}, t)\right).
\end{aligned}$$

Upon canceling out the discount factor $\exp(-\delta t)$ and separating out the control dependence from the Hamiltonian, the HJB equation is

$$
0 = V_t^*(\mathbf{x}, t) - \delta V^*(\mathbf{x}, t) + p_0 x_2 + r_1 x_1 (1 - x_1/K_1) V_{x_1}^*(\mathbf{x}, t) + r_2 x_2 V_{x_2}^*(\mathbf{x}, t)
$$
$$
+ \frac{1}{2}\sigma_1^2 x_1^2 V_{x_1, x_1}^*(\mathbf{x}, t) + \frac{1}{2}\sigma_2^2 x_2^2 V_{x_2, x_2}^*(\mathbf{x}, t)
$$
$$
+ \sum_{i=1}^{n_1} \lambda_{i,1}(t) \left(V^*((1 + v_{i,1})x_1, x_2, t) - V^*(\mathbf{x}, t) \right) \tag{11.12}
$$
$$
+ \sum_{i=1}^{n_2} \lambda_{i,2}(t) \left(V^*(x_1, (1 + v_{i,2})x_2, t) - V^*(\mathbf{x}, t) \right)
$$
$$
+ \widehat{S}^*(\mathbf{x}, t),
$$

where the control switching term has the quadratic form

$$
\widehat{S}(\mathbf{x}, u_1, t) \equiv p_1 q_1 u_1 x_1 x_2 - c_1 u_1 - c_2 u_1^2 - q_1 u_1 x_1 V_{x_1}^*(\mathbf{x}, t),
$$

including only the control dependent terms. The interior critical point of $\widehat{S}(x, u, t)$ with respect to the control u is the regular optimal control,

$$
u_1^{(reg)}(\mathbf{x}, t) = \frac{0.5}{c_2} \left(\left(p_1 x_2 - V_{x_1}^*(\mathbf{x}, t) \right) q_1 x_1 - c_1 \right), \tag{11.13}
$$

since $c_2 > 0$, with the regular control being easily computed in terms of the gradient $V_x^*(\mathbf{x}, t)$ due to the quadratic cost assumption. As in the case of many applications, the control is constrained as in (11.11), so the constrained optimal control is the composite **bang-regular-bang** control function,

$$
u_1^*(\mathbf{x}, t) = \begin{cases} U_1^{(min)}, & u_1^{(reg)}(\mathbf{x}, t) \leq U_1^{(min)} \\ u_1^{(reg)}(\mathbf{x}, t), & U_1^{(min)} \leq u_1^{(reg)}(\mathbf{x}, t) \leq U_1^{(max)} \\ U_1^{(max)}, & U_1^{(max)} \leq u_1^{(reg)}(\mathbf{x}, t) \end{cases}. \tag{11.14}
$$

The temporal side condition for the backward HJB equation (11.12) is the final condition $V^*(\mathbf{x}, t_f) = 0$ in the absence of any terminal conditions for the first quadrant of state space and the natural corner condition

$$
V^*(0, 0, t) = - \left(c_1 + c_2 U_1^{(min)} \right) U_1^{(min)} (1 - \exp(-\delta(t_f - t)))/\delta
$$

at the origin $(0, 0)$ for $0 < t < t_f$ since $U_1^{(min)} \geq 0$. On the edge $(x_1, 0)$ for $x_1 > 0$, the boundary condition is similar to solving the pure jump optimal resource HJB equation of Subsection 11.1.1 except that there is an additional diffusion term. On the edge $(0, x_2)$ for $x_2 > 0$, the boundary condition also involves solving an even less similar HJB equation since in this case the deterministic inflationary growth is exponential rather than logistic.

In [116], data of the International Pacific Halibut Commission (IPHC) annual reports [149] are used since the catch and price data were readily available over a long period.

Other data came from Clark [57]. The hybrid extrapolated-predictor-corrector Crank–Nicolson method similar to that described in the previous section and in Section 8.1 was used. The major result was that large inflationary increases had a very strong effect on the optimal return but not on the optimal effort.

Another multidimensional optimal harvesting problem can be found in the Lake Michigan salmon-alewife predator-prey model of Hanson in [106], where the alewife suffered large-scale die-offs every several years. The model was also mixed economically, since the salmon are fished recreationally while the alewife were fished in a commercial fishery, now disbanded. Multidimentional visualization and parallel processing for renewable resources are developed by Practico et al. [228] and Hanson et al. [113].

11.2 Stochastic Biomedical Applications

Variability plays an important role in medicine. Discussing critical care, Buchman [45] emphasizes that variability is **normal** for the individual patient and that illness is often accompanied by loss of individual variability. For instance, Boker et al. [37] find a variable ventilator improved lung function during surgery and recovery more than a controlled constant ventilator. Priplata et al. [231] find that input noise enhances balance, particulary for the elderly. Ashkenazy et al. [14] present a stochastic model to portray the variation in an individual's gait showing that variability changes with maturation and aging. Moss et al. [211] find increased sensitivity in detecting threshold levels with stochastic noise for stochastic resonance to occur for nonlinear neural systems during information processing.

Swan [261] presents many applications of optimal control to biomedicine in his book, but the emphasis is on deterministic compartment or ODE models. One chapter is on cancer therapy control and another is on drug administration control. Murray's [213, 214] two volumes on models of mathematical biology have information on cancer and other models but no real optimal control models.

According to Steel [255] and Goldie and Coldman [100], stochastic effects play an important role in the stages of development of cancer, the subsequent growth, and the invasiveness of tumors or the more liquid lymphomas. Mutations can be induced by environmental chemical agents or ionizing radiation, while spontaneous mutations are more rare, usually without obvious cause [100].

11.2.1 Diffusion Approximation of Tumor Growth and Tumor Doubling Time Application

Tumor Growth Branching Process

Sometimes approximating a discrete stochastic process by a diffusion process can be useful. Hanson and Tier [118] present an example for a branching process for modeling the growth of tumor cells. This discrete model is then approximated as a diffusion process for the purposes of analysis and computation.

Let B_i be the branching process, in the ith generation for $i = 1, 2, 3, \ldots$, such that there are three possible transitions in the time-interval $(t, t + \Delta t)$ for generation i,

$$B_i = \begin{cases} 0 & \text{if cell death} \\ 1 & \text{if no cell change} \\ 2 & \text{if cell division} \end{cases}, \tag{11.15}$$

similar to a birth-death model but with a middle state of no change. This yields a total cancer cell count change from $N(t)$ at time t to

$$N(t + \Delta t) = \sum_{i=1}^{N(t)} B_i \tag{11.16}$$

with the cell count change in $(t, t + \Delta t)$ being

$$\Delta N(t) = \sum_{i=1}^{N(t)} B_i - N(t) = \sum_{i=1}^{N(t)} (B_i - 1).$$

The B_i are assumed to be IID random variables with basic conditional moments that are dependent on $N(t)$, i.e., density dependent,

$$\mathrm{E}[B_i \mid N(t) = n] = m(n; \Delta t)$$

and

$$\mathrm{Var}[B_i \mid N(t) = n] = v(n; \Delta t).$$

The higher moments

$$\mathrm{E}[(B_i - m(N(t), \Delta t)^k \mid N(t) = n] = m_k(n; \Delta t)$$

will also be needed to demonstrate that they will be small for $k > 3$.

Thus, the basic conditional moments of the tumor cell count change $\Delta N(t)$ are

$$\mathrm{E}[\Delta N(t) \mid N(t) = n] = \sum_{i=1}^{n} \mathrm{E}[B_i \mid N(t) = n] - n = n(m(n; \Delta t) - 1)$$

and

$$\begin{aligned}
\mathrm{Var}[\Delta N(t) \mid N(t) = n] &= \mathrm{E}[(\Delta N(t) - n(m(n; \Delta t) - 1))^2 \mid N(t) = n] \\
&= \mathrm{E}\left[\sum_{i=1}^{n} (B_i - m(n; \Delta t))^2 \mid N(t) = n \right] \\
&= \sum_{i=1}^{n} \sum_{j=1}^{n} \mathrm{E}[(B_i - m(n; \Delta t))(B_j - m(n; \Delta t)) \mid N(t) = n] \\
&= \sum_{i=1}^{n} \mathrm{E}[(B_i - m(n; \Delta t))^2 \mid N(t) = n] = nv(n; \Delta t),
\end{aligned}$$

where the usual diagonalization technique has been used to apply the IID property of the B_i.

Diffusion Approximation of the Tumor Branching Process

Using some additional assumptions, a diffusion approximation will be constructed. Suppose T is some reference generation time such as the threshold for detection so

$$\tau = t/T$$

is a new scaled time, and let a new scaled stochastic process be

$$X(\tau) = N(t)/T$$

since the tumor will grow roughly as the number of generations. For the model to be consistent with these scalings, the basic moments have to be refined so that the changes in $X(\tau)$ are small for small changes in τ. *The basic idea of the diffusion approximation is that it will not work well unless the order of the mean state changes are the same order as the time changes*, i.e., $E[\Delta X(\tau)] = O(\Delta \tau)$ and the variance of the state changes are of the same order. Hence, let the infinitesimal mean be of the *near-replacement* form

$$m(N(t), \Delta t) = 1 + (m_1(X(\tau)) + o(1))\Delta\tau \text{ as } \Delta\tau \to 0,$$

where $m_1(x)$ is a function to be specified, and let the infinitesimal variance be of the form

$$v(N(t), \Delta t) = (v_1(X(\tau)) + o(1))T\Delta\tau \text{ as } \Delta\tau \to 0,$$

where $v_1(x)$ is a function to be specified. In addition, the higher moments should satisfy the form

$$m_k(N(t), \Delta t) = o(\Delta\tau) \text{ as } \Delta\tau \to 0.$$

First for the diffusion approximation, the infinitesimal moments of $X(\tau)$ with $\Delta X(\tau) = \Delta N(t)/T$ are computed as in (7.64)–(7.65),

$$\mu(x) = \lim_{\Delta\tau \to 0} \frac{E[\Delta X(\tau) \mid X(\tau) = x]}{\Delta\tau} = \lim_{\Delta\tau \to 0} \frac{E[\Delta N(T\tau)/T \mid N(T\tau) = Tx]}{\Delta\tau}$$

$$= \lim_{\Delta\tau \to 0} \frac{x(m(Tx, T\Delta\tau) - 1)}{\Delta\tau} = \lim_{\Delta\tau \to 0} [x(m_1(x) + o(1))] = xm_1(x) \quad (11.17)$$

and

$$\sigma^2(x) = \sigma_{1,1}(x) = \lim_{\Delta\tau \to 0} \frac{\text{Var}[\Delta X(\tau) \mid X(\tau) = x]}{\Delta\tau}$$

$$= \lim_{\Delta\tau \to 0} [x(v_1(x) + o(1))] = xv_1(x). \quad (11.18)$$

In addition, the vanishing of higher central moment condition (7.66) when $k = 3$ is used (since any $k \geq 3$ can be used) in place of the probabilistic continuity condition (7.63) by invoking a form of the Chebyshev inequality (7.67). Hence,

$$\lim_{\Delta\tau \to 0} \frac{E[|\Delta X(\tau) - x(m(Tx, T\Delta\tau) - 1)|^3 \mid X(\tau) = x]}{\Delta\tau} = \lim_{\Delta\tau \to 0} \frac{n \cdot m_3(Tx, T\Delta\tau)}{T^3\Delta\tau}$$

$$= \lim_{\Delta\tau \to 0} \frac{x \cdot o(\Delta\tau)}{T^2\Delta\tau} = 0,$$

completing the verification of the diffusion approximation and going substantially beyond the justification in [118].

For this particular application, the deterministic growth is chosen to be the **Gompertz growth model** [255, 100]

$$\mu(x) = x m_1(x) = \mu_1 x \ln(K/x), \tag{11.19}$$

where μ_1 is a constant growth coefficient and K is the carrying capacity or saturation level. Note that the Gompertz growth is singular as $x \to 0^+$ in that its derivative is unbounded as $x \to 0^+$ since $d(\mu_1 x \ln(k/x))/dx = -\mu_1 \ln(ex/K) \to +\infty$. However, the Gompertz model is often used in analyzing cancer experiments, although other models are also used, such as the simpler exponential growth model in shorter clinical trials [100]. In addition, the infinitesimal variance is taken to be purely linear, i.e.,

$$\sigma^2(x) = x v_1(x) = \sigma_1 x,$$

where $\sigma_1 > 0$ is a constant.

In summary, the backward operator in this time homogeneous case is

$$\mathcal{B}_{x_0}[u](x_0) = \frac{1}{2}\sigma_1 x u''(x_0) + \mu_1 x_0 \ln(K/x_0) u'(x_0). \tag{11.20}$$

Expected Tumor Doubling Time

The interest here is in the tumor doubling time, so suppose the tumor start is at the observed size c and then find the time it takes the tumor to double in size to $X(t) = 2c$. However, due to the stochastic nature of cancer, the tumor could become extinct, $X(t) = 0$, before it doubles in size. Hence, the proper problem is one of conditional probabilities, with the condition that the tumor doubles before it becomes extinct.

First consider the exit time at $2c$ starting at the general size $x_0 > 0$ at time t_0,

$$\tau_e^{(2c)}(x_0, t_0) = \inf_t [\, t \mid X(t) = 2c, X(s) \in (0^+, 2c), t_0 \le s < t, X(t_0) = x_0], \tag{11.21}$$

so the backward formulation of Subsection 7.7.1 can be used with variable x_0, here with $b = 2c$. Again let the exit time distribution function be

$$\Phi_{\tau_e^{(2c)}(x_0, t_0)}(t) = \mathrm{Prob}[\tau_e^{(2c)}(x_0, t_0) < t]$$

with corresponding density $\phi_{\tau_e^{(2c)}(x_0, t_0)}(t)$ and let the ultimate probability of exit at $X(t) = 2c$ be

$$\Phi_e^{(2c)}(x_0, t_0) = \int_0^\infty \phi_{\tau_e^{(2c)}(x_0, t_0)}(t) dt.$$

Consequently, the final answer will be the expected doubling time

$$\Phi_{\mathrm{dbl}}(c) = \Phi_e^{(2c)}(c, 0),$$

eventually using the initial values $x_0 = c$ and $t_0 = 0$.

Now let $u = u_0(x_0) = \Phi_e^{(2c)}(x_0, 0)$, and this satisfies the homogeneous backward equation

$$\mathcal{B}_{x_0}[u_0](x_0) = \frac{1}{2}\sigma_1 x u_0''(x_0) + \mu_1 x_0 \ln(K/x_0)u_0'(x_0) = 0 \qquad (11.22)$$

from (11.20) in particular and (7.59) in general, but with boundary conditions

$$u_0(0^+) = 0 \quad \text{and} \quad u_0(2c) = 1,$$

since an exit at $X(0) = 0^+$ is excluded under the conditioning and an exit at $X(t) = 2c$ is a certain conditional exit. Equation (11.22) is integrable in u and $x_0 > 0$ by using an integrating factor or its inverse called the **Wronskian** (also called the diffusion **scale density**),

$$W(x_0) \equiv \exp\left(-2\int^{x_0} \frac{\mu(x)}{\sigma^2(x)}dx\right) = \exp\left(-2\frac{\mu_1}{\sigma_1}\int^{x_0} \ln\left(\frac{K}{x}\right)dx\right)$$

$$= \exp\left(-2\frac{\mu_1 x_0}{\sigma_1}\ln\left(\frac{K}{ex_0}\right)\right) = (\beta_1 x_0)^{\gamma_1 x_0} \qquad (11.23)$$

here for the Gompertz model, where $\gamma_1 = 2\mu_1/\sigma_1$ and $\beta_1 = e/K > 0$. Thus, (11.22) simplifies to

$$(u_0'/W)'(x_0) = 0.$$

Thus, after two integrations and boundary condition substitutions lead to the solution of the boundary value problem, we have

$$\Phi_e^{(2c)}(x_0, 0) = u_0(x_0) = \frac{\int_{0^+}^{x_0} W(x)dx}{\int_{0^+}^{2c} W(y)dy}. \qquad (11.24)$$

Since as $x \to 0^+$, $W(x) = (\beta_1 x)^{\gamma_1 x} \sim 1 + \gamma_1 x \ln(\beta_1 x)$ and then

$$\int_{0^+}^{x} dy W(y) \sim x + 0.5\gamma_1 x^2(\ln(\beta_1 x) - 0.5),$$

$W(x)$ is integrable as $x \to 0^+$ so that (11.24) is well defined, all other points on $(0, 2c)$ being obviously regular or nonsingular points. Thus, setting $x_0 = c$ as the initial size gives the ultimate probability of a tumor doubling in size, $\Phi_e^{(2c)}(c, 0)$. More results by way of numerical and asymptotic approximations are given in [118].

The expected doubling time from (7.61) is

$$T_e^{(2c)}(c) = M_e^{(2c)}(c)/\Phi_e^{(2c)}(c, 0), \qquad (11.25)$$

normalizing the first moment from (7.60), which here is

$$M_e^{(2c)}(x_0) \equiv \int_0^{+\infty} t\phi_{\tau_e^{(2b)}(x_0, 0)}(t)dt$$

for general initial size x_0 and satisfying the backward equation from (7.62)

$$\mathcal{B}_{x_0}\left[M_e^{(2c)}\right](x_0) = -\Phi_e^{(2c)}(x_0, 0). \qquad (11.26)$$

The backward equation for the moment is easier to solve than the one derived for the expected time quotient (11.25) since the quotient leads to a much more complicated equation. The boundary conditions are homogeneous,

$$M_e^{(2c)}{}'(0^+) = 0 \quad \text{and} \quad M_e^{(2c)}{}'(2c) = 0,$$

but for different reasons, the first because 0^+ is the excluded exit and the second because it means an instant exit.

The solution can again utilize the Wronskian as a reciprocal integrating factor such that

$$(u_0'/W)'(x_0) = -2V(x_0)u_0(x_0),$$

where

$$V(x) \equiv \frac{1}{\sigma^2(x)W(x)} = \frac{1}{\sigma_1 x (\beta_1 x)^{\gamma_1}},$$

here for the Gompertz model, is called the **speed density**. As $x \to 0+$,

$$V(x) \sim \frac{1}{\sigma_1 x}(1 - \gamma_1 x \ln(\beta_1 x))$$

so that for $0 < \epsilon \ll x \ll 1$,

$$\int_\epsilon^x dy \, V(y) \sim \sigma^{-1}(\ln(x/\epsilon) + \gamma_1 \epsilon \ln(\beta_1 \epsilon) + 1))$$

and is not integrable as $\epsilon \to 0^+$. The integrability of both $W(x)$ and $V(x)$, as well as that of some other functions, plays a role in the Feller classification of boundaries for the Kolmogorov equations in one-dimension [31, 163]. Since a boundary is called a **regular boundary** if both $W(x)$ and $V(x)$ are integrable as the boundary point is approached from the interior of the domain, then 0^+ is a nonregular or **singular boundary** [163].

After two integrations, substitution of the boundary conditions to eliminate constants of integration and some manipulation of the integral forms, the solution of (11.26) can be written in the form

$$u_1(x_0) = 2(1 - u_0(x_0)) \int_{0^+}^{x_0} dy \, W(y) \int_y^{2c} dz \, V(z) u_0(z)$$

$$- 2u_0(x_0) \int_{x_0}^{2c} dy \, W(y) \int_y^{2c} dz \, V(z) u_0(z), \tag{11.27}$$

provided the integrals exist. Letting $x_0 = c$, the expected doubling time is given by the formula in (11.25) or more simply by

$$T_e^{(2c)}(c) = u_1(c)/u_0(c).$$

The multiple integral form of the solution (11.27) is too complicated to analyze further here, but additional numerical and asymptotic results are given by Hanson and Tier [118], including deterministic results. The application in [118] is based upon Fortner plasmacytoma data of Simpson-Herren and Lloyd [249]. The presentation here is somewhat different since it needed to be consistent with the notation and analytical formulation of this text.

Related formulation and results for other optimal stopping problems are some extinction problems for stochastic populations. They are examined for both diffusion in [117] and Poisson noise in [120, 122].

11.2.2 Optimal Drug Delivery to Brain PDE Model

In many applications, the control problem is formulated in terms of PDEs, not ODEs, since the problem depends on spatial variations and not just time variations. The ODE-driven control problem is usually called **lumped parameter control**, sometimes arising from **compartmental models** lumping the spatial variables so that a PDE is not used, while the PDE-driven control model is called **distributed parameter control**. The parameters in this latter case refer to the spatial variables in the background of the control problem. The mathematical background to this problem can be found in Online Section A.5 or in Gunzberger [102] in much more detail for flow problems.

Cancer drug delivery to eliminate or reduce tumors is usually based upon expensive sets of experiments using animal and later human subjects to determine a fixed dose size and dose period to fit general patient, tumor, and drug characteristics. Brain tumors can be very invasive and deadly, especially gliomas [262, 214]. When possible, most of the mass of the tumor is removed (also called resectioned), but drug chemotherapy or radiotherapy is used in an attempt to kill any remaining cancer cells, including mobile metastases [81]. Gliomas can also be very diffusive [214], so reaction-diffusion equations may be used to model the drug delivery to the brain [262, 214, 93]. However, these reaction-diffusion investigations are only studies of the behavior of the solutions. No control of the drug delivery is involved. In this subsection, the paper of Chakrabarty and Hanson [49] on the control of reaction diffusion equations for optimal drug delivery to the brain is briefly summarized.

Optimal Control Problem for Drug Delivery Reaction-Diffusion Equations

Consider a reaction-diffusion model of a three-state system consisting of tumor cells, normal cells, and cancer drug concentration in a brain. Let $y_1(\mathbf{x}, t)$ be the density of remaining tumor cells, $y_2(\mathbf{x}, t)$ be the density of normal cells, and $y_3(\mathbf{x}, t)$ be the concentration of the drug at time t in time horizon $[0, t_f]$ and position $\mathbf{x}$ in the brain domain $\mathcal{D}_x$. Let $\mathbf{y}(\mathbf{x}, t) = [y_3(\mathbf{x}, t)]_{3 \times 1}$ be the global state vector.

The tumor cell density satisfies the coupled reaction-diffusion equation

$$\frac{\partial y_1}{\partial t}(\mathbf{x}, t) = D_1 \nabla_x^2 [y_1](\mathbf{x}, t) + a_1 y_1 g_1(y_1) - (\alpha_{1,2} y_2 + \kappa_{1,3} y_3) y_1 \tag{11.28}$$

and the normal cells satisfy a similar equation,

$$\frac{\partial y_2}{\partial t}(\mathbf{x}, t) = D_2 \nabla_x^2 [y_3](\mathbf{x}, t) + a_2 y_2 g_2(y_2) - (\alpha_{2,1} y_1 + \kappa_{2,3} y_3) y_2, \tag{11.29}$$

where D_i is the diffusion coefficient for the ith state, $a_i y_i g_i(y_i)$ is the growth law for the ith state, the interaction coefficient $\alpha_{i,j} > 0$ signifies a constant death rate of tissue of state i due to tissue state j, and the coefficient $\kappa_{i,3} > 0$ denotes a constant death rate due to the drug. For concreteness, the growth terms are taken to be logistic, i.e., $a_i y_i g_i(y_i) = a_i y_i (1 - y_i / K_i)$, where a_i is a constant intrinsic growth coefficient and $K_i > 0$ is a constant carrying-capacity or saturation level. Thus, there can be a strong interaction between the tumor and normal tissues, but the drug interaction is unidirectional. The drug concentration $y_3(\mathbf{x}, t)$ diffuses, gets absorbed, and is controlled according to this reaction diffusion equation,

$$\frac{\partial y_3}{\partial t}(\mathbf{x}, t) = D_3 \nabla_x^2 [y_3](\mathbf{x}, t) + a_3 y_3 g_3(y_3) + u(\mathbf{x}, t), \tag{11.30}$$

where $a_3 y_3 g_3(y_3)$ is the drug absorption loss term and $u(\mathbf{x}, t)$ is the drug input control variable. For simplicity, the absorption term is taken to be exponential decay, so $a_3 y_3 g_3(y_3) = a_3 y_3$, where $a_3 < 0$ is the negative of the absorption coefficient and is assumed constant.

The vector reaction-diffusion PDE form merging (11.28), (11.29), and (11.30) corresponding to (A.138) is

$$\frac{\partial \mathbf{y}}{\partial t}(\mathbf{x}, t) = D\nabla_x^2[\mathbf{y}](\mathbf{x}, t) + \mathbf{B}(\mathbf{y}(\mathbf{x}, t), \mathbf{x}, t) + A\mathbf{u}(\mathbf{x}, t), \tag{11.31}$$

where $D = [D_i \delta_{i,j}]_{3\times 3}$ is the diffusion coefficient,

$$\begin{aligned}
\mathbf{B}(\mathbf{y}(\mathbf{x}, t), \mathbf{x}, t) = {} & \left(a_1 y_1(1 - y_1) - (\alpha_{1,2} y_2 + \kappa_{1,3} y_3) y_2\right) \mathbf{e}_1 \mathbf{e}_1^\top \\
& + \left(a_2 y_2(1 - y_2) - (\alpha_{2,1} y_1 + \kappa_{2,3} y_3) y_2\right) \mathbf{e}_2 \mathbf{e}_2^\top \\
& + a_3 y_1 \mathbf{e}_3 \mathbf{e}_3^\top
\end{aligned} \tag{11.32}$$

is the bilinear reaction term with unit vectors $\mathbf{e}_k = [\delta_{i,k}]_{3\times 1}$ for $k = 1:3$, $A = \mathbf{e}_3 \mathbf{e}_3^\top$ is the unit drug control coefficient, and the drift term does not appear since $C \equiv 0$ here. The initial conditions for the vector PDE (11.31) are the vector

$$\mathbf{y}(\mathbf{x}, 0) = \mathbf{y}_0(\mathbf{x}) \text{ for } \mathbf{x} \in \mathcal{D}_x \tag{11.33}$$

and the boundary condition is a no-flux condition,

$$-(\widehat{\mathbf{n}}^\top \nabla_x)[\mathbf{y}](\mathbf{x}, t) = \mathbf{0}, \tag{11.34}$$

where $\widehat{\mathbf{n}} = \widehat{\mathbf{n}}(\mathbf{x}, t)$ is the outward normal to the boundary $\partial \mathcal{D}_x$.

An objective in space-time is the minimization of the quadratic costs form

$$\begin{aligned}
V[\mathbf{y}, \mathbf{u}] = {} & \frac{1}{2} \int_{t_0}^{t_f} dt \int_{\mathcal{D}_x} d\mathbf{x} \left(\mathbf{y}^\top Q\mathbf{y} + (\mathbf{u} - \mathbf{u}_0)^\top R(\mathbf{u} - \mathbf{u}_0)\right)(\mathbf{x}, t) \\
& + \frac{1}{2} \int_{\mathcal{D}_x} d\mathbf{x} \left(\mathbf{y}^\top S\mathbf{y}\right)(\mathbf{x}, t_f),
\end{aligned} \tag{11.35}$$

which is a slight variation in the control of the form (A.139), where the quadratic coefficients are $R = r_3 \mathbf{e}_3 \mathbf{e}_3^\top$ for the tumor burden cost, $S = s_1 \mathbf{e}_1 \mathbf{e}_1^\top$ for the drug delivery costs, and $Q = q_1 \mathbf{e}_1 \mathbf{e}_1^\top + q_3 \mathbf{e}_3 \mathbf{e}_3^\top$ for the terminal costs, while the target threshold control value is $\mathbf{u}_0 = u_{0,3} \mathbf{e}_3$.

Hamiltonian Variational Formulation

The optimization problem above has three sets of constraints: the dynamics (11.31), the initial condition (11.33), and the boundary condition (11.34) and so requires three Lagrange multipliers $\boldsymbol{\lambda}(\mathbf{x}, t)$, $\boldsymbol{\mu}(\mathbf{x}, t)$, and $\boldsymbol{\nu}(\mathbf{x})$ (without t since $t = 0$ for the initial condition), respectively, to form the **pseudo-Hamiltonian** as in (A.140):

$$\mathcal{H}(\mathbf{y}, \mathbf{u}, \boldsymbol{\lambda}, \boldsymbol{\mu}, \boldsymbol{\nu}) = V[\mathbf{y}, \mathbf{u}] + \int_{t_0}^{t_f} dt \int_{\mathcal{D}_x} d\mathbf{x}\, \boldsymbol{\lambda}^{\top}\big(\mathbf{y}_t - D\nabla_x^2[\mathbf{y}] - \mathbf{B} - A\mathbf{u}\big)(\mathbf{x}, t)$$

$$+ \int_{t_0}^{t_f} dt \int_{\partial\mathcal{D}_x} d\boldsymbol{\Gamma}\, \boldsymbol{\mu}^{\top}\big(-\big(\widehat{\mathbf{n}}^{\top}\nabla_x\big)[\mathbf{y}]\big)(\mathbf{x}, t) \qquad (11.36)$$

$$+ \int_{\mathcal{D}_x} d\mathbf{x}\, \boldsymbol{\nu}^{\top}\big(\mathbf{y}(\mathbf{x}, 0^+) - \mathbf{y}_0(\mathbf{x})\big).$$

The main idea is that the Lagrange multipliers extend the three-vector state space to an extended six-vector state space

$$\mathbf{z}(\mathbf{x}, t) \equiv \{\mathbf{y}(\mathbf{x}, t), \mathbf{u}(\mathbf{x}, t), \boldsymbol{\lambda}(\mathbf{x}, t), \boldsymbol{\mu}(\mathbf{x}, t), \boldsymbol{\nu}(\mathbf{x})\}$$

to make the variations $\delta\mathbf{z}(\mathbf{x}, t)$ about $zbf^*(\mathbf{x}, t)$ in the extended objective systematic. Hence,

$$\mathcal{H}(\mathbf{z}^*(\mathbf{x}, t) + \delta\mathbf{z}(\mathbf{x}, t)) = \mathcal{H}(\mathbf{z}^*(\mathbf{x}, t)) + \delta\mathcal{H}(\mathbf{z}^*(\mathbf{x}, t), \delta\mathbf{z}(\mathbf{x}, t)) + \mathrm{O}(|\delta\mathbf{z}|^2(\mathbf{x}, t)),$$

assuming that $\mathbf{z}^*(\mathbf{x}, t)$ exists and is a unique optimal solution under sufficient differentiability assumptions on $\mathcal{H}(\mathbf{z}(\mathbf{x}, t))$. Critical to these assumptions is that the perturbation of the nonlinear reaction term $\mathbf{B}(\mathbf{y}, \mathbf{x}, t)$ has a quadratic approximation, but that is trivial for this application since $\mathbf{B}$ is quadratic in $\mathbf{y}$.

Skipping the details contained in Subsection A.5.2, something very similar to the first variation $\delta\mathcal{H}(\mathbf{z}^*(\mathbf{x}, t), \delta\mathbf{z}(\mathbf{x}, t))$ in (A.142) is found. Setting the coefficients of $\delta\boldsymbol{\lambda}^{\top}(\mathbf{x}, t)$, $\delta\boldsymbol{\nu}^{\top}(\mathbf{x})$, and $\delta\boldsymbol{\mu}^{\top}(\mathbf{x}, t)$ (only for $\mathbf{x} \in \mathcal{D}_x$), respectively, to zero confirms that the PDE (11.31), initial condition (11.33), and boundary condition (11.34) hold with the optimal state $\mathbf{y}^*(\mathbf{x}, t)$ replacing the state $\mathbf{y}(\mathbf{x}, t)$ of the original problem.

The final boundary value PDE problem for the optimal adjoint state $\boldsymbol{\lambda}^*(\mathbf{x}, t)$ comes from setting the coefficients for $\delta\mathbf{y}^{\top}(\mathbf{x}, t_f)$, $\delta\mathbf{y}^{\top}(\mathbf{x}, t_f)$ and $\delta\mathbf{y}^{\top}(\mathbf{x}, t)$ (only for $\mathbf{x} \in \mathcal{D}_x$), respectively, to zero, producing

$$\big(\boldsymbol{\lambda}_t^* + \nabla_x^2[D\boldsymbol{\lambda}^*] - \nabla_y[\mathbf{B}^{\top}]^*\boldsymbol{\lambda}^* - Q\mathbf{y}^*\big)(\mathbf{x}, t) = \mathbf{0}, \quad \mathbf{x} \in \mathcal{D}_x, \quad t \in [0, t_f) \qquad (11.37)$$

with final condition

$$\big(\boldsymbol{\lambda}^* + S\mathbf{y}^*\big)(\mathbf{x}, t_f) = \mathbf{0}, \quad \mathbf{x} \in \mathcal{D}_x \qquad (11.38)$$

and boundary condition

$$\big(\widehat{\mathbf{n}}^{\top}\nabla_x\big)[D\boldsymbol{\lambda}^*](\mathbf{x}, t) = \mathbf{0}, \quad \mathbf{x} \in \partial\mathcal{D}_x, \quad t \in (0, t_f), \qquad (11.39)$$

which is the corresponding no-flux condition in backward form.

Setting the coefficient of $\delta\mathbf{u}(\mathbf{x}, t)$ to zero leads to

$$R(\mathbf{u}^*(\mathbf{x}, t) - \mathbf{u}_0^*(\mathbf{x}, t)) = A^{\top}\boldsymbol{\lambda}(\mathbf{x}, t),$$

which reduces to

$$u_3^*(\mathbf{x}, t) = u_{0,3}^*(\mathbf{x}, t) + \lambda_3^*(\mathbf{x}, t)/r_3, \quad \mathbf{x} \in \mathcal{D}_x, \quad t \in [0, t_f). \qquad (11.40)$$

There are other optimality conditions that interrelate the Lagrange multipliers,

$$\boldsymbol{\nu}^*(\mathbf{x}) = \boldsymbol{\lambda}^*(\mathbf{x}, 0^+) \text{ for } \mathbf{x} \in \mathcal{D}_x$$

and

$$\boldsymbol{\mu}^*(\mathbf{x}, t) = D\boldsymbol{\lambda}^*(\mathbf{x}, t), \quad \mathbf{x} \in \partial\mathcal{D}_x, \quad t \in [0, t_f),$$

which will not be needed in the computations.

Forward-Backward Computational Iterations

The presence of nonlinear reaction terms in the forward state equation (11.31) using $\mathbf{y}^*(\mathbf{x}, t)$ with $\mathbf{u}^*(\mathbf{x}, t)$ and in the corresponding backward costate equation (11.37) for $\boldsymbol{\lambda}^*(\mathbf{x}, t)$ make computational methods essential. The computational method of Chakrabarty and Hanson [49, 50, 51] employs a forward state integration of (11.31) and a backward integration of (11.37) with sufficient iterations until the norm of the iteration difference is small enough. The forward equation (11.31) is independent of the costate $\boldsymbol{\lambda}^*(\mathbf{x}, t)$ but depends on the optimal control $\mathbf{u}^*(\mathbf{x}, t)$ which is a critical objective to be determined, so a starting guess for $\mathbf{u}^*(\mathbf{x}, t)$ is needed to start the forward integration, until a backward itegration generates a better guess using (11.40). On the other hand, the backward equation (11.37) depends strongly on the state distribution $\mathbf{y}^*(\mathbf{x}, t)$ as well as on its final values from (11.38), so that iterations, each consisting of a **double-shot** of both a forward interation followed by a backward iteration, are required for reasonable accuracy. This double shot method is similar to the **opposite directions** multiple shooting method of Hackbusch [104] for parabolic equations. Gunzberger [102] calls many such methods **one-shot** methods and gives a more rigorous justification of them.

To keep the computational presentation manageable, let the forward and backward PDEs be represented in the more compact notation

$$\mathbf{y}_t^*(\mathbf{x}, t) = \mathcal{F}(\mathbf{x}, t, \mathbf{y}^*(\mathbf{x}, t), \mathbf{u}^*(\mathbf{x}, t)),$$
$$\mathbf{0} = \boldsymbol{\lambda}_t^*(\mathbf{x}, t) + \mathcal{G}(\mathbf{x}, t, \boldsymbol{\lambda}^*(\mathbf{x}, t), \mathbf{y}^*(\mathbf{x}, t))$$

with general vector functions $\mathcal{F}$ and $\mathcal{G}$ for the forward and backward equations, respectively. Let the space vector $\mathbf{x}$ be replaced by the discrete representation

$$\mathbf{x_j} \equiv [x_{j_i,1} + (j_i - 1) \cdot \Delta x_i]_{3 \times 1},$$

where Δx_i is the step size in the ith direction, $\mathbf{j} = [j_i]_{3 \times 1}$, where, $j_i = 1{:}M_i$ nodes per direction, $i = 1{:}3$. Let the time t be replaced by the forward discretization

$$t_k \equiv k\Delta t$$

for $k = 0{:}K$ time-steps where Δt is the forward time-step size, $t_0 = 0$, and $t_K = t_f$. The backward discrete time will be of the form $t_k^{(b)} \equiv t_f - k\Delta t = (K - k)\Delta t = t_{K-k}$. The corresponding discretization of the dependent vectors will be

$$\mathbf{y}(\mathbf{x_j}, t_k) \simeq \mathbf{Y_{j,k}}, \boldsymbol{\lambda}(\mathbf{x_j}, t_k) \simeq \boldsymbol{\Lambda_{j,k}} \quad \text{and} \quad \mathbf{u}(\mathbf{x_j}, t_k) \simeq \mathbf{U_{j,k}}.$$

The numerical procedure used is the Crank–Nicolson method for second order accuracy in both space and time but modified with additional extrapolation, prediction, and correction techniques to accommodate nonlinear terms and multidimensions. The forward and backward discrete versions are written as

$$\mathbf{Y}_{\mathbf{j},k+1}^{(\gamma+1,\ell)} = \mathbf{Y}_{\mathbf{j},k}^{(\ell)} + \Delta t \mathcal{F}_{\mathbf{j},k+0.5}^{(\gamma,\ell)}, \tag{11.41}$$

$$\mathbf{\Lambda}_{\mathbf{j},k-1}^{(\gamma+1,\ell)} = \mathbf{\Lambda}_{\mathbf{j},k}^{(\ell)} + \Delta t \mathcal{G}_{\mathbf{j},k-0.5}^{(\gamma,\ell)} \tag{11.42}$$

for $\gamma = 0:n_c$ corrections ($\gamma = 0$ is the prediction step) in each time-step k until a relative stopping criterion for corrections in the tumor cell state component $Y_{1,\mathbf{j},k+1}^{(\gamma+1,\ell)}$ is satisfied,

$$\left\| Y_{1,\mathbf{j},k+1}^{(\gamma+1,\ell)} - Y_{1,\mathbf{j},k+1}^{(\gamma,\ell)} \right\| < \text{tol}_y \left\| Y_{1,\mathbf{j},k+1}^{(\gamma,\ell)} \right\| \tag{11.43}$$

for every state index $\mathbf{j}$ for $k = 0:K-1$ and during all double-shot iterations $\ell = 1:L$, provided $\|Y_{1,\mathbf{j},k+1}^{(\gamma,\ell)}\| \neq 0$. The general notation means that

$$\mathcal{F}_{\mathbf{j},k+0.5}^{(\gamma,\ell)} = \mathcal{F}\left(\mathbf{x_j}, t_{k+0.5}, \mathbf{Y}_{\mathbf{j},k+0.5}^{(\gamma,\ell)}, \mathbf{U}_{\mathbf{j},k+0.5}^{(\gamma,\ell)}\right)$$

and similarly for $\mathcal{G}_{\mathbf{j},k-0.5}^{(\gamma,\ell)}$. The relative tolerance in $Y_{1,\mathbf{j},k}^{(\gamma,\ell)}$ is tol_y. The Crank–Nicolson midpoint values are ordinarily approximated by the average

$$\mathbf{Y}_{\mathbf{j},k+0.5}^{(\gamma,\ell)} \simeq 0.5 \left(\mathbf{Y}_{\mathbf{j},k+1}^{(\gamma,\ell)} + \mathbf{Y}_{\mathbf{j},k}^{(\ell)}\right)$$

for $k = 0:K-1$ and

$$\mathbf{\Lambda}_{\mathbf{j},k-0.5}^{(\gamma,\ell)} \simeq 0.5 \left(\mathbf{\Lambda}_{\mathbf{j},k-1}^{(\gamma,\ell)} + \mathbf{\Lambda}_{\mathbf{j},k}^{(\ell)}\right)$$

for $k = K:-1:1$, where $\mathbf{Y}_{\mathbf{j},k}^{(\ell)}$ and $\mathbf{\Lambda}_{\mathbf{j},k}^{(\ell)}$ are the final corrections for each time-step k given shot ℓ, consistent with the second order Crank–Nicolson accuracy and implicitness reduction. A similar form is used for $\mathbf{U}_{\mathbf{j},k+0.5}^{(\gamma,\ell)}$. Second order central finite differences are used for all derivatives and based upon $\mathbf{Y}_{\mathbf{j},k+0.5}^{(\gamma,\ell)}$ or $\mathbf{\Lambda}_{\mathbf{j},k-0.5}^{(\gamma,\ell)}$.

The final stopping criterion for the convergence of the double-shot iterations $\ell = 2:L$ is the pair of norms

$$\left\| U_{3,\mathbf{j},k}^{(\ell)} - U_{3,\mathbf{j},k}^{(\ell-1)} \right\| < \text{tol}_u \left\| U_{3,\mathbf{j},k}^{(\ell-1)} \right\| \quad \text{and} \quad \left\| Y_{1,\mathbf{j},k}^{(\ell)} - Y_{1,\mathbf{j},k}^{(\ell-1)} \right\| < \text{tol}_y \left\| Y_{1,\mathbf{j},k}^{(\ell-1)} \right\|, \tag{11.44}$$

where the norm is over all $\mathbf{j}$ and k for $\ell = 2:L$ until satisfied, provided $\|U_{3,\mathbf{j},k}^{(\ell-1)}\| \neq 0$ and $\|Y_{1,\mathbf{j},k}^{(\ell-1)}\| \neq 0$, where $\text{tol}_u > 0$ and $\text{tol}_y > 0$ are some specified tolerances.

The treatment of the bilinear reaction term (11.32) requires careful consideration to accommodate the usual linear framework of the Crank–Nicolson method. Since this term has the pure bilinear form,

$$\mathbf{B}(\mathbf{y}, \mathbf{x}, t) = \widehat{B}(\mathbf{y})\mathbf{y},$$

in this application, this quasi-linear approximation is very appropriate,

$$\widehat{B}\left(\mathbf{Y}_{\mathbf{j},k+0.5}^{(\gamma,\ell)}\right)\mathbf{Y}_{\mathbf{j},k+0.5}^{(\gamma,\ell)} \simeq \widehat{B}\left(\mathbf{Y}_{\mathbf{j},k+0.5}^{(\gamma-1,\ell)}\right)\mathbf{Y}_{\mathbf{j},k+0.5}^{(\gamma,\ell)},$$

in the forward equation for corrections $\gamma \geq 1$ and time-steps $k \geq 1$.

Another special treatment needed is that of the no-flux boundary condition since central differences are inappropriate at the boundary, but backward and forward differences of the same second order accuracy work very well, e.g.,

$$\mathbf{0} = -\left((\widehat{\mathbf{n}}^\top\nabla_x)[\mathbf{Y}^*]\right)_{\mathbf{j},k}^{(\gamma,\ell)} \simeq -\frac{\left(3\mathbf{Y}_{\mathbf{j},k}^{(\gamma,\ell)} - 4\mathbf{Y}_{\mathbf{j}-\widehat{\mathbf{n}},k}^{(\ell)} + \mathbf{Y}_{\mathbf{j}-2\widehat{\mathbf{n}},k}^{(\gamma,\ell)}\right)}{\left(2|\widehat{\mathbf{n}}^\top\Delta\mathbf{x}|\right)},$$

$$\mathbf{0} = \left((\widehat{\mathbf{n}}^\top\nabla_x)[(\boldsymbol{\Lambda})^*]\right)_{\mathbf{j},k}^{(\gamma,\ell)} \simeq +\frac{\left(3\boldsymbol{\Lambda}_{\mathbf{j},k}^{(\gamma,\ell)} - 4\boldsymbol{\Lambda}_{\mathbf{j}-\widehat{\mathbf{n}},k}^{(\gamma,\ell)} + \boldsymbol{\Lambda}_{\mathbf{j}-2\widehat{\mathbf{n}},k}^{(\gamma,\ell)}\right)}{\left(2|\widehat{\mathbf{n}}^\top\Delta\mathbf{x}|\right)},$$

respectively, where $\widehat{\mathbf{n}} \equiv \widehat{\mathbf{n}}_{\mathbf{j},k}$, $\Delta\mathbf{x} = [\Delta x_i]_{3\times1} > \mathbf{0}$, and, e.g.,

$$\mathbf{Y}_{\mathbf{j}-\widehat{\mathbf{n}},k}^{(\gamma,\ell)} = \mathbf{Y}^{(\gamma,\ell)}(\mathbf{x_j} - |\widehat{\mathbf{n}}^\top\Delta\mathbf{x}|\widehat{\mathbf{n}}, t_k).$$

A sample output of the computations shown in Figure 11.1 shows significant decrease in tumor size in one space dimension for a 5-day drug treatment trial. For information on the parameters used see Chakrabarty and Hanson [49]. For the corresponding two-dimensional space model of drug delivery see [50] and for a three-dimensional model see [51]. For more detail, see Chakrabarty's thesis [48].

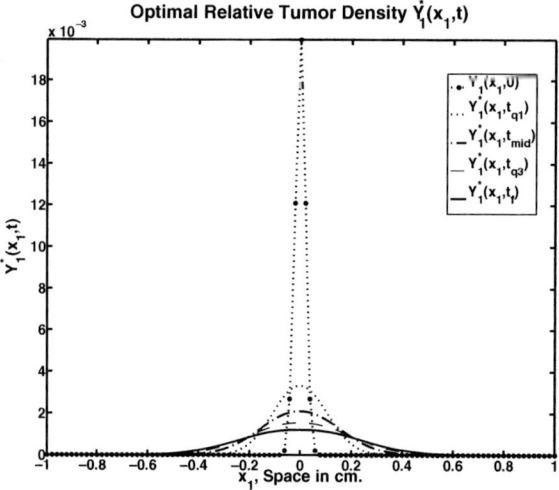

Figure 11.1. *Optimal tumor density* $Y_1^*(x_1, t)$ *in the one-dimensional case with time as a parameter rounded at quartile values* $\{0, t_{q_1} = t_f/4, t_{mid} = t_f/2, t_{q_3} = 3t_f/4, t_f\}$, *where* $t_f = 5$ *days. The total tumor density integral is reduced by 29% in the 5-day simulated drug treatment trial.*

Suggested References for Further Reading

- Anderson and Sutinen, 1984 [9]

- Ashkenazy et al., 2002 [14]

- Bharucha-Reid, 1960 [31]

- Boker et al., 2004 [37]

- Buchman, 2004 [45]

- Chakrabarty and Hanson, 2005a [49], 2005b [50], and 2006 [51]

- Clark, 1976 and 1990 [57]

- Clark and Lamberson, 1982 [58]

- Engelhard, 2000 [81]

- Goel and Dyn, 1974 [99]

- Goldie and Coldman, 1998 [100]

- Gunzberger, 2003 [102]

- Hackbusch, 1978 [104]

- Hanson, 1987 [106] and 1996 [109]

- Hanson et al., 1993 [113]

- Hanson and Ryan, 1988 [114] and 1998 [116]

- Hanson and Tier, 1981 [117] and 1982 [118]

- Hanson and Tuckwell, 1978 [119], 1981 [120], 1983 [121], and 1997 [122]

- Hennemuth et al., 1980 [139]

- Karlin and Taylor, 1975 [162, I] and 1981 [163, II]

- Kamien and Schwartz, 1981 [159]

- Kushner and Dupuis, 2001 [179]

- Ludwig, 1974 [187] and [188]

- Mangel, 1985 [193]

- Moss et al., 2004 [211]

- Murray, 2002 [213, I] and 2003 [214, II]

- Nisbet and Gurney, 1982 [219]

- Practico et al., 1992 [228]

- Priplata et al., 2002 [231]

- Ryan and Hanson, 1958 [241] and 1986 [242]

- Simpson-Herren and Lloyd, 1970 [249]

- Steel, 1977 [255]

- Swan, 1984 [261]

- Swanson, 1999 [262]

- Taylor and Karlin, 1998 [265]

- Tuckwell, 1989 [269] and 1995 [270]

Chapter 12

Applied Guide to Abstract Theory of Stochastic Processes

Mathematicians are like Frenchmen: Whatever you say to them they translate into their own language and forthwith it is something entirely different.

—Johann Wolfgang von Goethe (1749–1832) at http://www.quotationspage.com/quotes/22893.html

Since the mathematicians have invaded the theory of relativity, I do not understand it myself anymore.

—Albert Einstein (1879–1955) at http://en.wikiquote.org/wiki/Mathematics

Martingale (1589): Any of several systems of betting in which a player increases the stake usually by doubling each time a bet is lost.

—Merriam-Webster's Online Dictionary, definition 3 at http://www.m-w.com/dictionary/martingale

Martingales are treated because of their great importance, but they are not used as a tool in this book.

—William (Willy) Feller (1906–1970), p. 209 in [85]

The concept of martingales is due to P. Lévy, but it was J. L. Doob who realized its unexpected potential and developed the theory.

—William (Willy) Feller (1906–1970), p. 210 in [85]

Our view of Brownian motion never focused too closely on the underlying measure space, and, by and large, we have profited from keeping a respectful distance.

—J. Michael Steele, p. 218 in [256]

This chapter briefly introduces more of the abstract analytical methods, such as measure theoretic methods, martingale methods, Radon–Nikodým derivatives, Girsanov's theorem, Itô processes, Lévy processes, characteristic functions and exponents, Lévy–Klintchine formula, jump-diffusion process comparisons, and other topics from the applied point of view as a bridge to more abstract methods.

The purpose of this chapter is to supply some insightful and useful background to make the more abstract literature on stochastic processes and control more accessible to general students in applied mathematics, statistics, computer science, applied science, and engineering. The overall approach in this book is designed to start from the common calculus and analysis constructs of entry-level graduate students in these applied areas by evolving these constructs to those of applied stochastic processes and control, much as genes have evolved by small but powerful changes. However, students still need to understand the important results that come from using more abstract methods.

The applied motivation is to solve problems with a combination of analytical and computational methods. These problems may have great complexity in terms on nonlinearities in the state and other dependencies. It is necessary to train both students and researchers from a broad range of areas in science and engineering to solve large-scale problems. In the abstract approach the emphasis is not necessarily to solve applied problems, but to prove existence, uniqueness, and convergence, often in very abstract language. However, sometimes the conditions of the proofs are too restrictive, so as to exclude many complex and large-scale applications. Proofs as such are not given in this chapter, but some formal applied derivations are given and readers can refer to the list of references at the end of the chapter for more rigorous treatments.

12.1 Very Basic Probability Measure Background

To keep things simple and concise, it will be necessary to compromise on completeness but keep sufficient detail for a coherent story. The notation will be somewhat different from the usual, if there is such a thing as usual notation, so that we can avoid conflict with the stochastic process notation where possible. The symbols are selected so that they are related to what the quantity signifies, where possible.

12.1.1 Mathematical Measure Theory Basics

The starting point will be some notions of measure theory and its algebra, called σ-algebra. Measure theory provides an abstract generalization of integration theory including expectations and distributions that are based on counts, intervals, areas, volumes, and mass to that of general sets. The ultimate goal is **probability measure**, but the presentation begins with the foundations in the more general mathematical measure theory.

Measure σ-Algebra Definition

Let $\mathcal{U}$ be a nonempty set called the **universe**, but it really is only the principal set of interest. Let Σ be a collection of subsets on $\mathcal{U}$.

Definition 12.1. Σ *is a σ-algebra* if

- $\emptyset \in \Sigma$, *i.e., the empty set $\emptyset$ is included.*

- $\mathcal{U} \in \Sigma$, *i.e., the universe $\mathcal{U}$ is included.*

- *The set $\mathcal{S} \in \Sigma \implies \mathcal{S}^c \in \Sigma$, i.e., its complement $\mathcal{S}^c$ with respect to $\mathcal{U}$ is included too, i.e., verifying that $\mathcal{S} \cup \mathcal{S}^c = \mathcal{U}$.*

- *If $\{\mathcal{S}_i \in \Sigma : i = 1 : n\}$ is a sequence of subsets, then the union $\bigcup_{i=1}^{n} \mathcal{S}_i \in \Sigma$, i.e., additive closure under unions.*

*If so, then $\{\mathcal{U}, \Sigma\}$ is called a **measurable space**.*

Often the symbol Ω is used for the general universe $\mathcal{U}$ and the symbol $\mathcal{F}$ is used for the σ-algebra Σ. Recall that the **union** of two sets

$$\mathcal{S}_1 \cup \mathcal{S}_2 = \{\text{points } X : X \in \mathcal{S}_1 \text{ OR } X \in \mathcal{S}_2\},$$

the logical operator symbol **(OR)** playing an important role when translated to probability measures.

A **Borel set** $\Sigma = \mathcal{B} = \mathcal{B}(\mathbb{R}^{n_x})$ is the σ-algebra of open sets on $\mathcal{U} = \mathbb{R}^{n_x}$, so $\mathcal{B}(\mathbb{R}^{n_x})$ automatically contains all closed sets of $\mathbb{R}^{n_x}$ by complementarity.

Measure Definition

Definition 12.2. *The measure $\mathcal{M}$ is a function on the measurable space $\{\mathcal{U}, \Sigma\}$ that maps $\Sigma \longrightarrow [0, \infty)$, such that*

- $\mathcal{M}(\emptyset) = 0$, *i.e., the empty set $\emptyset$ has **measure zero**.*

- *If for any subset $\mathcal{S} \in \Sigma$, then $\mathcal{M}(\mathcal{S}) \geq 0$, i.e., **nonnegativity**, as in mass.*

- *If $\{\mathcal{S}_i \in \Sigma : i = 1, 2, \ldots,\}$ is any countable sequence of **disjoint subsets** (i.e., $\mathcal{S}_i \bigcap \mathcal{S}_j = \emptyset$, $i \neq j$, the intersection is empty), then the measure of the union is the sum of the measures*

$$\mathcal{M}\left(\bigcup_{i=1}^{\infty} \mathcal{S}_i\right) = \sum_{i=1}^{\infty} \mathcal{M}(\mathcal{S}_i), \tag{12.1}$$

*i.e., **countable additivity**, as in preserving mass under partitioning.*

The triplet $\{\mathcal{U}, \Sigma, \mathcal{M}\}$ is called a **measure space**. Often the symbol μ is used for the general measure symbol $\mathcal{M}$ used here, but the former conflicts with the use of μ as the mean or drift in this book. Recall that the **intersection** of two sets

$$\mathcal{S}_1 \cap \mathcal{S}_2 = \{\text{points } X : X \in \mathcal{S}_1 \text{ AND } X \in \mathcal{S}_2\},$$

the logical operator symbol **AND** playing an important role when translated to probability measures.

The nonnegativity measure property $\mathcal{M}(\mathcal{S}) \geq 0$ means that **positive measure** has been defined. Positive meaures, among other things, facilitate convergence proofs, i.e., monotone convergence. However, if for any subset $\mathcal{S} \in \Sigma$ and $\mathcal{M}(\mathcal{S}) \leq 0$, then $\mathcal{M}(\mathcal{S})$

would be a **negative measure** and negative measure may be needed for some applications in spite of the added awkwardness of the proofs.

Lebesgue Measure Introduction

If the set S is measurable, the $\mathcal{M}(S)$ is called the **total mass of the set**, e.g., if S is an interval $[a, b]$, then it is the length $(b - a)$; if a rectangle $[a, b] \times [c, d]$, then it is the area $(b - a) \cdot (d - c)$; or if a cube $[a, b] \times [a, b] \times [a, b]$, then it is the volume $(b - a)^3$. The closed intervals $[a, b]$, open intervals (a, b), and semiopen intervals $[a, b)$ or $(a, b]$ have the same measure or mass or length of $(b - a)$ since they differ only by **points of zero measure**.

In general, a **Lebesgue measure** is a measure on an n_x dimensional space of real vectors, so the universe is $\mathcal{U} = \mathbb{R}^{n_x}$, where a representative set is a hypercube

$$S = (\mathbf{a}, \mathbf{b}) \equiv (a_1, b_1) \times (a_2, b_2) \times \cdots \times (a_{n_x}, b_{n_x})$$

such that $-\infty < a_i < b_i < +\infty$, and the measure has the form

$$\mathcal{M}(S) = \prod_{i=1}^{n_x} (b_i - a_i).$$

Alternatively,

$$\mathcal{M}(S) = \int_S d\mathbf{x}.$$

Lebesgue measure is a special case of Borel measure specialized to $\mathbb{R}^{n_x}$.

Often, the infinitesimal hypercube measure between vector positions from $\mathbf{x}$ to $\mathbf{x} + d\mathbf{x}$ is abbreviated as

$$\mathcal{M}(d\mathbf{x}) = \mathcal{M}((\mathbf{x}, \mathbf{x} + d\mathbf{x}))$$

for compact notation, letting $d\mathbf{x}$ represent the vector-interval set $(\mathbf{x}, \mathbf{x} + d\mathbf{x})$. This also recognizes the translation invariance of the measure of a generalized interval $(\mathbf{x}, \mathbf{x} + d\mathbf{x})$, since its generalized length $\prod_{i=1}^{n_x} dx_i$ is independent of the interval start at $\mathbf{x}$.

Dirac Measures

Another measure that complements the Lebesgue measure is the **Dirac measure** δ_x for some point in $\mathcal{U}$, having the properties that for some set $S \subseteq \mathcal{U}$,

$$\delta_x(S) = \begin{cases} 1, & x \in S \\ 0, & x \notin S \end{cases}. \tag{12.2}$$

This is the set version of the **Dirac delta function** and apparently has the same basic definition as the indicator function $\mathbf{1}_{x \in S}$, except without the measure infrastructure.

Counting Measures

For Poisson processes and other discrete applications, there are also counting measures, i.e., when

$$\mathcal{M}(S) = N(S) \equiv \textit{number of elements in set } S. \tag{12.3}$$

This includes the points of zero measure that **do count**.

Some Properties of Measures

- The measure space $\{\mathcal{U}, \Sigma, \mathcal{M}\}$ is **finite** if $\mathcal{M}(\mathcal{U}) < \infty$ and real.

- The measure space $\{\mathcal{U}, \Sigma, \mathcal{M}\}$ is σ**-finite** if there exists a countable sequence of measurable sets $\{\mathcal{S}_i \in \Sigma : i = 1, 2, \ldots,\}$ such that $\mathcal{M}(\mathcal{S}_i) < \infty$ and real for all i, i.e., sets of finite measure, and

$$\mathcal{U} = \bigcup_{i=1}^{\infty} \mathcal{S}_i,$$

  the union of a countable number of sets of finite measure. Note that σ-finite is not necessarily finite, since the set of real intervals $[i, i + 1]$ has unit measure which is finite (a Lebesgue measure), but their union is the real line, $\mathcal{U} = \mathbb{R}^1$, which is infinite, so $\mathcal{U}$ is σ-finite while not finite.

- The measure $\mathcal{M}$ is a **monotonic function** since if measurable sets $\mathcal{S}_1$ and $\mathcal{S}_2$ are ordered $\mathcal{S}_1 \subseteq \mathcal{S}_2$, then $\mathcal{M}(\mathcal{S}_1) \le \mathcal{M}(\mathcal{S}_2)$.

- If $\{\mathcal{S}_i \in \Sigma : i = 1, 2, \ldots,\}$ is any countable sequence of subsets that are **not necessarily disjoint**, then the measure of the union is bounded only by the sum of the measures,

$$\mathcal{M}\left(\bigcup_{i=1}^{\infty} \mathcal{S}_i\right) \le \sum_{i=1}^{\infty} \mathcal{M}(\mathcal{S}_i),$$

  unlike the lack of redundancies of disjoint sets given in (12.1).

- If $\{\mathcal{S}_i \in \Sigma : i = 1, 2, \ldots,\}$ is any countable sequence of subsets that are **forward nested** so that $\mathcal{S}_i \subseteq \mathcal{S}_{i+1}$, then the limit of the **measure of the union** has the limiting measure

$$\mathcal{M}\left(\bigcup_{i=1}^{\infty} \mathcal{S}_i\right) = \lim_{n \to \infty} \mathcal{M}(\mathcal{S}_n),$$

  noting that $\mathcal{M}(\mathcal{S}_i \cup \mathcal{S}_{i+1}) = \mathcal{M}(\mathcal{S}_{i+1})$.

- If $\{\mathcal{S}_i \in \Sigma : i = 1, 2, \ldots,\}$ is any countable sequence of subsets that are **backward nested** so that $\mathcal{S}_{i+1} \subseteq \mathcal{S}_i$, then the limit of the **measure of the intersection** has the limiting measure

$$\mathcal{M}\left(\bigcap_{i=1}^{\infty} \mathcal{S}_i\right) = \lim_{n \to \infty} \mathcal{M}(\mathcal{S}_n),$$

  noting that $\mathcal{M}(\mathcal{S}_i \cap \mathcal{S}_{i+1}) = \mathcal{M}(\mathcal{S}_{i+1})$.

- A **null set** $\mathcal{N} \in \Sigma$ is a measurable set such that $\mathcal{M}(\mathcal{N}) = 0$, a **negligible set** is a subset of a null set, and a measure $\mathcal{M}$ is **complete** if every negligible set is measurable. A σ-algebra Σ can always be completed by adding any missing null sets to it.

- A property P holds **almost everywhere (a.e.)** if the set of elements $\mathcal{S}$ in Σ for which the property does not hold is a null set, i.e., $\mathcal{S} = \mathcal{N}$ is a set with measure zero such that $\mathcal{M}(\mathcal{N}) = 0$.

- Given the measure space $\{\mathcal{U}, \Sigma, \mathcal{M}_1\}$, another measure $\mathcal{M}_2$ on the measurable space $\{\mathcal{U}, \Sigma\}$ is **absolutely continuous (a.c.)** with respect to $\mathcal{M}_1$ if for any measurable set $\mathcal{S} \in \Sigma$

$$\mathcal{M}_1(\mathcal{S}) = 0 \Longrightarrow \mathcal{M}_2(\mathcal{S}) = 0.$$

Absolute continuity is written symbolically as $\mathcal{M}_2(\mathcal{S}) \prec \mathcal{M}_1(\mathcal{S})$ (or as $\mathcal{M}_2(\mathcal{S}) \ll \mathcal{M}_1(\mathcal{S})$, but this conflicts with asymptotic notation). This property permits defining the ratio $\mathcal{M}_2(\mathcal{S})/\mathcal{M}_1(\mathcal{S})$ for comparison between two measures of a set.

If $\mathcal{M}_2(\mathcal{S}) \prec \mathcal{M}_1(\mathcal{S})$ and $\mathcal{M}_1(\mathcal{S}) \prec \mathcal{M}_2(\mathcal{S})$, i.e., both are mutually absolutely continuous with respect to the other, then the measures $\mathcal{M}_1$ and $\mathcal{M}_2$ are said to be **equivalent** ($\mathcal{M}_1(\mathcal{S}) \overset{\text{a.c.}}{\equiv} \mathcal{M}_1(\mathcal{S})$). As Cont and Tankov [60] suggest, the term **equivalence** is perhaps misleading and should be called something like **comparable**.

Many of these properties are needed for proofs of existence and convergence, as well as for constructing stochastic processes.

Measurable Functions

A prerequisite that a function f is integrable is that f is a **measurable function**.

Definition 12.3. *Given two measurable spaces, $(\mathcal{U}_1, \Sigma_1)$ and $(\mathcal{U}_2, \Sigma_2)$, a mapping of the function f from $\mathcal{U}_1$ to $\mathcal{U}_2$ is **measurable** with respect to (Σ_1, Σ_2) if the **inverse (preimage)** $f^{-1}(\mathcal{S}_2) \in \Sigma_1$ for all $\mathcal{S}_2 \in \Sigma_2$, i.e., there is an $\mathcal{S}_1 \in \Sigma_1$ such that $f(\mathcal{S}_1) = \mathcal{S}_2$.*

Just as in Riemann integration for general Riemann integrable functions, where the integral is built up from the limit of finite Riemann sums, the integral with respect to a measurable function is built up from sums of step functions called a **simple function**.

Definition 12.4.

- *A **simple function** is a finite linear combination of set indicator functions $\{\mathbf{1}_{x \in \mathcal{S}_i}\}$ of measurable sets $\mathcal{S}_i$ for $i = 1:n$ on a measurable space $(\mathcal{U}, \Sigma)$ with real coefficients (could also be complex) c_i, having the form*

$$f(x) = \sum_{i=1}^{n} c_i \mathbf{1}_{x \in \mathcal{S}_i},$$

where $x \in \mathcal{U}$.

- *The **integral with respect to the measure $\mathcal{M}$** for such a simple function is*

$$\mathcal{I}_{\mathcal{M}}[f] = \sum_{i=1}^{n} c_i \mathcal{M}(\mathcal{S}_i),$$

provided all the measures $\mathcal{M}(\mathcal{S}_i)$ are finite, i.e., providing the analogy to the Riemann sums holds.

- *For a general, positive measurable function f, **integrability** can be extended to f by **comparison to simple measurable functions** on $\mathcal{U}$, such as*

$$\mathcal{I}_{\mathcal{M}}[f] = \sup_{g}\left\{\mathcal{I}_{\mathcal{M}}[g]: \; g(x) = \sum_{i=1}^{n} c_i \mathbf{1}_{x \in \mathcal{S}_i}, \; g(x) \leq f(x), \; x \in \mathcal{U}\right\},$$

*provided $\mathcal{I}_{\mathcal{M}}[f]$ is finite. For functions that change sign, i.e., signed functions, the **positive-negative decomposition** $f(x) = f_+(x) - f_-(x)$ with the $f_{\pm}(x) \equiv (|f|(x) \pm f(x))/2$ for $x \in \mathcal{U}$, such that*

$$\mathcal{I}_{\mathcal{M}}[f] = \mathcal{I}_{\mathcal{M}}[f_+] - \mathcal{I}_{\mathcal{M}}[f_-],$$

provided the $\mathcal{I}_{\mathcal{M}}[f_{\pm}]$ are finite. (The positive-negative decomposition is used in Chapter 8 for numerical up-winding to ensure stability.)

- *If $\mathcal{M}$ is a Lebesgue measure, then the Lebesgue of the measure function f on $\mathcal{S} \in \mathcal{U}$ can be written*

$$\mathcal{I}_{\mathcal{M}}[f] = \int_{\mathcal{S}} f(x)\mathcal{M}(dx) = \int_{\mathcal{U}} \mathbf{1}_{x \in \mathcal{S}}\, f(x)\mathcal{M}(dx),$$

recalling that dx symbolizes the set $(x, x + dx)$.

- ***Monotone Convergence Theorem***
  Given the measure space $(\mathcal{U}, \mathbf{\Sigma}, \mathcal{M})$ if $\{f_n(x), \; f_n(x) \geq 0 \text{ for } n = 1, 2, \dots, \}$ is a countable sequence of one-dimensional (nonnegative) measurable functions on $\mathcal{U}$ that is a.e. monotone increasing and converging pointwise to $f(x)$ a.e., then we have

$$\lim_{n \to \infty} \int_{\mathcal{U}} f_n(x)\mathcal{M}(dx) = \int_{\mathcal{U}} f_n(x)\mathcal{M}(dx).$$

This basic convergence theorem leads to several others.

12.1.2 Change of Measure: Radon–Nikodým Theorem and Derivative

The abstract analogue of the change of variables, chain rule, and Jacobian techniques for the Riemann or Riemann–Stieltjes integral is the change of measures and the Radon–Nikodým derivative.

Theorem 12.5. *Radon–Nikodým Change of Measures.*
Given the measure space $\{\mathcal{U}, \mathbf{\Sigma}, \mathcal{M}_1\}$ with σ-finite measure $\mathcal{M}_1$, if $\mathcal{M}_2$ is a finite measure that is absolutely continuous with respect to $\mathcal{M}_1$ ($\mathcal{M}_2 \prec \mathcal{M}_1$), then there exists a measurable real function $\mathbb{D}(x) > 0$ for $x \in \mathcal{U}$ such that for each measurable set $\mathcal{S} \in \mathbf{\Sigma}$

$$\mathcal{M}_2(\mathcal{S}) = \mathcal{I}_{\mathcal{M}_1}[\mathbb{D}\,\mathbf{1}_{* \in \mathcal{S}}] = \int_{\mathcal{U}} \mathbb{D}(x)\mathbf{1}_{x \in \mathcal{S}}d\mathcal{M}_1(x) = \int_{\mathcal{S}} \mathbb{D}(x)d\mathcal{M}_1(x), \qquad (12.4)$$

where $d\mathcal{M}_i(x) = \mathcal{M}_i(dx)$ is equivalent notation for $i = 1:2$. The function $\mathbb{D}$ is the **Radon–Nikodým derivative** *of $\mathcal{M}_2$ with respect to $\mathcal{M}_1$, i.e.,*

$$\mathbb{D}(x) = \frac{d\mathcal{M}_2}{d\mathcal{M}_1}(x) \quad \text{or} \quad \mathbb{D}(\mathcal{S}) = \frac{d\mathcal{M}_2}{d\mathcal{M}_1}(\mathcal{S}). \tag{12.5}$$

Further, if η is integrable with respect to the measure $\mathcal{M}_2$, then

$$\mathcal{I}_{\mathcal{M}_2}[\eta] = \int_{\mathcal{U}} \eta(x)d\mathcal{M}_2(x) = \int_{\mathcal{U}} \eta(x)\frac{d\mathcal{M}_2(x)}{d\mathcal{M}_1(x)}d\mathcal{M}_1(x)$$

$$= \mathcal{I}_{\mathcal{M}_2}[\eta\,\mathbb{D}] = \int_{\mathcal{U}} \eta(x)\mathbb{D}(x)\mathcal{M}_1(x),$$

i.e., using the Radon–Nikodým derivative in a measure-theoretic chain rule.

Thus, the Radon–Nikodým derivative is the analogue of the Jacobian of the transformation (9.56) in an integral change of variables and leads to the absolutely continuous measure chain rule, symbolically substituting for $\mathbb{D}$,

$$d\mathcal{M}_2 = \frac{d\mathcal{M}_2}{d\mathcal{M}_1}d\mathcal{M}_1.$$

If $d\mathcal{M}_2$ and $d\mathcal{M}_1$ are mutually absolutely continuous, i.e., equivalent ($\mathcal{M}_1(\mathcal{S}) \overset{\text{a.c.}}{\equiv} \mathcal{M}_1(\mathcal{S})$), the Radon–Nikodým derivatives are mutual reciprocals,

$$\frac{d\mathcal{M}_1}{d\mathcal{M}_2} = 1 \Big/ \frac{d\mathcal{M}_2}{d\mathcal{M}_1},$$

formally justified by common null sets.

See the probability measure Examples 12.13 on p. 377 illustrating applied-oriented calculations for Radon–Nikodým derivatives in Subsection 12.2.1.

12.1.3 Probability Measure Basics

Since the probability distribution function for the real random variable $\mathbf{X}$ on the real set $\mathcal{S} \subseteq \mathbb{R}^{n_x}$ has the property that

$$\Phi_{\mathbf{X}}(\mathcal{S}) = \text{Prob}[\mathbf{X} \in \mathcal{S}] \in [0, 1],$$

it is a natural candidate for a measure and the density $\phi_{\mathbf{X}}(\mathbf{x})$ could play the role of the Radon–Nikodým derivative. According to convention, we reset the universe as $\mathcal{U} = \Omega$, the σ-algebra as $\Sigma = \mathcal{F}$, and the measure as $\mathcal{M} = \mathbb{P}$. For the jump part of jump-diffusions, counting or jump measures will also be needed.

Definition 12.6. *Probability Measure.*
*A **probability space** $(\Omega, \mathcal{F}, \mathbb{P})$ is a measure space with elements $\omega \in \Omega$ called **sample points** of random events in the **sample space** Ω, elements $\mathcal{F}_i \in \mathcal{F}$ called random **events**, and any **probability measure** $\mathbb{P}$ on the measurable space $(\Omega, \mathcal{F})$ has total mass of one, i.e., $\mathbb{P}(\Omega) = 1$.*

Summarizing the Kolmogorov axioms [33] *of a probability space* $(\Omega, \mathcal{F}, \mathbb{P})$,

- $\mathbb{P}(\emptyset) = 0$ *and* $\mathbb{P}(\Omega) = 1$.

- $\mathbb{P}(\mathcal{S}) \geq 0$ *for all* $\mathcal{S} \in \Omega$.

- $\mathbb{P}(\cup_{i=1}^{\infty} \mathcal{S}_i) = \cup_{i=1}^{\infty} \mathbb{P}(\mathcal{S}_i)$, *assuming the* $\{\mathcal{S}_i\}$ *are disjoint and countable, i.e., there is* **countable additivity**, *so that if* $\mathcal{S} \cup \mathcal{S}^c = \Omega$, *then the complementarity property also holds,* $\mathbb{P}(\mathcal{S}^c) = \mathbb{P}(\Omega) + \mathbb{P}(\mathcal{S})$.

- *If* $\mathcal{S}_2 \subseteq \mathcal{S}_2$ *and* $\mathbb{P}(\mathcal{S}_1) = 0$, *then* $\mathbb{P}(\mathcal{S}_2) = 0$, *i.e., the probability space is* **complete**.

Some additional properties and nomenclature follow.

- *The* $\omega \in \Omega$ *are also called scenarios as well as* **outcomes***, the* **underlying** *or* **background random variables***, e.g., like the mark variable of the compound Poisson process or Poisson random measure.*

- *An event set* $\mathcal{S}$ *with probability* $\mathbb{P}(\mathcal{S}) = 1$ *is said to happen* **almost surely** **(a.s.)** *or* **with probability one (w.p.o.)***, equivalent to* **almost everywhere (a.e.)** *for mathematical measures. If an event* $\mathcal{S}$ *has probability* $\mathbb{P}(\mathcal{S}) = 0$, *the event is said to be* **impossible***.*

- *Given a probability space* $(\Omega, \mathcal{F}, \mathbb{P})$, *then a (real)* **random variable** $\mathbf{X}(\omega)$ *is a measurable mapping from* Ω *to* $\mathbb{R}^{n_x}$ *such that the inverse (preimage)* $\mathbf{X}^{-1}(\mathcal{S}) = \{\omega \in \Omega : \mathbf{X}(\omega) \in \mathcal{S}\}$ *is* $\mathcal{F}$-*measurable for Borel (open) sets* $\mathcal{S} \in \mathcal{B}(\mathbb{R}^{n_x})$, *i.e.,* $\mathbf{X}(\omega)$ *is the* **realization** $\mathbf{X}$ *upon event* ω. *If* f *is a (real) measurable function, then* $f(\mathbf{X}(\omega))$ *will also be a random variable.*

- *If the problem involves only a single probability measure* $\mathbb{P}$ *for the single random variable* ω, *then we can write in more usual notation,*

$$X \equiv \omega, \quad \text{Prob}[X \in \mathcal{S}] = \text{Pr}[X \in \mathcal{S}] \equiv \mathbb{P}(\mathcal{S}),$$

i.e., the probability measure is the distribution $\Phi_\omega(\mathcal{S}) = \mathbb{P}(\mathcal{S})$ *for* $\mathcal{S} \subseteq \Omega$.

- *In general, if* $\mathbf{X} = \mathbf{X}(\omega) \in \mathbb{R}^{n_x}$ *for* $\omega \in \Omega$, *then let* $\omega \in \mathcal{S}_\omega \subseteq \Omega, \mathbf{X}(\omega) \in \mathcal{S}_{\mathbf{X}} = \mathbf{X}(\mathcal{S}_\omega)$, *and assuming the preimage* $\mathcal{S}_\omega = \mathbf{X}^{-1}(\mathcal{S}_{\mathbf{X}})$ *exists, then the* **distribution** *of* $\mathbf{X}$ *is the probability measure*

$$\Phi_{\mathbf{X}}(\mathcal{S}_{\mathbf{X}}) = \mathbb{P}(\mathbf{X}^{-1}(\mathcal{S}_{\mathbf{X}})),$$

so $\Phi_{\mathbf{X}}(\mathbf{x}) = \mathbb{P}(\{\omega \ni \mathbf{X} \leq \mathbf{x}\})$, *the inequality* $(\mathbf{X} \leq \mathbf{x})$ *meant element-wise.*

- *The* **expectation** *for a measurable real function* f *of* $X \in \mathbb{R}^{n_x}$ *with* $\omega \in \Omega$ *is then*

$$\text{E}[f(\mathbf{X})] = \int_{\Omega} f(\mathbf{X}(\omega))\mathbb{P}(d\omega) = \int_{\Omega} f(\mathbf{X}(\omega))d\mathbb{P}(\omega) = \int_{\mathbb{R}^{n_x}} f(\mathbf{x})\Phi_{\mathbf{X}}(d\mathbf{x}),$$

provided f *is absolutely integrable,*

$$\int_{\Omega} |f(\mathbf{X}(\omega))|\mathbb{P}(d\omega) < \infty,$$

noting that the $d\omega$ argument of $\mathbb{P}$ is an abbreviation for the interval $(\omega, \omega + d\omega)$ and that $d\mathbb{P}(\omega)$ and $\mathbb{P}(d\omega)$ will be assumed to be equivalent notation.

- ***Almost sure equivalence:*** *Let $\mathbf{X}_1(\omega)$ and $\mathbf{X}_2(\omega)$ be two random variables for $\omega \in \Omega$; then $\mathbf{X}_1 \overset{\text{a.s.}}{=} \mathbf{X}_2$ if*

$$\mathbb{P}(\{\omega \in \Omega, \mathbf{X}_1(\omega) = \mathbf{X}_2(\omega)\}) = 1.$$

*(See also the prior definition of **almost surely**.)*

- ***Equivalence in distribution:*** *Let $\mathbf{X}_1(\omega)$ and $\mathbf{X}_2(\omega)$ be two random variables for $\omega \in \Omega$. If the distribution satisfies*

$$\Phi_{\mathbf{X}_1} = \Phi_{\mathbf{X}_2},$$

*then $\mathbf{X}_1(\omega)$ and $\mathbf{X}_2(\omega)$ are called **equal in distribution** and we write*

$$\mathbf{X}_1 \overset{\text{dist}}{=} \mathbf{X}_2.$$

*(Also called **equal in law** or **identically distributed**; the notation $\mathbf{X}_1 \overset{\text{d}}{=} \mathbf{X}_2$ is also used.)*

- *The set of n random variables $\{X_i\}$ are **independent** with respect to the measurable sets $\mathcal{S}_i$ for $i = 1:n$ if the probability of the union is the product of the probabilities*

$$\mathbb{P}\left(\bigcup_{i=1}^{n}\{X_i \in \mathcal{S}_i\}\right) = \prod_{i=1}^{n}\mathbb{P}(\{X_i \in \mathcal{S}_i\}),$$

where the underlying random variable ω has been suppressed. A more concrete and useful form as a distribution in the vector $\mathbf{X} = [X_i]_{n \times 1}$ is

$$\Phi_{\mathbf{X}}(\mathbf{x}) = \mathbb{P}\left(\bigcup_{i=1}^{n}\{X_i \leq x_i\}\right) = \prod_{i=1}^{n}\mathbb{P}(\{X_i \leq x_i\}) = \prod_{i=1}^{n}\Phi_{X_i}(x_i).$$

An immediate corollary is the multiplication rule for the expectation of a set of independent random variables,

$$\mathrm{E}\left[\prod_{i=1}^{n}X_i\right] = \prod_{i=1}^{n}\mathrm{E}[X_i],$$

assuming finite expectations, $\mathrm{E}[|X_i|] < \infty$ for $i = 1:n$.

For more background information, see Applebaum [12], Billingsley [32], Bingham and Kiesel [33], Cont and Tankov [60], Cyganowski, Kloeden, and Ombach [67], Øksendal [222], and Øksendal and Sulem [223].

Much of the further results, such as conditional expectations, follow the applied path in this book, except that matters like that of positivity and changes in sign have to be treated with care to account for particular abstract constructs and conditions that are designed to facilitate proofs rather than the wide variety of problem applications.

12.1.4 Stochastic Processes in Continuous Time on Filtered Probability Spaces

Since the emphasis of this book is on jump-diffusions, stochastic processes in continuous time are treated and the relatively simpler, but not simple, discrete time stochastic processes are omitted (see Pliska's [225] book or Bingham and Kiesel's chapter devoted to discrete time processes) [33, Chapter 3]. The main additional difficulty treating stochastic processes in continuous time is extending the notion of a single probability space to a family of probability spaces over the continuous time variable t which often has infinite range.

Definition 12.7. *Filtered Probability Space.*

- *Based upon a probability space $(\Omega, \mathcal{F}, \mathbb{P})$, a **filtration** is a family of increasing σ-algebras*

$$\mathbb{F} = \{\mathcal{F}_t : t \geq 0; \ \mathcal{F}_s \subseteq \mathcal{F}_t, \ 0 \leq s \leq t < \infty\}$$

*and the extended space $(\Omega, \mathcal{F}, \mathbb{P}, \mathbb{F})$ is called a **filtered probability space**. The sub-σ-algebra $\mathcal{F}_t$ represents the known information of the system on $(0, t]$ at time t.*

- *The usual filtration conditions (with jump-diffusions in mind) are*

  - *The initial sub-σ-algebra $\mathcal{F}_0$ consists of all the $\mathbb{P}$-null-sets of $\mathcal{F}$.*

  - *The filtration $\mathbb{F}$ is **right-continuous with left limits (RCLL, or cádlág in French)**, i.e., $\mathcal{F}_t = \mathcal{F}_{t^+} = \lim_{\epsilon \to 0^+} \mathcal{F}_{t+\epsilon}$ for the RC part and $\mathcal{F}_{t^-} = \lim_{\epsilon \to 0^+} \mathcal{F}_{t-\epsilon}$ for the LL part exist. The **jump** in the sub-σ-algebra at time t is $[\mathcal{F}]_t = \mathcal{F}_{t^+} - \mathcal{F}_{t^-}$. If only continuous processes such as diffusions are under consideration, then **right-continuity (RC)** is sufficient.*

Definition 12.8. *Stochastic Process.*

- *Given the filtered probability space $(\Omega, \mathcal{F}, \mathbb{P}, \mathbb{F})$, a **stochastic process in continuous time** $\mathbf{X} = \{\mathbf{X}(t) : t \geq 0\}$ and X is $\mathcal{F}_t$-**adapted** to the filtration $\mathbb{F}$ if $\mathbf{X}(t)$ is $\mathcal{F}_t$-measurable $(\mathbf{X}(t) \in \mathcal{F}_t)$ for each t.*

- *The **natural filtration** for the stochastic process $\mathbf{X}(t)$ can be written as*

$$\mathcal{F}_{t,\mathbf{X}} = \widehat{\sigma}(\mathbf{X}(s), \ 0 \leq s \leq t)$$

with $\widehat{\sigma}$ signifying the σ-field of $\mathbf{X}(t)$, or more loosely the information or history of the process $\mathbf{X}(t)$ up until time t.

- *Including the dependence on the underlying random variable, $\omega \in \Omega$, the $\mathbf{X}(t; \omega)$ defines a random function of time, called the **sample path** and is a mapping from $[0, t] \times \Omega$ to $\mathbb{R}^{n_x}$. Usually, $\mathbf{X}(t; \omega)$ is denoted by $\mathbf{X}_t(\omega)$ or just X_t; however, in this book real subscripts are reserved to denote partial derivatives, except for algebraic quantities like $\mathcal{F}_t$ that are not genuine functions.*

- *If X is adapted, i.e., $\mathcal{F}_t$-**adapted** to $\mathbb{F}$, for $t \geq 0$, then the conditional expectation satisfies*

$$\mathrm{E}[X(t) \mid \mathcal{F}_t] \overset{\mathrm{a.s.}}{=} X(t)$$

*since $X(t)$ is known from $\mathcal{F}_t$ (recall the symbol $\overset{a.s.}{=}$ denotes **equals almost surely**). Saying that X or $X(t)$ is $\mathcal{F}_t$-**adapted** to $\mathbb{F}$ means the same as saying that $X(t)$ is **nonanticipating**.*

- *Two stochastic processes X_1 and X_2 are the **same with respect to a set of finite-dimensional distributions** if for some positive integer n and discrete time points $\{t_i : i = 1 : n\}$, the random vectors $\mathbf{X}_j = [X_{i,j}]_{n \times 1}$ for $j = 1 : 2$ have the same n-dimensional distribution, corresponding to the stochastic processes X_j for $j = 1 : 2$, respectively.*

12.1.5 Martingales in Continuous Time

Martingales are processes with the property that the best predictor of the process future value is the present value given present knowledge, i.e., it represents a fair game of gambling, rather than a favorable or unfavorable one.

Definition 12.9. *Martingale Properties in Continuous Time.*

- *Given a filtered probability space $(\Omega, \mathcal{F}, \mathbb{P}, \mathbb{F})$ and $\mathcal{F}_t$-adapted process $X(t)$ on $[0, T]$, $T < \infty$, then $X(t)$ is a **martingale** if*

$$\mathrm{E}[X(t) \mid \mathcal{F}_s] \overset{a.s.}{=} X(s), \quad t > s \geq 0, \tag{12.6}$$

provided $X(t)$ is absolutely integrable, $\mathrm{E}[|X(t)|] < \infty$ on $[0, T]$, i.e., the best predictor of $X(t)$ with respect to the filter $\mathcal{F}_s$ is $X(s)$.

- *If instead of (12.6),*

$$\mathrm{E}[X(t) \mid \mathcal{F}_s] \overset{a.s.}{\leq} X(s), \quad t > s \geq 0,$$

*then $X(t)$ is a **supermartingale**, but if*

$$\mathrm{E}[X(t) \mid \mathcal{F}_s] \overset{a.s.}{\geq} X(s), \quad t > s \geq 0,$$

*then $X(t)$ is a **submartingale**. (The submartingale corresponds to the favorable game and the supermartingale corresponds to the unfavorable game, provided $X(t) - X(s)$ represents the gain.)*

- *Two martingales $\mathcal{M}_1(t)$ and $\mathcal{M}_2(t)$ which are also equivalent or mutually absolutely continuous measures, i.e., $\mathcal{M}_1(t) \overset{a.c.}{\equiv} \mathcal{M}_2(t)$, are called **equivalent martingale measures (EMM)** and they play an important role in mathematical finance.*

Examples 12.10. *Diffusion, Jump, and other Martingales.*
For this set of examples, the time-interval $[0, T]$ as well as the coefficients will be finite, so there is no question that the stochastic processes will be absolutely integrable.

1. *Let $X(t)$ be a $\{\mu_0, \sigma_0\}$-constant coefficient, **diffusion process** with SDE*

$$dX(t) = \mu_0 dt + \sigma_0 dW(t),$$

and it is of interest to know for what values of μ_0 is $X(t)$ a martingale, a super-martingale, or a submartingale.

The solution by integrating over $[s, t]$ is

$$X(t) = X(s) + \mu_0(t - s) + \sigma_0(W(t) - W(s)),$$

noting that $(W(t) - W(s)) \overset{\text{dist}}{=} W(t - s)$ by the stationary property and is independent of $W(s)$ and $\text{E}[W(t - s)|W(s)] = 0$, so with $\mathcal{F}_t = \widehat{\sigma}(X(r), \ 0 \le r \le t)$, the natural filtration for $X(t)$,

$$\text{E}[X(t) \mid \mathcal{F}_s] = X(s) + \mu_0(t - s),$$

$0 \le s < t$. *Hence, $X(t)$ is a martingale if $\mu_0 = 0$ (the case of the zero-mean infinitesimal diffusion, denoted by $d\widehat{X}(t) = \sigma_0 dW(t)$), a supermartingale if $\mu_0 < 0$, or a submartingale if $\mu_0 > 0$. Alternatively, the translated process*

$$\widetilde{X}(t) \equiv X(t) - \mu_0 t \overset{\text{dist}}{=} \widetilde{X}(s) + \sigma_0 W(t - s)$$

is a martingale.

2. *Let $X(t)$ be a $\{\mu_0, \sigma_0\}$-constant coefficient, **geometric diffusion process**,*

$$dX(t) = X(t)(\mu_0 dt + \sigma_0 dW(t)),$$

which has the Itô calculus solution,

$$X(t) \overset{\text{dist}}{=} X(s) \exp((\mu_0 - \sigma_0^2/2)(t - s) + \sigma_0 W(t - s)),$$

so

$$\text{E}[X(t) \mid \mathcal{F}_s] = X(s) \exp(\mu_0(t - s)),$$

$0 \le s < t$. *Again, $X(t)$ is a martingale if $\mu_0 = 0$, a supermartingale if $\mu_0 < 0$, or a submartingale if $\mu_0 > 0$. Alternatively, the scaled process*

$$\widetilde{X}(t) \equiv \exp(-\mu_0 t) X(t) \overset{\text{dist}}{=} \widetilde{X}(s) \exp(\sigma_0(W(t - s) - \sigma_0(t - s)/2))$$

*is a martingale, or more specifically an **exponential martingale** [22], and the scaling corresponds to the Girsanov transformation of $W(t)$ that will be discussed in Subsection 12.2.2.*

3. *Let $X(t)$ be a **simple Poisson process** $P(t)$ with additional drift and constant coefficients, $\{\mu_0, \nu_0, \lambda_0\}$,*

$$dX(t) = \mu_0 dt + \nu_0 dP(t),$$

where $\text{E}[dP(t)] = \lambda_0 dt = \text{Var}[dP(t)]$. The solution is

$$X(t) = X(s) + \mu_0(t - s) + \nu_0(P(t) - P(s)),$$

where $(P(t) - P(s)) \overset{\text{dist}}{=} P(t - s)$ and the conditional expectation is

$$\text{E}[X(t) \mid \mathcal{F}_s] = X(s) + (\mu_0 + \lambda_0 \nu_0)(t - s),$$

so $X(t)$ is a martingale if $\mu_0 = -\lambda_0 v_0$ (the zero-mean infinitesimal jump process, denoted by $d\widehat{X}(t) = v_0 d\widehat{P}(t)$, using the zero mean Poisson $d\widehat{P}(t) \equiv dP(t) - \lambda_0 dt$, a supermartingale if $\mu_0 < -\lambda_0 v_0$, or a submartingale if $\mu_0 > -\lambda_0 v_0$. Alternatively, the translated process

$$\widetilde{X}(t) \equiv X(t) - (\mu_0 + \lambda_0 v_0)t \overset{dist}{=} \widetilde{X}(s) + v_0\widehat{P}(t-s)$$

is a martingale.

4. *Let $X(t)$ be a **compound Poisson process** with additional drift and constant coefficients, $\{\mu_0, v_0, \lambda_0, \mu_Q\}$,*

$$dX(t) = \mu_0 dt + v_0 \sum_{i=P(t)+1}^{P(t)+dP(t)} Q_i,$$

where $\mathrm{E}[dP(t)] = \lambda_0 dt = \mathrm{Var}[dP(t)]$ and the Q_i are IID random marks with mean μ_Q and variance σ_Q^2 which will not be needed. (Also note that the zero-one law has not been applied to $dP(t)$.) The solution is

$$X(t) \overset{dist}{=} X(s) + \mu_0(t-s) + v_0 \sum_{i=1}^{P(t-s)} Q_i$$

and the conditional expectation, by iterated conditional expectations between the Poisson counting process and the marks, is

$$\mathrm{E}[X(t) \mid \mathcal{F}_s] = X(s) + (\mu_0 + \lambda_0 v_0 \mu_Q)(t-s),$$

so $X(t)$ is a martingale if $\mu_0 = -\lambda_0 v_0 \mu_Q$ (the zero-mean infinitesimal compound Poisson, $d\widehat{X}(t) = v_0\mu_Q d\widehat{P}(t) + v_0 \sum_{i=P(t)+1}^{P(t)+dP(t)} \widehat{Q}_i$, where $\widehat{P}(t) \equiv (P(t) - \lambda_0 t)$ and $\widehat{Q}_i = (Q_i \quad \mu_Q)$, a supermartingale if $\mu_0 < -\lambda_0 v_0 \mu_Q$, or a submartingale if $\mu_0 > -\lambda_0 v_0 \mu_Q$. The alternative process

$$\widetilde{X}(t) = X(t) - (\mu_0 + \lambda_0 v_0 \mu_Q)t = \widetilde{X}(s) + v_0\mu_Q \widehat{P}(t-s) + v_0 \sum_{i=1}^{P(t-s)} \widehat{Q}_i$$

is a martingale, such that the difference $\widetilde{X}(t) - \widetilde{X}(s)$ is a linear combination of zero-mean random processes, or variables as in the case of $\widehat{Q}_i$, counting only the jumps in $(s, t]$.

5. *As an exercise the reader can find the similar martingale properties as a function of the additional drift for the **geometric jump-diffusion** problem with constant coefficients,*

$$dX(t) = X(t)\left(\mu_0 dt + \sigma_0 dW(t) + v_0 \sum_{i=P(t)+1}^{P(t)+dP(t)} (\exp(Q_i) - 1)\right),$$

where again the marks are IID with mean μ_Q and variance σ_Q^2, with the amplitude in the log-ready exponential form.

6. *The simplest, but trivial, example is the **constant process** $X(t) = c_0$ for $t \geq 0$, i.e., $dX(t) = 0$, so $X(t)$ is a martingale since $\mathrm{E}[X(t)|\mathcal{F}_s] = c_0 = X(s)$ for $s < t$.*

7. *Another example is the **closed martingale** that is constructed from an absolutely integrable random variable Y, independent of t, on the filtered probability space, such that a stochastic process is defined as*

$$X(t) \equiv \mathrm{E}[Y \mid \mathcal{F}_t], \quad t \geq 0.$$

*Thus, by the **tower law** [22, p. 34], [209, Rule 6, p. 72],*

$$\mathrm{E}[X(t) \mid \mathcal{F}_s] = \mathrm{E}[\mathrm{E}[Y \mid \mathcal{F}_t] \mid \mathcal{F}_s] = \mathrm{E}[Y \mid \mathcal{F}_s] \equiv X(s)$$

for $s < t$, since the conditioning on $\mathcal{F}_t$ followed by the conditioning on $\mathcal{F}_s$ is the same as the original conditioning on $\mathcal{F}_s$, i.e., dependence is on the smaller of the conditioning filters.

12.1.6 Jump-Diffusion Martingale Representation

Martingale representations are heavily relied upon for hedging in financial applications. There are many versions of martingale representation in the literature. Some have useful and elementary presentations. Many are restricted to diffusions except for a mention of jump processes. See Baxter and Rennie [22], Duffie [75], Glasserman [97], Øksendal [222], and Steele [256]. Here, a form of the martingale repesentation theorem is given for marked-jump-diffusion processes following Applebaum [12] and, particularly, Runggaldier [239]. Their formulation, after Jacod and Shiryaev [153] and Kunita and Watanabe [173], respectively, uses Poisson random measure $\mathcal{P}(\mathbf{dt}, \mathbf{dq})$ defined beginning in (5.1) on p. 130 and whose integrals are related to compound Poisson processes (5.6) on the mark-sample-space $\mathcal{Q}$ by

$$\int_{\mathcal{Q}} h(t, q)\mathcal{P}(\mathbf{dt}, \mathbf{dq}) = \sum_{i=P(t)+1}^{P(t)+dP(t)} h(T_i^-, Q_i) \tag{12.7}$$

without using the zero-one law for $dP(t)$, where the T_i^- are the prejump-times and the Q_i are the IID sampled marks, but are often found in martingale form by using the centered or mean-zero Poisson random measure

$$\widetilde{\mathcal{P}}(\mathbf{dt}, \mathbf{dq}) \equiv \mathcal{P}(\mathbf{dt}, \mathbf{dq}) - \mathrm{E}[\mathcal{P}(\mathbf{dt}, \mathbf{dq})] = \mathcal{P}(\mathbf{dt}, \mathbf{dq}) - \phi_Q(q; t)dq\lambda(t)dt,$$

where $\Phi_Q(dq; t) = \phi_Q(q; t)dq$ is jump-amplitude probability measure and $\lambda(t)$ is the Poisson jump-rate. The mean-zero relationship corresponding to the original relationship (12.7) is then

$$\int_{\mathcal{Q}} h(t, q)\widetilde{\mathcal{P}}(\mathbf{dt}, \mathbf{dq}) = \sum_{i=P(t)+1}^{P(t)+dP(t)} h(T_i^-, Q_i) - \mathrm{E}_Q[h(t, Q)]\lambda(t)dt, \tag{12.8}$$

where $\mathrm{E}_Q[h(t, Q)] = \int_{\mathcal{Q}} h(t, q)\phi_Q(q; t)dq$.

Theorem 12.11. *Marked-Jump-Diffusion Martingale Representation Theorem.*
Given the Wiener process $W(t)$ and compound Poisson triplet

$$\{dP(t), \lambda(t), \phi_Q(q; t)\}$$

or a Poisson random measure $\mathcal{P}(\mathbf{dt}, \mathbf{dq})$ on the sigma field

$$\mathbb{F} = \mathcal{F}_t^{(W, P, Q)} = \widehat{\sigma}\{W(s), P(t), \mathcal{S}_Q, \mathcal{S}_N : 0 \le s \le t, \ \mathcal{S}_Q \in \mathcal{Q}, \ \mathcal{S}_N \in \mathcal{N}_1\},$$

where $\mathcal{N}_1$ is the collection of null-sets of $\mathbb{P}$. Then, any $(\mathbb{P}, \mathbb{F})$-martingale $\mathcal{M}(t)$ has the two alternate representions

$$\mathcal{M}(t) = \mathcal{M}(0) + \int_0^t \Gamma^{(D)}(s)dW(s) + \int_0^t \int_Q \Gamma^{(MJ)}(s, q)\widetilde{\mathcal{P}}(\mathbf{ds}, \mathbf{dq})$$

$$= \mathcal{M}(0) + \int_0^t \Gamma^{(D)}(s)dW(s) + \sum_{i=1}^{P(t)} \Gamma^{(MJ)}\left(T_i^-, Q_i\right) \qquad (12.9)$$

$$- \mathrm{E}_Q\left[\Gamma^{(MJ)}(t, Q)\right]\Lambda(t),$$

where $\Gamma^{(D)}(t)$ is a predictable (measurable with respect to $\mathbb{P}$), square-integrable process, while $\Gamma^{(MJ)}(t, q)$ is a $\mathcal{F}_t^{(W, P, Q)}$-predictable, Q-marked process, such that

$$\mathrm{E}_Q\left[\Gamma^{(MJ)}(t, Q)\right] = \int_Q \Gamma^{(MJ)}(t, q)\phi_Q(q; t)dq < \infty$$

and $\Lambda(t) \equiv \int_0^t \lambda(s)ds$ is the mean jump-count.

The martingale representation theorem is used in Subsection 12.2.2 for two versions of Girsanov's stochastic process transformation theorem, one for the diffusion process alone, i.e., without the Poisson terms in (12.9), and another for marked-jump-diffusion processes using the full form in (12.9).

The martingale approach may be a favored approach to solving SDE problems, but Heath and Schweizer [136] show the equivalence of the martingale and PDE approaches for a number of financial applications. The Feynman–Kac formula (see (7.71) on p. 216 or see the appendix of Duffie [75, Appendix E] for more background) is used to solve the corrresponding PDE problem that is derived from the SDE.

12.2   Change in Probability Measure: Radon–Nikodým Derivatives and Girsanov's Theorem

12.2.1   Radon–Nikodým Theorem and Derivative for Change of Probability Measure

Here, a version of the Radon–Nikodým theorem, Theorem 12.5, and derivative is formulated especially for probability measures and expectations. The abstract analogue of the change of distribution corresponding to a change in random variables is presented in (B.5) for the distribution and (B.6) for the density on pp. B3–B4 in Online Appendix B.

Theorem 12.12. Radon–Nikodým Change of Probability Measures.
Given a filtered probability space $(\Omega, \mathcal{F}, \mathbb{P}, \mathbb{F})$ *with* σ*-finite measure* $\mathbb{P}_1$, *if* $\mathbb{P}_2$ *is a finite measure that is mutually absolutely continuous with* $\mathbb{P}_1$ *(equivalent,* $\mathbb{P}_2 \overset{a.c.}{\equiv} \mathbb{P}_1$*), then there exists a positive measurable real function*

$$\mathbb{D}(x) = \frac{d\mathbb{P}_2}{d\mathbb{P}_1}(x) \quad \text{or} \quad \mathbb{D}(\mathcal{S}) = \frac{d\mathbb{P}_2}{d\mathbb{P}_1}(\mathcal{S}) \tag{12.10}$$

called the **Radon–Nikodým derivative** *of* $\mathbb{P}_2$ *with respect to* $\mathbb{P}_1$ *for* $x \in \Omega$ *such that for each measurable set* $\mathcal{S} \in \mathcal{F}$

$$\mathbb{P}_2(\mathcal{S}) = E_{\mathbb{P}_1}[\mathbb{D}(X)\mathbf{1}_{X \in \mathcal{S}}] = \int_{\Omega} \mathbb{D}(x)\mathbf{1}_{x \in \mathcal{S}} d\mathbb{P}_1(x) = \int_{\mathcal{S}} \mathbb{D}(x) d\mathbb{P}_1(x), \tag{12.11}$$

where $d\mathbb{P}_i(x) = \mathbb{P}_i(dx)$ *is equivalent notation for* $i = 1:2$.
Further, if η *is absolutely integrable with respect to the measure* $\mathbb{P}_2$, *then*

$$E_{\mathbb{P}_2}[\eta(X)] = \int_{\Omega} \eta(x) d\mathbb{P}_2(x) = \int_{\Omega} \eta(x) \frac{d\mathbb{P}_2(x)}{d\mathbb{P}_1(x)} d\mathbb{P}_1(x)$$

$$= E_{\mathbb{P}_2}[\eta(X)\mathbb{D}(X)] = \int_{\Omega} \eta(x)\mathbb{D}(x)\mathbb{P}_1(x),$$

i.e., using the Radon–Nikodým derivative in a measure-theoretic chain rule.

Thus, the Radon–Nikodým derivative is the analogue of the Jacobian of the transformation (9.56) in an integral change of variables and leads to the absolutely continuous measure chain rule, symbolically substituting for g,

$$d\mathbb{P}_2 = \frac{d\mathbb{P}_2}{d\mathbb{P}_1} d\mathbb{P}_1.$$

If $d\mathbb{P}_2$ and $d\mathbb{P}_1$ are mutually absolutely continuous, i.e., equivalent ($\mathbb{P}_1(\mathcal{S}) \overset{a.c.}{\equiv} \mathbb{P}_1(\mathcal{S})$), the Radon–Nikodým derivatives are mutual reciprocals,

$$\frac{d\mathbb{P}_1}{d\mathbb{P}_2} = 1 \Big/ \frac{d\mathbb{P}_2}{d\mathbb{P}_1},$$

formally justified by common null sets.

Examples 12.13. Radon–Nikodým Derivative Calculations.

- *Normal distributions:*
  Suppose a transformation from a standard normal distribution with density

$$\phi_1(x) = \exp\left(-x^2/2\right)\Big/\sqrt{2\pi}$$

  to a mean-μ*, variance-*σ^2 *normal distribution with density*

$$\phi_2(x) = \exp\left(-(x - \mu)^2/\left(2\sigma^2\right)\right)\Big/\sqrt{2\pi\sigma^2}.$$

The change in measure coincides with a change of drift and a change of scale. Thus, $\mathbb{P}_1(x) = \Phi_1(x)$ is the first probability measure, and the second is

$$\mathbb{P}_2(x) = \Phi_2(x) = \int_{-\infty}^{x} \mathbb{D}(y)\phi_1(y)dy = \int_{-\infty}^{x} \phi_2(y)dy$$

or $\phi_2(x) = \mathbb{D}(x)\phi_1(x)$ upon differentiating according to the fundamental theorem of integral calculus, and the Radon–Nikodým derivative is

$$\mathbb{D}(x) = \frac{d\mathbb{P}_2(x)}{d\mathbb{P}_1(x)} = \frac{d\Phi_2(x)}{d\Phi_1(x)} = \frac{\phi_2(x)}{\phi_1(x)} = \frac{\exp\left(-(x-\mu)^2/(2\sigma^2)\right)\big/\sqrt{2\pi\sigma^2}}{\exp\left(-x^2/2\right)\big/\sqrt{2\pi}}$$

$$= \frac{1}{\sigma}\exp\left(-\frac{(1-\sigma^2)x^2 - 2\mu x + \mu^2}{2\sigma^2}\right). \tag{12.12}$$

Hence, under measure $\mathbb{P}_1$ the random variable X has mean 0 and variance 1, but under measure $\mathbb{P}_2$ the random variable X has mean μ and variance σ^2.

If $\sigma = 1$, then there is only a change of drift and the Radon–Nikodým derivative is simpler:

$$\mathbb{D}(x) = \frac{d\mathbb{P}_2(x)}{d\mathbb{P}_1(x)} = \exp\left(\frac{\mu(2x-\mu)}{2}\right).$$

The more general form (12.12), formally justified here, can be transformed to the form in a proposition of Cont and Tankov [60, Proposition 9.7, p. 306] for two diffusion or Brownian motion processes, both denoted by $X = X(T)$, with parameters $\mu \to \mu_j T$ for the drifts and $\sigma^2 \to \sigma_1^2 T = \sigma^2 T = \sigma_2^2 T$ for a common variance on $(\Omega, \mathcal{F}_T, \mathbb{P}_j, \mathbb{F})$ for $j = 1{:}2$. Hence, using the fact the Radon–Nikodým derivative is the ratio of the two densities,

$$\mathbb{D}(X(T), T) = \frac{d\mathbb{P}_2(X(T), T)}{d\mathbb{P}_1(X(T), T)}$$

$$= \frac{\exp\left(-(X(T)-\mu_2 T)^2/\left(2\sigma_2^2 T\right)\right)\big/\sqrt{2\pi\sigma_2^2 T}}{\exp\left(-(X(T)-\mu_1 T)^2/\left(2\sigma_1^2 T\right)\right)\big/\sqrt{2\pi\sigma_1^2 T}}$$

$$= \exp\left(\frac{2(\mu_2-\mu_1)X(T) - (\mu_2^2-\mu_1^2)T}{2\sigma^2}\right). \tag{12.13}$$

This corrects an error in [60, Proposition 9.7, p 306]. They also convert this to the Cameron–Martin theorem form by letting $X(T) = \mu_1 T + \sigma W_1(T)$ in the notation here, so

$$\mathbb{D}(T) = \frac{d\mathbb{P}_2(T)}{d\mathbb{P}_1(T)} = \exp\left(\frac{2(\mu_2-\mu_1)\sigma W_1(T) - (\mu_2-\mu_1)^2 T}{2\sigma^2}\right), \tag{12.14}$$

which is correct in [60, following Proposition 9.7, p. 306].

- **Sets of independent random variables:**
  Let $\mathbf{X} = [X_i]_{n \times 1}$ be a set of n independent random variables with vector mean $\boldsymbol{\mu}^{(1)} = [\mu_i^{(1)}]_{n \times 1}$ and variance vector $\mathbf{V}^{(1)} = [\sigma_i^{(1)}]_{n \times 1}$ with product density

  $$\phi^{(1)}(\mathbf{x}) = \prod_{i=1}^{n} \phi_i^{(1)}(x_i)$$

  due to the independence property. The relationship between the measure, the distribution

  $$\Phi^{(1)}(\mathbf{x}) = \text{Prob}_{\mathbb{P}_1}[\mathbf{X} \leq \mathbf{x}],$$

  and the density can be written formally as

  $$\frac{d\mathbb{P}_1(\mathbf{x})}{d\mathbf{x}} = \left(\prod_{i=1}^{n} \frac{\partial}{\partial x_i} \right) \Phi^{(1)}(\mathbf{x}) = \phi^{(1)}(\mathbf{x}),$$

  where $\mathbf{X} \leq \mathbf{x}$ means $X_i \leq x_i$ for $i = 1:n$ and $d\mathbf{x} = \prod_{i=1}^{n} dx_i$ is the infinitesimal n-dimensional Euclidean measure, not a vector differential.

  Let there be a function $\mathbb{D}(\mathbf{x})$ that generates a second distribution or measure,

  $$\Phi^{(2)}(\mathbf{x}) = \text{Prob}_{\mathbb{P}_2}[\mathbf{X} \leq \mathbf{x}] = \left(\prod_{i=1}^{n} \int_{-\infty}^{x_i} dy_i \phi_i^{(2)}(y_i) \right)$$

  $$= \left(\prod_{i=1}^{n} \int_{-\infty}^{x_i} dy_i \phi_i^{(1)}(y_i) \right) \mathbb{D}(\mathbf{y}),$$

  so

  $$\frac{d\mathbb{P}_2(\mathbf{x})}{d\mathbf{x}} = \left(\prod_{i=1}^{n} \frac{\partial}{\partial x_i} \right) \Phi^{(2)}(\mathbf{x}) = \prod_{i=1}^{n} \int_{-\infty}^{x_i} dy_i \phi_i^{(2)}(x_i) = \phi_i^{(2)}(\mathbf{x})$$

  $$= \mathbb{D}(\mathbf{x}) \prod_{i=1}^{n} \int_{-\infty}^{x_i} dy_i \phi_i^{(1)}(x_i) = \mathbb{D}(\mathbf{x})\phi^{(1)}(\mathbf{x}).$$

  Solving produces

  $$\mathbb{D}(\mathbf{x}) = \frac{d\mathbb{P}_2(\mathbf{x})}{d\mathbb{P}_1(\mathbf{x})} = \frac{\phi_i^{(2)}(\mathbf{x})}{\phi^{(1)}(\mathbf{x})} = \prod_{i=1}^{n} \frac{\phi_i^{(2)}(x_i)}{\phi_i^{(1)}(x_i)}. \tag{12.15}$$

  This result is important for stochastic processes $X(t)$ for $t \in [0, T]$ since a Radon–Nikodým derivative cannot be computed for a random variable over an infinite-dimensional interval, but it is possible to sample $X(t)$ at sample times $t_i = (i-1)T/n$ using $X_i = X(t_i)$ for $i = 1:n$, assuming the process of interest has independent increments.

As a more concrete example, suppose that the X_i have a standard normal distribution, i.e., IID with $\mu_i^{(1)} = 0$ and $(\sigma_i^{(1)})^2 = 1$, and a nonstandard distribution is sought with mean $\mu_i^{(2)} = \mu_i$ and $(\sigma_i^{(2)})^2 = \sigma_i^2$. Then using (12.12),

$$\mathbb{D}(\mathbf{x}) = \frac{d\mathbb{P}_2(\mathbf{x})}{d\mathbb{P}_1(\mathbf{x})} = \frac{1}{\prod_{j=1}^n \sigma_j} \exp\left(-\sum_{i=1}^n \left(\frac{(1 - \sigma_i^2)x_i^2 - 2\mu_i x_i + \mu_i^2}{2\sigma_i^2}\right)\right). \quad (12.16)$$

This example is similar to one in Glasserman [97], except there $\sigma_i \equiv 1$.

- **Poisson distribution, a discrete analogy:**
  Next consider a Poisson cumulative distribution with parameter Λ_1 for the discrete variable N_1,

$$\Phi_n^{(1)} = \text{Prob}[N_1 < n] = e^{-\Lambda_1} \sum_{k=0}^n \frac{\Lambda_1^n}{n!},$$

  which has increment (discrete derivative analogue)

$$\Delta\Phi_{n-1}^{(1)} \equiv \Phi_n^{(1)} - \Phi_{n-1}^{(1)} = e^{-\Lambda_1} \frac{\Lambda_1^n}{n!},$$

  the numerical forward difference notation, corresponding to a discrete density and consistent with Itô rules. The change of measure from variable N_1 with parameter Λ_1 to variable N_2 with parameter Λ_2 is given by

$$\Phi_n^{(2)} = \text{Prob}[N_2 < n] = e^{-\Lambda_2} \sum_{k=0}^n \frac{\Lambda_2^k}{k!} = e^{-\Lambda_1} \sum_{k=0}^n \mathbb{D}_k \frac{\Lambda_1^k}{k!}$$

  with the Radon–Nikodým discrete derivative satisfying

$$\Delta\Phi_{n-1}^{(2)} = e^{-\Lambda_2} \frac{\Lambda_2^n}{n!} = \mathbb{D}_n e^{-\Lambda_1} \frac{\Lambda_1^n}{n!},$$

  and solving yields

$$\mathbb{D}_n = \frac{\Delta\mathbb{P}_2(n-1)}{\Delta\mathbb{P}_1(n-1)} = \frac{\Delta\Phi_{n-1}^{(2)}}{\Delta\Phi_{n-1}^{(2)}} = e^{\Lambda_1 - \Lambda_2} \left(\frac{\Lambda 2}{\Lambda 1}\right)^n$$

$$= e^{\Lambda_1 - \Lambda_2 + n \ln(\Lambda_2/\Lambda_1)}. \quad (12.17)$$

  Thus, with the change in measure from $\mathbb{P}_1$ to $\mathbb{P}_2$, the mean or average jump-count changes from Λ_1 to Λ_2.

- **Poisson distribution with fixed-size jumps:**
  Now, consider a Poisson distribution for discrete variable N_1 with parameter Λ_1 and constant jump size $v_1 \neq 0$, so

$$X = v_1 N_1.$$

  Given the primary measure

$$\mathbb{P}_1(x) = \text{Prob}[X \leq x] = \text{Prob}[N_1 \leq x/v_1] = e^{-\Lambda_1} \sum_{k=0}^\infty \frac{\Lambda_1^k}{k!} \mathbf{1}_{\{k \leq x/v_1\}},$$

a change in measure with parameters $\{\Lambda_2, v_2\}$ is sought such that

$$\mathbb{P}_2(x) = \mathrm{Prob}[X \leq x] = \mathrm{Prob}[N_2 \leq x/v_2] = e^{-\Lambda_2} \sum_{k_2=0}^{\infty} \frac{\Lambda_2^{k_2}}{k_2!} \mathbf{1}_{\{k_2 \leq x/v_2\}}$$

$$= e^{-\Lambda_1} \sum_{k_1=0}^{\infty} \frac{\Lambda_1^{k_1}}{k_1!} \mathbf{1}_{\{k_1 \leq x/v_1\}} \mathbb{D}_{k_1}.$$

In lieu of a proper derivative for the indicator functions $\mathbf{1}_{\{k_j \leq x/v_j\}}$ for $j = 1:2$, consider the increment at $x = (n-1)v_2$,

$$\Delta \mathbb{P}_2((n-1)v_2) = \mathbb{P}_2((n-1)v_2 + \Delta x) - \mathbb{P}_2((n-1)v_2)$$

$$= e^{-\Lambda_2} \sum_{k_2=0}^{\infty} \frac{\Lambda_2^{k_2}}{k_2!} \mathbf{1}_{\{n-1 < k_2 \leq n-1 + \Delta x/v_2\}}$$

$$= e^{-\Lambda_1} \sum_{k_1=0}^{\infty} \frac{\Lambda_1^{k_1}}{k_1!} \mathbf{1}_{\{(n-1)v_2/v_1 < k_1 \leq (n-1)v_2/v_1 + \Delta x/v_1\}} \mathbb{D}_{k_1}.$$

Aside from coupling the potential Radon–Nikodým discrete derivatives $\mathbb{D}_{k_1}$, as Cont and Tankov [60, Proposition 9.5, p. 303] state, the two measures will not be equivalent since their null sets will in general not coincide unless the jump sizes are the same, $v_2 = v_1$.

Thus, with $v_2 = v_1$ and $\Delta x = v_1$ for a semiopen unit step $(n-1, n]$, the new measure increment becomes

$$\Delta \mathbb{P}_2((n-1)v_1) = \mathbb{P}_2(nv_1) - \mathbb{P}_2((n-1)v_1)$$

$$= e^{-\Lambda_2} \sum_{k_2=0}^{\infty} \frac{\Lambda_2^{k_2}}{k_2!} \mathbf{1}_{\{n-1 < k_2 \leq n\}} = e^{-\Lambda_2} \frac{\Lambda_2^n}{n!}$$

$$= e^{-\Lambda_1} \sum_{k_1=0}^{\infty} \frac{\Lambda_1^{k_1}}{k_1!} \mathbf{1}_{\{n-1 < k_1 \leq n\}} \mathbb{D}_{k_1} = e^{-\Lambda_1} \frac{\Lambda_1^n}{n!} \mathbb{D}_n,$$

thus obtaining the same Radon–Nikodým discrete derivative as in the previous unit step example (12.17)

$$\mathbb{D}_n = \frac{\Delta \mathbb{P}_2(n-1)}{\Delta \mathbb{P}_1(n-1)} = e^{\Lambda_1 - \Lambda_2 + n \ln (\Lambda_2/\Lambda_1)}. \tag{12.18}$$

*Note that although the original measures $\mathbb{P}_j(nv_j)$ are RCLL as they should be, inherited from the indicators $\mathbf{1}_{\{k \leq n\}}$, the increment $\Delta \mathbb{P}_1((n-1)v_1)$ is **left continuous, right limits (LCRL)** due to the indicator increments $\mathbf{1}_{\{n-1 < k_2 \leq n\}}$, but they precisely allow the selection of just the nth jump term in the Poisson distribution sum since the indicator increments are closed at n and open at $n-1$.*

This Poisson distribution example is an applied justification of the proposition in Cont and Tankov [60, Proposition 9.5, p. 303] for two Poisson processes $n = N_j = P(T) =$

$N(T)$ with parameters $\Lambda_j = \lambda_j T$ on $(\Omega, \mathcal{F}_T, \mathbb{P}_j, \mathbb{F})$ for $j = 1:2$, i.e.,

$$\frac{\Delta \mathbb{P}_2(N(T) - 1)}{\Delta \mathbb{P}_1(N(T) - 1)} = e^{(\lambda_1 - \lambda_2)T + N(T) \ln (\lambda_2/\lambda_1)}, \qquad (12.19)$$

but only for the same size, $v_2 = v_1$, which has an explicit form as given here.

12.2.2   Change in Measure for Stochastic Processes: Girsanov's Theorem

There are many versions of Girsanov's theorem for changing a probability measure to change the drift of a stochastic diffusion process, and some of these variants are not very distinguishable from the Radon–Nikodým theorem. Here, a modification of Runggaldier's version [239] (see also Brémaud [43] for even more details) will be followed since it has been found to be the most useful, the Radon–Nikodým derivative being relatively easy to calculate and coming with an extension to jump-diffusions. The application of this theorem is determining the measure change for a relative change $\gamma(t)$ of the drift from $\mu_1(t)$ to a drift $\mu_2(t)$ appropriate for the problem of interest, e.g., the change of the drift coefficient $\mu_1(t) = \mu$ in the Black–Scholes [34] method to the current market rate $\mu_2(t) = r$.

Diffusion Girsanov Transformations

Let the reference $\mathbb{P}_1$-SDE for a state diffusion process $X(t)$ be

$$dX(t) = \mu_1(t)dt + \sigma(t)dW_1(t) \qquad (12.20)$$

with time-dependent coefficients $\{\mu_1(t), \sigma(t)\}$, whose integrabilities are implied by the following Girsanov diffusion theorem, on a finite time-interval $[0, T]$ on the filtered probability space $(\Omega, \mathcal{F}_t, \mathbb{P}_1, \mathbb{F})$ with $W_1(t)$ being a $\mathbb{P}_1$-Wiener process. In addition, let the target $\mathbb{P}_2$-SDE objective for this state diffusion process $X(t)$ be

$$dX(t) = \mu_2(t)dt + \sigma(t)dW_2(t) \qquad (12.21)$$

with the same volatility $\sigma(t)$ but changed to drift $\mu_2(t)$, integrability also implied, on the finite time-interval $[0, T]$ on the filtered probability space $(\Omega, \mathcal{F}_t, \mathbb{P}_2, \mathbb{F})$ with $W_1(t)$ being a corresponding $\mathbb{P}_2$-Wiener process.

Theorem 12.14. *Girsanov's Theorem for Changing the Probability Measure of a Diffusion Process to Change the Drift.*
Let $(\Omega, \mathcal{F}_t, \mathbb{P}_1, \mathbb{F})$ be a filtered probability space with $\mathbb{F} = \cup_t \mathcal{F}_t$, symbolically over t. Let $\gamma(t)$ be a square integrable predictable (measurable with respect to $\mathbb{P}_1$, i.e., knowable given $\mathcal{F}_t$) drift process

$$\int_0^t \gamma^2(s)ds < \infty$$

for all $t \in [0, T]$. Then, the Radon–Nikodým derivative $\mathbb{D}(t)$ at time t for the process $X(t)$ is given by the martingale representation (12.9)

$$d\mathbb{D}(t) = \mathbb{D}(t)\gamma(t)dW_1(t), \quad \mathbb{D}(0) \overset{\text{w.p.o.}}{=} 1, \qquad (12.22)$$

supposing that $E_{\mathbb{P}_1}[\mathbb{D}(t)] = 1$ and there exists a second probability measure $\mathbb{P}_2$ on $\mathbb{F}$ that is equivalent to $\mathbb{P}_1$ (mutually absolutely continuous, $\mathbb{P}_2 \overset{a.c.}{\equiv} \mathbb{P}_1$), such that

$$d\mathbb{P}_2 = \mathbb{D}(t)d\mathbb{P}_1$$

and

$$dW_2(t) = dW_1(t) - \gamma(t)dt, \qquad (12.23)$$

where $W_1(t)$ is a $\mathbb{P}_1$-Wiener process as in (12.20) while $W_2(t)$ is a $\mathbb{P}_2$-Wiener process as in (12.21).

The Radon–Nikodým derivative is explicitly given by

$$\mathbb{D}(t) = \frac{d\mathbb{P}_2(t)}{d\mathbb{P}_1(t)} = \exp\left(\int_0^t \gamma(s)\left(dW_1(s) - \frac{1}{2}\gamma(s)ds\right)\right) \qquad (12.24)$$

and the relative drift change is

$$\gamma(t) = \frac{\mu_2(t) - \mu_1(t)}{\sigma(t)}. \qquad (12.25)$$

If the filtration

$$\mathbb{F} = \mathcal{F}_t^{(W_1)} = \widehat{\sigma}\{W_1(s),\ \mathcal{S}_N : 0 \le s \le t,\ \mathcal{S}_N \in \mathcal{N}_1\},$$

where $\mathcal{N}_1$ is the collection of null-sets of $\mathbb{P}_1$, then conversely every probability measure $\mathbb{P}_2 \overset{a.c.}{\equiv} \mathbb{P}_1$ has the same Radon–Nikodým derivative structure.

Substituting (12.23), the Wiener process shift, into the original SDE, we obtain

$$dX(t) = \mu_1(t)dt + \sigma(t)dW_1(t) = (\mu_1(t) + \sigma(t)\gamma(t))dt + \sigma(t)dW_2(t),$$

thus comparison to the second SDE $\mu_2(t) = \mu_1(t) + \sigma(t)\gamma(t)$ and (12.25) for $\gamma(t)$ is immediate, given common volatilities $\sigma_1(t) = \sigma(t) = \sigma_2(t)$.

Upon applying the Itô stochastic chain rule to solve the $\mathbb{D}$-SDE (12.22),

$$d\ln(\mathbb{D}(t)) \overset{dt}{=} \frac{\gamma(t)\mathbb{D}(t)dW_1(t)}{\mathbb{D}(t)} - \frac{(\gamma(t)\mathbb{D}(t))^2 dt}{2\mathbb{D}^2(t)} = \gamma(t)\left(dW_1(t) - \frac{1}{2}\gamma(t)dt\right),$$

integrating with $\mathbb{D}(0) = 1$, and inverting the logarithm, the answer for $\mathbb{D}(t)$ in (12.24) follows. Note that the assumption of common volatility is essential for obtaining the simple linear SDE in $\mathbb{D}(t)$ given in (12.22), since from just one of the independent example terms i in (12.16) it is seen that there is a quadratic term in x of the ith exponent unless $\sigma_i^{(2)} = \sigma_i = \sigma_i^{(1)} = 1$, the common σ in this example. Hence, this Girsanov theorem is quite simple and special. The crudely derived constant coefficient case in (12.14), as an example for Radon–Nikodým derivatives, can be properly recovered from the Girsanov form (12.24) by setting $t = T$ and replacing the time-dependent coefficients by constants, i.e., $\mu_j(s) \to \mu_j$ for $j = 1{:}2$ and $\sigma(s) \to \sigma$.

Note that the relative drift shift (12.25), being state-independent, is also the same for the linear diffusion case,

$$dX(t) = X(t)\left(\mu_1(t)dt + \sigma(t)dW_1(t)\right), \tag{12.26}$$

which is important for applications in finance. This is a linear SDE for the **geometric Brownian motion (GBM) or multiplicative diffusion noise** of the Black–Scholes–Merton [34, 201] option pricing model, while the reference SDE (12.26) for Theorem 12.14 is for **arithmetic Brownian motion (ABM) or additive diffusion noise** of the historic 1900 Bachelier [16] model. The multiplicative model is better for compounded effects, while the additive model is better for strictly cumulative effects. It is well known that the multiplicative model can be transformed into an additive one by the logarithmic transformation using Itô rules,

$$d\ln(X(t)) = \left(\mu_1 - \sigma^2(t)/2\right)dt + \sigma(t)dW_1(t). \tag{12.27}$$

Since the diffusion coefficient shift, $\sigma^2(t)/2$, of the drift would be the same for the GBM target model (2) as for the GBM reference model (1), it is clear that the diffusion Girsanov transformation of the drift will be the same as for the ABM model, i.e.,

$$\gamma^{(\mathrm{GBM})}(t) = \frac{\mu_2(t) - \mu_1(t)}{\sigma(t)}. \tag{12.28}$$

Marked-Jump-Diffusion Girsanov Transformations

Now consider the case of marked-jump-diffusions or compound-jump-diffusions. Let the reference $\mathbb{P}_1$-SDE for a state marked-jump-diffusion process $X(t)$ be

$$dX(t) = \mu_1(t)dt + \sigma(t)dW_1(t) + \int_{\mathcal{Q}_1} h_1(t,q)\mathcal{P}_1(\mathbf{dt},\mathbf{dq}) \tag{12.29}$$

with $\mathbb{P}_1$-Wiener process $W_1(t)$, $\mathbb{P}_1$-Poisson process $P_1(t)$, $\mathrm{E}_{\mathbb{P}_1}[dP_1(t)] = \lambda_1(t)dt$ defines the jump-rate, integrable time-dependent coefficients $\{\mu_1(t), \sigma(t), \lambda_1(t)\}$, (time, mark)-dependent jump-amplitude $h_1(t,q)$, whose integrability is implied by the following theorem, $\mathbb{P}_1$-Poisson jump-times T_i, and IID sample marks Q_i distributed with density $\phi_Q^{(1)}(q;t)$ on the filtered probability space $(\Omega, \mathcal{F}_t, \mathbb{P}_1, \mathbb{F})$ with $W_1(t)$ over on a finite time-interval $[0, T]$. Several forms of the Poisson measure integrals,

$$\int_{\mathcal{Q}_1} h_1(t,q)\mathcal{P}_1(\mathbf{dt},\mathbf{dq}) = \sum_{k=P_1(t)+1}^{P_1(t)+dP_1(t)} h_1(T_k^-, Q_k) \stackrel{\mathrm{dt}}{=} h_1(t, Q)dP_1(t),$$

will be used here, sometimes one form being more convenient than the other.

In addition, let the target $\mathbb{P}_2$-SDE objective for this state marked-jump-diffusion process $X(t)$ be

$$dX(t) = \mu_2(t)dt + \sigma(t)dW_2(t) + \int_{\mathcal{Q}_2} h_2(t,q)\mathcal{P}_2(\mathbf{dt},\mathbf{dq}) \tag{12.30}$$

with $\mathbb{P}_2$-Wiener process $W_2(t)$, $\mathbb{P}_2$-Poisson process $P_2(t)$, $\mathrm{E}_{\mathbb{P}_2}[dP_2(t)] = \lambda_2(t)dt$ defines the jump-rate, the same volatility $\sigma(t)$ but changed to drift $\mu_2(t)$ and changed jump-rate $\lambda_2(t)$, (time, mark)-dependent jump-amplitude $h_2(t, q)$, integrability also implied, $\mathbb{P}_2$-Poisson jump-times T_i, and IID sample marks Q_i distributed with density $\phi_Q^{(2)}(q; t)$ on the finite time-interval $[0, T]$ on the filtered probability space $(\Omega, \mathcal{F}_t, \mathbb{P}_2, \mathbb{F})$.

The following theorem follows Runggaldier [239, Theorem 2.4] but is also presented more in the notation of this book.

Theorem 12.15. *Girsanov's Theorem for Changing the Probability Measure of a Jump-Diffusion Process to Change the Drift, Jump-Rate, and Mark-Density.*
Let $(\Omega, \mathcal{F}_t, \mathbb{P}_1, \mathbb{F})$ be a filtered probability space on the finite time-interval $[0, T]$ with the mark space $\mathcal{Q} = \mathbb{R}$ and the (jump-rate, mark-density) characteristics

$$(\lambda_1(t), \Phi_Q(\mathbf{dq}; t)) \stackrel{\text{gen}}{=} (\lambda_1(t), \phi_Q(q; t)dq).$$

Let $\gamma^{(\mathrm{D})}(t)$ be the square integrable diffusion drift change given in (12.25)

$$\gamma^{(\mathrm{D})}(t) = \frac{\mu_2(t) - \mu_1(t)}{\sigma(t)} \tag{12.31}$$

of Theorem 12.14. Let $\gamma^{(\mathrm{J})}(t)$ be a nonnegative, $\mathcal{F}_t$-predictable jump-rate user-defined scaling process such that

$$\Lambda_2(t) \equiv \int_0^t \lambda_2(s)ds = \int_0^t \gamma^{(\mathrm{J})}(s)\lambda_1(s)ds < \infty$$

for all $t \in [0, T]$, i.e., the transformed mean jump-count is finite, and let $\gamma^{(\mathrm{M})}(q; t)$ be a nonnegative, $\mathcal{F}_t$-predictable, $\mathcal{Q}$-space dependent mark-distribution user-defined scaling process such that

$$\int_{\mathcal{Q}_2} \phi_Q^{(2)}(q; t)dq = \int_{\mathcal{Q}_1} \gamma^{(\mathrm{M})}(q; t)\phi_Q^{(1)}(q; t)dq = 1,$$

i.e., transformed mark-space probability is conserved.
Let

$$\mathbb{D}(t) = \mathbb{D}^{(\mathrm{D})}(t)\mathbb{D}^{(\mathrm{MJ})}(t),$$

where the diffusion martingale representation factor $\mathbb{D}^{(\mathrm{D})}(t)$ is given in (12.24) with stochastic differential in (12.22) and the marked-jump factor $\mathbb{D}^{(\mathrm{MJ})}(t)$ is given by the marked-jump martingale representation (12.9),

$$d\mathbb{D}^{(\mathrm{MJ})}(t) = \mathbb{D}^{(\mathrm{MJ})}(t) \int_{\mathcal{Q}_1} \left(\gamma^{(\mathrm{J})}(t)\gamma^{(\mathrm{M})}(q; t) - 1\right) \widehat{\mathcal{P}}(\mathbf{dt}, \mathbf{dq}), \tag{12.32}$$

subject to the side condition

$$\mathrm{E}_{\mathbb{P}_1}[\mathbb{D}^{(\mathrm{MJ})}] = 1,$$

where

$$\widehat{\mathcal{P}}(\mathbf{dt}, \mathbf{dq}) \equiv \mathcal{P}(\mathbf{dt}, \mathbf{dq}) - \mathrm{E}[\mathcal{P}(\mathbf{dt}, \mathbf{dq})] = \mathcal{P}(\mathbf{dt}, \mathbf{dq}) - \left(\gamma^{(\mathrm{J})}(t) - 1\right)\lambda_1(t)dt,$$

so the solution to the SDE in (12.32) *is*

$$
\begin{aligned}
\mathbb{D}^{(MJ)}(t) &= \exp\left(\int_0^t \left(\int_{\mathcal{Q}_1} \ln\left(\gamma^{(J)}(s)\gamma^{(M)}(q;s)\right) - \left(\gamma^{(J)}(s) - 1\right)\lambda_1(s)ds\right)\right) \\
&= \exp\left(\int_0^t -\left(\gamma^{(J)}(s) - 1\right)\lambda_1(s)ds\right)\prod_{k=1}^{P_1(t)}\gamma^{(J)}(T_k^-)\gamma^{(M)}(Q_k;T_k^-).
\end{aligned}
\tag{12.33}
$$

The transformed quantities are

$$
dW_2(t) = dW_1(t) - \gamma^{(D)}(t)dt, \tag{12.34}
$$

$$
\lambda_2(t) = \gamma^{(J)}(t)\lambda_1(t), \tag{12.35}
$$

$$
\phi_Q^{(2)}(q;t) = \gamma^{(M)}(q;t)\phi_Q^{(1)}(q;t). \tag{12.36}
$$

Thus, the explicit form of the marked-jump-diffusion Radon–Nikodým derivative is

$$
\begin{aligned}
\mathbb{D}(t) &= \frac{d\mathbb{P}_2(t)}{d\mathbb{P}_1(t)} \\
&= \exp\left(\int_0^t \left(\gamma^{(D)}(s)\left(dW_1(s) - \gamma^{(D)}(s)ds/2\right) - \left(\gamma^{(J)}(s) - 1\right)\lambda_1(s)ds\right)\right) \\
&\quad \cdot \prod_{i=1}^{P_1(t)}\gamma^{(J)}(T_k^-)\gamma^{(M)}(Q_k;T_k^-).
\end{aligned}
\tag{12.37}
$$

If the filtration

$$
\mathbb{F} = \mathcal{F}_t^{(W_1,P_1,Q)} = \widehat{\sigma}\{W_1(s), P_1(t), \mathcal{S}_Q, \mathcal{S}_N : 0 \le s \le t, \ \mathcal{S}_Q \in \mathcal{Q}_1, \ \mathcal{S}_N \in \mathcal{N}_1\},
$$

where $\mathcal{N}_1$ *is the collection of null-sets of* $\mathbb{P}_1$, *then conversely every probability measure* $\mathbb{P}_2 \overset{\text{a.c.}}{\equiv} \mathbb{P}_1$ *has the same Radon–Nikodým derivative structure.*

Note that the Wiener process W_1 is independent of the marked Poisson process double (P_1, Q), but the mark random variables Q are only conditionally independent of P_1, and that condition is that there exists a jump of the state X in time, so the factoring $\mathbb{D}(t) = \mathbb{D}^{(D)}(t)\mathbb{D}^{(MJ)}(t)$ into only two parts makes sense. Also, using the product form of Itô's stochastic chain rule,

$$
\begin{aligned}
d\mathbb{D}(t) &= \mathbb{D}^{(MJ)}(t)d\mathbb{D}^{(D)}(t) + \mathbb{D}^{(D)}(t)d\mathbb{D}^{(MJ)}(t) + d\mathbb{D}^{(D)}(t)d\mathbb{D}^{(MJ)}(t) \\
&\overset{\text{dt}}{=} \mathbb{D}^{(MJ)}(t)d\mathbb{D}^{(D)}(t) + \mathbb{D}^{(D)}(t)d\mathbb{D}^{(MJ)}(t) \\
&= \mathbb{D}(t)\left(\gamma^{(D)}(t)dW_1(t) + \int_{\mathcal{Q}_1}\left(\gamma^{(J)}(t)\gamma^{(M)}(q;t) - 1\right)\widehat{\mathcal{P}}_1(\mathbf{dt},\mathbf{dq})\right).
\end{aligned}
$$

Since (12.32) for $\mathbb{D}^{(MJ)}$ is linear, in formal dt-precision notation, we have

$$
d\ln\left(\mathbb{D}^{(MJ)}(t)\right) = -\left(\gamma^{(J)}(t) - 1\right)\lambda_1(t)dt + \ln\left(\gamma^{(J)}(t)\gamma^{(M)}(q;t)\right)dP_1(t).
$$

If dP_1 jumps, then the jump is given by

$$
\left[\mathbb{D}^{(MJ)}\right](t) = \left(\gamma^{(J)}(t)\gamma^{(M)}(q;t) - 1\right)dP_1(t)
$$

and the jump of the logarithm is

$$\left[\ln\left(\mathbb{D}^{(MJ)}\right)\right](t) = \left(\ln\left(\mathbb{D}^{(MJ)} + \left(\gamma^{(J)}(t)\gamma^{(M)}(q;t) - 1\right)\mathbb{D}^{(MJ)}\right) - \ln\left(\mathbb{D}^{(MJ)}\right)\right)dP_1(t)$$
$$= \ln\left(\gamma^{(J)}(t)\gamma^{(M)}(q;t) - 1\right)dP_1(t),$$

so

$$\mathbb{D}^{(MJ)}(t) = \exp\left(-\int_0^t (\gamma^{(J)}(s) - 1)\lambda_1(s)ds + \sum_{i=1}^{P_1(t)} \ln\left(\gamma^{(J)}(t)\gamma^{(M)}(q;t)\right)\right)$$

$$= \exp\left(-\int_0^t \left((\gamma^{(J)}(s) - 1)\lambda_1(s)ds + \int_{\mathcal{Q}_1} \ln\left(\gamma^{(J)}(s)\gamma^{(M)}(q;s)\right)\mathcal{P}_1(\mathbf{ds}, \mathbf{dq})\right)\right). \quad (12.38)$$

Finally, combining equations (12.22), (12.24), (12.32), and (12.38) along with converting the exponential of a sum to a product yields the result (12.37) for $\mathbb{D}(t)$ for the marked-jump-diffusion change from measure $\mathbb{P}_1$ to $\mathbb{P}_2$ according to the recipe (12.34) to (12.36).

For the geometric or linear marked-jump-diffusion

$$dX(t) = X(t)\left(\mu_1(t)dt + \sigma(t)dW_1(t) + \sum_{i=P_1(t)+1}^{P_1(t)+dP_1(t)} h_1(T_i, Q_i)\right), \quad (12.39)$$

the logarithmic change of variable can transform the geometric model to an arithmetic one like (12.29),

$$d\ln(X(t)) = \left(\mu_1(t) - \sigma^2(t)/2\right)dt + \sigma(t)dW_1(t) + \sum_{i=P_1(t)+1}^{P_1(t)+dP_1(t)} \ln\left(h_1(T_i, Q_i) + 1\right). \quad (12.40)$$

Again assuming a common volatility $\sigma(t)$, the Itô rule diffusion coefficient shift of the drift coefficient will be common in both target (2) and reference (1) models, and the jump-rate becomes $\lambda_1(t)$ and jump-amplitude distribution is unchanged, then the Girsanov tranformation triplet becomes

$$dW_2(t) = dW_1(t) - \gamma^{(D)}(t)dt,$$

$$\lambda_2(t) = \gamma^{(J)}(t)\lambda_1(t), \quad (12.41)$$

$$\phi_Q^{(2)}(q;t) = \gamma^{(M)}(t)\phi_Q^{(1)}(q;t)$$

and will be preserved for the geometric case.

Also, see Øksendal and Sulem's Lévy process book [223] for a combined jump-rate and mark distribution scaling, with some financial examples.

Example 12.16. *Two-State, Two-Noise Model Girsanov Application.*
To determine both the diffusive scaling $\gamma^{(D)}(t)$ and the jump scaling $\gamma^{(J)}(t)$, at least two states (assets in financial applications) are needed to handle two sources of random noise. Following a financial example of Runggaldier [239], let $X_1(t)$ and $X_2(t)$ be two states with the same jump-diffusion noise, $W_1(t)$ and $P_1(t)$, but the jump-amplitude is assumed to be deterministic in magnitude given a Poisson jump-time, so $\gamma^{(M)}(q;t) \equiv 1$ here since there are no marks in the problem. The SDE dynamics are given by

$$dX_i(t) = X_i(t)\left(\mu_i(t)dt + \sigma_i(t)dW_1(t) + \nu_i(t)dP_1(t)\right) \quad (12.42)$$

for $i = 1:2$ and where $\mathbb{E}[dP_1(t)] = \lambda_1(t)dt$.

Let the second measure transformed noise be given by

$$dW_2(t) = dW_1(t) - \gamma^{(D)}(t)dt, \quad dP_2(t) = dP_1(t) - \gamma^{(J)}(t)\lambda_1(t)dt.$$

Hence, the dynamics are transformed to

$$dX_i(t) = X_i(t)\left(\left(\mu_i(t) + \gamma^{(D)}(t)\sigma_i(t) + \gamma^{(J)}(t)\nu_i(t)\lambda_1(t)\right)dt + \sigma_i(t)dW_2(t) + \nu_i(t)dP_2(t)\right)$$

such that the common Radon–Nikodým derivative from (12.37) is

$$\mathbb{D}(t) = \exp\left(\int_0^t \left(\gamma^{(D)}(s)\left(dW_1(s) - \gamma^{(D)}(s)ds/2\right) - \left(\gamma^{(J)}(s) - 1\right)\lambda_1(s)ds\right) \right.$$
$$\left. + \gamma^{(J)}(t)P_1(t)\right), \tag{12.43}$$

depending only on the given, common noise, and so yields an equivalent martingale measure $\mathbb{P}_2(t)$ *transformed from* $\mathbb{P}_1(t)$.

For convenience in applications, a scaling of the state $\widetilde{X}_i(t) = X_i(t)/B(t)$ *is introduced using a deterministic process*

$$dB(t) = r(t)B(t)dt,$$

$B(0) > 0$ *and* $r(t) \geq 0$, *that in finance would be called discounting if* $B(t)$ *were a riskless asset like a zero-coupound bond such that* $B(t)$ *is called the numeraire, the most common one. Thus by the chain rule,*

$$d\widetilde{X}_i(t) = \left(d^{(cont)}X_i(t)\right)\big/ B(t) - X_i(t)dB(t)/B^2(t) + [X_i/B](t)dP_1(t)$$
$$= \widetilde{X}_i(t)\left(\left(\mu_i(t) + \gamma^{(D)}(t)\sigma_i(t) + \gamma^{(J)}(t)\nu_i(t)\lambda_1(t) - r(t)\right)dt + \sigma_i(t)dW_2(t) + \nu_i(t)P_2(t)\right).$$

Selecting the common diffusion scaling $\gamma^{(D)}(t)$ *for both* $X_i(t)$ *for* $i = 1{:}2$ *consequently obtains the common jump scaling* $\gamma^{(J)}(t)$, *so*

$$\gamma^{(D)}(t) = \left(r(t) - \mu_i(t) - \gamma^{(J)}(t)\nu_i(t)\lambda_1(t)\right)\big/ \sigma_i(t)$$

for $l = 1 : 2$. *Solving simultaneously for the two scalings produces solutions explicit in the given parameters,*

$$\gamma^{(J)}(t) = \frac{\sigma_1(t)(r(t) - \mu_2(t)) - \sigma_2(t)(r(t) - \mu_1(t))}{(\sigma_1(t)\nu_2(t) - \sigma_2(t)\nu_1(t))\lambda_1(t)} \tag{12.44}$$

and

$$\gamma^{(D)}(t) = \frac{\nu_1(t)(\mu_2(t) - r(t)) + \nu_2(t)(r(t) - \mu_1(t))}{(\sigma_1(t)\nu_2(t) - \sigma_2(t)\nu_1(t))}, \tag{12.45}$$

provided $(\sigma_1(t)\nu_2(t) - \sigma_2(t)\nu_1(t)) \neq 0$ *and* $\gamma^{(J)}(t)\lambda_1(t) > 0$. *This produces a unique martingale measure, and in finance the measure uniqueness is required for the completeness of the market* [239, 134]. *The presence of an infinite number of mark IID random variables requires an infinite number of states or assets to exactly show uniqueness of the transformed martingale measure* $\mathbb{P}_2(t)$.

Refer to Runggaldier's jump-diffusion handbook article [239] for more information and examples on the multidimensional case, Poisson random measure formulation, and financial applications.

12.3   Itô, Lévy, and Jump-Diffusion Comparisons

12.3.1   Itô Processes and Jump-Diffusion Processes

Many authors—Bingham and Kiesel [33], Duffie [75], Glasserman [97], Hull [148], Merton [203], Mikosch [209], Øksendal [222], and others—mostly refer to Brownian motion or Wiener-driven processes with Wiener scaling by a factor $\sigma(t)$ and translated by drift $\mu(t)$,

$$dX(t) = \mu(t)dt + \sigma(t)dW(t), \tag{12.46}$$

at least as a basic definition of an **Itô process**. Some, such as Glasserman [97], Hull [148], Merton [203], and Mikosch [209], would explicitly allow the composite interpretation of the coefficient functions in the basic definition (12.46) to include dependence on the state $X(t)$, such that $\mu(t) = f(X(t), t), \sigma(t) = g(X(t), t)$, and

$$dX(t) = f(X(t), t)dt + g(X(t), t)dW(t). \tag{12.47}$$

Others extend the basic Itô process (12.46) to include (12.47) by application of the Itô chain rule using a transformation like $\widehat{X}(t) = F(X(t), t)$ to obtain

$$d\widehat{X}(t) = f(X(t), t)dt + g(X(t), t)dW(t),$$

where

$$f(X(t), t) = \left(F_t + \mu(t)F_x + \frac{1}{2}\sigma^2(t)F_{xx} \right)(X(t), t)$$

and

$$g(X(t), t) = \sigma(t)F_x(X(t), t).$$

Thus, the state dependent formula (12.47) will be taken as an acceptable **definition of the Itô process**.

However, in his stochastic differential equation classic 1951 memoir [150], Itô also correctly includes jumps in his discussion of simple Markov processes. Itô referred to simple Markov processes, specified by a stochastic differential equation, which for general Poisson noise with distributed jump-amplitudes might be called stochastic integral differential equations,

$$dX(t) = f(X(t), t)dt + g(X(t), t)dW(t) + \int_Q h(X(t), t, q)\mathcal{P}(dt, dq), \tag{12.48}$$

in our notation, or preferably by a stochastic integral equation,

$$X(t) = X(t_0) + \int_{t_0}^t \left[f(X(s), s)ds + g(X(s), s)dW(s) \right. $$
$$\left. + \int_Q h(X(s), s, q)\mathcal{P}(ds, dq) \right], \tag{12.49}$$

again in our notation. Hence, there is a historical basis for calling the jump-diffusion processes, that are the focus of this book, **Itô processes**.

Still others, for instance, Tavella and Randall [264], refer to a jump-diffusion process as a superposition of an Itô process and a Poisson jump process, while Øksendal and Sulem [223] refer to a similar combination as an Itô–Lévy process, but see the next subsection on Lévy processes for the differences between jump-diffusion and Lévy processes. Applebaum [12] and others more precisely call diffusion processes like (12.47) **Itô diffusion processes**.

Although diffusion processes are easier to treat since they have continuous sample paths, jump processes and jump-diffusion processes have discontinuous sample paths and that makes it relatively more difficult for proving theorems. Some of the most significant changes occur with jumps, such as extreme financial crashes and natural disasters.

Hence, according to the more of less standard Itô process usage (12.47),

$$\textit{Itô processes} \subset \textit{Jump-diffusion processes}. \tag{12.50}$$

12.3.2   Lévy Processes and Jump-Diffusion Processes

Lévy processes are essentially jump-diffusion processes but are extended to processes with infinite jump-rates. There have been many recent efforts to study and apply Lévy processes, such as the CGMY model [46], the VG model [47], and the NIG model [243]. Sometimes the term non-Gaussian processes is used as in Barndorff–Nielsen and Shepherd (GIG model) [20] but may not necessarily mean strict Lévy processes. There are also several recent books on Lévy processes such as that of Applebaum [12], and some of these books use jump processes or jump-diffusions in the titles, such as those of Cont and Tankov [60] and Øksendal and Sulem [223]. As with other abstract concepts, there are many different definitions of a Lévy process, and some attempt has been made to merge them within the spirit of this book.

Definition 12.17. *Basic Lévy Process Conditions.*
*A **Lévy process** satisfies the following conditions:*

- ***RCLL stochastic process**: $\{\mathbf{X}(t),\ t \geq 0\}$ on the probability space $(\Omega, \mathcal{F}, \mathbb{P})$ with values in $\mathbb{R}^{n_x}$. (The term **cádlág** means RCLL in French but is used in English probability texts, too.)*

- ***Initial condition**: $\mathbf{X}(0) \overset{\text{a.s.}}{=} \mathbf{0}$.*

- ***Independent increments**: For every partition $0 = t_0 < t_1 < t_2 < \cdots < t_{n_t} < \infty$, the increments*

$$\Delta\mathbf{X}(t_j) \equiv \mathbf{X}(t_{j+1}) - \mathbf{X}(t_j) \text{ for } j = 0{:}n - 1 \tag{12.51}$$

are independent.

- ***Stationary increments**: Together with independence,*

$$\Delta\mathbf{X}(t_j) \overset{\text{dist}}{=} \mathbf{X}(\Delta t_j), \tag{12.52}$$

where $\Delta t_j \equiv t_{j+1} - t_j$.

• **Stochastic continuity:** *The increments of* $\mathbf{X}(t)$ *satisfy*

$$\lim_{\Delta t \to 0} \text{Prob}[\mathbf{X}(t + \Delta t) - \mathbf{X}(t)| \geq \epsilon] = 0, \quad \forall\, \epsilon > 0 \text{ and } t \geq 0. \quad (12.53)$$

All but the last condition (12.53) are standard for the processes dealt with here when the **coefficients are constant**, so it is usually sufficient to show stochastic continuity. (*Note that continuous in probability is not the same as continuous.*) However, when the process coefficients are not constant, then the process will in general not be stationary, as the Lévy condition (12.52) requires. For many real problems, such as in financial markets, the time-dependence of process coefficients is important (for instance, see Hanson and Westman [127]), so (12.52) will not be valid for these problems. It is clear that the IID Wiener vector process $\mathbf{W}(t)$ or the Wiener driven vector Gaussian process with constant coefficients

$$\mathbf{G}(t) = \boldsymbol{\mu}_0 t + \sigma_0 \mathbf{W}(t)$$

and the Poisson vector process $\mathbf{P}(t)$ with constant jump-rates $\boldsymbol{\lambda}(t) = \boldsymbol{\lambda}_0$ will all be Lévy processes, as well as any linear combination that is a simple constant coefficient jump-diffusion n_x-vector process,

$$\mathbf{X}(t) = \boldsymbol{\mu}_0 t + \sigma_0 \mathbf{W}(t) + v_0 \mathbf{P}(t),$$

where $\sigma_0 \in \mathbb{R}^{n_x \times n_w}$ and $v_0 \in \mathbb{R}^{n_x \times n_p}$ consistent with IID $\mathbf{W}(t) \in \mathbb{R}^{n_w}$ and IID $\mathbf{P}(t) \in \mathbb{R}^{n_p}$. Adding the compound Poisson process to the combination will be discussed in the sequel.

There are some preliminary definitions that are important for further properties of Lévy processes.

Definition 12.18. *Infinitely Divisible Distribution.*
A probability distribution $\Phi_{\mathbf{X}}$ *on* $\mathbb{R}^{n_x}$ *is infinitely divisible if for each positive integer n there exists a set of IID random variables* $S\mathbf{Y}_j$ *for* $j = 1:n$ *such that the sum*

$$\mathbf{S}_n = \sum_{j=1}^{n} \mathbf{Y}_j \overset{\text{dist}}{=} \mathbf{X},$$

where $\mathbf{X}$ *has distribution* $\Phi_{\mathbf{X}}$.

Infinite divisibility can be related to the central limit theorem and is closely connected to Lévy processes via compound Poisson processes as follows [60].

Proposition 12.19. *Lévy Processes and Infinite Divisibility.*
Let $\mathbf{X}(t)$ *be a Lévy process for* $t \geq 0$ *on* $\mathbb{R}^{n_x}$; *then for every t, X(t) has an infinitely divisible distribution. Conversely, if* Φ *is an infinitely divisible distribution, then there exists a Lévy process* $\mathbf{X}(t)$ *with the distribution* Φ.

The compound Poisson process is included firmly as a Lévy process by the following result proved in Cont and Tankov [60].

Proposition 12.20. *Compound Poisson Processes as Lévy Processes.*
*The process $CP(t)$ for $t \geq 0$ is a **compound Poisson process**, i.e.,*

$$CP(t) = \sum_{j=1}^{P(t)} Q_i, \qquad (12.54)$$

*where $P(t)$ is a **simple Poisson process** with constant rate λ_0 and the Q_j are IID random jump-amplitudes with common distribution $\Phi_Y(y)$ such that λ_0 is independent of Q_i*

if and only if

*$CP(t)$ is a Lévy process and its sample paths are **piecewise constant functions**.*

Characteristic Functions and Lévy Characteristic Exponents

Definition 12.21. *Characteristic Function.*
The characteristic function of a random vector $\mathbf{X}$ on $\mathbb{R}^{n_x}$ is the complex-valued function

$$\mathcal{C}_{\mathbf{X}}(\mathbf{z}) \equiv \mathrm{E}_{\mathbf{X}} \left[\exp\!\left(i\mathbf{z}^{\top} \mathbf{X} \right) \right] \qquad (12.55)$$

for all $\mathbf{z} \in \mathbb{R}^{n_x}$ and i is the imaginary unit.

Clearly, the characteristic function of a **continuous random variable $\mathbf{X}$** is the **Fourier transform** of the density of $\mathbf{X}$, i.e.,

$$\mathcal{C}_{\mathbf{X}}(\mathbf{z}) = \int_{\mathbb{R}^{n_x}} e^{i\mathbf{z}^{\top}\mathbf{x}} \phi_{\mathbf{X}}(\mathbf{x}) d\mathbf{x},$$

while if X is a **discrete scalar random variable** with distribution given by the countable sequence of probabilities $\pi_k = \mathrm{Prob}[X = k]$, then the characteristic function is the discrete Fourier transform,

$$\mathcal{C}_X(z) = \sum_{k=0}^{\infty} e^{izk} \cdot \pi_k.$$

This is the basic random vector definition, but here the interest will be in the same definition when the random vector is a function of time t, i.e., a stochastic process $\mathbf{X}(t)$,

$$\mathcal{C}_{\mathbf{X}(t)}(\mathbf{z}) \equiv \mathrm{E}_{\mathbf{X}(t)} \left[\exp\!\left(i\mathbf{z}^{\top} \mathbf{X}(t) \right) \right].$$

One of the most important features of a Lévy process is that the characteristic function has relatively simple form [60, 12], as follows.

Proposition 12.22. *Lévy Characteristic Functions and Exponents.*
If $\mathbf{X}(t)$ *is a Lévy process for* $t \geq 0$ *on* $\mathbb{R}^{n_x}$, *then there exists a real-valued continuous function* $\eta_{\mathbf{X}(t)}(\mathbf{z})$ *of the characteristic vector* $\mathbf{z} \in \mathbb{R}^{n_x}$, *called the **characteristic exponent**, such that*

$$\mathcal{C}_{\mathbf{X}(t)}(\mathbf{z}) = \mathrm{E}_{\mathbf{X}(t)}\left[\exp\left(i\mathbf{z}^\top \mathbf{X}(t)\right)\right] = \exp\left(it\eta_{\mathbf{X}(t)}(\mathbf{z})\right). \tag{12.56}$$

However, for nonstationary problems without the Lévy stationarity condition (12.53), then it would be expected that in general the exponent would not be linear in t,

$$\mathcal{C}_{\mathbf{X}(t)}(\mathbf{z}) = \exp\left(i\overline{\eta}_{\mathbf{X}(t)}(\mathbf{z}; t)\right).$$

It is well known that Fourier transforms, and the characteristic function, are mainly useful for constant coefficients, with few exceptions.

Examples 12.23. *Characteristic Functions and Exponents of Lévy Processes.*

• *Standard Wiener process* $W(t)$ *on* $\mathbb{R}$:

$$\mathcal{C}_{W(t)}(z) = \mathrm{E}_{W(t)}\left[e^{izW(t)}\right] = \frac{1}{\sqrt{2\pi t}}\int_{-\infty}^{\infty} e^{izw} e^{-w^2/(2t)}\,dw$$

$$= e^{-(tz)^2/(2t)}\frac{1}{\sqrt{2\pi t}}\int_{-\infty}^{\infty} e^{-(w - itz)^2/(2t)}\,dw = e^{-tz^2/2}$$

using the completing the square technique, so the Lévy characteristic exponent is

$$\eta_{W(t)}(z) = -\frac{1}{2}z^2. \tag{12.57}$$

• *IID Wiener vector process* $\mathbf{W}(t)$ *on* $\mathbb{R}^{n_w}$ *with* $\mathrm{Cov}[\mathbf{W}(t), \mathbf{W}^\top(t)] = tI_{n_w}$:

$$\mathcal{C}_{\mathbf{W}(t)}(\mathbf{z}) = \mathrm{E}_{\mathbf{W}(t)}\left[e^{i\mathbf{z}^\top \mathbf{W}(t)}\right] = \prod_{j=1}^{n_w}\frac{1}{\sqrt{2\pi t}}\int_{-\infty}^{\infty} e^{iz_j w_j} e^{-w_j^2/(2t)}\,dw_j$$

$$= \prod_{j=1}^{n_w}\mathcal{C}_{W_j(t)}(z_j) = \exp\left(-t\sum_{j=1}^{n_x} z_j^2/2\right) = \exp\left(-t|\mathbf{z}|^2/2\right),$$

so the Lévy characteristic exponent is

$$\eta_{\mathbf{W}(t)}(\mathbf{z}) = -\frac{1}{2}|\mathbf{z}|^2. \tag{12.58}$$

• *IID Gaussian vector process* $\mathbf{G}(t) = \boldsymbol{\mu}_0 t + \sigma_0 \mathbf{W}(t)$ *on* $\mathbb{R}^{n_x}$ *with* $\mathrm{Cov}[\mathbf{W}(t), \mathbf{W}^\top(t)]$
$= t I_{n_w}$, *constant* $\boldsymbol{\mu}_0 \in \mathbb{R}^{n_x}$, *and constant* $\sigma_0 \in \mathbb{R}^{n_x \times n_w}$:

$$\mathcal{C}_{\mathbf{G}(t)}(\mathbf{z}) = \mathrm{E}_{\mathbf{W}(t)}\left[e^{i\mathbf{z}^\top(\boldsymbol{\mu}_0 t + \sigma_0 \mathbf{W}(t))}\right]$$

$$= e^{i\mathbf{z}^\top\boldsymbol{\mu}_0 t}\prod_{k=1}^{n_w}\frac{1}{\sqrt{2\pi t}}\int_{-\infty}^{\infty}\exp\left(i\sum_{j=1}^{n_w} z_j \sigma_{0,j,k} w_k\right)e^{-w_k^2/(2t)}dw_k$$

$$= \exp\left(it\mathbf{z}^\top\boldsymbol{\mu}_0 - \sum_{j=1}^{n_x} z_j \sum_{\ell=1}^{n_x} z_\ell \sum_{k=1}^{n_w}\sigma_{0,j,k}\sigma_{0,\ell,k}/2\right)$$

$$= \exp\left(it\mathbf{z}^\top\boldsymbol{\mu}_0 - t\mathbf{z}^\top(\sigma_0\sigma_0^\top)\mathbf{z}\right),$$

so the Lévy characteristic exponent is

$$\eta_{\mathbf{G}(t)}(\mathbf{z}) = i\mathbf{z}^\top\boldsymbol{\mu}_0 - \frac{1}{2}\mathbf{z}^\top(\sigma_0\sigma_0^\top)\mathbf{z}/2. \tag{12.59}$$

• *Simple Poisson process* $P(t)$ *on* $\mathbb{R}$ *with constant jump-rate* λ_0:

$$\mathcal{C}_{P(t)}(z) = \mathrm{E}_{P(t)}\left[e^{izP(t)}\right] = e^{-\lambda_0 t}\sum_{k=0}^{\infty}\frac{(\lambda_0 t)^k}{k!}e^{izk}$$

$$= e^{-\lambda_0 t}\sum_{k=0}^{\infty}\frac{\left(\lambda_0 t e^{iz}\right)^k}{k!} = e^{-\lambda_0 t + \lambda_0 t e^{iz}} = e^{\lambda_0 t\,(e^{iz} - 1)},$$

so the Lévy characteristic exponent is

$$\eta_{P(t)}(z) = \lambda_0(e^{iz} - 1). \tag{12.60}$$

• *Centered or martingale form of Poisson process* $\widetilde{P}(t) \equiv P(t) - \lambda_0 t$ *on* $\mathbb{R}$ *with constant jump-rate* λ_0:

$$\mathcal{C}_{\widetilde{P}(t)}(z) = \mathrm{E}_{P(t)}\left[e^{iz(P(t) - \lambda_0 t)}\right] = e^{-\lambda_0 t}\sum_{k=0}^{\infty}\frac{(\lambda_0 t)^k}{k!}e^{iz(k - \lambda_0 t)}$$

$$= e^{-\lambda_0 t(1 + iz)}\mathcal{C}_{P(t)}(z) = e^{\lambda_0 t(e^{iz} - 1 - iz)},$$

so the Lévy characteristic exponent is

$$\eta_{\widetilde{P}(t)}(z) = \lambda_0(e^{iz} - 1 - iz). \tag{12.61}$$

- **Simple Poisson vector process $\mathbf{P}(t)$ on $\mathbb{R}^{n_p}$ with independent components and constant jump-rate vector $\boldsymbol{\lambda}_0 = [\lambda_{0,j}]_{n_p \times 1}$:**

$$\mathcal{C}_{\mathbf{P}(t)}(\mathbf{z}) = \mathrm{E}_{\mathbf{P}(t)}\left[\exp\left(i\mathbf{z}^\mathsf{T}\mathbf{P}(t)\right)\right] = \prod_{j=1}^{n_p} e^{-\lambda_{0,j}t} \sum_{k_j=0}^{\infty} \frac{(\lambda_{0,j}t)^{k_j}}{k_j!} e^{iz_j k_j}$$

$$= \prod_{j=1}^{n_p} \exp\left(\lambda_{0,j}t\left(\exp(iz_j) - 1\right)\right) = \exp\left(t\sum_{j=1}^{n_p} \lambda_{0,j}\left(\exp(iz_j) - 1\right)\right)$$

$$= \exp\left(tn_p\left(\overline{\lambda_0 \exp(iz)} - \overline{\lambda_0}\right)\right),$$

where $\overline{\lambda_0} \equiv \sum_{j=1}^{n_p} \lambda_{0,j}/n_p$ and $\overline{\lambda_0 \exp(iz)} \equiv \sum_{j=1}^{n_p} \lambda_{0,j}\exp(iz_j)/n_p$, so the Lévy characteristic exponent is

$$\eta_{\mathbf{P}(t)}(\mathbf{z}) = n_p\left(\overline{\lambda_0 \exp(iz)} - \overline{\lambda_0}\right). \tag{12.62}$$

- **Simple compound Poisson process $CP(t) = \sum_{\ell=1}^{P(t)} Q_\ell$ on $\mathbb{R}$ with constant jump-rate λ_0 and IID jump-amplitudes Q_ℓ with distribution $\Phi_Q(q)$:**

$$\mathcal{C}_{CP(t)}(z) = \mathrm{E}_{P(t),Q}\left[e^{izX(t)}\right] = e^{-\lambda_0 t} \sum_{k=0}^{\infty} \frac{(\lambda_0 t)^k}{k!} \mathrm{E}_Q\left[\exp\left(iz\sum_{\ell=1}^{k} Q_\ell\right)\right]$$

$$= e^{-\lambda_0 t} \sum_{k=0}^{\infty} \frac{(\lambda_0 t)^k}{k!} \prod_{\ell=1}^{k} \mathrm{E}_Q\left[\exp(izQ_\ell)\right]$$

$$= e^{-\lambda_0 t} \sum_{k=0}^{\infty} \frac{(\lambda_0 t)^k}{k!} \mathrm{E}_Q^k\left[\exp(izQ)\right] = \exp\left(\lambda_0 t\left(\mathrm{E}_Q\left[\exp(izQ)\right] - 1\right)\right)$$

using the iterated conditional expectation technique and IID, so the Lévy characteristic exponent, substituting $\mathrm{E}_Q\left[\exp(izQ)\right] = \mathcal{C}_Q(z)$, is

$$\eta_{CP(t)}(z) = \lambda_0(\mathcal{C}_Q(z) - 1) \tag{12.63}$$

and the simple Poisson process exponent is recovered if $Q_\ell \overset{\text{w.p.o.}}{=} 1 \; \forall \ell \geq 1$.

- **Vector compound Poisson process $\mathbf{CP}(t) = \sum_{\ell=1}^{P(t)} \mathbf{Q}_\ell$ on $\mathbb{R}^{n_x}$ with constant jump-rate λ_0 and IID vector jump-amplitudes $\mathbf{Q}_\ell$ with distribution $\Phi_{\mathbf{Q}}(\mathbf{q})$:** *Note that the*

$\mathbf{Q}_\ell$ are IID as vectors not necessarily as components; thus

$$\mathcal{C}_{\mathbf{CP}(t)}(\mathbf{z}) = \mathrm{E}_{P(t),\mathbf{Q}} \left[e^{i\mathbf{z}^\top \mathbf{CP}(t)} \right] = e^{-\lambda_0 t} \sum_{k=0}^\infty \frac{(\lambda_0 t)^k}{k!} \mathrm{E}_\mathbf{Q} \left[\exp\left(i\mathbf{z}^\top \sum_{\ell=1}^k \mathbf{Q}_\ell \right) \right]$$

$$= e^{-\lambda_0 t} \sum_{k=0}^\infty \frac{(\lambda_0 t)^k}{k!} \prod_{\ell=1}^k \mathrm{E}_\mathbf{Q} \left[\exp\left(i\mathbf{z}^\top \mathbf{Q}_\ell \right) \right]$$

$$= e^{-\lambda_0 t} \sum_{k=0}^\infty \frac{(\lambda_0 t)^k}{k!} \mathrm{E}_\mathbf{Q}^k \left[\exp\left(i\mathbf{z}^\top \mathbf{Q}_\ell \right) \right]$$

$$= \exp\!\left(\lambda_0 t \left(\mathrm{E}_\mathbf{Q} \left[\exp\left(i\mathbf{z}^\top \mathbf{Q}_\ell \right) \right] - 1 \right) \right)$$

using the iterated conditional expectation technique and IID again, so the Lévy characteristic exponent, substituting $\mathrm{E}_\mathbf{Q}\left[\exp(i\mathbf{z}^\top \mathbf{Q})\right] = \mathcal{C}_\mathbf{Q}(\mathbf{z})$, is

$$\eta_{\mathbf{CP}(t)}(z) = \lambda_0 (\mathcal{C}_\mathbf{Q}(\mathbf{z}) - 1). \tag{12.64}$$

Lévy–Klintchine Jump-Diffusion Formula

In these previous examples, the ingredients for the fundamental theorem of the **Lévy–Klintchine representation formula** specialized to jump-diffusion processes have been derived, based on the vector Gaussian process exponent result in (12.59) and the vector compound Poisson process exponent result in (12.64).

Theorem 12.24. *Lévy–Klintchine Formula for Jump-Diffusion Processes.*
Let $\mathbf{X}(t)$ be the jump-diffusion process on $\mathbb{R}^{n_x}$ for $t \geq 0$,

$$\mathbf{X}(t) = \mathbf{X}(0) + \boldsymbol{\mu}_0 t + \sigma_0 \mathbf{W}(t) + \sum_{\ell=1}^{P(t)} \mathbf{Q}_\ell \tag{12.65}$$

with Lévy characteristic triplet $(\sigma_0 \sigma_0^\top, \lambda_0 \Phi_\mathbf{Q}(d\mathbf{q})dt, \boldsymbol{\mu}_0)$, where $\boldsymbol{\mu}_0 \in \mathbb{R}^{n_x}$ is a constant, $\sigma_0 \in \mathbb{R}^{n_x \times n_w}$ is a constant, $\mathbf{W}(t) \in \mathbb{R}^{n_w}$ is a vector Wiener process, and $\mathbf{P}(t) \in \mathbb{R}$ is a simple Poisson process with constant and finite jump-rate $\lambda_0 \in \mathbb{R}$ and compounded with IID vector jump-amplitudes $\mathbf{Q}_\ell \in \mathbb{R}^{n_x}$ with distribution $\Phi_\mathbf{Q}(\mathbf{q})$. The random triplet $(\mathbf{W}(t), P(t), Q)$ is independent variables, except that the jump-amplitude Q requires the existence of a jump of the Poisson process.

Then, the characteristic function with $\mathbf{z} \in \mathbb{R}^{n_x}$ for the initial condition translated process

$$\mathbf{Y}(t) \equiv \mathbf{X}(t) - \mathbf{X}(0) \tag{12.66}$$

is

$$\mathcal{C}_{\mathbf{Y}(t)}(\mathbf{z}) = \mathrm{E}_{\mathbf{Y}(t)}\!\left[\exp(i\mathbf{z}^\top \mathbf{Y}(t)\right] = \exp\left(t\eta_{\mathbf{Y}(t)}(\mathbf{z}) \right),$$

where the Lévy characteristic exponent is

$$\eta_{\mathbf{Y}(t)}(\mathbf{z}) = i\boldsymbol{\mu}_0 t - \frac{1}{2}\mathbf{z}^\top \sigma_0 \sigma_0^\top \mathbf{z} + \lambda_0 \int_{\mathbb{R}^{n_x}} \left(\exp(i\mathbf{z}^\top \mathbf{q}) - 1 \right) \phi_\mathbf{Q}(\mathbf{q})d\mathbf{q}. \tag{12.67}$$

Except for the technical details, the Lévy characteristic exponent result (12.67) follows from (12.59) for $\mathbf{G}(t)$ and from (12.64) for $\mathbf{CP}(t)$ by the independence properties between $\mathbf{G}(t)$ and $(P(t), Q)$ and by iterative conditional expectation between $P(t)$ and Q that is conditioned on the existence of a jump as for (12.64). Thus,

$$
\begin{aligned}
\mathcal{C}_{\mathbf{Y}(t)}(\mathbf{z}) &= \mathrm{E}_{\mathbf{Y}(t)}\big[\exp(i\mathbf{z}^{\top}\mathbf{Y}(t))\big] \\
&= \mathrm{E}_{\mathbf{W}(t)}\big[\exp(i\mathbf{z}^{\top}\mathbf{G}(t))\big] \cdot \mathrm{E}_{P(t),\mathbf{Q}}\big[\exp(i\mathbf{z}^{\top}\mathbf{CP}(t))\big] \\
&= \mathcal{C}_{\mathbf{G}(t)}(\mathbf{z}) \cdot \mathcal{C}_{\mathbf{P}(t)}(\mathbf{z}) \\
&= \exp\big(t\eta_{\mathbf{G}(t)}(\mathbf{z})\big) \cdot \exp\big(t\eta_{\mathbf{CP}(t)}(\mathbf{z})\big) \\
&= \exp\big(t\big(\eta_{\mathbf{G}(t)}(\mathbf{z}) + \eta_{\mathbf{CP}(t)}(\mathbf{z})\big)\big)
\end{aligned}
$$

so substituting (12.59) and (12.64) and expanding the expectations leads directly to the main result (12.67). It should be noted that embedded in this derivation is the semigroup property [12, 60] of the characteristic function in the case of constant coefficients.

In the case of the geometric or linear jump-diffusion process (5.42) with constant rate coefficients for $X(t) \in \mathbb{R}$ with SDE,

$$
dX(t) = X(t)\left(\mu_0 dt + \sigma_0 dW(t) + \sum_{k=P(t)+1}^{P(t)+dP(t)} \big(e^{Q_k} - 1\big)\right),
$$

the solution is exponential via a logarithmic change of variable technique,

$$
X(t) = X(0)\exp\left((\mu_0 - \sigma_0^2/2)t + \sigma_0 W(t) + \sum_{k=1}^{P(t)} Q_k\right) \tag{12.68}
$$

with $X(0) > 0$, and is obviously not a Lévy process due to the exponential time-dependence, without further transformation to fit the characteristic exponent form in (12.67).

Corollary 12.25. *Lévy–Klintchine Transformed Geometric Jump-Diffusion Formula. Assuming the hypotheses of Theorem 12.24, except that $n_x = 1$, $n_w = 1$, and that the Lévy characteristic triplet is $(\sigma_0^2, \lambda_0\Phi_Q(dq)dt, \mu_0 - \sigma_0^2)$, then the characteristic function with $z \in \mathbb{R}$ of the logarithmic-translated process $Y(t)$,*

$$
Y(t) \equiv \ln(X(t)/X(0)), \tag{12.69}
$$

corresponding to the geometric process (12.68), is

$$
\mathcal{C}_{Y(t)}(z) = \mathrm{E}_{Y(t)}[\exp(izY(t))] = \exp\big(t\eta_{Y(t)}(z)\big),
$$

where the Lévy characteristic exponent is

$$
\eta_{Y(t)}(z) = i(\mu_0 - \sigma_0^2/2)t - \frac{1}{2}\sigma_0^2 z^2 + \lambda_0 \int_{\mathbb{R}} (\exp(izq) - 1)\phi_Q(q)dq. \tag{12.70}
$$

Lévy–Klintchine Lévy Process Formula, Including Infinite Rate Processes

So far the jump-rate λ_0 has been assumed to be constant and either explicitly or implicitly finite in this subsection on Lévy processes. However, the infinite jump-rate is a distinguishing feature of Lévy processes so that, in general, it is not valid to write the jump-rate symbol λ_0 in Lévy process formulas. Instead, it is necessary to refer to the number of jumps rather than to the jump-rate.

Recall the definition (B.178) of the **jump function** on p. B62 of a process,

$$[\mathbf{X}](t) \equiv \mathbf{X}(t) - \mathbf{X}(t-),$$

written here for RCLL vector processes. (Caution: In some of the literature $\Delta\mathbf{X}(t)$ is used but can be confused with the analytic or numerical time increment, $\Delta\mathbf{X}(t) \equiv \mathbf{X}(t+\Delta t) - \mathbf{X}(t)$.) At points where $\mathbf{X}(t)$ is continuous, then $[\mathbf{X}](t) = 0$.

Definition 12.26. *Number of Jumps of a Process, Poisson Random Measure, and Lévy Measure. The number of jumps in the open set $\mathcal{S}$, assuming a bounded number of jumps and excluding zero jumps ($0 \notin \mathcal{S}$) on the interval $(0, t]$, is*

$$\mathcal{P}((0, t], \mathcal{S}) = \sum_{s \in (0,t]} \mathbf{1}_{[\mathbf{X}](s) \in \mathcal{S}}. \tag{12.71}$$

Here, $\mathcal{P}((0, t], \mathcal{S})$ is the Poisson random or jump measure [223]. The differential form is denoted by $\mathcal{P}(dt, d\mathbf{q}) = \mathcal{P}((t, t + dt], (\mathbf{q}, \mathbf{q} + d\mathbf{q}))$, as previously used in Chapter 5. An alternate form [232] uses a sequence of stopping or jump-times,

$$T_{k+1}(\mathcal{S}) = \inf\{t \mid t > T_k(\mathcal{S}), [X](t) \in \mathcal{S}\}; \quad T_0(\mathcal{S}) \equiv 0,$$

such that

$$\mathcal{P}((0, t], \mathcal{S}) = \sum_{k=1}^{\infty} \mathbf{1}_{T_k(\mathcal{S}) \leq t}.$$

The zero-mean (centered or martingale) form is denoted by

$$\widetilde{\mathcal{P}}(dt, \mathbf{q}) = \mathcal{P}(dt, d\mathbf{q}) - \nu^{(L)}(d\mathbf{q})dt, \tag{12.72}$$

*where now $\nu^{(L)}(d\mathbf{q})dt = \mathrm{E}[\mathcal{P}(dt, d\mathbf{q})]$ and $\nu^{(L)}$ is called the **Lévy measure** in general.*

A fundamental tool for separating out the large jumps in the presence of infinite jump-rates is the following decomposition after the concise form of Øksendal and Sulem [223].

Theorem 12.27. *Lévy–Itô Decomposition.*
Let $0 \leq R < \infty$ be a jump-amplitude cut-off. Then a Lévy process $\mathbf{X}^{(L)}(t)$ on $\mathbb{R}^{n_x}$ has the decomposition

$$\mathbf{X}^{(L)}(t) = \widetilde{\boldsymbol{\mu}}_{0,R}t + \sigma_0\mathbf{W}(t) + \int_{|\mathbf{q}|<R} \mathbf{q}\widetilde{\mathcal{P}}(t, d\mathbf{q}) + \int_{|\mathbf{q}| \geq R} \mathbf{q}\mathcal{P}(t, d\mathbf{q}), \tag{12.73}$$

where $\mathbf{W}(t) \in \mathbb{R}^{n_w}$ is an independent vector Wiener process, $\widetilde{\boldsymbol{\mu}}_{0,R} \in \mathbb{R}^{n_x}$ is a constant adjusted with R from the original drift $\boldsymbol{\mu}_0 \in \mathbb{R}^{n_x}$, and $\sigma_0 \in \mathbb{R}^{n_x \times n_w}$ is a constant.

In particular, the Lévy–Itô decomposition states that the Lévy process is, as is the jump-diffusion, decomposable into a continuous process and a discontinuous process,

$$\mathbf{X}^{(L)}(t) = \mathbf{X}^{(cont)}(t) + \mathbf{X}^{(discont)}(t);$$
$$\mathbf{X}^{(cont)}(t) = \widetilde{\boldsymbol{\mu}}_{0,R}t + \sigma_0\mathbf{W}(t);$$
$$\mathbf{X}^{(discont)}(t) = \mathbf{X}^{(L)}(t) - \mathbf{X}^{(cont)}(t).$$

One consequence of this Lévy–Itô decomposition is another fundamental result [223, 232], as follows.

Theorem 12.28. *Lévy–Klintchine Representation Formula for Lévy Processes.*
Let $\mathbf{X}^{(L)}(t)$ be a Lévy process for $t \geq 0$ with Lévy measure $v^{(L)}$ on $\mathbb{R}^{n_x}$, given constants $\widetilde{\boldsymbol{\mu}}_{0,R} \in \mathbb{R}^{n_x}$ and $\sigma_0 \in \mathbb{R}^{n_x \times n_w}$. Then the jump-count satisfies

$$\int_{\mathbb{R}^{n_x}} \min(|\mathbf{q}|^2, R)v^{(L)}(d\mathbf{q}) < \infty$$

and the characteristic function on $\mathbf{z} \in \mathbb{R}^{n_x}$ for $\mathbf{X}(t) = \mathbf{X}^{(L)}(t)$ is

$$\mathcal{C}_{\mathbf{X}(t)}(\mathbf{z}) = \mathrm{E}_{\mathbf{X}(t)}\big[\exp\big(i\mathbf{z}^\top\mathbf{X}(t)\big)\big] = \exp\big(t\eta_{\mathbf{X}(t)}(\mathbf{z})\big),$$

where the Lévy characteristic exponent is

$$\eta_{\mathbf{X}(t)}(\mathbf{z}) = i\widetilde{\boldsymbol{\mu}}_{0,R}\,t - \tfrac{1}{2}\mathbf{z}^\top\sigma_0\sigma_0^\top\mathbf{z}$$
$$+ \int_{|\mathbf{q}|<R} \big(\exp\big(i\mathbf{z}^\top\mathbf{q}\big) - 1 - i\mathbf{z}^\top\mathbf{q}\big)\, v^{(L)}(d\mathbf{q}) \qquad (12.74)$$
$$+ \int_{|\mathbf{q}|\geq R} \big(\exp\big(i\mathbf{z}^\top\mathbf{q}\big) - 1\big)\, v^{(L)}(d\mathbf{q}).$$

Conversely, given constants $\widetilde{\boldsymbol{\mu}}_{0,R} \in \mathbb{R}^{n_x}$ and $\sigma_0 \in \mathbb{R}^{n_x \times n_w}$, along with the Lévy measure $v^{(L)}$ on $\mathbb{R}^{n_x}$ such that the jump-count satisfies

$$\int_{\mathbb{R}^{n_x}} \min(|\mathbf{q}|^2, R)v^{(L)}(d\mathbf{q}) < \infty,$$

then there exists a Lévy process $\mathbf{X}^{(L)}$ that is unique in distribution such that the Lévy characteristic is (12.74) for $\mathbf{z} \in \mathbb{R}^{n_x}$.

Note that the extra linear term $i\mathbf{z}^\top\mathbf{q}$ in the first or inner integral of (12.74) is related to the zero-mean Poisson $\widetilde{P}(t)$ form iz found in (12.61) but not in (12.60) for $P(t)$.

Although jump-process time-dependent coefficients, like drift and volatility coefficients, do not strictly satisfy the stationary increment condition (12.52) for a Lévy process, Øksendal and Sulem [223] define **Levy–driven processes** which satisfy the Lévy–Itô decomposition formula (12.73) but not the constant coefficient condition. For example, analogous to the **Wiener-driven Itô process** (12.47), there is the **Lévy-driven Itô–Lévy**

process [223, Theorem 1.14, p. 6] on $\mathbb{R}$ with time-random coefficients

$$dX(t) = \widetilde{\mu}_{0,R}(t;\omega)dt + \sigma_0(t;\omega)dW(t)$$
$$+ \int_{|q|<R} h(t,q;\omega)\widetilde{\mathcal{P}}(dt,dq) + \int_{|q|\geq R} h(t,q;\omega)\mathcal{P}(dt,dq) \tag{12.75}$$

for some $R \in [0,\infty)$, $(\widetilde{\mu}_{0,R}(t;\omega), \sigma_0(t;\omega), h(t,q\omega)$ are integrable functions, and ω is some background random variable.

The **Lévy-driven geometric Lévy process** [223, Example 1.15, p. 7] is similarly defined,

$$dX(t) = X(t)\bigg(\widetilde{\mu}_{0,R}(t;\omega)dt + \sigma_0(t;\omega)dW(t)$$
$$+ \int_{|q|<R} h(t,q;\omega)\widetilde{\mathcal{P}}(dt,dq) + \int_{|q|\geq R} h(t,q;\omega)\mathcal{P}(dt,dq) \bigg), \tag{12.76}$$

where, in addition, the jump-amplitude $h(t,q;\omega) \geq -1$ preserves positivity assuming $X(0) > 0$, with more potential uses in financial applications.

In general, these processes are special cases of what Øksendal and Sulem [223, Theorem 1.19, p. 10] call **Lévy diffusions** governed by **Lévy stochastic differential equations**,

$$d\mathbf{X}(t) = \widetilde{\mu}(t,\mathbf{X}(t))dt + \sigma(t,\mathbf{X}(t))d\mathbf{W}(t)$$
$$+ \int_{|q|<R} h(t,q;\omega)\widetilde{\mathcal{P}}(dt,dq) + \int_{|q|\geq R} h(t,\mathbf{X}(t),q)\mathcal{P}(dt,dq), \tag{12.77}$$

where $0 \leq t \leq T$, $\mathbf{X} \in \mathbb{R}^{n_x}$, $\widetilde{\mu} \in \mathbb{R}^{n_x}$, $\mathbf{W} \in \mathbb{R}^{n_w}$, $\sigma \in \mathbb{R}^{n_x \times n_w}$, $\mathcal{P} \in \mathbb{R}^{n_p}$, $\mathbf{Q} \in \mathbb{R}^{n_x}$, and $h \in \mathbb{R}^{n_x \times n_p}$ subject to the usual linear growth and Lipschitz continuity conditions.

For many other Lévy process models, including models which push the limits of the assumptions here, see Applebaum [12, Subsection 5.4.7, p. 286ff].

Concluding this subsection like the last, the relative size of Lévy processes is compared to that of jump-diffusions. According to the strict Lévy process definition leading to a restriction to constant coefficients,

$$\left\{ \begin{array}{c} \textit{constant coefficient} \\ \textit{jump-diffusion processes} \end{array} \right\} \subset \textit{Lévy processes}, \tag{12.78}$$

since ordinarily jump-diffusions based upon Poisson processes do not allow for infinite jump-rates on $[0,t]$. However, if the infinite jump activity is controlled, then

$$\left\{ \begin{array}{c} \textit{finite jump-rate} \\ \textit{Lévy processes} \end{array} \right\} \subset \textit{jump-diffusion processes}, \tag{12.79}$$

since jump-diffusions in general include variable coefficients and nonlinear terms.

If the comparison is made to the Lévy-driven processes discussed by Øksendal and Sulem [223] and summarized here, then

$$\{\textit{jump-diffusion processes}\} \subset \textit{Lévy-driven processes} \tag{12.80}$$

due to the inclusion of infinite jump-rates with nonlinear and time-dependent coefficients in Lévy-driven processes.

12.4   Exercise

1. Similar to Examples 12.10 on p. 372, find the martingale properties as a function of the additional drift for the **geometric jump-diffusion** problem with constant coefficients

$$dX(t) = X(t) \left(\mu_0 dt + \sigma_0 dW(t) + \nu_0 \sum_{i=P(t)+1}^{P(t)+dP(t)} \left(e^{Q_i} - 1\right) \right),$$

where again the marks are IID with mean μ_Q and variance σ_Q^2.

Suggested References for Further Reading

- Applebaum, 2004 [12]

- Bain, 2006 [17]

- Baxter and Rennie, 1996 [22]

- Billingsley, 1986 [32]

- Bingham and Kiesel, 2004 [33]

- Bossaerts, 2002 [41]

- Brémaud, 1981 [43]

- Cont and Tankov, 2004 [60]

- Cyganowski, Kloeden, and Ombach, 2002 [67]

- Doob, 1953 [70]

- Duffie, 1992 [75]

- Gihman and Skorohod, 1972 [95]

- Glasserman, 2003 [97]

- Harrison and Pliska, 1981, [133]

- Harrison and Pliska, 1983, [134]

- Heath and Schweizer, 2000 [136]

- Hull, 2000 [148]

- Itô, 1951 [150]

- Karatzas and Shreve, 1998 [161]

- Karlin and Taylor, 1981 [163]

- Klebaner, 1998 [165]

- Mikosch, 1998 [209]

- Neftci, 2000 [217]

- Øksendal, 1998 [222]

- Øksendal and Sulem, 2005 [223]

- Pliska, 1997 [225]

- Protter, 2004 [232]

- Runggaldier, 2003 [239]

- Rogers and Williams, 2000 [236]

- Shreve, 2004 [248]

- Steele, 2001 [256]

- Yong and Zhou, 1999 [290]

Bibliography

[1] M. L. ABELL AND J. P. BRASELTON, *The Maple V Handbook*, Academic Press, New York, NY, 1994.

[2] M. ABRAMOWIITZ AND I. A. STEGUN, EDS., *Handbook of Mathematical Functions with Formulas, Graphs, and Mathematical Tables*, Applied Mathematics Series 55, National Bureau of Standards, Washington, DC, 1964.

[3] R. M. ABU-SARIS AND F. B. HANSON, *Computational Suboptimal Filter for a Class of Wiener-Poisson Driven Stochastic Process*, Dyn. and Control, vol. 7, no. 3, 1997, pp. 279–292.

[4] N. AHMED AND K. TEO, *Optimal Control of Distributed Parameter Systems*, North–Holland, New York, NY, 1981.

[5] Y. AïT-SAHALIA, *Disentangling Diffusion from Jumps*, J. Fin. Econ., vol. 74, 2004, pp. 487–528.

[6] T. G. ANDERSEN, L. BENZONI, AND J. LUND, *An Empirical Investigation of Continuous-Time Equity Return Models*, J. Fin., vol. 57, no. 3, 2002, pp. 1239–1284.

[7] B. D. O. ANDERSON AND J. B. MOORE, *Optimal Filtering*, Prentice–Hall, Englewood Cliffs, NJ, 1979.

[8] B. D. O. ANDERSON AND J. B. MOORE, *Optimal Control: Linear Quadratic Methods*, Prentice–Hall, Englewood Cliffs, NJ, 1990.

[9] P. ANDERSEN AND J. G. SUTINEN, *Stochastic Bioeconomics: A Review of Basic Methods and Results*, Marine Res. Econ., vol. 1, 1982, pp. 1–10.

[10] H. L. ANDERSON, *Metropolis, Monte Carlo Method, and the MANIAC*, Los Alamos Science, Fall 1986, pp. 96–107; http://www.fas.org/sgp/othergov/doe/lanl/pubs/numbe14.htm.

[11] C. A. AOURIR, D. OKUYAMA, C. LOTT, AND C. EGLINTON, *Exchanges—Circuit Breakers, Curbs, and Other Trading Restrictions*, 2007, http://invest-faq.com/articles/exch-circuit-brkr.html.

[12] D. APPLEBAUM, *Lévy Processes and Stochastic Calculus*, Cambridge University Press, Cambridge, UK, 2004.

[13] L. ARNOLD, *Stochastic Equations: Theory and Applications*, John Wiley, New York, NY, 1974.

[14] Y. ASHKENAZY, J. M. HAUSDORFF, P. C. IVANOV, AND H. E. STANLEY, *A Stochastic Model of Human Gait Dynamics*, Phys. A, vol. 316, 2002, pp. 662–670.

[15] M. ATHANS AND P. L. FALB, *Optimal Control: An Introduction to the Theory and Its Applications*, McGraw-Hill, New York, NY, 1966.

[16] L. BACHELIER, *Théorie de la Spéculation*, Annales de l'Ecole Normale Supérieure, vol. 17, 1900, pp. 21–86. English translation by A. J. Boness in *The Random Character of Stock Market Prices*, P. H. Cootner, ed., MIT Press, Cambridge, MA, 1967, pp. 17–78.

[17] A. BAIN, *Stochastic Calculus and Stochastic Filtering*, 2007. http://www.chiark.greenend.org.uk/~alanb/.

[18] C. A. BALL AND W. N. TOROUS, *On Jumps in Common Stock Prices and Their Impact on Call Option Prices*, J. Finance, vol. 40, 1985, pp. 155–173.

[19] M. S. BARTLETT, *Stochastic Equations: Theory and Applications*, 3rd ed., Cambridge University Press, Cambridge, UK, 1978.

[20] O. E.. BARNDORFF-NIELSEN AND N. SHEPHERD, *Non-Gaussian Ornstein-Uhlenbeck-Based Models and Some of Their Uses in Financial Economics*, J. Roy. Stat. Soc., ser. B, vol. 3, part 2, 2001, pp. 167–241.

[21] T. BASAR, *Twenty-Five Seminal Papers in Control*, IEEE Control Syst. Mag., vol. 20, no. 1, Feb. 2000, pp. 69–70.

[22] M. BAXTER AND A. RENNIE, *Financial Calculus: An Introduction to Derivative Pricing*, Cambridge University Press, Cambridge, UK, 1996.

[23] I. BEICHL AND F. SULLIVAN, *The Metropolis Algorithm*, Computing in Sci. and Engineering, vol. 2, no. 1, 2000, pp. 65–69.

[24] D. J. BELL AND D. H. JACOBSON, *Singular Optimal Control Problems*, Academic Press, New York, NY, 1994.

[25] R. E. BELLMAN, *Dynamic Programming*, Princeton University Press, Princeton, NJ, 1957.

[26] R. E. BELLMAN, *Adaptive Control Processes: A Guided Tour*, Princeton University Press, Princeton, NJ, 1961.

[27] R. BELLMAN AND R. KALABA, *Selected Papers on Mathematical Trends in Control Theory*, Dover Publications, New York, NY, 1964.

[28] C. M. BENDER AND S. A. ORSZAG, *Advanced Mathematical Methods for Scientists and Engineers*, McGraw-Hill, New York, NY, 1978.

[29] D. S. BERNSTEIN, *Feedback Control: An Invisible Thread in the History of Technology*, IEEE Control Systems Magazine, vol. 22, no. 2, April 2002, pp. 53–68.

[30] J. T. BETTS, *Practical Methods for Optimal Control Using Nonlinear Programming*, SIAM, Philadelphia, PA, 2001.

[31] A. T. BHARUCHA-REID, *Elements of the Theory of Markov Processes and Their Applications*, McGraw-Hill, New York, NY, 1960.

[32] P. BILLINGSLEY, *Probability and Measures*, 2nd ed., John Wiley, New York, NY, 1986.

[33] N. H. BINGHAM AND R. KIESEL, *Risk-Neutral Valuation: Pricing and Hedging of Financial Derivatives*, Springer-Verlag, New York, NY, 2004.

[34] F. BLACK AND M. SCHOLES, *The Pricing of Options and Corporate Liabilities*, J. Political Economy, vol. 81, 1973 (Spring), pp. 637–659.

[35] F. BLACK, *How We Came Up with the Option Formula*, J. Portfolio Mgmt., vol. 15, winter 1989, pp. 4–8.

[36] G. A. BLISS, *Lectures on the Calculus of Variations*, University of Chicago Press, Chicago, IL, 1946.

[37] A. BOKER, C. J. HABERMAN, L. GIRLING, R. P. GUZMAN, G. LOURIDAS, J. R. TANNER, M. CHEANG., M. MATH, B. W. MAYCHER, D. D. BELL, AND G. J. DOAK, *Variable Ventilation Improves Perioperative Lung Function in Patients Undergoing Abdominal Aortic Aneurysmectomy*, Anesthesiology, vol. 100, no. 3, 2004, pp. 608–616.

[38] G. E. P. BOX AND M. E. MULLER, *A Note on the Generation of Random Normal Deviates*, Ann. Math. Stat., vol. 29, 1958, pp. 610–611.

[39] P. BOYLE, *Options: A Monte Carlo Approach*, J. Fin. Econ., vol. 4, 1977, pp. 323–338.

[40] P. BOYLE, M. BROADIE, AND P. GLASSERMAN, *Monte Carlo Methods for Security Pricing*, J. Econ. Dyn. and Control, vol. 21, 1997, pp. 1267–1321.

[41] P. BOSSAERTS, *The Paradox of Asset Pricing*, Princeton University Press, Princeton, NJ, 2002.

[42] P. W. BRIDGEMAN, *Dimensional Analysis*, Yale University Press, New Haven, CT, 1963.

[43] P. BRÉMAUD, *Point Processes and Queues: Martingale Dynamics*, Springer-Verlag, New York, NY, 1981.

[44] A. E. BRYSON AND Y. C. HO, *Applied Optimal Control*, John Wiley, New York, NY, 1975.

[45] T. G. BUCHMAN, *Nonlinear Dynamics, Complex Systems, and the Pathobiology of Critical Illness*, Curr. Opin. Crit. Care, vol. 10, no. 5, 2004, pp. 378–382.

[46] P. CARR, H. GEMAN, D. B. MADAN, AND M. YOR, *Stochastic Volatility for Lévy Processes*, Math. Fin., vol. 13, no. 3, 3003, pp. 345–382.

[47] P. CARR AND D. B. MADAN, *Option Valuation Using the Fast Fourier Transform*, J. Comp. Fin., vol. 2, 1999, pp. 61–73.

[48] S. P. CHAKRABARTY, *Optimal Control of Drug Delivery to Brain Tumors Using a Distributed Parameters Deterministic Model*, Ph.D. Thesis in Mathematics, University of Illinois, Chicago, IL, 2006, pp. 1–126.

[49] S. P. CHAKRABARTY AND F. B. HANSON, *Optimal Control of Drug Delivery to Brain Tumors for a Distributed Parameters Model*, in Proceedings of the 2005 American Control Conference, American Automatic Control Council, Dayton, OH, 2005, pp. 973–978.

[50] S. P. CHAKRABARTY AND F. B. HANSON, *Optimal Control of Drug Delivery to Brain Tumors for a PDE Driven Model Using the Galerkin Finite Element Method*, in Proceedings of the 44th IEEE Conference on Decision and Control and European Control Conference, 2005, pp. 1613–1618.

[51] S. P. CHAKRABARTY AND F. B. HANSON, *Cancer Drug Delivery in Three Dimensions for a Distributed Parameter Control Model Using Finite Elements*, in Proceedings of the 45th IEEE Conference on Decision, 2006, pp. 2088–2093; available at http://www.math.uic.edu/˜hanson/pub/CDC2006/cdc06spcfbhweb.pdf.

[52] G. CHICHILNISKY, *Fischer Black: The Mathematics of Uncertainty*, Notices of the AMS, vol. 43, no. 3, 1996, pp. 319–322.

[53] S.-L. CHUNG AND F. B. HANSON, *Optimization Techniques for Stochastic Dynamic Programming*, in Proceedings of the 29th IEEE Conference on Decision and Control, vol. 4, 1990, pp. 2450–2455.

[54] S.-L. CHUNG AND F. B. HANSON, *Parallel Optimizations for Computational Stochastic Dynamic Programming*, in Proceedings of the 1990 International Conference on Parallel Processing, Pennsylvania State University Press, University Park, PA, vol. 3: Algorithms and Applications, P.-C. Yew, ed., 1990, pp. 254–260.

[55] S.-L. CHUNG, F. B. HANSON, AND H. H. XU, *Parallel Stochastic Dynamic Programming: Finite Element Methods*, Linear Algebra Appl., vol. 172, 1992, pp. 197–218.

[56] E. ÇINLAR, *Introduction to Stochastic Processes*, Prentice-Hall, Englewood Cliffs, NJ, 1975.

[57] C. W. CLARK, *Mathematical Bioeconomics: The Optimal Management of Renewable Resources*, 1st and 2nd eds., John Wiley, New York, NY, 1976, 1990.

[58] C. W. CLARK AND R. LAMBERSON, *An Economic History and Analysis of Pelagic Whaling*, Marine Policy, vol. 6, 1982, pp. 103–120.

[59] J. E. COHEN, *Mathematics Is Biology's Next Microscope, Only Better; Biology Is Mathematics' Next Physics, Only Better*, PLoS Biology, vol. 2, no. 12, 2004, pp. 2017–2023.

[60] R. CONT AND P. TANKOV, *Financial Modelling with Jump Processes*, Chapman & Hall/CRC, Boca Raton, FL, 2004.

[61] E. T. COPSON, *Asymptotic Expansions*, Cambridge University Press, Cambridge, UK, 1965.

[62] J. M. COURTAULT, Y. KABANOV, B. BRU, P. CRÉPEL, I. LEBON, AND A. L. MARCHAND, *Louis Bachelier on the Centenary of Théorie De La Spéculation*, Math. Fin., vol. 10, no. 3, 2000, pp. 341–353.

[63] D. R. COX AND H. D. MILLER, *The Theory of Stochastic Processes*, Chapman and Hall, London, UK, 1965.

[64] J. C. COX AND M. RUBINSTEIN, *Options Markets*, Prentice-Hall, Englewood Cliffs, NJ, 1985.

[65] S. CYGANOWSKI, L. GRÜNE, AND P. KLOEDEN, *Maple for Jump-Diffusion Stochastic Differential Equations in Finance*, in Programming Languages and Systems in Computational Economics and Finance, S. S. Nielsen, ed., Kluwer Academic Publishers, Amsterdam, 2002, pp. 233–269; available at http://www.uni-bayreuth.de/departments/math/~lgruene/papers/jumpfin.html.

[66] S. CYGANOWSKI AND P. KLOEDEN, *Maple Schemes for Jump-Diffusion Stochastic Differential Equations*, in Proceedings of the 16th IMACS World Congress, Lausanne, 2000, M. Deville and R. Owens, eds., International Association of Mathematics and Computer Simulation, Piscataway, NJ, CD-ROM Paper 216-9, pp. 1–16; available at http://www.math.uni-frankfurt.de/~numerik/maplestoch/jumpdiff.pdf.

[67] S. CYGANOWSKI, P. KLOEDEN, AND J. OMBACH, *Elementary Probability to Stochastic Differential Equations with Maple*, Springer-Verlag, New York, NY, 2002.

[68] B. N. DATTA, *Numerical Linear Algebra and Applications*, Brooks/Cole, New York, NY, 1995.

[69] J. DONGARRA AND F. SULLIVAN, *Guest Editor's Introduction: The Top Ten Algorithms*, Computing in Sci. and Engineering, vol. 2, no. 1, 2000, pp. 22–23.

[70] J. L. DOOB, *Stochastic Processes*, John Wiley, New York, NY, 1953.

[71] G. D. DOOLEN AND JOHN HENDRICKS, *Monte Carlo at Work*, Los Alamos Science, Special Issue Dedicated to S. Ulam, 1987, pp. 142–143. http://www.fas.org/sgp/othergov/doe/lanl/pubs/00326867.pdf.

[72] P. DORATO, C. ABDALLAH, AND V. CERONE, *Linear-Quadratic Control: An Introduction*, Prentice-Hall, Englewood Cliffs, NJ, 1995.

[73] J. DOUGLAS, JR., AND T. DUPONT, *Galerkin Methods for Parabolic Equations*, SIAM J. Numer. Anal., vol. 7, 1970, pp. 575–626.

[74] J. DOUGLAS, JR., *Time Step Procedures of Nonlinear Parabolic PDEs*, in Mathematics of Finite Elements and Applications, MAFELAP, J. Whiteman, ed., Academic Press, London, 1979, pp. 289–304.

[75] D. DUFFIE, *Dynamic Asset Pricing Theory*, Princeton University Press, Princeton, NJ, 2001.

[76] D. DÜVELMEYER, *Inkorrektheitsphnomene und Regularisierung bei der Parameterschtzung für Jump-Diffusions-Prozesse*, Dissertation, Facultät für Mathematik, Technische Universität Chemnitz, Chemnitz, 2005; http://www-usercgi.tuchemnitz.de/~dana/diss/php.

[77] P. DYER AND S. R. MCREYNOLDS, *The Computation and Theory of Optimal Control*, Academic Press, New York, NY, 1970.

[78] E. B. DYKIN, *Markov Processes* I *and* II, Academic Press, New York, NY, 1965.

[79] R. ECKHARDT, *Ulam, John von Neumann, and the Monte Carlo Method*, Los Alamos Science, Special Issue Dedicated to S. Ulam, 1987, pp. 131–137. http://www.fas.org/sgp/othergov/doe/lanl/pubs/00326867.pdf.

[80] P. EMBRECHTS, C. KLÜPPELBERG, AND T. MIKOSCH, *Modelling Extremal Events for Insurance and Finance*, Springer-Verlag, New York, NY, 2003.

[81] H. H. ENGELHARD, *Brain Tumors and the Blood-Brain Barrier*, in Neuro-Oncology: The Essentials, Thieme Medical Publishers, Inc., New York, 2000, pp. 49–53.

[82] J. D. ESARY, F. PROSCHAN, AND D. W. WALKUP, *Association of Random Variables, with Applications*, Ann. Math. Statist., vol. 38, 1967, pp. 1466–1474.

[83] M. EVANS, N. HASTINGS, AND B. PEACOCK, *Statistical Distributions*, 3rd ed., John Wiley, New York, NY, 2000.

[84] W. FELLER, *An Introduction to Probability Theory and Its Application*, vol. 1, 3rd ed., John Wiley, New York, NY, 1968.

[85] W. FELLER, *An Introduction to Probability Theory and Its Application*, vol. 2, 2nd ed., John Wiley, New York, NY, 1971.

[86] W. H. FLEMING AND R. W. RISHEL, *Deterministic and Stochastic Optimal Control*, Springer-Verlag, New York, NY, 1975.

[87] J. J. FLORENTIN, *Optimal Control of Systems with Generalized Poisson Inputs*, J. Basic Engr., ASME Trans., Ser. D, vol. 85, no. 2, pp. 217–221.

[88] G. E. FORSYTHE, M. A. MALCOLM, AND C. MOLER, *Computer Methods for Mathematical Computations*, Prentice-Hall, Englewood Cliffs, NJ, 1977.

[89] B. FRIEDMAN, *Principles and Techniques of Applied Mathematics*, John Wiley, New York, NY, 1956.

[90] W. H. FLEMING, ED., *Future Directions in Control Theory: A Mathematical Perspective*, SIAM, Philadelphia, PA, 1988.

[91] R. M. MURRAY, ED., *Control in an Information Rich World: Report of the Panel on Future Directions in Control, Dynamics, and Systems*, SIAM, Philadelphia, PA, 2003.

[92] T. C. GARD, *Introduction to Stochastic Differential Equations*, Marcel-Dekker, New York, NY, 1988.

[93] R. A. GATENBY AND E. T. GAWLINSKI, *A Reaction-Diffusion Model of Cancer Invasion*, Cancer Research, vol. 56, 1996, pp. 5745–5753.

[94] S. B. GERSHWIN, *Manufacturing Systems Engineering*, Prentice-Hall, Englewood Cliffs, NJ, 1994.

[95] I. I. GIHMAN AND A. V. SKOROHOD, *Stochastic Differential Equations*, Springer-Verlag, New York, NY, 1972.

[96] I. I. GIHMAN AND A. V. SKOROHOD, *Controlled Stochastic Processes*, Springer-Verlag, New York, NY, 1979.

[97] P. GLASSERMAN, *Monte Carlo Methods in Financial Engineering*, Springer-Verlag, New York, NY, 2003.

[98] P. W. GLYNN AND W. WHITT, *The Asymptotic Efficiency of Simulation Estimators*, Oper. Res., vol. 40, no. 3, 1992, pp. 505–520.

[99] N. S. GOEL AND N. RICHTER-DYN, *Stochastic Models in Biology*, Springer-Verlag, NY, 2003.

[100] J. H. GOLDIE AND A. J. COLDMAN, *Drug Resistance in Cancer: Mechanisms and Models*, Cambridge University Press, Cambridge, UK, 1998.

[101] M. S. GREWAL AND A. P. ANDREWS, *Kalman Filtering: Theory and Practice*, Prentice-Hall, Englewood Cliffs, NJ, 1993.

[102] M. D. GUNZBURGER, *Perspectives in Flow Control and Optimization*, SIAM, Philadelphia, PA, 2003.

[103] R. HABERMAN, *Elementary Applied Partial Differential Equations with Fourier Series and Boundary Value Problems*, Prentice-Hall, Englewood Cliffs, NJ, 1983.

[104] W. HACKBUSCH, *A Numerical Method for Solving Parabolic Equations with Opposite Orientations*, Computing, vol. 20, 1978, pp. 229–240.

[105] J. M. HAMMERSLEY AND D. C. HANDSCOMB, *Monte Carlo Methods*, Methuen, London, UK, 1964.

[106] F. B. HANSON, *Bioeconomic Model of the Lake Michigan Alewife Fishery*, Canadian Journal of Fisheries and Aquatic Sciences, vol. 44, suppl. 2, 1987, pp. 298–305.

[107] F. B. HANSON, *Stochastic Dynamic Programming: Advanced Computing Constructs*, in Proceedings of the 28th IEEE Conference on Decision and Control, vol. 1, 1989, pp. 901–903.

[108] F. B. HANSON, *Computational Dynamic Programming on a Vector Multiprocessor*, IEEE Trans. Autom. Control, vol. 36, no. 4, 1991, pp. 507–511.

[109] F. B. HANSON, *Computational Stochastic Dynamic Programming*, in Stochastic Digital Control System Techniques, Control and Dynamic Systems: Advances in Theory and Applications, vol. 76, C. T. Leondes, ed., Academic Press, New York, NY, 1996, pp. 103–162.

[110] F. B. HANSON, *Local Supercomputing Training in the Computational Sciences Using Remote National Centers*, Future Generation Computer Systems: Special Issue on Education in the Computational Sciences, vol. 19, 2003, pp. 1335–1347.

[111] F. B. HANSON, *Computational Stochastic Control: Basic Foundations, Complexity and Techniques*, in Proceedings of the 42nd IEEE Conference on Decision and Control, 2003, pp. 3024–3029.

[112] F. B. HANSON AND K. NAIMIPOUR, *Convergence of Numerical Method for Multistate Stochastic Dynamic Programming*, in Proceedings of the International Federation of Automatic Control, 12th World Congress, vol. 9, International Federation of Automatic Control, Laxenburg, Austria, 1993, pp. 501–504.

[113] F. B. HANSON, C. J. PRATICO, M. S. VETTER, AND H. H. XU, *Multidimensional Visualization Applied to Renewable Resource Management*, in Proceedings of the Sixth SIAM Conference on Parallel Processing for Scientific Computing, vol. II, R. F. Sincovec et al., eds., SIAM, Philadelphia, PA, 1993, pp. 1033–1036.

[114] F. B. HANSON AND D. RYAN, *Optimal Harvesting with Density Dependent Random Effects*, Natural Resource Modeling, vol. 2, 1988, pp. 439–455.

[115] F. B. HANSON AND D. RYAN, *Mean and Quasideterministic Equivalence for Linear Stochastic Dynamics*, Math. Biosci., vol. 93, 1989, pp. 1–14.

[116] F. B. HANSON AND D. RYAN, *Optimal Harvesting with Both Population and Price Dynamics*, Math. Biosci., vol. 148, 1998, pp. 129–146.

[117] F. B. HANSON AND C. TIER, *An Asymptotic Solution of the First Passage Problem for Singular Diffusion in Population Biology*, SIAM J. Appl. Math., vol. 40, 1981, pp. 113–132.

[118] F. B. HANSON AND C. TIER, *A Stochastic Model of Tumor Growth*, Math. Biosci., vol. 61, 1982, pp. 73–100.

[119] F. B. HANSON AND H. C. TUCKWELL, *Persistence Times of Populations with Large Random Fluctuations*, Theor. Population Biol., vol. 14, 1978, pp. 46–61.

[120] F. B. HANSON AND H. C. TUCKWELL, *Logistic Growth with Random Density Independent Disasters*, Theor. Population Biol., vol. 19, 1981, pp. 1–18.

[121] F. B. HANSON AND H. C. TUCKWELL, *Diffusion Approximations for Neuronal Activity Including Reversal Potentials*, J. Theor. Neurobiology, vol. 2, 1983, pp. 127–153.

[122] F. B. HANSON AND H. C. TUCKWELL, *Population Growth with Randomly Distributed Jumps*, J. Math. Biol., vol. 36, no. 2, 1997, pp. 169–187.

[123] F. B. HANSON AND J. J. WESTMAN, *Optimal Consumption and Portfolio Policies for Important Jump Events: Modeling and Computational Considerations*, in Proceedings of the American Control Conference, American Automatic Control Council, Dayton, OH, 2001, pp. 4456–4661.

[124] F. B. HANSON AND J. J. WESTMAN, *Optimal Consumption and Portfolio Control for Jump-Diffusion Stock Process with Log-Normal Jumps*, in Proceedings of the American Control Conference, American Automatic Control Council, Dayton, OH, July 2002, pp. 4256–4261; for corrected version see ftp://ftp.math.uic.edu/pub/Hanson/ACC02/acc02webcor.pdf.

[125] F. B. HANSON AND J. J. WESTMAN, *Stochastic Analysis of Jump-Diffusions for Financial Log-Return Processes*, in Stochastic Theory and Control, Proceedings of a Workshop held in Lawrence, Kansas, Oct. 18–20, 2001, Lecture Notes in Control and Information Sciences, vol. 280, B. Pasik-Duncan, ed., Springer-Verlag, New York, pp. 169–184, 2002; for corrected version see http://www.math.uic.edu/~hanson/pub/KU02/ku02hwfmcor.pdf.

[126] F. B. HANSON AND J. J. WESTMAN, *Jump-Diffusion Stock Return Models in Finance: Stochastic Process Density with Uniform-Jump Amplitude*, in Proceedings of the 15th International Symposium on Mathematical Theory of Networks and Systems, D. S. Gilliam and J. Rosenthal, eds., University of Notre Dame, South Bend, IN, http://www.nd.edu/~mtns, Aug. 2002, pp. 1–7.

[127] F. B. HANSON AND J. J. WESTMAN, *Portfolio Optimization with Jump–Diffusions: Estimation of Time-Dependent Parameters and Application*, in Proceedings of the 41st IEEE Conference on Decision and Control, Dec. 2002, pp. 377–382.

[128] F. B. HANSON AND J. J. WESTMAN, *Jump-Diffusion Stock-Return Model with Weighted Fitting of Time-Dependent Parameters*, in Proceedings of the American Control Conference, American Automatic Control Council, Dayton, OH, 2003, pp. 4869–4874.

[129] F. B. HANSON, J. J. WESTMAN, AND Z. ZHU, *Maximum Multinomial Likelihood Estimation of Market Parameters for Stock Jump-Diffusion Models*, in Mathematics of Finance: Proceedings of the 2003 AMS-IMS-SIAM Joint Summer Research Conference on Mathematics of Finance, AMS Contemporary Mathematics, vol. 351, G. Yin and Q. Zhang, eds., 2004, pp. 155–169.

[130] F. B. HANSON AND J. J. WESTMAN, *Optimal Portfolio and Consumption Policies Subject to Rishel's Important Jump Events Model: Computational Methods*, IEEE

Trans. Autom. Control, Special Issue on Stochastic Control Methods in Financial Engineering, vol. 48, no. 3, 2004, pp. 326–337.

[131] F. B. HANSON AND G. YAN, *American Put Option Pricing for Stochastic-Volatility, Jump-Diffusion Models*, in Proceedings of the 2007 American Control Conference, American Automatic Control Council, Dayton, OH, pp. 1–6; http://www.math.uic.edu/~hanson/pub/GYan/ACC07fhgywebpub.pdf.

[132] F. B. HANSON AND Z. ZHU, *Comparison of Market Parameters for Jump-Diffusion Distributions Using Multinomial Maximum Likelihood Estimation*, in Proceedings of the 42nd IEEE Conference on Decision and Control, Dec. 2004, pp. 3919–3924.

[133] J. M. HARRISON AND S. R. PLISKA, *Martingales and Stochastic Integrals in the Theory of Continuous Trading*, Stochastic Proc. Appl., vol. 11, 1981, pp. 215–260.

[134] J. M. HARRISON AND S. R. PLISKA, *A Stochastic Calculus Model of Continuous Trading: Complete Markets*, Stochastic Proc. Appl., vol. 15, 1983, pp. 313–316.

[135] M. B. HAUGH AND A. W. LO, *Computational Challenges in Portfolio Management*, Computing in Sci. and Engineering, May/June 2000, pp. 54l–559.

[136] D. HEATH AND M. SCHWEIZER, *Martingales Versus PDEs in Finance: An Equivalence Result with Examples*, J. Appl. Prob., vol. 37, 2000, pp. 947–957.

[137] J. W. HELTON AND O. MERINO, *Classical Control Using H^∞ Methods: An Introduction to Design*, SIAM, Philadelphia, PA, 1998.

[138] J. W. HELTON AND O. MERINO, *Classical Control Using H^∞ Methods: Theory, Optimization, and Design*, SIAM, Philadelphia, PA, 1998.

[139] R. C. HENNEMUTH, J. E. PALMER, AND B. E. BROWN, *A Statistical Description of Recruitment in Eighteen Selected Fish Stocks*, J. Northw. Alt. Fish Sci., vol. 1, 1980, pp. 101–111.

[140] D. J. HIGHAM, *An Algorithmic Introduction to Numerical Simulation of Stochastic Differential Equations*, SIAM Rev., vol. 43, no. 3, 2001, pp. 525–546.

[141] D. J. HIGHAM, *An Introduction to Financial Option Valuation: Mathematics, Stochastics and Computation*, Cambridge University Press, Cambridge, UK, 2004.

[142] D. J. HIGHAM, *Black-Scholes for Scientific Computing Students*, Comput. Sci. Engrg., vol. 6, no. 6, Nov./Dec. 2004, pp. 72–79.

[143] D. J. HIGHAM AND N. J. HIGHAM, *MATLAB Guide*, 2nd ed., SIAM, Philadelphia, PA, 2005.

[144] D. J. HIGHAM AND P. E. KLOEDEN, *Maple and MATLAB for Stochastic Differential Equations in Finance*, in Programming Languages and Systems in Computational Economics and Finance, S. S. Neilsen, ed., Kluwer, Dordrecht, The Netherlands, 2002, pp. 233–270. http://www.maths.strath.ac.uk/~aas96106/algfiles.html.

[145] D. J. HIGHAM AND P. E. KLOEDEN, *Numerical Methods for Nonlinear Stochastic Differential Equations with Jumps*, Numer. Math., vol. 101, 2005, pp. 101–119.

[146] D. J. HIGHAM AND P. E. KLOEDEN, *Convergence and Stability of Implicit Methods for Jump-Diffusion Systems*, Int. J. Numer. Anal. Model., vol. 3, 2006, pp. 125–140.

[147] D. J. HIGHAM, X. MAO, AND A. M. STUART, *Strong Convergence of Euler-Type Methods for Nonlinear Stochastic Differential Equations*, SIAM J. Numer. Anal., vol. 40, no. 3, 2002, pp. 1041–1063.

[148] J. C. HULL, *Options, Futures, & Other Derivatives*, 4th ed., Prentice-Hall, Englewood Cliffs, NJ, 2000.

[149] INTERNATIONAL PACIFIC HALIBUT COMMISSION, *Annual Reports*, Seattle, WA, 1984–1985.

[150] K. ITÔ, *On Stochastic Differential Equations*, Mem. Amer. Math. Soc., no. 4, 1951, pp. 1–51.

[151] P. JÄCKEL, *Monte Carlo Methods in Finance*, John Wiley, New York, NY, 2002.

[152] D. H. JACOBSON AND D. Q. MAYNE, *Differential Dynamic Programming*, American Elsevier, New York, NY, 1970.

[153] J. JACOD AND A. N. SHIRYAEV, *Limit Theorems for Stochastic Processes*, Springer-Verlag, Berlin, 1987.

[154] R. A. JARROW AND E. R. ROSENFELD, *Jump Risks and the Intertemporal Capital Asset Pricing Model*, J. Business, vol. 57, no. 3, 1984, pp. 337–351.

[155] A. H. JAZWINSKI, *Stochastic Processes and Filtering Theory*, Academic Press, New York, NY, 1970.

[156] P. JORION, *On Jump Processes in the Foreign Exchange and Stock Markets*, Rev. Fin. Studies, vol. 88, no. 4, 1989, pp. 427–445.

[157] R. E. KALMAN, *Contributions to the Theory of Optimal Control*, Bol. Soc. Mat. Mexicana, vol. 5, 1960, pp. 102–119.

[158] M. H. KALOS AND P. A. WHITLOCK, *Monte Carlo Methods*, Volume I: Basics, John Wiley and Sons, New York, NY, 1986.

[159] M. I. KAMIEN AND N. L. SCHWARTZ, *Dynamic Optimization: The Calculus of Variations and Optimal Control in Economics and Management*, North-Holland, New York, NY, 1981.

[160] I. KARATZAS, J. P. LEHOCZKY, S. P. SETHI, AND S. E. SHREVE, *Explicit Solution of a General Consumption/Investment Problem*, Math. Oper. Res., vol. 11, 1986, pp. 261–294. (Also available in Sethi [245, Chapter 2].)

[161] I. KARATZAS AND S. E. SHREVE, *Methods of Mathematical Finance*, Springer-Verlag, New York, NY, 1998.

[162] S. Karlin and H. M. Taylor, *A First Course in Stochastic Processes*, 2nd ed., Academic Press, New York, NY, 1975.

[163] S. Karlin and H. M. Taylor, *A Second Course in Stochastic Processes*, Academic Press, New York, NY, 1981.

[164] D. E. Kirk, *Optimal Control Theory: An Introduction*, Prentice-Hall, Englewood Cliffs, NJ, 1970. (Reprinted by Dover Publications, Mineola, NY, 2004.)

[165] F. C. Klebaner, *Introduction to Stochastic Calculus with Applications*, Imperial College Press, London, UK, 1998.

[166] P. E. Kloeden and E. Platen, *Numerical Solution of Stochastic Differential Equations*, Springer-Verlag, New York, NY, 1992.

[167] P. E. Kloeden, E. Platen, and H. Schurz, *Numerical Solution of SDE Though Computer Experiments*, Springer-Verlag, New York, NY, 1994.

[168] P. Kokotović, H. K. Khalil, and J. O'Reilly, *Singular Perturbation Methods in Control: Analysis and Design*, Academic Press, New York, NY, 1986.

[169] A. N. Kolmogorov and S. V. Fomin, *Introductory Real Analysis*, translated from the Russian and edited by R. A. Silverman, Prentice-Hall, Englewood Cliffs, NJ, 1970. (Reprinted by Dover Publications, Mineola, NY, 1975.)

[170] S. G. Kou, *A Jump Diffusion Model for Option Pricing*, Management Sci., vol. 48, 2002, pp. 1086–1101.

[171] S. G. Kou and H. Wang, *Option Pricing Under a Double Exponential Jump Diffusion Model*, Management Science, vol. 50, no. 9, 2004, pp. 1178–1192.

[172] K. Koya, H. Kimura, and M. Kawato, *Neural Mechanisms of Learning and Control*, IEEE Control Systems Magazine, vol. 21, no. 4, Aug. 2001, pp. 42–54.

[173] H. Kunita and S. Watanabe, *On Square-Integrable Martingales*, Nagoya Math. J., vol. 30, 1967, pp. 209–245.

[174] H. J. Kushner, *Stochastic Stability and Control*, Academic Press, New York, NY, 1967.

[175] H. J. Kushner, *A Survey of Some Applications of Probability and Stochastic Control Theory to Finite Difference Methods for Degenerate Elliptic and Parabolic Equations*, SIAM Rev., vol. 18, 1976, pp. 545–577.

[176] H. J. Kushner, *Numerical Methods for Stochastic Control Problems in Continuous Time*, SIAM J. Control Optim., vol. 28, 1990, pp. 999–1048.

[177] H. J. Kushner, *Jump-Diffusions with Controlled Jumps: Existence and Numerical Methods*, J. Math. Anal. Appl., vol. 249, no. 1, 2000, pp. 179–198.

[178] H. J. KUSHNER AND G. DIMASI, *Approximation for Functionals and Optimal Control Problems on Jump Diffusion Processes*, J. Math. Anal. Appl., vol. 63, no. 3, 1978, pp. 772–800.

[179] H. J. KUSHNER AND P. G. DUPUIS, *Numerical Methods for Stochastic Control Problems in Continuous Time*, 2nd ed., Springer-Verlag, New York, NY, 2001.

[180] H. J. KUSHNER AND D. J. JARVIS, *Large-Scale Computations for High Dimension Control Systems*, in Proceedings of the 33rd IEEE Conference on Decision and Control, vol. 1, 1994, pp. 461–465.

[181] H. J. KUSHNER AND G. G. YIN, *Stochastic Approximation Algorithms, Recursive Algorithms and Applications*, Springer-Verlag, New York, NY, 2003.

[182] R. E. LARSON, *A Survey of Dynamic Programming Computational Procedures*, IEEE Trans. Autom. Control, vol. AC-16, 1967, pp. 767–774.

[183] G. P. LEPAGE, *A New Algorithm for Adaptive Multidimensional Integration*, J. Comp. Phys., vol. 27, no. 2, 1978, pp. 192–203.

[184] F. L. LEWIS, *Optimal Estimation with an Introduction to Stochastic Control Theory*, John Wiley, New York, NY, 1986.

[185] M. J. LIGHTHILL, *Introduction to Fourier Analysis and Generalised Functions*, Cambridge University Press, Cambridge, UK, 1964.

[186] A. LIPTON, *Mathematical Methods for Foreign Exchange: A Financial Engineer's Approach*, World Scientific, Singapore, 2001.

[187] D. LUDWIG, *Stochastic Population Theories*, Springer-Verlag, New York, NY, 1974.

[188] D. LUDWIG, *Persistence of Dynamical Systems Under Random Perturbations*, SIAM Rev., vol. 17, 1975, pp. 605–640.

[189] D. LUDWIG, *Optimal Harvesting of a Randomly Fluctuating Resource. I: Application of Perturbation Methods*, SIAM J. Appl. Math., vol. 37, 1979, pp. 166–184.

[190] D. LUDWIG AND J. M. VARAH, *Optimal Harvesting of a Randomly Fluctuating Resource. II: Numerical Methods and Results*, SIAM J. Appl. Math., vol. 37, 1979, pp. 185–205.

[191] Y. MAGHSOODI, *Mean Square Efficient Numerical Solution of Jump-Diffusion Stochastic Differential Equations*, Sankhyā, ser. A, vol. 58, no. 1, 1996, pp. 25–47.

[192] Y. MAGHSOODI AND C. J. HARRIS, *In-Probability Approximation and Simulation of Nonlinear Jump-Diffusion Stochastic Differential Equations*, IMA J. Math. Control Inform., vol. 4, 1996, pp. 65–92.

[193] M. MANGEL, *Decision and Control in Uncertain Resource Systems*, Academic Press, New York, NY, 1985.

[194] M. MARITON, *Jump Linear Systems in Automatic Control*, Marcel Dekker, New York, NY, 1990.

[195] G. MARSAGLIA AND T. A. BRAY, *A Convenient Method for Generating Normal Variables*, SIAM Rev., vol. 6, 1964, pp. 260–264.

[196] D. Q. MAYNE, *A Second–Order Gradient Method for Determining Optimal Control of Non–Linear Discrete Time Systems*, Int. J. Control, vol. 3, 1966, pp. 85–95.

[197] D. Q. MAYNE, *Differential Dynamic Programming — A Unified Approach to the Optimization of Dynamical Systems*, Control and Dynamical Systems: Advances in Theory and Applications, vol. 10, C. T. Leondes, ed., Academic Press, New York, NY, 1973, pp. 179–254.

[198] R. C. MERTON, *Lifetime Portfolio Selection Under Uncertainty: The Continuous-Time Case*, Rev. Econ. Stat., vol. 51, 1969, pp. 247–257. (Also available in Merton [203, Chapter 4].)

[199] R. C. MERTON, *Optimum Consumption and Portfolio Rules in a Continuous-Time Model*, J. Econ. Theory, vol. 3, no. 4, 1971, pp. 373–413. (Also available in Merton [203, Chapter 5].)

[200] R. C. MERTON, *Eratum*, J. Econ. Theory, vol. 6, no. 2, 1973, pp. 213–214.

[201] R. C. MERTON, *Theory of Rational Option Pricing*, Bell J. Econ. Mgmt. Sci., vol. 4, 1973 (Spring), pp. 141–183. (Also available in Merton [203, Chapter 8].)

[202] R. C. MERTON, *Option Pricing When Underlying Stock Returns are Discontinuous*, J. Fin. Econ., vol. 3, 1976, pp. 125–144. (Also available in Merton [203, Chapter 9].)

[203] R. C. MERTON, *Continuous Time Finance*, Blackwell Publishers, Cambridge, MA, 1992.

[204] R. C. MERTON AND M. S. SCHOLES, *Fischer Black*, J. Finance, vol. 50, no. 5, 1996, pp. 1359–1369.

[205] W. C. MESSNER AND D. M. TILBURY, *Control Tutorials for MATLAB and Simulink: User's Guide*, Addison-Wesley, Reading, MA, 2002. (See additional information in http://www.engin.umich.edu/group/ctm/.)

[206] N. METROPOLIS, *The Beginning of the Monte Carlo Method*, Los Alamos Science, Special Issue Dedicated to S. Ulam, 1987, pp. 125–130. http://www.fas.org/sgp/othergov/doe/lanl/pubs/00326866.pdf.

[207] N. METROPOLIS, A. N. ROSENBLUTH, M. N. ROSENBLUTH, A. H. TELLER, AND E. TELLER, *Equation of State Calculation by Fast Computing Machines*, J. Chem. Phys., vol. 21, no. 6, 1953, pp. 1087–1092.

[208] N. METROPOLIS AND S. ULAM, *The Monte Carlo Method*, J. Amer. Stat. Assoc., vol. 44, no. 247, 1949, pp. 335–341.

[209] T. MIKOSCH, *Elementary Stochastic Calculus: With Finance in View*, World Scientific, Singapore, 1998.

[210] C. MOLER ET AL., *Using MATLAB*, version 6, Mathworks, Natick, MA, 2000.

[211] F. MOSS, L. M. WARD, AND W. G. SANNITA, *Stochastic Resonance and Sensory Information Processing*, Clinical Neurophysiology, vol. 115, 2004, pp. 267–281.

[212] G. I. MURPHY, *Clupeoids*, in Fish Population Dynamics, J. A. Gulland, ed., John Wiley, New York, NY, 1977, pp. 283–308.

[213] J. D. MURRAY, *Mathematical Biology, I: An Introduction*, Springer-Verlag, New York, NY, 2002.

[214] J. D. MURRAY, *Mathematical Biology, II: Spatial Models and Biomedical Applications*, Springer-Verlag, New York, NY, 2003.

[215] R. M. MURRAY, K. J. ASTRÖM, R. W. BROCKETT, AND G. STEIN, *Future Directions in Control in an Information Rich World: A Summary of the Report of the Panel on Future Directions in Control, Dynamics, and Systems*, IEEE Control Systems Magazine, vol. 23, no. 2, April 2003, pp. 20–33. http://www.cds.caltech.edu/~murray/cdspanel/.

[216] K. NAIMIPOUR AND F. B. HANSON, *Convergence of a Numerical Method for the Bellman Equation of Stochastic Optimal Control with Quadratic Costs and Constrained Control*, Dyn. and Control, vol. 3, no. 3, 1993, pp. 237–259.

[217] S. N. NEFTCI, *Introduction to the Mathematics of Financial Derivatives*, 2nd ed., Academic Press, New York, NY, 2000.

[218] H. NIEDERREITER, *Random Number Generation and Quasi-Monte Carlo Methods*, SIAM, Philadelphia, PA, 1992.

[219] R. M. NISBET AND W. S. C. GURNEY, *Modelling Fluctuating Populations*, John Wiley, New York, NY, 1982.

[220] N. S. NISE, *Control Systems Engineering*, John Wiley, New York, NY, 2000.

[221] J. NOCEDAL AND S. J. WRIGHT, *Numerical Optimization*, Springer-Verlag, New York, NY, 1999.

[222] B. ØKSENDAL, *Stochastic Differential Equations: An Introduction with Applications*, 5th ed., Springer-Verlag, New York, NY, 1998.

[223] B. ØKSENDAL AND A. SULEM, *Applied Stochastic Control of Jump Diffusions*, Springer-Verlag, Berlin, 2005.

[224] E. PARZEN, *Stochastic Processes*, Holden-Day, San Francisco, CA, 1962.

[225] S. R. PLISKA, *Introduction to Mathematical Finance: Discrete Time Models*, Blackwell Publishers, Cambridge, MA, 1997.

[226] L. S. PONTRYAGIN, V. G. BOLTYANSKII, R. V. GAMKRELIDZE, AND E. F. MISHCHENKO, *The Mathematical Theory of Optimal Processes*, Wiley-Interscience Publishers, New York, NY, 1962.

[227] E. POLAK, *An Historical Survey of Computational Methods in Optimal Control*, SIAM Rev., vol. 15, 1973, pp. 553–584.

[228] C. J. PRATICO, F. B. HANSON, H. H. XU, D. J. JARVIS, AND M. S. VETTER, *Visualization for the Management of Renewable Resources in an Uncertain Environment*, in Proceedings of Supercomputing '92, 1992, pp. 258–266 and p. 853.

[229] W. H. PRESS AND G. R. FARRAR, *Recursive Stratified Sampling for Multidimensional Monte Carlo Integration*, Computers in Physics, vol. 4, no. 2, 1990, pp. 190–195.

[230] W. H PRESS, S. A. TEUKOLSKY, W. T. VETTERLING, AND B. P. FLANNERY, *Numerical Recipes in C++: The Art of Scientific Computing*, 2nd ed., Cambridge University Press, Cambridge, UK, 2002.

[231] A. PRIPLATA, J. NIEMI, M. SALEN, J. HARRY, L. A. LIPSITZ, AND J. J. COLLINS, *Noise-Enhanced Human Balance Control*, Phys. Rev. Let., vol. 89, no. 23, 2002, pp. 238101–238104.

[232] P. PROTTER, *Stochastic Integration and Differential Equations: A New Approach*, Springer-Verlag, Berlin, 2004.

[233] D. W. REPPERGER, *Celebrating the 100th Anniversary of Controlled, Sustained, and Powered Air Flight,* in IEEE Control Systems Magazine, vol. 23, no. 6, 2003, pp. 12–16.

[234] R. C. SMITH AND M. A. DEMETRIOU, EDS., *Research Directions in Distributed Parameter Systems*, SIAM, Philadelphia, PA, 2003.

[235] R. RISHEL, *Modeling and Portfolio Optimization for Stock Prices Dependent on External Events,* in Proceedings of the 38th IEEE Conference on Decision and Control, 1999, pp. 2788–2793.

[236] L. C. G. ROGERS AND D. WILLIAMS, *Diffusions, Markov Processes and Martingales*, Cambridge University Press, Cambridge, UK, 2000.

[237] S. M. ROSS, *Stochastic Processes*, John Wiley, New York, NY, 1983.

[238] S. M. ROSS, *Introduction to Probability Models*, 7th ed., Academic Press, New York, NY, 2000.

[239] W. J. RUNGGALDIER, *Jump-Diffusion Models*, in Handbook of Heavy Tailed Distributions in Finance, S. T. Rachev, ed., Handbooks in Finance, Elsevier/North-Holland, New York, NY, 2003, pp. 169–209.

[240] J. RUST, *Do People Behave According to Bellman's Principle of Optimality?*, Hoover Institution Working Paper E-92-10, 1994, pp. 1–65.

[241] D. RYAN AND F. B. HANSON, *Optimal Harvesting with Exponential Growth in an Environment with Random Disasters and Bonanzas*, Math. Biosci., vol. 74, 1985, pp. 37–57.

[242] D. RYAN AND F. B. HANSON, *Optimal Harvesting of a Logistic Population in an Environment with Stochastic Jumps*, J. Math. Biol., vol. 24, 1986, pp. 259–277.

[243] T. H. RYDBERG, *The Normal Inverse Gaussian Lévy Process: Simulation and Approximation*, Comm. Stat. Stoch. Models, vol. 13, 1997, pp. 887–910.

[244] Z. SCHUSS, *Theory and Applications of Stochastic Differential Equations*, John Wiley, New York, NY, 1980.

[245] S. P. SETHI, *Optimal Consumption and Investment with Bankruptcy*, Kluwer Academic Publishers, Boston, MA, 1997.

[246] S. P. SETHI AND M. TAKSAR, *A Note on Merton's "Optimum Consumption and Portfolio Rules in a Continuos-Time Model"*, J. Econ. Theory, vol. 46, no. 2, 1988, pp. 395–401. (Also available in Sethi [245, Chapter 3].)

[247] S. P. SETHI AND Q. ZHANG, *Hierarchical Decision Making in Stochastic Manufacturing Systems*, Birkhäuser, Boston, MA, 1994.

[248] S. E. SHREVE, *Stochastic Calculus Models for Finance* II: *Continuous-Time Models*, Springer-Verlag, New York, NY, 2004.

[249] L. SIMPSON-HERREN AND H. H. LLOYD, *Kinetics Parameters and Growth Curves for Experimental Tumor Systems*, Cancer Chemo. Reps., vol. 54, 1970, pp. 143–174.

[250] C. SINDERMANN, *Principal Diseases of Marine Fish and Shellfish*, Academic Press, New York, NY, 1970.

[251] I. N. SNEDDON, *Elements of Partial Differential Equations*, McGraw-Hill, New York, NY, 1957.

[252] D. L. SNYDER AND M. I. MILLER, *Random Point Processes in Time and Space*, 2nd ed., Springer-Verlag, New York, NY, 1991.

[253] I. M. SOBOL', *The Distribution of Points in a Cube and the Approximate Evaluation of Integrals*, Comp. Math. and Math. Physics, vol. 7, no. 4, 1967, pp. 86–112.

[254] H. W. SORENSEN, ED., *Kalman Filtering: Theory and Application*, IEEE Press, New York, NY, 1985.

[255] G. G. STEEL, *Growth Kinetics of Tumors*, Clarendon Press, Oxford, UK, 1977.

[256] J. M. STEELE, *Stochastic Calculus and Financial Applications*, Springer-Verlag, New York, NY, 2001.

[257] R. B. STEIN, *Nerve and Muscle: Membranes, Cells, and Systems*, Plenum Press, New York, NY, 1980.

[258] R. STENGEL, *Stochastic Optimal Control: Theory and Application*, Dover Publications, New York, NY, 1994.

[259] S. STOJANOVIC, *Computational Financial Mathematics Using Mathematica: Optimal Trading in Stocks and Options*, Birkhäuser, Boston, MA, 2002.

[260] R. L. STRATONOVICH, *A New Representation for Stochastic Integrals and Equations*, SIAM J. Control, vol. 4, 1966, pp. 362–371.

[261] G. W. SWAN, *Applications of Optimal Control in Biomedicine*, Marcel Dekker, New York, NY, 1984.

[262] K. R. SWANSON, *Mathematical Modeling of the Growth and Control of Tumors*, Ph.D. Thesis, University of Washington, Seattle, 1999.

[263] A. E. TAYLOR AND W. R. MANN, *Advanced Calculus*, 2nd ed., Xerox College Publishing, Lexington, MA, 1972.

[264] D. TAVELLA AND C. RANDALL, *Pricing Financial Instruments: The Finite Difference Method*, John Wiley, New York, NY, 2000.

[265] H. M. TAYLOR AND S. KARLIN, *An Introduction to Stochastic Modeling*, 3rd ed., Academic Press, New York, NY, 1998.

[266] C. TIER AND F. B. HANSON, *Persistence in Density Dependent Stochastic Populations*, Math. Biosci., vol. 53, 1981, pp. 89–117.

[267] H .C. TIJMS, *Stochastic Modelling and Analysis: A Computational Approach*, John Wiley, New York, NY, 1986.

[268] D. M. TILBURY AND W. C. MESSNER, *Control Tutorials for Software Instruction over the World Wide Web*, IEEE Trans. Automatic Control, vol. 42, no. 4, 1999, pp. 237–246.

[269] H .C. TUCKWELL, *Stochastic Processes in the Neurosciences*, SIAM, Philadelphia, PA, 1989.

[270] H .C. TUCKWELL, *Elementary Applications of Probability Theory*, Chapman and Hall, London, UK, 1995.

[271] M. TURELLI, *Random Environments and Stochastic Calculus*, Theor. Population Biol., vol. 12, 1977, pp. 140–178.

[272] W. T. VETTERLING, S. A. TEUKOLSKY, W. H PRESS, AND B. P. FLANNERY, *Numerical Recipes: Example Book* (C++), 2nd ed., Cambridge University Press, Cambridge, UK, 2002.

[273] J. VON NEUMANN, *Various Techniques Used in Connection with Random Digits*, in Applied Mathematics Series, vol. 12, U. S. National Bureau of Standards, 1951, pp. 36–38.

[274] J. J. WESTMAN AND F. B. HANSON, *The LQGP Problem: A Manufacturing Application*, in Proceedings of the American Control Conference, 1997, pp. 566–570.

[275] J. J. WESTMAN AND F. B. HANSON, *Nonlinear State Dynamics: Computational Methods and Manufacturing Example*, International Journal of Control, 2000, vol. 73, pp. 464–480.

[276] J. J. WESTMAN AND F. B. HANSON, *State Dependent Jump Models in Optimal Control*, in Proceedings of the 38th IEEE Conference on Decision and Control, 1999, pp. 2378–2383.

[277] J. J. WESTMAN AND F. B. HANSON, *Nonlinear State Dynamics: Computational Methods and Manufacturing Example*, Int. J. Control, vol. 73, 2000, pp. 464–480.

[278] J. J. WESTMAN AND F. B. HANSON, *MMS Production Scheduling Subject to Strikes in Random Environments*, in Proceedings of the American Control Conference, American Automatic Control Council, Dayton, OH, 2000, pp. 2194–2198.

[279] J. J. WESTMAN, F. B. HANSON, AND E. K. BOUKAS, *Optimal Production Scheduling for Manufacturing Systems with Preventive Maintenance in an Uncertain Environment*, in Proceedings of the American Control Conference, American Automatic Control Council, Dayton, OH, 2001, pp. 1375–1380.

[280] N. WIENER, *Differential Space*, J. Math. Phys., vol. 2, 1923, pp. 132–174.

[281] N. WIENER, *Generalized Harmonic Analysis*, Acta Math., vol. 55, 1930, pp. 117–258.

[282] P. WILMOTT, S. HOWISON AND J. DEWYNNE, *The Mathematics of Financial Derivatives: A Student Introduction*, Cambridge University Press, Cambridge, UK, 1996.

[283] P. WILMOTT, *Derivatives: The Theory and Practice of Financial Engineeing*, vols. 1 & 2, John Wiley, New York, NY, 2000.

[284] P. WILMOTT, *Paul Wilmott on Quantitative Finance*, John Wiley, New York, NY, 1998.

[285] S. WOLFRAM, *Mathematica: A System for Doing Mathematics by Computer*, Addison-Wesley, Reading, MA, 1988.

[286] W. M. WONHAM, *Random Differential Equations in Control Theory*, in Probabilistic Methods in Applied Mathematics, vol. 2, Academic Press, New York, NY, 1970, pp. 131–212.

[287] YAHOO! FINANCE, *Historical Prices, S & P 500 Index Symbol ^GSPC*, June 2007, http://finance.yahoo.com/q/hp?s=%5EGSPC.

[288] G. YAN, *Option Pricing for a Stochastic-Volatility Jump-Diffusion Model*, Ph.D. Thesis in Mathematics, University of Illinois, Chicago, IL, 2006, pp. 1–128.

[289] G. YAN AND F. B. HANSON, *Option Pricing for a Stochastic-Volatility Jump-Diffusion Model with Log-Uniform Jump-Amplitudes*, Proceedings of the 2006 American Control Conference, American Automatic Control Council, Dayton, OH, 2006, pp. 2989–2994.

[290] J. YONG AND X. Y. ZHOU, *Stochastic Controls: Hamiltonian Systems and HJB Equations*, Springer-Verlag, New York, NY, 1999.

[291] Z. ZHU, *Option Pricing and Jump-Diffusion Models*, Ph.D. Thesis in Mathematics, University of Illinois, Chicago, IL, 2005, pp. 1–165.

[292] Z. ZHU AND F. B. HANSON, *A Monte-Carlo Option-Pricing Algorithm for Log-Uniform Jump-Diffusion Model* in Proceedings of the 44th IEEE Conference on Decision and Control and European Control Conference, 2005, pp. 5221–5226.

[293] Z. ZHU AND F. B. HANSON, *Optimal Portfolio Application with Double-Uniform Jump Model*, in Control Theory Applications in Financial Engineering and Manufacturing, International Series in Operations Research and Management Sciences, Springer/Kluwer, New York, NY, 2006, pp. 331–358.

Index

Note: Page numbers preceded by an A, B, or C refer to pages in the corresponding appendices found online at www.siam.org/books/dc13

$\sim$, asymptotic, B50

./, element-wise division, 229

.$*$, element-wise multiplication, 184

$\mathbf{1}_S$, indicator function, 239, 262, 272

$\mathbf{1}_{x \in A}$, indicator function, B53

$(2k-1)!!$, double factorial, 7

$(f * g)(x)$, convolution, B31

$\Delta P(t)$, Poisson process increment, 12

$\Delta P(t; Q)$, marked Poisson increment, 132

$\Delta W(t)$, Wiener process increment, 3

$\Delta \Lambda(t)$, jump count increment, 20

Δ_h, h-jump, 194

$\delta P(t; Q)$, marked Poisson differential, 131, 132

$\delta(x)$, Dirac delta function, B53

$\delta_{i,j}$, Kronecker delta, B19

$\delta_R(x)$, right-continous (RC) delta function, B58

$\eta_3[X]$, skewness coefficient, B5

$\eta_4[X]$, kurtosis coefficient, B5

$\Gamma(x)$, gamma function, 6, B51

$\Lambda(t)$, time-dependent jump-count, 20

λ, Poisson jump rate, 2, B19

$\lambda(t)$, time-dependent jump-rate, 20

μ, mean, B4

$\Phi(x)$, probability distribution, B2, B3

$\Phi_{dP(t)}(k; \lambda dt)$, differential simple Poisson distribution, 13

$\Phi_e(t; \mu)$, exponential distribution, B15

$\Phi_{G(t)}(x)$, Gaussian distribution, B10

Φ_{ln}, lognormal distribution, B11

$\Phi_n(x; \mu, \sigma^2)$, normal distribution, B8

$\Phi_{P(t)}(k; \Lambda(t))$, temporal Poisson distribution, 21

$\Phi_{P(t)}(k; \lambda t)$, simple Poisson distribution, 13

$\Phi_u(x; a, b)$, uniform distribution, B5

$\Phi_{X,Y}(x, y)$, continuous joint distribution, B21

$\Phi_{X,Y}(x, y_j)$, mixed joint distribution, B21

$\Phi_{\Delta P(t)}(k; \Delta \Lambda(t))$, temporal Poisson increment distribution, 21

$\Phi_{\Delta P(t)}(k; \lambda \Delta t)$, simple Poisson increment distribution, 13

$\Phi_{\Delta T_j | T_j}(\Delta t)$, interjump time distribution, 23

$\Phi_{dP(t)}(k; d\lambda(t))$, differential Poisson distribution, 22

$\Phi_{dP(t)}(k; \lambda(t)dt)$, differential Poisson distribution, 21

$\Phi_{X|Y}(x|y)$, conditional distribution, B25

$\phi(x)$, probability density, B3

$\phi_{dW(t)}(w)$, Wiener process differential density, 4

$\phi_e(t; \mu)$, exponential density, B15

ϕ_{ln}, lognormal distribution, B11

$\phi_n(x; \mu, \sigma^2)$, normal density, B8

ϕ_{sn}, secant-normal density, 156

$\phi_u(x; a, b)$, uniform density, B5

$\phi_{W(t)}(w)$, Wiener process density, 3

$\phi_{X,Y}(x, y)$, continuous joint distribution, B21

$\phi_{X,Y}(x, y_j)$, hybrid joint density-distribution, B22

$\phi_{\Delta W(t)}(w)$, Wiener process increment density, 4

$\phi_{X|Y}(x|y)$, conditional density, B27

π_k, discrete distribution, B17

$\pi_{X,Y}(x_i, y_j)$, discrete joint distribution, B21

$\pi_{X,Y}(x_i, y_j)$, joint discrete distribution, B24

$\pi_{X|Y}(x|y)$, conditional discrete distribution, B25

$\rho_{i,j}$, correlation coefficient, 159, B45

σ, standard deviation, B4

σ-algebra, 363

σ^2, variance, B4

$\tau_e(x_0, t_0)$, exit time, 208

A^{-1}, inverse, B41

$A^\top$, transpose, B40

$F_\mathbf{x}(\mathbf{x})$, gradient, B43

$[F](X(t), t)$, jump in F, 103

$\int h(P)dP$, jump-integral, 68

$\int h(X, t)dP$, jump-integral, 70

$\mathcal{B}_x$, backward operator, 193, 194, 198

$\binom{n}{k}$, binomial coefficient, B47

$\mathcal{F}_x$, forward operator, 198

$\frac{\partial F}{\partial \mathbf{x}}(\mathbf{x})$, gradient, B43

$\overset{dt}{=}$, equals in dt-precision, 42

$\overset{dt}{\underset{ms}{=}}$, equals in mean square dt-precision, 42

$\overset{gen}{=}$, generalized equality, B54

$\overset{ims}{=}$, Itô mean square equals, 37, 63

$\overset{ims}{\underset{sym}{=}}$, symbolically equals in Itô mean square, 42

$\| A \|_p$, matrix p-norm, B41

$\| \mathbf{x} \|$, vector norm, B41

$f^{(m)}(x)$, mth order derivative, B63

$k = m_1 : m_2$
  colon notation, B17
  definition, B17
  loop notation, B7, B17, B63

$\lfloor x \rfloor$, floor function, B57

$\nabla_\mathbf{x}[F](\mathbf{x})$, gradient, B43

$x!$, factorial function, 6, B51

a.e. (almost everywhere), 365

ABM (arithmetic Brownian motion), 384

absolute extrema, B68

absolutely continuous, 366

absorbing boundary conditions, 205

acceptance-rejection method, 14, 270

adapted, 41

additive
  Brownian motion (ABM), 291
  noise, 94, 95, 291

adjoint
  formal, 201
  operator, 198
  state, A3, A7

algebra
  matrix, B39

algorithm
  Box–Muller, 273

almost
  everywhere (a.e.), 365
  surely (a.s.), 369

amplitude
  jump, 104

analysis, B1
  matrix, B39

antithetic variates, 279

applications
  biology, 339
  biomedicine, 347
  financial engineering, 287
  financial mathematics, 287
  mathematical biology, 339

approximation
  diffusion, 202, 349
  normal, B19, B48
  piecewise-constant, 47
  Poisson, B48
  Taylor, B63
  Taylor remainder, B63

arbitrage
  profits, 296

arithmetic Brownian motion (ABM), 384

asymptotic
  $\sim$, B50
  expansion, B50
  notation, B49
    O, B49
    o, B50
    ord, B50
    $\sim$, B50
  sequence, B50

backward
  equation, 193
  Euler's method, 33, 257
  finite difference (BFD), 228
  Kolmogorov equation, 196
  operator, 193, 194, 196, 198
bang control, A20
bang-bang control, 344
bang-regular-bang control, 342, 346
bankruptcy, 317
Bayes' rule, B70
Bellman
  principle of optimality, 172, 175,
    A30
Bellman's curse of dimensionality, 230
Bernoulli
  distribution, 20
  trials, B47
BFD (backward finite difference), 228
bias, 262
big oh, O, B49
binomial
  coefficient, B47
  distribution, B47, B73
    Bernoulli trials, B47
    normal approximation, B48
    Poisson approximation, B48
  expansion, B47
  expectation, B48
  $\binom{n}{k}$, B47
  $p(f_1, f_2; \pi_1, \pi_2)$, B47
  theorem, B47
biomass, 340, 344
biomedical
  applications, 347
  variability, 347

bivariate normal
  density, B46
  distribution, B46, B73
Black–Scholes
  fraction, 298
  Merton fraction, 298
  model, 288
  option pricing PDE, 300
  PDE, 290
Black–Scholes–Merton model, 291
Bolza form, A2
bond, 321
boundary
  absorbing, 322
  extrema, B68
  regular, 352
  singular, 352
boundary condition
  absorbing, 205
  derivative, 227
  no-flux, 227
  reflecting, 206
boundary point, A2
Box–Muller algorithm, 273
brain tumors, 353
branching process, 347
Brownian motion, 2
  additive, 291
  geometric, 97, 289
budget equation, 297

calculus of variations, A3–A5, A7
  fundamental lemma, 201, A6
  fundamental theorem, A6
call option, 288, 292
cancer
  drug, 353
  growth, 347
carrying capacity, 340
catchability, 340
Cauchy–Binet formula, B42
Cauchy–Schwarz inequality, 49, B69
centered Poisson process
  characteristic function, 394
central
  finite difference (CFD), 223, 228,
    229

limit theorem, B39
moments
   fourth, B5
   kurtosis, B5
   second, B4
   skew, B5
   third, B5
   variance, B4
CFD (central finite difference), 223,
   228, 229
chain rules
   diffusion $F(X(t), t)$, 98
   $G(W(t))$, 85
   $G(W(t), t)$, 86
   $\mathcal{H}(P(t), t)$, 101, 102
   Itô, 85
   jump $F(X(t), t)$, 103
   jump-diffusion, 106
change of index, 36
Chapman–Kolmogorov equation, 205
characteristic function, 392, B33
   centered Poisson process, 394
   compound Poisson process, 395
   Gaussian process, 394
   jump-diffusion process, 396
   Poisson process, 394
   Wiener process, 393
chattering control, 184
cheap control, 344
Chebyshev inequality, 38, 214, 349, B69
closed-loop control, A24
coefficient
   binomial, B47
   correlation, 159, B45
   diffusion, B10
   excess kurtosis, B5
   kurtosis, B5
   skewness, B5
   variation, 111
coefficient of variation, 111, 209
colon notation, $k = m_1 : m_2$, B17
colored noise, 120
completing the square, 91, 110, B13,
   B30, B73
complexity
   computational, A12

compound Poisson process, 131, 135,
   136, 392
   characteristic function, 395
computational
   complexity, A12
   cost, A12
   method
      stochastic dynamic
         programming, 219
      stochastic simulations, 241
concave, A9
$\text{cond}_p[A]$, condition number, B41
condition
   Legendre–Clebsch, A8
   necessary, A3
   number, B41
   sufficient for optimum, A8
conditional
   density, B27
   distribution, B25
   exit, 212
   expectation, B27
   infinitesimal moments, 141
   probability, B25
confidence interval, 264
confidence level, 264
conjunct, 199, 201
conservation
   probability, B6, B15, B17
constant of integration
   genuine, 90
consumption
   optimal, 325
   regular, 325
   wealth, 321
continuity, B61
continuous
   absolutely, 366
control, 169
   bang, A20
   bang-bang, 344
   bang-regular-bang, 342, 346, A13
   bang-singular-bang, A21
   closed-loop, A24
   corner, A20
   deterministic, A1

distributed parameter, 353, A34
drug delivery, 353
feedback, A24
impulse, A21
law, A24
normal, A8
open loop, A24
optimal, 169, A2, A4
reaction-diffusions, 353
regular, A3, A8, A9, A12
stochastic, 169
variates, 281
convergence
hierarchy, 37
mean, 37
mean square, 37
probability, 37
convex, A9
function, B70
convolution, B30
density, B31
normal densities, B72
uniform densities, B72
corner, A20
correlation, 159
coefficient, B45
$\rho_{i,j}$, 159, B45
bounds, B45
cosine transform, B34
cost
Bolza form, A2
instantaneous, A2
Lagrangian, A2
objective, A2
running, A2
terminal, A2
costate, A3, A7
counting
measure, 364
process, 2, 11
Cov[X, Y], covariance, B23
covariance, B23
Cov[X, Y], B23
$P(t)$, 16
$W(t)$, 4
Crank–Nicolson method, 223

critical point, A2, A8, B68
crude Monte Carlo method, 274
cumulative Poisson count, 269
current value, 342, 345
curse of dimensionality, 230
CV (coefficient of variation), 111
$\mathcal{C}_X(u)$, characteristic function, B33
$\mathcal{C}_X(z)$, characteristic function, 392

DDE (differential-difference equation),
        27
decomposition
Lévy–Itô, 398
decomposition rule
integration, A29
minimization, A29
delta
function
$\delta(x)$, B53
$\delta_R(x)$, B58
Dirac, B53
Kronecker, B19
hedge, 312
hedging, 290, 300
portfolio, 290
delta-correlated, 122
Gaussian white noise, 122
Poisson white noise, 122
delta-density, 25
$\phi_{dW(t)}(w)$, 25
demographic stochasticity, 97
density
$\phi_{\Delta T_j | T_j}(\Delta t)$, 23
$\phi_{\Delta W(t)}(w)$, 4
$\phi_{dW(t)}(w)$, 4
$\phi_e(t; \mu)$, B15
ϕ_{ln}, B11
$\phi_u(x; a, b)$, B5
$\phi_n(\mathbf{x}; \boldsymbol{\mu}, \boldsymbol{\Sigma})$, B45
$\phi_n(x; \mu, \sigma^2)$, B8
$\phi_{T_j | T_{j-1}}(\Delta t)$, 23
$\phi_{W(t)}(w)$, 3
$\phi(x)$, B3
$\phi_{X,Y}(x, y)$, B21
$\phi_{X|Y}(x|y)$, B27
bivariate normal, B46

conditional, B27
convolution, B31
differential Wiener process, 4
exponential, B15
joint, B21
lognormal, B11
marginal, B22
  $\phi_X(x)$, B23
  $\phi_Y(y)$, B23
multidimensional, 205
multivariate normal, B45
normal, B8
  product, B71
Poisson process, B58
probability, B3
secant-normal, 156, 157, 320
  ϕ_{sn}, 156
sum, B31
total probability, B29
transition, 197
triangular, 320
uniform, B5
Wiener process, 3
Wiener process increment, 4
density-distribution
  hybrid
    $\phi_{X,Y}(x, y_j)$, B22
derivative
  boundary condition, 227
  $f^{(m)}(x)$, B63
  partial, 86
Det[A], determinant, B41
determinant, B41
deterministic
  control, A1
  dynamic programming, A30
  integration
    Riemann integral, 33
  optimal control, A1
  process, B59
differential, B61
  mean square, 42
  process
    $dW(t)$, 2
    Wiener, 2
    Wiener density, 4

differential-difference equation, 27
diffusion
  approximation, 202, 213, 233, 347,
    349
  coefficient, 94, B10
  differential $dW(t)$, 2
  equation, B10
  increment $\Delta W(t)$, 3
  process, 1
diffusion, $W(t)$, 1
diffusion-dominated, 227
dimensional analysis, 304
dimensionless groups, 304
Dirac delta function, B53
discontinuity
  jump, B61
discontinuous process
  Poisson, 12
discount factor, 190
discount rate, 322, 331
  cumulative, 322, 331
discrete
  distribution, B17
    π_k, B17
  Hessian, 230
distributed parameter systems (DPS),
  353, A34
distribution
  $p_k(\Lambda)$, B18
  $\Phi(x)$, B2, B3
  $\Phi_{dP(t)}(k; \lambda dt)$, 13
  $\Phi_e(t; \mu)$, B15
  $\Phi_{G(t)}(x)$, B10
  Φ_{ln}, B11
  $\Phi_n(x; \mu, \sigma^2)$, B8
  $\Phi_{P(t)}(k; \Lambda(t))$, 21
  $\Phi_{P(t)}(k; \lambda t)$, 13
  $\Phi_u(x; a, b)$, B5
  $\Phi_{X,Y}(x, y)$, B21
  $\Phi_{X,Y}(x, y_j)$, B21
  $\Phi_{\Delta P(t)}(k; \Delta\Lambda(t))$, 21
  $\Phi_{\Delta P(t)}(k; \lambda\Delta t)$, 13
  $\Phi_{\Delta T_j | T_j}(\Delta t)$, 23
  $\Phi_{dP(t)}(k; d\Lambda(t))$, 22
  $\Phi_{dP(t)}(k; \lambda(t)dt)$, 21
  $\Phi_{X|Y}(x|y)$, B25

π_k, B17
$\pi_{X,Y}(x_i, y_j)$, B21, B24
$\pi_{X|Y}(x|y)$, B25
Bernoulli, 20
binomial, B47, B73
bivariate normal, B46, B73
conditional, B25
discrete, B17
exponential, B14, B15
Gaussian, B8, B10
independent random variable, B24
infinitely divisible, 391
joint, B20
lognormal, 97, 98, 112, B11
marginal, B22
    $\Phi_X(x)$, B22
    $\pi_X(x_i)$, B22
    $\pi_Y(y_j)$, B22
multinomial, B46
multivariate, B45
multivariate normal, B45
normal, B8
piecewise quadratic, B72
Poisson, 13, B18
Poisson process, B19
probability, B2
proper, B2
right-continuous, B59
sum, B31
total probability, B29
transition, 197
triangular, B72
uniform, 319, B5
$d\Lambda(t)$, jump count differential, 20
dot product, B40
    double $A : B$, 163
double factorial, 7
    $(2k - 1)!!$, 7
double-dot product
    $A : B$, 163, 178, 179, 181
doubling time, 350
DPS (distributed parameter systems),
    353, A34
$dP(t)$, Poisson process differential, 2
$dP(t; Q)$, compound Poisson process
    differential, 134

drift, 94, 104, B10
drug delivery, 353
$dW(t)$, Wiener process differential, 2
dynamic programming, 177, A30
    Bellman's, A28
    deterministic, A28
Dynkin's
    equation, 208
    formula, 193, 208

eigenvalue, B43
element-wise
    division ./, 229
    multiplication, .∗, A26
environmental stochasticity, 97
equation
    Bernoulli, 326
    growth, 340
    HJB (Hamilton–Jacobi–Bellman),
        176
    logistic, 340, 344
equivalence
    quasi-deterministic, 111, 146
equivalent
    martingale measures, 372
    measure, 366
erf, error function, B8
erfc, complementary error function, B8
Euler's
    explicit method, 33
    method, 33, 244
        jump-adapted, 257
Euler–Maruyama method, 244
European option, 288, 292
$E[X]$, expectation, B4, B18
exercise price, 288, 292
exit
    conditional, 212
    time, 208
    times, 206
expansion
    asymptotic, B50
    binomial, B47
    multinomial, B49

expectation, B4, B18, B23
  μ, B4
  binomial, B48
  conditional, B27
    $\mathcal{P}$, 140
  continuous random variables, B4
  discrete set, B18
  $dP(t; Q)$, 134
  $E[X]$, B4, B18
  $E[X|Y]$, B27
  $E_{X,Y}[f(X, Y)]$, B23
  iterated, B28
  joint random variables, B23
  $\mathcal{P}(\mathbf{dt}, \mathbf{dq})$, 133
  total probability, B29
expectation-integration interchange, 56
expected
  exit times, 206
  rate of return, 293
  return, 293
expiration
  condition, 299
  time, 288, 292
exponential
  $\Phi_e(t; \mu)$, B15
  $\phi_e(t; \mu)$, B15
  density, B15
  distribution, B14, B15
  growth, 340, 346
  random variables, 268
  RNG, 268
  series, B19
    Poisson distribution, B19
  stochastic, 88
exponentially distributied
  interjump time, 13
extrema, B67
  boundary, B68
  local, B68
  relative, B68
extremal principle, B67
$E[X|Y]$, conditional expectation, B27
$E_{X,Y}[f(X, Y)]$, joint expectation, B23

factorial
  function, 6, B51
  $x!$, 6

fat tails, 309
feedback
  control, A24
  gain, A24
Feynman–Kac formula, 216
FFD (forward finite difference), 228
final
  condition, 172, 299, A7
  value problem (FVP), 299
financial engineering
  applications, 287
financial mathematics
  applications, 287
finite
  difference
    backward, 228
    central, 223, 228, 229
    forward, 228
    upwind, 228
  difference methods, 220
  element method, 228
first passage time, 206
floor function, $\lfloor x \rfloor$, B57
Fokker–Planck equation, 199
formula
  Cauchy–Binet, B42
  Dynkin's, 193
  Feynman–Kac, 216
  Lévy–Klintchine, 396
  Stirling's, B52
  Taylor's, B63
forward
  equation, 193
  finite difference (FFD), 228
  integration, 33
  Kolmogorov equation, 198
  operator, 198, 201
Fourier transform, B33
fourth central moment, B5
fractional sampling, 278
Fubini's theorem, B31
function
  $(2k - 1)!!$, 7
  $\Gamma(x)$, B51
  $\delta_R(x)$, B58

Dirac delta, B53
double factorial, 7
erf, B8
erfc, B8
factorial, 6, B51
floor, B57
gamma, 6, B51
generalized, B52
$H(x)$, B31, B52
impulse, B53
indicator, B53
jump, 71, B62
min, 4, 6, 9
of integration, 89
step, B52
$x!$, 6, B51
$\lfloor x \rfloor$, B57
functional PDE, 178
FVP (final value problem), 299
fzero, MATLAB zero finder, A11

gain
feedback, A24
gamma
function, 6, B51
Gauss-statistics rules, 224
Gaussian
$G(t)$, B10
$\Phi_{G(t)}(x)$, B10
distribution, B8, B10
process, 391, B8, B10
characteristic function, 394
white noise, 120
GBM (geometric Brownian motion), 97,
289, 384
general Markov SDEs, 129
generalized
equality, B54
$\overset{gen}{=}$, B54
function, B52
right-continuous step-function, B57
generating function, B34
geometric
Brownian motion (GBM), 97, 289,
384
Lévy process, 400

Girsanov
example, 387
Girsanov's theorem, 382
diffusion, 382
jump-diffusion, 385
global
extrema, B68
maximum, B67
minimum, B67
Gompertz model, 350
gradient, B43
matrix-vector product, B43
quadratic form, B44
Greenspan process, 328
growth
logistic, 340, 346

h-jump, 194, 196
Hamilton's
equations, A2, A3
Hamilton–Jacobi–Bellman (HJB), 176
Hamiltonian, A3
current value, 342, 345
distributed parameter, A35
present value, 342, 345
harvest effort, 340
Heaviside step function, B52
Hessian
discrete, 230
matrix, B44
quadratic form, B44
hist, MATLAB histogram, B6
histogram, B6
HJB (Hamilton–Jacobi–Bellman)
equation, 176
homogeneous options, 298
$H(x)$, step function, B31, B52, B53
$H_R(x)$, RC step function, B57
hybrid
density-distribution
$\phi_{X,Y}(x, y_j)$, B22
stochastic system, B59

identity, B40
IID (independent identically
distributed), B37
ill-conditioned, B42

importance sampling, 259, 274
   VEGAS, 276
impulse
   control, A21
   function, B53
increment, B61
   independent, B60
independent, B24
   increments, 3, 132, B60
      $\Delta P(t)$, 12
      Poisson process, 12
      Wiener process, 3
   random variable, B24
indicator function, B53
   $\mathbf{1}_{x \in A}$, B53
   $\mathbf{1}_S$, 239, 262, 272
inequality
   Cauchy, B41
   Cauchy–Schwarz, 49, B69
   Chebyshev, 38, 214, 349, B69
   Jensen's, B70
   Schwarz, 49, B69
   triangular, B41
infinitely divisible distribution, 391
infinitesimal mean
   diffusion, 94
   $E[dX(t) \mid X(t) = x]$, 106
   jump exponent, 105
   jump-diffusion, 106
infinitesimal variance
   diffusion, 94
   jump exponent, 105
   jump-diffusion, 107
   $Var[dX(t) \mid X(t) = x]$, 107
instantaneous cost, A2
integral
   Itô, 34, 40
   Riemann–Stieltjes, 34
   Stieltjes, 34
integration
   additive decomposition rule, A29
   Itô, 34, 40
   Riemann, 32
   rule
      left rectangular, 33
      midpoint, 33

      right rectangular, 33
      trapezoidal, 60
   Stieltjes, 34
interarrival time, 13, B20
   $T_{k+1} - T_k$, B20
interjump time, 13
   exponentially distributed, 13
   simulation, 14
intrinsic growth rate, 340
inverse, B41
   method, 268
   transformation method, 268, B15
inverse Poisson method, 269
i-PWCA (piecewise constant
      approximations), 47
isometry
   Itô, 51, 77
   martingale, 51, 77
iterated
   expectation, 149, B28
   probability, B26
Itô
   formula, 86
   lemma, 86
   lemma with jumps, 109
   mean square differential, 42
   process, 389
   stochastic chain rule, 85, 86
   stochastic integral, 40
Itô–Lévy process, 400
Itô–Taylor expansion, 249

Jacobian, 274
Jensen's inequality, B70
JLQG (jump linear quadratic Gaussian),
      179, 184
joint
   density, B21
   $\phi_{X,Y}(x, y)$, B21
   distribution, B20
   $\Phi_{X,Y}(x, y)$, B21
   $\Phi_{X,Y}(x, y_j)$, B21
   $\pi_{X,Y}(x_i, y_j)$, B21
   probability, B21
   $Prob[X \leq x, Y \leq y]$, B21
   $Prob[X = x_i, Y = y_j]$, B21
   $Prob[X \leq x, Y = y_j]$, B21

jump, B62
$\Delta P(t)$, 12
amplitude, 104
condition, 333
counter
indicator function, 108
discontinuity, B61, B62
function, 71, B62
functions, 100
$[h](dP(t))$, 100
integral
$\int (P dP)(t)$, 65
number, 398
process, 1
$\Delta P(t)$, 12
$d P(t)$, 2
$P(t)$, 1
time, B20
time, T_j, 13
time, T_k, B20
jump-adapted Euler method, 257
jump-diffusion, 106
chain rule, 109
density, 112, 152
process, 81, 106
characteristic function, 396
state dependent, 139
jump-risk, 312
jump-time
simulation, 14

Kolmogorov
equation
backward, 193, 196, 203
forward, 193, 198, 203
multidimensional, 203
Kronecker delta, B19
$\delta_{i,j}$, B19
kurtosis, B5
coefficient, B5
$\eta_4[X]$, B5

Lagrange multiplier, A3
Lagrangian, A2
Laplace's method, B51

large numbers
law, B38
law
control, A24
large numbers, B38
power, A8
total probability, 111, B29
transformation of probabilities, B15
zero-one jump, 19
LCRL (left continuous, right limits), 381
Legendre–Clebsch conditions, A8
leptokurtic, 309, B5
Lévy
diffusions, 400
jump number, 398
process, 390, 400, B36
characteristic exponents, 393
characteristic functions, 393
definition, 390
geometric, 400
SDEs, 400
Lévy-driven process, 400
Lévy-driven processes, 399
Lévy–Itô decomposition, 398
Lévy–Klintchine
formula, 396
geometric jump-diffusion, 397
jump-diffusion, 396
Lévy process, 399
representation, 396
$\lim_{n\to\infty}^{\text{mean}}$, limit in the mean, 37
$\lim_{n\to\infty}^{\text{ms}}$, mean square limit, 37, 63
limit
in probability, 37
$\lim_{n\to\infty}^{prob}$, 37
in the mean, 37
$\lim_{n\to\infty}^{\text{mean}}$, 37
mean square, 37
$\lim_{n\to\infty}^{\text{ms}}$, 37
linear
discrete mark-jump-diffusion
expectation, 143
mark-jump
expectation, 143
mark-jump-diffusion
simulation, 143

quadratic
  problem, A22, A32
  SDEs
    diffusion transformations, 94
    jump-diffusion, 109
    jump transformations, 104
linear quadratic
  jump-diffusion
    problem, 179, 184
little oh, o, B50
local
  extrema, B68
  maximum, B68
  minimum, B68
local optimum, A2
log, MATLAB natural logarithm, 14
logistic
  equation, 340, 344
  growth, 340
lognormal, 97
  density, B11
    ϕ_{ln}, B11
  distribution, 98, 112, B11
    Φ_{ln}, B11
  mean/mode, B70
  mode, B70
  moment, B13
  random number generator, B14
loop notation, $k = m_1 : m_2$, B17
LQ (linear quadratic), A22, A32
LQGP (linear quadratic Gaussian
        problem), 179, 184
LQJD (linear quadratic jump-diffusion),
        179, 221
LQJD/U (LQJD in control only), 221,
        229
lumped parameter systems, 353, A34

$\mathcal{M}_1(\mathcal{S}) \overset{\text{a.c.}}{\equiv} \mathcal{M}_1(\mathcal{S})$, equivalent measure,
        366
$\mathcal{M}_2(\mathcal{S}) \prec \mathcal{M}_1(\mathcal{S})$, absolutely
        continuous, 366
machine epsilon, 15
marginal
  density, B22
    $\phi_X(x)$, B23

$\phi_Y(y)$, B23
  distribution, B22
    $\Phi_X(x)$, B22
    $\pi_X(x_i)$, B22
    $\pi_Y(y_j)$, B22
mark space, 130
mark-time Poisson
  process, 130, 139
Markov
  chain, B60
  chain approximation (MCA), 231
  Poisson process, 12
  process, 2, B59, B60
    stationary, B61
  $P(t)$, 12
  Wiener process, 3
martingale, 10, 24, 77, 372
  equivalent measures, 372
  methods, 361
  representation theorem, 376
  sub-, 372
  super-, 372
mathematical biology
  applications, 339
matrix
  algebra, B39
  analysis, B39
  equality, B39
  Hessian, B44
  identity, B39
  notation, B39
  positive definite, B44
  positive semidefinite, B44
maxima, B67
maximum
  local, B68
  necessary condition, A3, A15
  principle
    Pontryagin, A14
  relative, B68
  sufficient condition, A8
MCA (Markov chain approximation),
        231
  consistency conditions
    diffusion, 233
    jump-diffusion, 238

mean, B4, B18
   μ, B4
   continuous random variables, B4
   convergence, 37
   discrete set, B18
   $E[X]$, B18
   sample, B37
mean square
   convergence, 37
   deviation, B4
   differential, 42
     $(dP\,dW)(t)$, 72
     $(dP)^m(t)$, 78
     $dt\,dP(t)$, 72
     $(dt)^k((dP)^m(dW)^n)(t)$, 74
     $(dW)^2(t)$, 42
     $(W\,dW)(t)$, 42
   limit, 37
     $\lim^{\mathrm{ms}}_{n\to\infty}$, 37
     $\mathcal{P}(\mathbf{dt}, \mathbf{dq})$, 136
   random measure
     $\mathcal{P}(\mathbf{dt}, \mathbf{dq})$, 136
measurable
   function, 366
   space, 363
measure, 362, 363
   absolutely continuous, 366
   counting, 364
   equivalence, 366
   space, 363
   total mass, 364
Merton fraction
   Black–Scholes, 298
   volatility, 298
mesokurtic, 309
method
   acceptance-rejection, 14, 270
   backward Euler's, 33, 257
   change of index, 36
   characteristics, 302
   completing the square, 91, 110,
     B13, B30, B73
   computational, 219
   Crank–Nicolson, 223
   Euler's, 33, 244
   Euler–Maruyama, 244

finite element, 228
forward integration, 33
inverse, 268
inverse transformation, B15
Laplace's, B51
left rectangular rule, 33
midpoint rectangular rule, 33
Milstein's, 248, 249
rejection, 14
right rectangular rule, 33
simulations, 241
stochastic Euler, 244
stochastic theta, 254, 257
tangent-line, 33
trapezoidal rule, 60
Metropolis algorithm, 258, 259
midpoint rule, 33, 223
Milstein's method, 248, 249
min, minimum, 4, 9
minima, B67
minimization decomposition rule, A29
minimum
   function
     min, 4, 9
   local, B68
   necessary condition, A3, A15
   principle
     Pontryagin, A14
   relative, B68
   sufficient condition, A8
MISER sampling code, 279
m_n, sample mean, B37
mode, B70
   most likely value, B70
model
   Black–Scholes, 288
   Black–Scholes–Merton, 291
   Gompertz, 350
   option pricing, 288
moment
   log-jump-diffusion, 147
   lognormal, B13
Monte Carlo
   acceptance-rejection, 270
   antithetic variates, 279
   basic method, 260

computational costs ratio, 267
confidence interval, 264
control variates, 281
efficiency, 267
estimator, 261, 275
importance sampling, 274
inverse method, 268
methods, 258
MISER, 279
stratified sampling, 276
variance ratio, 267
VEGAS, 276
most likely value, B70
multinomial
    distribution, B46
        multinomial expansion, B49
    expansion theorem, B49
    joint probability, B47
    $p(\mathbf{f}; \boldsymbol{\pi})$, B47
multiplicative noise, 94, 96, 97, 291, 329
multivariate
    distribution, B45
    normal
        density, B45
        distribution, B45

necessary conditions, A3
neutral risk, A9
no-flux boundary condition, 227
noise, 120
    additive, 94, 95
    colored, 120
    multiplicative, 94, 96, 97, 329
    white, 120
    white Gaussian, 120
    white Poisson, 120
    word, 94
nonanticipatory, 41
nondifferentiability, 9
    $W(t)$, 9
non-Gaussian process, 390
nonsmooth process, B63
nonsmoothness, 9, B61
nonstationary Poisson process, 20
norm, B41
    matrix, B41

normal
    approximation, B19, B48
    control, A8
    density, B8
        $\phi_n(x; \mu, \sigma^2)$, B8
        product, B71
    distribution, B8
        $\Phi_n(x; \mu, \sigma^2)$, B8
        skewless, B9
    random number generator (RNG),
        B9
    variate RNG, 273
null set, 365

O, big oh, B49
o, little oh, B50
ODE (ordinary differential equation)
        problem, 302
    deterministic, 302
    stochastic, 302
open loop control, A24
operator
    adjoint, 198
    backward, 193, 194, 196, 198
    forward, 198, 201
optimal
    control, 169
    portfolio, 321
    sampling allocation, 279
optimality
    principle, A29
optimum
    condition
        sufficient, A8
    necessary condition, A3
option
    Black–Scholes pricing PDE, 300
    call, 288, 292
    European, 288, 292
    exercise price, 288, 292
    expiration time, 288, 292
    pricing
        Bachelier, 291
    pricing model, 288
    pricing PDE, 299

put, 288, 292
   risk-neutral, 312
ord, same order, B50
orderliness, 19
orthogonal, B43
orthonormal, B43

parabolic mesh ratio, 227
   CFD, 227
   UFD, 228
partial derivatives, 86
partial differential equation (PDE), B10
   diffusion, B10
   driven dynamics, 353, A34
   option pricing, 299
   problem, 299
   stochastic dynamic programming,
      179
partial sum, B63
partial sum, S_n, B63
PDE (partial differential equation), B10
$\mathcal{P}(\mathbf{dt}, \mathbf{dq})$, Poisson
   random measure, 130
performance index, A2
$p(\mathbf{f}; \boldsymbol{\pi})$, multinomial joint probability,
   B47
$p(f_1, f_2; \pi_1, \pi_2)$, binomial distribution,
   B47
PIDCP (partial integral differential
      complementary problem), 317
PIDE (partial integral differential
      equation), 178, 179
piecewise-constant, 47
   approximation (PWCA), 47
   assumption, 47
$p_k(\Lambda)$, Poisson distribution, B18
$p_k(\lambda t)$, simple Poisson distribution, B19
plant function, A2
platokurtic, B5
plot, MATLAB plot function, 14
point
   interior critical, B68
   process, 11
   stationary, B68
Poisson
   approximation, B48

compound process, 392
counting process, 11, B29
distributed, 13
distribution, 13, B18, B19
   exponential series, B19
   $p_k(\Lambda)$, B18
distribution, $p_k(\lambda t)$, B19
inverse, 269
jump-rate, λ, B19
mark space, 130
point process, 11
$P(t)$, 1
random measure, 130, 398
   notation, 132
random measure, $\mathcal{P}(\mathbf{dt}, \mathbf{dq})$, 130
simple process, 392
white noise, 120
Poisson distribution
   Poisson random measure
      $\mathcal{P}(\Delta\mathbf{t}_i, \Delta\mathbf{q}_j)$, 133
Poisson process, 1, B18
   $\Delta P(t)$, 12
   $dP(t)$, 2
   $P(t)$, 1
   characteristic function, 394
   compound, 129, 131
   count, $\Delta\Lambda(t)$, 20
   density, B58
   differential, 2
   discontinuity, 12
   distribution, 13
   increments $\Delta P(t)$, 12
   interarrival time, 13, B20
   jump-rate, $\lambda(t)$, 20
   jump-time, 13, B20
      simulation, 14
   marked point process, 129
   Markov, 12
   nonstationary, 20
   orderliness, 19
   right-continuity, 12
   space-time, 129
   state-dependent, 129
   stationary process, 12
   temporal, 20
   time-inhomogeneous, 20

unit jumps, 12
zero-one jump law, 19
Pontryagin, A14
optimum principles, A15
portfolio
budget equation, 297
delta, 290
self-financing, 296
zero aggregate, 296
power law, A8
preliminaries, B1
present value, 342, 345
principle
maximum, A15
minimum, A15
of optimality, 172, A29
optimum, A15
Prob, probability, B2, B3, B12, B17,
B21, B25, B26
probability, B1
Prob, B2, B3, B12, B17, B21, B25,
B26
conditional, B25
conservation, B2, B6, B15, B17
convergence, 37
density, B3
$\phi(x)$, B3
distribution, B2
$\Phi(x)$, B2
inversion, 97, 111, B12
joint, B21
measure, 362, 368
nonnegativity, B17
iterated, B26
transition, 205, B60, B61
problem
canonical LQJD, 184
eigenvalue, B43
JLQG, 179, 184
linear quadratic, A22, A32
linear quadratic jump-diffusion,
179, 184
LQGP, 179, 184
LQJD, 179, 184
LQJD in control only, 180
LQJD/U, 180

process
branching, 347
Brownian motion, 2
compound Poisson, 392
continuous, B62
counting, 2
deterministic, B59
diffusion, 1
diffusion, $W(t)$, 1
discontinuous, 12, B62
Gaussian, 391, B8, B10
geometric Poisson, 329
Greenspan, 328
Itô, 389
Itô–Lévy, 400
jump, 1
jump, $P(t)$, 1
jump-diffusion, 81, 106
Lévy, 390, B36
Lévy-driven, 400
Markov, 2, B59, B60
stationary, B61
non-Gaussian, 390
nonsmooth, B63
Poisson, 1, 2, B18
compound, 135, 136
mark-time, 130
space-time, 130
quasi-deterministic, 328–330
random, 1, B59
right-continuous, B62
simple Poisson, 392
smooth, B63
stochastic, 1, B59
Wiener, 1, 2
product
dot, B40
double-dot, 163, 178, 179, 181,
B67
matrix-matrix, B40
matrix-vector, B40
trace, B67
programming
dynamic, 177, A30
stochastic dynamic, 177, A30
proper distribution, B2

proportional sampling, 277
pseudo-Hamiltonian, A30
pseudo-random number generator, B6
$P(t)$, Poisson process, 1, 2, B19
$P(t)$-stochastic integration, 63
$P(t; Q)$, marked Poisson process, 131
put option, 288, 292
PWCA (piecewise-constant
    approximation), 47, 64

quadratic form
    symmetry, B44
quasi-deterministic
    approximation, 343
    equivalence, 111, 146, 147
    process, 328–330

Radon–Nikodým
    derivative, 367, 377
    theorem, 367, 377
rand
    MATLAB uniform RNG, 14, B6
    'state', B8
randn
    MATLAB normal RNG, 5, B9
    'state', 5
random, 1
    bonanza, 106
    disaster, 106
    process, 1, B59
    seed, B8
    state, B8
    stochastic, 1
    variable
        continuous, B2
        discrete, B17
        IID (independent identically
          distributed), B37
        independent, B24
        sum, B30, B37, B71
random number generator (RNG), B6
    diffusion, 5
    exponential, B16
    jump-diffusion additive noise, 113
    linear diffusion SDE, 115
    linear jump SDE, 115

linear jump-diffusion SDE, 113
linear mark-jump-diffusion, 158,
    256
lognormal, B14
normal, B9
Poisson jump-time, 14
pseudo, B6
rand, B8
randn, B9
RNG, B6
seed, B8
simple Poisson process, 14
state, B8
uniform, B8
Wiener process, 5
x=exp(mu+sigma*randn),
    B14
zero-one jump law, 14
RC (right-continuous), B57
RCLL (right-continuous, left limits),
    371, 381, 398
reaction-diffusion equations, 353
recursive stratified sampling, 279
reflecting boundary conditions, 206
reflection principle, 206
regular boundary, 352
regular control, A3, A8, A9, A12
regular point, A2
rejection method, 14
    zero-one jump law, 14
relative extrema, B68
representation
    Lévy–Klintchine, 396
return
    expected rate, 293
Riccati equation, A24
    matrix, 188, A28, A33
    scalar, A24
Riemann
    integral, 33
    integration, 32
    sum, 33
Riemann–Stieltjes integral, 34
right-continuous (RC), B57
    distribution, B59
    Poisson process, 12

process, B62
    Poisson, 12
    $P(t)$, 12
risk, 293
    aversion, 325, 332
    more risky, 293, 336
    neutral, A9
    riskfree, 293, 294
    riskier, 293, 336
    riskiness measure, 293
    riskless, 294
risk-neutral, 312
    option, 312
risky asset, 293
RMS (root mean square), 2
RNG (random number generator), B6
    exponential, 268
    normal variate, 273
rule
    Bayes', B70
    left rectangular, 33
    midpoint, 33
    right rectangular, 33
    trapezoidal, 60
running cost, A2

sample
    mean, 261, 275, B37
        m_n, B37
    variance, B37
    variance, 275, B37
        s_n^2, B37
        unbiased estimate, B37
        unbiased estimate $\widehat{s}_n^2$, B37
sampling
    fractional, 278
    importance, 274
        VEGAS, 276
    optimal allocation, 279
    proportional, 277
    recursive stratified, 279
        MISER, 279
    stratified, 276
scale density, 351
Schwarz inequality, 49, B69
    $E[|\,XY\,|]$, 49
    expectation, 49

SDE (stochastic differential equation)
    backward Euler, 257
    diffusion, 94
    Euler's method, 244
    general Markov, 129
    jump, 104
    jump-diffusion, 106
    mark-time Poisson, 129
    Milstein's method, 248, 249
    multidimensional, 129
    problem, 299
    simulation, 241
    space-time Poisson, 129
    state-dependent, 129, 141
    Taylor expansion, 249
    theta method, 254
SDP (stochastic dynamic programming),
    176
secant-normal density, 156, 157
second central moment, B4
self-financing portfolio, 296
self-similar solution, 304
sequence
    asymptotic, B50
series
    exponential, B19
sgn (sign function), 184
shadow price, 343
simple function, 366
simulations
    SDEs, 241
simulation
    $(dW)^2(t)$, 43
    diffusion, 5
    Euler–Maruyama SDE, 244
    jump-diffusion additive noise, 113
    linear diffusion SDE, 115
    linear jump SDE, 115
    linear jump-diffusion SDE, 113
    linear mark-jump-diffusion, 143,
        158, 256
    Milstein SDE, 249
    Poisson jump-time, 14
    simple Poisson process, 14
    Wiener, 45

Wiener process, 5
    zero-one jump law, 14
Simulink, A12
sine transform, B34
singular boundary, 352
singular point, A2
singular solution, A19
skewness, 309
    coefficient, B5
    $\eta_3[X]$, B5
    negative, 309
smooth process, B63
S_n, partial sum, B63
s_n^2, sample variance, B37
$\widehat{s}_n^2$, sample variance unbiased estimate,
    B37
space-time Poisson
    noise, 129
    process, 130, 139
speed density, 352
split-step backward Euler method, 257
sqrt, MATLAB square root function, 5
stability criteria, 227
standard deviation, B4
    σ, B4
standard Wiener process, 3
    $W(t)$, 3
state
    space, B59
state-dependent SDEs, 141
stationary, B61
    Markov process, B61
    point, A8, B68
    process, 3
        Markov, B61
        Poisson process, 12
        $P(t)$, 12
        Wiener process, 3
        $W(t)$, 3
step function
    Heaviside, B52
    $H(x)$, B31, B52
    $H_R(x)$, B57
    $U(x; a, b)$, 15, 113, 121, 122

Stieltjes
    integral, 34, B3
    $\int f(X(t), t) dX(t)$, 34
Stirling's formula, B52
stochastic
    calculus, 81
        general, 129
    chain rule
        diffusion $F(X(t), t)$, 98
        $G(W(t))$, 85
        $G(W(t), t)$, 86
        Itô, 85, 86
        state-dependent, 142
    diffusion integration, 34, 57
    diffusion process, 1, 31
    dynamic programming, 169, 177
        PDE, 179
        PIDE, 179
    Euler's method, 244
    integral
        $\int (dW)^2(t)$, 41
        $\int (P dP)(t)$, 65
        $\int (W dW)(t)$, 41
        mean, 48
    integration, 31, 63
        $\int f(W(t), t) dt$, 33
        $\int f(X(t), t) dt$, 34
        $\int g(W(t), t) dt$, 40
        $\int h dP$, 77
        $\overset{ims}{=}$, 63
        diffusion, 31, 57
        Euler's method, 40, 57
        forward integration, 40, 57
        Itô, 34, 40, 57
        jump, 63, 77
        left rectangular rule, 40, 57
        $\lim_{n \to \infty}^{ms}$, 63
        midpoint rule, 56
        $P(t)$, 63, 74
        Stratonovich, 56
        t, 33, 34
        trapezoidal, 60
        $W(t)$, 34
    jump
        integral, 63
        integral expectation, 74

integral rule, 74
integration, 63, 77
process, 1, 63
jump-diffusion processes, 81
natural exponential, 88
optimal control, 169
Poisson jump
    integration, 63
process, 1, B59
    continuous-time, 1
    diffusion, 1, 31
    jump, 1, 63
processes
    abstract theory, 361
    jump-diffusion, 81
random, 1
system
    hybrid, B59
Taylor expansion, 249
theta method, 254, 257
stochastic calculus
    fundamental theorem
        $d[\int(g(W)dW)(t)]$, 45
        $d(\int(h(P)dP)(t))$, 64
        $\int dG(W(t)$, 46
        $\int d\mathcal{H}(P(t))$, 64
        $\int(G'(W)dW)(t)$, 85
        Itô's, 85
stochasticity
    demographic, 97
    environmental, 97
stock fraction, 321
stopping times, 206
stratified sampling, 276
    recursive, 279
Stratonovich
    stochastic integration, 56
        $\int(WdW)(t)$, 56
        mean, 56
substitution test, 90
submartingale, 372
supertriangular numbers, 68
SVJD (stochastic-volatility,
        jump-diffusion), 317

tangent-line method, 33
Taylor approximation, B63
    jump, B67
    remainder, B63
Taylor expansion
    stochastic, 249
Taylor's formula, B63
temporal Poisson process, 20
terminal cost, A2
test
    substitution, 90
theorem
    binomial, B47
    central limit, B39
    Fubini's, B31
    Girsanov's, 382
    gradient peel, B44
    martingale representation, 376
    multinomial expansion, B49
third central moment, B5
time
    doubling, 350
    interarrival, B20
    jump, B20
time-homogeneous, B61
time-inhomogeneous, 20
times
    exit, 206
    first passage, 206
    stopping, 206
T_j, jump-time, 13
total probability
    density, B29
    distribution, B29
    expectation, B29
    law, 111, B29
trace, B40
Trace[A], trace, B40
transformation of probabilities law, B15
transition
    density, 197
    distribution, 197
    probability, 233, B60, B61
        matrix, B61
transpose, B40
transversality, A8

trapezoidal rule, 60
triangular number, 67
$n(n + 1)/2$, 67
t-stochastic integration, 33
tumor growth, 347

UFD (upwinded finite difference), 228
unbiased estimate, 262, B37
$\widehat{s_n^2}$, B37
uniform
density, B5
$\phi_u(x; a, b)$, B5
distribution, B5
$\Phi_u(x; a, b)$, B5
random number generator
seed, B8
state, B8
upwinded finite difference (UFD), 228
utility
consumption, 322
CRRA (constant relative
risk-aversion), 325, 332
HARA (hyperbolic absolute
risk-aversion), 317, 325
marginal, 325
power, 325, 332
terminal, 322
$U(x; a, b)$, step function, 15, 113, 121,
122

Var[X], variance
discrete case, B4, B18
variance, B4, B18
σ^2, B4
continuous random variables, B4
discrete set, B18
$dP(t; Q)$, 134
product, B69
sample, B37
sum, B69
Var[X], B4, B18
variance-expectation identity, B69

variation
coefficient, 209
vector notation, B39
VEGAS algorithm, 276
volatility, 94, 293
fraction, 298
smile, 309
volatility-risk, 312

wealth
consumption, 321
equation, 321, 331
weighted sampling, 259
well-conditioned, B42
white noise, 120
delta-correlated, 122
Gaussian, 120
Poisson, 120
Wiener process, 1
characteristic functions, 393
continuity, 3
density, 3
differential, 4
density, 4
differential $dW(t)$, 2
increment, 4
density, 4
independent increments, 3
Markov, 3
nondifferentiability, 9
nonsmoothness, 9
stationary process, 3
with probability one (w.p.o.), 2, 369
Wronskian, 351
$W(t)$-stochastic integration, 34
$W(t)$, Wiener process, 1, 2

zero-one jump law, 19, 20
approximate, 18
Bernoulli distribution, 20
$dP(t)$, 19
mean square limit, 73
Poisson process, 19